Masahito Hayashi

Quantum Information

Masahito Hayashi

Quantum Information

An Introduction

With 14 Figures and 10 Tables

Springer

Masahito Hayashi
Japan Science and Technology Agency
201 Daini Hongo White Bldg
5-28-3, Hongo, Bunkyo-ku
Tokyo 113-0033, Japan
e-mail: masahito@qci.jst.go.jp

Library of Congress Control Number: 2006923433

ISBN-10 3-540-30265-4 Springer Berlin Heidelberg New York
ISBN-13 978-3-540-30265-0 Springer Berlin Heidelberg New York

Springer is a part of Springer Science+Business Media

springer.com

Printed in Germany

Typesetting: Protago-TEX-Production GmbH, Berlin
Production: LE-TEX Jelonek, Schmidt & Vöckler GbR, Leipzig
Cover design: eStudio Calamar S.L., F. Steinen-Broo, Pau/Girona, Spain

Printed on acid-free paper 57/3100/YL 5 4 3 2 1 0

To my parents

Preface

This book is the revised English edition of the Japanese book *Introduction to Quantum Information Theory*, which systematically describes quantum information theory and was originally published by Saiensu-sha, Tokyo, Japan in May 2003. The study of information processing based on the physical principles of quantum mechanics was initiated in the 1960s. Recently, such quantum information processing has demonstrated experimentally, and its theoretical aspects have been examined more deeply and mathematically. The research field addressing the relevant theory is called Quantum Information Theory, and is now being studied by many researchers from various viewpoints.

However, only Holevo's book *Probabilistic and Statistical Aspects of Quantum Theory*, which was published back in 1980 (English version in 1982), places a heavy emphasis on the mathematical foundation of quantum information theory. Several books concerning quantum information science have been published since the late 1990s. However, they treat quantum computation, the physical aspects of quantum information, or the whole of quantum information science and are not mainly concerned with quantum information theory. Therefore, it seemed to me that many researchers would benefit from an English book on quantum information theory, and so I decided to publish the English version of my book. I hope that it will make a contribution to the field of quantum information theory.

This book was written as follows. First, the author translated the original Japanese version in cooperation with Dr. Tim Barnes. Next, the book was revised through the addition of many new results to Chaps. 8–10 and a historical note to every chapter. Several exercises were also added, so that the English version has more than 330 exercises. Hence, I take full responsibility for the content of this English version. In this version, theorems and lemmas are displayed along with the names of the reserchers who contributed them. However, when the history of the theorems and lemmas is not so simple, they are displayed without the contributing researchers' names and their histories are explained in a historical note at the end of the given chapter.

I am indebeted to Prof. Masanao Ozawa and Dr. Tohya Hiroshima for their feedback on the Japanese version, which been incorporated into the English version. I am also grateful to (in alphabetical order) Dr. Giulio Chiribella, Mr. Motohisa Fukuda, Prof. Richard Gill, Dr. Michael Horodecki, Dr. Satoshi Ishizaka, Dr. Paolo Perinotti, Dr. Toshiyuki Shimono, and Dr. Andreas Winter, for reviewing the technical aspects of the English version. Further, Dr. Tomohisa Hayakawa, Mr. Daichi Isami, Mr. Takashi Okajima, Mr. Tomotake Sasaki, Mr. Taiji Suzuki, Mr. Fuyuhiko Tanaka, and Mr. Ken'ichiro Tanaka used the draft of the English version in their seminar and verified its contents. Miss Rika Abe commented on the nontechnical parts of the book. Further, Mr. Motohisa Fukuda helped me in compiling the references. I would like to express my appreciation for their cooperation.

I also would like to thank Prof. Hiroshi Imai of the University of Tokyo and the people associated with the ERATO Quantum Computation and Information Project for providing the research environments for this English version. I would like to express my gratitude to Dr. Glenn Corey and editorial staffs of Springer for good excellent editing process. I would also like to thank Dr. Claus E. Ascheron of Springer Science+Business Media for his encouragement and patience during the preparation of the manuscript.

Tokyo, February 2006 *Masahito Hayashi*

Preface to Japanese Version

This textbook attempts to describe quantum information theory, which is a presently evolving field. It is organized so that the reader can understand its contents with very elementary prior knowledge. This research field has been developed by many researchers from various backgrounds and has matured rapidly in the last 5 years.

Recently, many people have considered that more interdisciplinary activities are needed in the academic world. Hence, education and research must be performed and evaluated on a wide scope. However, since the extreme segmentation of each research area has increased the difficulty of interdisciplinary activities. On the other hand, quantum information theory can in some sense form a bridge between several fields because it deals with topics in a variety of disciplines including physics and information science. Hence, it can be expected to contribute in some way to removing the segmentation of its parent fields. In fact, information science consists of subfields such as computer science, mathematical statistics, and Shannon's information theory. These subfields are studied in separate contexts.

However, in quantum information theory, we must return to the fundamentals of the topic, and there are fewer boundaries among the different fields. Therefore, many researchers now transcend these boundaries.

Given such a starting point, the book was written to enable the reader to efficiently attain the interdisciplinary knowledge necessary for understanding quantum information theory. This book assumes only that the reader has knowledge of linear algebra, differential and integral calculus, and probability/statistics at the undergraduate level. No knowledge of quantum mechanics is assumed.

Some of the exercises given in the text are rather difficult. It is recommended that they be solved in order to acquire the skills necessary for tackling research problems. Parts of the text contain original material that does not appear elsewhere. Comments will be given for such parts.

The author would like to thank Prof. Hiroshi Imai of the University of Tokyo, Prof. Shun-ichi Amari of the Brain Science Institute at RIKEN, Prof. Kenji Ueno of Kyoto University, and the people associated with the ERATO Quantum Computation and Information Project, the Brain Science Institute at RIKEN, and the Department of Mathematics at Kyoto University for providing me with the means to continue my research. The author also wishes to thank Prof. Hiroshi Nagaoka of the University of Electro-Communications, Prof. Akio Fujiwara of Osaka University, Prof. Keiji Matsumoto of the National Institute of Informatics, and Dr. Tomohiro Ogawa of the University of Tokyo for helpful discussions and advice. This text would not have been possible without their enlightening discussions.

I also received valuable comments from Prof. Alexander Holevo of the Steklov Mathematical Institute, Prof. Masanao Ozawa of Tohoku University, Dr. Ryutaroh Matsumoto of the Tokyo Institute of Technology, Dr. Fumiaki Morikoshi of NTT, Dr. Yodai Watanabe of RIKEN, and Dr. Mitsuru Hamada, Dr. Yoshiyuki Tsuda, Dr. Heng Fan, Dr. Xiangbin Wang, and Mr. Toshiyuki Shimono of the ERATO Quantum Computation and Information Project regarding the contents of this text. They have also earned a debt of gratitude. I would also like to thank Mr. Kousuke Hirase of Saiensu-sha for his encouragement and patience during the preparation of the manuscript.

Tokyo, December 2003 *Masahito Hayashi*

Contents

Prologue

Invitation to Quantum Information Theory

Understanding the implications of recognizing matter and extracting information from it has been a long-standing issue in philosophy and religion. However, recently this problem has become relevant to other disciplines such as cognitive science, psychology, and neuroscience. Indeed, this problem is directly relevant to quantum mechanics, which forms the foundation of modern physics. In the process of recognition, information cannot be obtained directly from matter without any media. To obtain information, we use our five senses; that is, a physical medium is always necessary to convey information to us. For example, in vision, light works as the medium for receiving information. Therefore, observations can be regarded as information processing via a physical medium. Hence, this problem can be treated by physics. Of course, to analyze this problem, the viewpoint of information science is also indispensable because the problem involves, in part, information processing.

In the early 20th century, physicists encountered some unbelievable facts regarding observations (measurements) in the microscopic world. They discovered the contradictory properties of light, i.e., the fact that light has both wave- and particlelike properties. Indeed, light behaves like a collection of minimum energy particles called photons. In measurements using light, we observe the light after interactions with the target. For example, when we measure the position of the matter, we detect photons after interactions with them. Since photons possess momentum and energy, the speed of the object is inevitably disturbed.[1] In particular, this disturbance cannot be ignored when the mass of the measured object is small in comparison with the energy of the photon. Thus, even though we measure the velocity of an object after the measurement of its position, we cannot know the velocity of an object precisely because the original velocity has been disturbed by the first

[1] The disturbance of measurement is treated in more detail in the formulation of quantum mechanics in Chap. 7.

measurement. For the same reason, when we measure the velocity first, its position would be disturbed. Therefore, our naive concept of a "perfect measurement" cannot be applied, even in principle. In the macroscopic world, the mass of the objects is much larger than the momentum of the photons. We may therefore effectively ignore the disturbance by the collisions of the photons. Although we consider that a "perfect measurement" is possible in this macroscopic world, the same intuition cannot be applied to the microscopic world.

In addition to the impossibility of "perfect measurements" in the microscopic world, no microscopic particles have both a determined position and a determined velocity. This fact is deeply connected to the wave-particle duality in the microscopic world and can be regarded as the other side of the nonexistence of "perfect measurements."[2] Thus it is impossible to completely understand this microscopic world based on our macroscopic intuitions, but it is possible to predict probabilistically its measured value based on the mathematical formulation of quantum theory.

So far, the main emphasis of quantum mechanics has been on examining the properties of matter itself, rather than the process of *extracting information*. To discuss how the microscopic world is observed, we need a quantitative consideration from the viewpoint of "information." Thus, to formulate this problem clearly, we need various theories and techniques concerning information. Therefore, the traditional approach to quantum mechanics is insufficient. On the other hand, theories relating to information pay attention only to the data-processing rather than the extraction process of information. Therefore, in this quantum-mechanical context, we must take into account the process of obtaining information from microscopic (quantum-mechanical) particles. We must open ourselves to the new research field of *quantum information science*. This field is to be broadly divided into two parts: (1) *quantum computer science*, in which algorithms and complexity are analyzed using an approach based on computer science, and (2) *quantum information theory*, in which various protocols are examined from the viewpoint of information theory and their properties and limits are studied. Specifically, since quantum information theory focuses on the amount of accessible information, it can be regarded as the theory for quantitative evaluation of the process of *extracting information*, as mentioned above.

Since there have been few textbooks describing this field, the present textbook attempts to provide comprehensive information ranging from the fundamentals to current research. Quantum computer science is not treated in this book because it has been addressed in many other textbooks. Since quantum information theory forms a part of the basis of quantum computer science, this textbook may be useful for not only researchers in quantum information theory but also those in quantum computer science.

---

[2] The relation between this fact and nonexistence can be mathematically formulated by (7.27) and (7.30).

History of Quantum Information Theory

Let us briefly discuss the history of quantum information theory. Quantum mechanics was first formulated by Schrödinger (wave mechanics) and Heisenberg (matrix mechanics). However, their formulations described the dynamics of microscopic systems, but they had several unsatisfactory aspects in descriptions of measurements. Later, the equivalence between both formulations were proved. To resolve this point, von Neumann [408] established the formulation of quantum theory that describes measurements as well as dynamics based on operator algebra, whose essential features will be discussed in Chap. 1. However, in studies of measurements following the above researches, the philosophical aspect has been emphasized too much, and a quantitative approach to extracting information via measurements has not been examined in detail. This is probably because approaches to mathematical engineering have not been adopted in the study of measurements.

In the latter half of the 1960s, a Russian researcher named Stratonovich, who is one of the founders of stochastic differential equations, and two American researchers, Helstrom and Gordon, proposed a formulation of optical communications using quantum mechanics. This was the first historical appearance of quantum information theory. Gordon [148, 149], Helstrom [201], and Stratonovich [380] mainly studied error probabilities and channel capacities for communications. Meanwhile, Helstrom [381] examined the detection process of optical communication as parameter estimation. Later, many American and Russian researchers such as Holevo [212, 214], Levitin [264], Belavkin [34], Yuen [432], and Kennedy [431] also examined these problems.[3] In particular, Holevo obtained the upper bound of the communication speed in the transmission of a classical message via a quantum channel in his two papers [212, 214] published in the 1970s. Further, Holevo [215, 216], Yuen [432], Belavkin, and their coworkers also analyzed many theoretically important problems in quantum estimation.

Unfortunately, the number of researchers in this field rapidly decreased in the early 1980s, and this line of research came to a standstill. Around this time, Bennett and Brassard [35] proposed a quantum cryptographic protocol (BB84) using a different approach to quantum mechanical systems. Around the same time, Ozawa [327] gave a precise mathematical formulation of the state reduction in the measurement process in quantum systems.

In the latter half of the 1980s, Nagaoka investigated quantum estimation theory as a subfield of mathematical statistics. He developed the asymptotic theory of quantum-state estimation and quantum information geome-

[3] Other researchers during this period include Grishanin, Mityugov, Kuriksha, Liu, Personick, Lax, Lebedev, Forney [116] in the United States and Russia. Many papers were published by these authors; however, an accurate review of all of them is made difficult by their lack of availability. In particular, while several Russian papers have been translated into English, some of them have been overlooked despite their high quality. For details, see [202, 216].

try [304]. This research was continued by many Japanese researchers, including Fujiwara, Matsumoto, and the present author in the 1990s [119, 120, 126, 127, 129, 165–168, 170, 171, 178, 182, 281, 282, 284–286]. For this history, see Hayashi [194].

In the 1990s, in the United States and Europe several researchers started investigating quantum information processing, e.g., quantum data compression, quantum teleportation, superdense coding, another quantum cryptographic protocol (B92), etc. [36, 38, 39, 43, 245, 360]. In the second half of the 1990s, the study of quantum information picked up speed. In the first half of the 2000s, several information-theoretic approaches were developed, and research has been advancing at a rapid pace.

We see that progress in quantum information theory has been achieved by connecting various topics. This text clarifies these connections and discusses current research topics starting with the basics.

Structure of the Book

Quantum information theory has been studied by researchers from various backgrounds. Their approach can be broadly divided into two categories. The first approach is based on information theory. In this approach, existing methods for information processing are translated (and extended) into quantum systems. The second approach is based on quantum mechanics.

In this text, four chapters are dedicated to examining problems based on the first approach, i.e., establishing information-theoretic problems. These are Chap. 3, "Quantum Hypothesis Testing and Discrimination of Quantum States," Chap. 4, "Classical Quantum Channel Coding (Message Transmission)," Chap. 6, "Quantum Information Geometry and Quantum Estimation," and Chap. 10, "Source Coding in Quantum Systems." Problems based on the second approach is treated in three chapters: Chap. 5, "State evolution and Trace-Preserving Completely Positive Maps," Chap. 7, "Quantum measurements and State Reduction," and Chap. 8, "Entanglement and Locality Restrictions."

Advanced topics in quantum communication such as quantum teleportation, superdense coding, quantum-state transmission (quantum error correction), and quantum cryptography are often discussed in quantum information theory. Both approaches are necessary for understanding these topics, which are covered in Chap. 9, "Analysis of Quantum Communication Protocols."

Some quantum-mechanical information quantities are needed to handle these problems mathematically, and these problems are covered in Sects. 2.2, 5.4, 5.5, 8.2, and 8.3. This allows us to touch upon several important information-theoretic problems using a minimum amount of mathematics. The book also includes 330 exercises together with short solutions (for simple problems, the solutions are omitted for brevity). Solving these problems should provide readers not only with knowledge of quantum information theory but also the necessary techniques for pursuing original research in the field.

Chapter 1 covers the mathematical formulation of quantum mechanics in the context of quantum information theory. It also gives a review of linear algebra. Chapter 2 summarizes classical information theory. This not only provides an introduction to the later chapters but also serves as a brief survey of classical information theory. Topics such as the large deviation principle for entropy, Fisher information, and information geometry are covered here. Quantum relative entropy, which is the quantum-theoretic measure of information, is also briefly discussed. This concludes the preparatory part of the text.

Chapter 3 covers quantum hypothesis testing and the discrimination of quantum states. This chapter serves to answer the question: If there are two states, which is the true state? The importance of this question may not at first be apparent. However, this problem provides the foundation for other problems in information theory and is therefore crucially important. Also, this problem provides the basic methods for quantum algorithm theory. Many of the results of this chapter will be used in subsequent chapters. In particular, the quantum version of Stein's lemma is discussed here; it can be used a basic tool for other topics. Furthermore, many of the difficulties associated with the noncommutativity of quantum theory can be seen here in their simplest forms. This chapter can be read after Chap. 1 and Sects. 2.1, 2.2, and A.3.

Chapter 4 covers classical quantum channel coding (message transmission). That is, we treat the tradeoff between the transmission speed and the error probability in the transmission of classical messages via quantum states. In particular, we discuss the channel capacity, i.e., the theoretical bound of the transmission rate when the error probability is 0, as well as its associated formulas. This chapter can be read after Chap. 1 and Sects. 2.1, 2.2, 3.4, 3.6, and 3.7.

Chapter 5 discusses the trace-preserving completely positive map, which is the mathematical description of state evolution in quantum systems. Its structure will be illustrated with examples in quantum two-level systems. We also briefly discuss the relationship between the state evolution and information quantities in quantum systems (the entropy and relative entropy). In particular, the part covering the formulation of quantum mechanics (Sects. 5.1 to 5.3) can be read after only Chap. 1.

Chapter 6 describes the relationship between quantum information geometry and quantum estimation. First, the inner product for the space of quantum states is briefly discussed. Next, we discuss the geometric structure naturally induced from the inner product. The theory of state estimation in quantum systems is then discussed by emphasizing the Cramér–Rao inequality. Most of this chapter can be read after Chaps. 1 and 2 and Sect. 5.1.

Chapter 7 covers quantum measurement and state reduction. First, it is shown that the state reduction due to a quantum measurement follows naturally from the axioms of the quantum systems discussed in Chap. 1. Next, we discuss the relationship between quantum measurement and the uncertainty relations. Finally, it is shown that under certain conditions it is possible, in principle, to perform a measurement such that the required information can be obtained while the state demolition is negligible. Readers who only wish to read Sects. 7.1 and 7.3 can read them after Chap. 1 and Sect. 5.1. Section 7.2 requires the additional background of Sect. 6.1.

Chapter 8 discusses the relationship between locality and entanglement, which are fundamental topics in quantum mechanics. First, we examine state operations when the condition of locality is imposed on quantum operations. Next, the information quantities related to entanglement are considered. The theory for distilling a completely entangled state from an incompletely entangled state is discussed. Information-theoretic methods play a central role in entanglement distillation. Further, quantification of entanglement is discussed from various viewpoints.

Chapter 9 delves deeply into topics in quantum channels such as quantum teleportation, superdense coding, quantum-state transmission (quantum error correction), and quantum key distribution based on the theory presented in previous chapters. These topics are very simple when noise is not present. However, if noise is present in a channel, these problems require the information-theoretic methods discussed in previous chapters. The relationship among these topics is also discussed. Further, the relation between channel capacities and entanglement theory is also treated.

Finally, Chap. 10 discusses source coding in quantum systems. We treat not only the theoretical bounds of quantum fixed-length source coding but also universal quantum fixed-/variable-length source coding, which does not depend on the form of the information source. The beginning part of this chapter, excepting the purification scheme, requires only the contents of Chapts. 1 and 2 and Sect. 5.1. In particular, in universal quantum variable-length source coding, a measurement is essential for determining the coding length. Hence this measurement causes the demolition of the state to be sent, which makes this a more serious problem. However, it can be solved by a measurement with negligible state demolition, which is described in Chap. 7. Then we treat quantum-state compression with mixed states and its several variants. The relations between these problems and entanglement theory are also treated. Further, we treat the relationships between the reverse capacities (reverse Shannon theorem) and these problems. Excluding

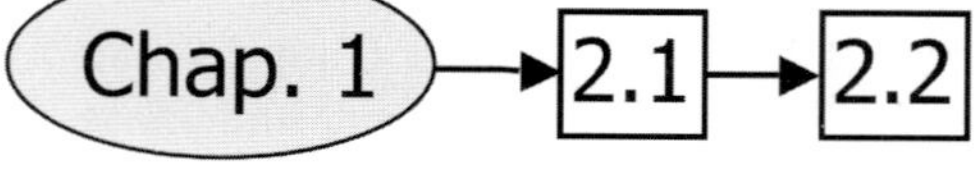

Fig. 1. Example

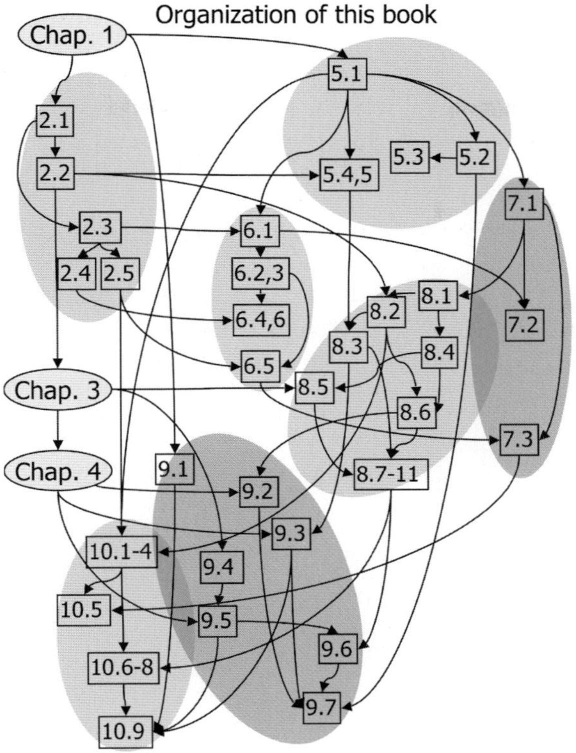

Fig. 2. Organization of this book

Sects. 10.6–10.9, this chapter can be read after Chap. 1 and Sects. 2.1, 2.2, 2.4, 4.1, and 5.1.

This text thus covers a wide variety of topics in quantum information theory. Quantum hypothesis testing, quantum-state discrimination, and quantum-channel coding (message transmission) have been discussed such that only a minimal amount of mathematics is needed to convey the essence of these topics. Prior to this text, these topics required the study of advanced mathematical theories for quantum mechanics, such as those presented in Chap. 5. Further, Chaps. 5 ("State Evolution and Trace Preserving Completely Positive Maps in Quantum Systems") and 7 ("Quantum Measurement and State Reduction") have been written such that they can be understood with only the background provided in Chap. 1. Therefore, this text should also be suitable for readers who are interested in either the information-theoretic aspects of quantum mechanics or the foundations of quantum mechanics.

Finally, the organization of this book is illustrated in Fig. 2. For example, Fig. 1 implies that that section requires the contents of Chap. 1 and Sect. 2.1 or only that of Sect. 2.1.

1

Mathematical Formulation of Quantum Systems

Summary. In this chapter, we cover the fundamentals of linear algebra and provide a mathematical formulation of quantum mechanics for use in later chapters. It is necessary to understand these topics since they form the foundation of quantum information processing discussed later. In the first section, we cover the fundamentals of linear algebra and introduce some notation. The next section describes the formulation of quantum mechanics. Further, we examine a quantum two-level system, which is the simplest example of a quantum-mechanical system. Finally, we discuss the tensor product and matrix inequalities. More advanced discussions on linear algebra are available in Appendix A.

Table 1.1. Denotations used in Chap. 1

$\bar{x}$	Complex conjugate of given number x
$\mathcal{S}(\mathcal{H})$	Set of density matrices of given Hilbert space $\mathcal{H}$
A^T	Transpose of given matrix A
A^*	Adjoint of given matrix A
$[X,Y]$	Communtator of given matrices A and B
$X \circ Y$	Symmetrized product of given matrices A and B
$\mathrm{P}_\rho^{\boldsymbol{M}}$	Probability distribution when measurement is $\boldsymbol{M}$ and state is ρ
ρ_{mix}	Completely mixed state
Tr_A	Partial trace concerning system $\mathcal{H}_A$
$\{X \geq 0\}$	Projection defined by (1.32)
$\kappa_{\boldsymbol{M}}$	Pinching of PVM $\boldsymbol{M}$ (1.11)
κ_X	Pinching of Hermitian matrix X (1.12)
$\kappa_{\boldsymbol{M}}$	Pinching of POVM $\boldsymbol{M}$ (1.13)

Quantum two-level system	
S_i	Pauli matrix (1.15)
$\rho_{\boldsymbol{x}}$	Stokes parameterization (1.16)

1.1 Quantum Systems and Linear Algebra

In order to treat information processing in quantum systems, it is necessary to mathematically formulate fundamental concepts such as quantum systems, measurements, and states. First, we consider the quantum system. It is described by a Hilbert space $\mathcal{H}$ (a finite- or infinite-dimensional complex vector space with a Hermitian inner product), which is called a *representation space*. Before considering other important concepts such as measurements and states, we give a simple overview of linear algebra. This will be advantageous because it is not only the underlying basis of quantum mechanics but is also as helpful in introducing the special notation used for quantum mechanics. In mathematics, a Hilbert space usually refers to an infinite-dimensional complex vector space with a Hermitian inner product. In physics, however, a Hilbert space also often includes finite-dimensional complex vector spaces with Hermitian inner products. This is because in quantum mechanics, the complex vector space with a Hermitian inner product becomes the crucial structure. Since infinite-dimensional complex vector spaces with Hermitian inner products can be dealt with analogously to the finite-dimensional case, we will consider only the finite-dimensional case in this text. Unless specified, the dimension will be labeled d.

The representation space of a given system is determined by a physical observation. For example, spin-$\frac{1}{2}$ particles such as electrons possess, an internal degree of freedom corresponding to "spin" in addition to their motional degree of freedom. The representation space of this degree of freedom is $\mathbb{C}^2$. The representation space of a one-particle system with no internal degrees of freedom is the set of all square integrable functions from $\mathbb{R}^3$ to $\mathbb{C}$. In this case, the representation space of the system is an infinite-dimensional space, which is rather difficult to handle. Such cases will not be examined in this text.

Before discussing the states and measurements, we briefly summarize some basic linear algebra with some emphasis on Hermitian matrices. This will be important particularly for later analysis. The Hermitian product of two vectors

$$u = \begin{pmatrix} u^1 \\ u^2 \\ \vdots \\ u^d \end{pmatrix}, \quad v = \begin{pmatrix} v^1 \\ v^2 \\ \vdots \\ v^d \end{pmatrix} \in \mathcal{H}$$

is given by

$$\langle u|v\rangle \stackrel{\text{def}}{=} \overline{u^1}v^1 + \overline{u^2}v^2 + \ldots + \overline{u^d}v^d \in \mathbb{C},$$

where the complex conjugate of a complex number x is denoted by $\overline{x}$. The norm of the vector is given by $\|u\| \stackrel{\text{def}}{=} \sqrt{\langle u|u\rangle}$. The inner product of the vectors satisfies the *Schwarz inequality*

$$\|u\|\|v\| \geq |\langle u|v\rangle| \,. \tag{1.1}$$

When a matrix

$$X = \begin{pmatrix} x^{1,1} & x^{1,2} & \dots & x^{1,d} \\ x^{2,1} & x^{2,2} & \dots & x^{2,d} \\ \vdots & \vdots & \ddots & \vdots \\ x^{d,1} & x^{d,2} & \dots & x^{d,d} \end{pmatrix} \tag{1.2}$$

satisfies the following condition

$$X = X^* \stackrel{\text{def}}{=} \begin{pmatrix} \overline{x^{1,1}} & \overline{x^{2,1}} & \dots & \overline{x^{d,1}} \\ \overline{x^{1,2}} & \overline{x^{2,2}} & \dots & \overline{x^{d,2}} \\ \vdots & \vdots & \ddots & \vdots \\ \overline{x^{1,d}} & \overline{x^{2,d}} & \dots & \overline{x^{d,d}} \end{pmatrix}, \tag{1.3}$$

it is called *Hermitian*. We also define the complex conjugate matrix $\overline{X}$ and its transpose matrix X^T as follows:

$$\overline{X} \stackrel{\text{def}}{=} \begin{pmatrix} \overline{x^{1,1}} & \overline{x^{1,2}} & \dots & \overline{x^{1,d}} \\ \overline{x^{2,1}} & \overline{x^{2,2}} & \dots & \overline{x^{2,d}} \\ \vdots & \vdots & \ddots & \vdots \\ \overline{x^{d,1}} & \overline{x^{d,2}} & \dots & \overline{x^{d,d}} \end{pmatrix}, \quad X^T \stackrel{\text{def}}{=} \begin{pmatrix} x^{1,1} & x^{2,1} & \dots & x^{d,1} \\ x^{1,2} & x^{2,2} & \dots & x^{d,2} \\ \vdots & \vdots & \ddots & \vdots \\ x^{1,d} & x^{2,d} & \dots & x^{d,d} \end{pmatrix}.$$

Then, a Hermitian matrix X satisfies $X^T = \overline{X}$. If a Hermitian matrix X satisfies $\langle u|Xu\rangle \geq 0$ for an arbitrary vector $u \in \mathcal{H}$, it is called *positive semidefinite* and denoted by $X \geq 0$. If $\langle u|Xu\rangle > 0$ for nonzero vectors u, X is called *positive definite*. The condition of positive semidefiniteness is equivalent to all the eigenvalues of a diagonalized Hermitian matrix X that are either zero or positive. As shown later, the trace of the product of two positive semidefinite matrices X and Y satisfies

$$\operatorname{Tr} XY \geq 0 \,. \tag{1.4}$$

However, in general, the product XY is not a Hermitian matrix. Note that although the matrix $XY + YX$ is Hermitian, it is generally not positive semidefinite.

We can regard each element $u \in \mathcal{H}$ as an element of the dual space $\mathcal{H}^*$ according to the correspondence between $\mathcal{H}$ and $\mathcal{H}^*$ given by the inner product. We denote the corresponding element of the dual space $\mathcal{H}^*$ by $\langle u|$, in accordance with the conventional notation in physics. If we wish to emphasize that u is an element not of $\mathcal{H}^*$ but of $\mathcal{H}$, we write $|u\rangle$. That is,

$$|u\rangle = \begin{pmatrix} u^1 \\ u^2 \\ \vdots \\ u^d \end{pmatrix} \in \mathcal{H}, \quad \langle u| = \left(\overline{u^1}\ \overline{u^2} \cdots \overline{u^d}\right) \in \mathcal{H}^* .$$

The Hermitian inner product $\langle u|v\rangle$ can also be considered as the matrix product of $\langle u|$ and $|v\rangle$. Note that this notation is used in this text even if the norm of v is not equal to 1. On the other hand, the opposite matrix product $|v\rangle\langle u|$ is a $d \times d$ matrix:

$$|v\rangle\langle u| = \begin{pmatrix} v^1 \\ v^2 \\ \vdots \\ v^d \end{pmatrix} \begin{pmatrix} \overline{u^1}\ \overline{u^2} \cdots \overline{u^d} \end{pmatrix} = \begin{pmatrix} v^1\overline{u^1} & v^1\overline{u^2} & \dots & v^1\overline{u^d} \\ v^2\overline{u^1} & v^2\overline{u^2} & \dots & v^2\overline{u^d} \\ \vdots & \vdots & \ddots & \vdots \\ v^d\overline{u^1} & v^d\overline{u^2} & \dots & v^d\overline{u^d} \end{pmatrix}. \tag{1.5}$$

Although $|Xv\rangle = X|v\rangle$, $\langle Xv| = \langle v|X^*$. Evidently, if matrix X is Hermitian, then $\langle u|Xv\rangle = \langle Xu|v\rangle$. This also equals $\operatorname{Tr}|v\rangle\langle u|X$, which is often denoted by $\langle u|X|v\rangle$. Using this notation, matrix X given by (1.2) may be written as $X = \sum_{i,j} x^{i,j}|u_i\rangle\langle u_j|$, where u_i is a unit vector whose ith element is 1 and remaining elements are 0.

A Hermitian matrix X may be transformed into the diagonal form U^*XU by choosing an appropriate unitary matrix U. Since $X = U(U^*XU)U^*$, we may write

$$X = \begin{pmatrix} u_1^1 & \dots & u_d^1 \\ \vdots & \ddots & \vdots \\ u_1^d & \dots & u_d^d \end{pmatrix} \begin{pmatrix} x^1 & & O \\ & \ddots & \\ O & & x^d \end{pmatrix} \begin{pmatrix} \overline{u_1^1} & \dots & \overline{u_1^d} \\ \vdots & \ddots & \vdots \\ \overline{u_d^1} & \dots & \overline{u_d^d} \end{pmatrix}. \tag{1.6}$$

Define d vectors $u_1, u_2, \dots, u_d$

$$u_i = \begin{pmatrix} u_i^1 \\ \vdots \\ u_i^d \end{pmatrix}.$$

Then the unitarity of U implies that $\{u_1, u_2, \dots, u_d\}$ forms an orthonormal basis. Using (1.5), the Hermitian matrix X may then be written as $X = \sum_i x^i|u_i\rangle\langle u_i|$. This process is called diagonalization. If X and Y commute, they may be written as $X = \sum_{i=1}^d x^i|u_i\rangle\langle u_i|$, $Y = \sum_{i=1}^d y^i|u_i\rangle\langle u_i|$ using the same orthonormal basis $\{u_1, u_2, \dots, u_d\}$. If X and Y do not commute, they cannot be diagonalized using the same orthonormal basis.

Furthermore, we can characterize positive semidefinite matrices using this notation. A matrix X is positive semidefinite if and only if $x^i \geq 0$ for arbitrary i. Thus, this equivalence yields inequality (1.4) as follows:

$$\operatorname{Tr} XY = \operatorname{Tr} \sum_{i=1} x^i|u_i\rangle\langle u_i|Y = \sum_{i=1} x^i \operatorname{Tr}|u_i\rangle\langle u_i|Y = \sum_{i=1} x^i\langle u_i|Y|u_i\rangle \geq 0\,.$$

We also define the commutator $[X,Y]$ and symmetrized product $X \circ Y$ of two matrices X and Y as[1]

$$[X,Y] \stackrel{\text{def}}{=} XY - YX, \quad X \circ Y \stackrel{\text{def}}{=} \frac{1}{2}(XY + YX).$$

Exercises

1.1. Show Schwarz's inequality (1.1) noting that $\langle u+rcv|u+rcv\rangle \geq 0$ for an arbitrary real number r, where $c = \langle v|u\rangle/|\langle v|u\rangle|$.

1.2. Suppose that k vectors $u_1,\ldots,u_k$ ($u_j = (u^{ij})$) satisfy $\langle u_{j'}|u_j\rangle = \sum_{i=1}^d \overline{u^{ij'}}u^{ij} = \delta_{j'j}(1 \leq j,j' \leq k)$, where $\delta_{j,j'}$ is defined as 1 when $j = j'$ and as 0 otherwise. Show that there exist $d-k$ vectors $u_{k+1},\ldots,u_d$ such that $\langle u_{j'}|u_j\rangle = \delta_{j'j}$, i.e., the matrix $U = (u^{ij})$ is unitary.

1.3. Let $X = \sum_{i,j} x^{i,j}|u_i\rangle\langle u_j|$ be a Hermitian positive semidefinite matrix. Show that the transpose matrix X^T is also positive semidefinite.

1.4. Let X_θ,Y_θ be functions that are matrix valued. Show that the derivative of the product $(X_\theta Y_\theta)'$ can be written as $(X_\theta Y_\theta)' = X_\theta Y'_\theta + X'_\theta Y_\theta$, where X'_θ, Y'_θ are the derivatives of X_θ, Y_θ, respectively. Show that the derivative of $\operatorname{Tr} X_\theta$, i.e., $(\operatorname{Tr} X_\theta)'$, is equal to $\operatorname{Tr}(X'_\theta)$.

1.5. Let U be a unitary matrix and X be a matrix. Show that the equation $(UXU^*)^* = UX^*U^*$ holds. Also give a counterexample of the equation $(UXU^*)^T = UX^TU^*$.

1.2 State and Measurement in Quantum Systems

To discuss information processing in quantum systems, we must first be able to determine the probability that every measurement value appears as an outcome of the measurement. Few standard texts on quantum mechanics give a concise and accurate description of the probability distribution of each measurement value. Let us discuss the fundamental framework of quantum theory so as to calculate the probability distribution of a measurement value.

In the spin-$\frac{1}{2}$ system discussed previously, when the direction of "spin" changes, the condition of the particle also changes. In quantum systems, a description of the current condition of a system, such as the direction of the spin, is called a *state*. Any state is described by a Hermitian matrix ρ called a *density matrix* or simply *density*:

$$\operatorname{Tr} \rho = 1, \quad \rho \geq 0. \tag{1.7}$$

[1] A vector space closed under commutation $[X,Y]$ is called a Lie algebra. A vector space closed under the symmetrized product $X \circ Y$ is called a Jordan algebra.

Since quantum systems are too microscopic for direct observation, we must perform some measurement in order to extract information from the system. Such a measurement is described by a set of Hermitian matrices $\boldsymbol{M} \stackrel{\text{def}}{=} \{M_\omega\}_{\omega \in \Omega}$ satisfying the following conditions:

$$M_\omega \geq 0, \quad \sum_{\omega \in \Omega} M_\omega = I,$$

where I denotes the identity matrix. The set $\boldsymbol{M} = \{M_\omega\}_{\omega \in \Omega}$ is called a *positive operator valued measure* (*POVM*). For readers who have read standard texts on quantum mechanics, note that M_ω is not restricted to projection matrices. When ω is continuous, the summation $\sum$ is replaced by an integration $\int$ on Ω. Here we denote the set of the measurement values ω by Ω and omit it if there is no risk of confusion. If ω takes values in a discrete set, then the number of elements in Ω will be denoted by $|\boldsymbol{M}|$.

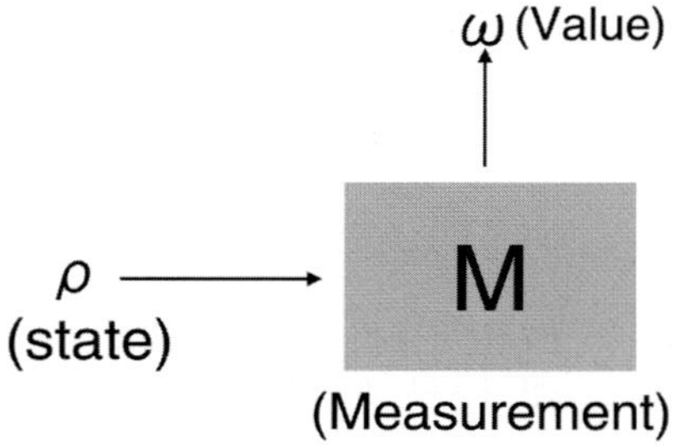

Fig. 1.1. Measurement scheme

The density matrix ρ and the POVM M form the mathematical representations of a state of the system and the measurement, respectively, in the following sense. If a measurement corresponding to $\boldsymbol{M} = \{M_\omega\}_{\omega \in \Omega}$ is perfomed on a system in a state corresponding to ρ, then the probability $\mathrm{P}_\rho^{\boldsymbol{M}}(\omega)$ of obtaining ω is[2]

$$\mathrm{P}_\rho^{\boldsymbol{M}}(\omega) := \operatorname{Tr} \rho M_\omega . \tag{1.8}$$

The above definition satisfies the axioms of a probability since $\operatorname{Tr} \rho M_\omega$ is positive and its summation becomes 1, as follows. The inequality $\operatorname{Tr} \rho M_\omega \geq 0$ can be verified by the fact that ρ and M_ω are both positive semidefinite. Furthermore, since

[2] In quantum mechanics, one often treats the state after the measurement rather than before it. The state change due to a measurement is called *state reduction*, and it requires more advanced topics than those described here. Therefore, we postpone its discussion until Sect. 7.1.

$$\sum_{\omega\in\Omega} \operatorname{Tr} \rho M_\omega = \operatorname{Tr} \rho \sum_{\omega\in\Omega} M_\omega = \operatorname{Tr} \rho I = \operatorname{Tr} \rho = 1 ,$$

we see that this is indeed a probability distribution. In this formulation, it is implicitly assumed that both the state and the measurement are reproducible (otherwise, it would be virtually impossible to verify experimentally (1.8)). For brevity, we shall henceforth refer to the system, state, and measurement by $\mathcal{H}$, ρ, and $\boldsymbol{M}$, respectively.

Let us now discuss the structure of a set of density matrices that shall be denoted by $\mathcal{S}(\mathcal{H})$. Consider a system that is in state ρ_1 with a probability λ and in state ρ_2 with a probability $1-\lambda$. Now, let us perform a measurement $\boldsymbol{M} = \{M_\omega\}$ on the system. The probability of obtaining the measurement value ω is given by

$$\lambda \operatorname{Tr} \rho_1 M_\omega + (1-\lambda) \operatorname{Tr} \rho_2 M_\omega = \operatorname{Tr}[(\lambda\rho_1 + (1-\lambda)\rho_2) M_\omega] . \tag{1.9}$$

The state of the system may be considered to be given by $\rho' = \lambda\rho_1 + (1-\lambda)\rho_2$. Thus, even by using (1.8) with this state and calculating the probability distributions of the measurement values, we will still be entirely consistent with the experiment. Therefore, we may believe that the state of the system is given by ρ'. This is called a probabilistic mixture (or incoherent mixture).

In quantum mechanics, a state $|u\rangle\langle u| \in \mathcal{S}(\mathcal{H})$ represented by a vector $u \in \mathcal{H}$ of norm 1 is called a *pure state.* This u is referred to as a state in the sense of $|u\rangle\langle u|$. The set of all vectors of norm 1 is written as $\mathcal{H}^1$. In contrast, when a state is not a pure state, it is called a *mixed state.* A pure state cannot be written as the probabilistic mixture of other states. However, all mixed states may be written as the probabilistic mixture of other states, such as pure states. For example, if the dimensionality of $\mathcal{H}$ is d, then $\frac{1}{d}I$ is a mixed state. In fact, it is called a *completely mixed state* and is written as ρ_{mix}.

On the other hand, the vector $|x\rangle = \sum_i x^i |u_i\rangle$ is a quantum-mechanical superposition of $u_1, \ldots, u_d$, where $u_1, \ldots, u_d$ are the orthonormal basis states of $\mathcal{H}$. Note that this is different from the probabilistic mixture discussed above. The probabilistic mixture is independent of the choice of the orthonormal basis. However, the quantum-mechanical superposition depends on the choice of the basis, which depends on the physical properties of the system under consideration.

When the operators M_ω in a POVM $\boldsymbol{M} = \{M_\omega\}$ are projection matrices, the POVM is called a *projection valued measure* (*PVM*) (only PVMs are examined in elementary courses in quantum mechanics). This is equivalent to $M_\omega M_{\omega'} = 0$ for different ω, ω'. Hermitian matrices are sometimes referred to as "observables" or "physical quantities." We now explain why.

Let the eigenvalues of a Hermitian matrix X be x^i, and the projection matrices corresponding to this eigenspace be $E_{X,i}$, i.e., $X = \sum_i x^i E_{X,i}$. The right-hand side of this equation is called the *spectral decomposition* of X. The decomposition $\boldsymbol{E}_X = \{E_{X,i}\}$ is then a PVM. When more than one eigenvector corresponds to a single eigenvalue, the diagonalization $X = \sum_{i=1}^d x^i |u_i\rangle\langle u_i|$

is not unique, while the spectral decomposition is unique. Using (1.8), we may calculate the expectation value and variance of a PVM $\boldsymbol{E}_X$ acting on a state ρ as $\operatorname{Tr} \rho X$ and $\operatorname{Tr} \rho X^2 - (\operatorname{Tr} \rho X)^2$, respectively. Note that these are expressed completely in terms of X and ρ. Therefore, we identify the Hermitian matrix X as PVM $\boldsymbol{E}_X$ and refer to it as the measurement for the Hermitian matrix X.

When two Hermitian matrices X, Y commute, we can use a common orthonormal basis $u_1, \ldots, u_d$ and two sets of real numbers $\{x_i\}, \{y_i\}$ to diagonalize the matrices

$$X = \sum_i x^i |u_i\rangle\langle u_i|, \qquad Y = \sum_i y^i |u_i\rangle\langle u_i| . \tag{1.10}$$

A measurement of X and Y may be performed simultaneously using the PVM $\{|u_i\rangle\langle u_i|\}_i$. Evidently, if all $X_1, \ldots, X_k$ commute, it is also possible to diagonalize them using a common basis.

In general, the elements M_ω of the PVM $\boldsymbol{M} = \{M_\omega\}$ and the state ρ do not necessarily commute. This noncommutativity often causes many difficulties in their mathematical treatment. To avoid these difficulties, we sometimes use the *pinching* map $\kappa_{\boldsymbol{M}}$ defined by

$$\kappa_{\boldsymbol{M}}(\rho) \stackrel{\text{def}}{=} \sum_\omega M_\omega \rho M_\omega . \tag{1.11}$$

This is because the pinching map $\kappa_{\boldsymbol{M}}$ modifies the state ρ such that the state becomes commutative with M_ω. Hence, the pinching map is an important tool for overcoming the difficulties associated with noncommutativities. We often treat the case when PVM M is expressed as the spectral decomposition of a Hermitian matrix X. In such a case, we use the shorthand κ_X instead of $\kappa_{\boldsymbol{E}_X}$. That is,

$$\kappa_X \stackrel{\text{def}}{=} \kappa_{\boldsymbol{E}_X} . \tag{1.12}$$

For a general POVM $\boldsymbol{M}$, we may define

$$\kappa_{\boldsymbol{M}}(\rho) \stackrel{\text{def}}{=} \sum_\omega \sqrt{M_\omega} \rho \sqrt{M_\omega} . \tag{1.13}$$

Note that this operation does not necessarily have the same effect as making the matrices commute. Further, to treat the data as well as the state, we define

$$\hat{\kappa}_{\boldsymbol{M}}(\rho) \stackrel{\text{def}}{=} \sum_\omega \sqrt{M_\omega} \rho \sqrt{M_\omega} \otimes |e_\omega\rangle\langle e_\omega| , \tag{1.14}$$

where $\{e_\omega\}$ is a CONS of another system.

Exercises

1.6. Show that when one performs a PVM $\boldsymbol{E}_X$ on a system in a state ρ, the expectation value and the variance are given by $\operatorname{Tr} \rho X$ and $\operatorname{Tr} \rho X^2 - (\operatorname{Tr} \rho X)^2$, respectively.

1.3 Quantum Two-Level Systems

A quantum system with a two-dimensional representation space is called a *quantum two-level system* or a *qubit*, which is the abbreviation for quantum bit. This is a particularly important special case for examining a general quantum system. The spin-$\frac{1}{2}$ system, which represents a particle with a total angular momentum of $\frac{1}{2}$, is the archetypical quantum two-level system. The electron is an example of such a spin-$\frac{1}{2}$ system. A spin-$\frac{1}{2}$ system precisely represents a specific case of angular momentum in a real system; however, it is sometimes referred to as any quantum system with two levels. In particle physics, one comes across a quantum system of isospin, which does not correspond to the motional degrees of freedom but to purely internal degrees of freedom.

Mathematically, they can be treated in the same way as spin-$\frac{1}{2}$ systems. In this text, since we are interested in the general structure of quantum systems, we will use the term quantum two-level system to refer generically to all such systems.

In particular, the Hermitian matrices S_0, S_1, S_2, S_3 given below are called *Pauli matrices*:

$$S_0 = I, \quad S_1 = \begin{pmatrix} 0 & 1 \\ 1 & 0 \end{pmatrix}, \quad S_2 = \begin{pmatrix} 0 & -i \\ i & 0 \end{pmatrix}, \quad S_3 = \begin{pmatrix} 1 & 0 \\ 0 & -1 \end{pmatrix}. \tag{1.15}$$

They will help to simplify the expressions of matrices. The density matrix can be parameterized by using the Pauli matrices:

$$\rho_{\boldsymbol{x}} = \frac{1}{2} \begin{pmatrix} 1 + x^3 & x^1 - x^2 i \\ x^1 + x^2 i & 1 - x^3 \end{pmatrix} = \frac{1}{2} \left(S_0 + \sum_{i=1}^{3} x^i S_i \right), \tag{1.16}$$

which is called Stokes parameterization.The range of $\boldsymbol{x} = (x^1, x^2, x^3)$ is the set $\{\boldsymbol{x} | \sum_{i=1}^{3} (x^i)^2 \leq 1\}^{\text{Ex. 1.7}}$. We often focus on the basis

$$e_0 \stackrel{\text{def}}{=} \begin{pmatrix} 1 \\ 0 \end{pmatrix}, \quad e_1 \stackrel{\text{def}}{=} \begin{pmatrix} 0 \\ 1 \end{pmatrix}$$

in the space $\mathbb{C}^2$. If there are several representation spaces $\mathcal{H}_A$, $\mathcal{H}_B$, etc. equivalent to $\mathbb{C}^2$, and we wish to specify the space of the basis e_0, e_1, we will write e_0^A, e_1^A, etc. In this case, the Pauli matrix S_i and the identity matrix will be denoted by S_i^A and I_A, respectively.

Next, we consider the measurements in a quantum two-level system. The measurement of the observable S_1 is given by the PVM $\boldsymbol{E}_{S_1} = \{E_1, E_{-1}\}$, where

$$E_1 = \frac{1}{2}\begin{pmatrix} 1 & 1 \\ 1 & 1 \end{pmatrix}, \quad E_{-1} = \frac{1}{2}\begin{pmatrix} 1 & -1 \\ -1 & 1 \end{pmatrix}.$$

Given a density matrix $\rho_{\boldsymbol{x}}$, the probability $\mathrm{P}_{\rho_{\boldsymbol{x}}}^{\boldsymbol{E}_{S_1}}(1)$ of obtaining the measurement value 1 is $\operatorname{Tr}\rho E_1 = \frac{1+x_1}{2}$. Similarly, the probability $\mathrm{P}_{\rho_{\boldsymbol{x}}}^{\boldsymbol{E}_{S_1}}(-1)$ of obtaining the measurement value -1 is $\frac{1-x_1}{2}$. A more detailed treatment of quantum two-level systems will be deferred until Sect. 5.3.

Exercises

1.7. Verify that the set $\boldsymbol{x} = (x^1, x^2, x^3)|\rho \geq 0\}$ is equal to $\{\boldsymbol{x}| \sum_{i=1}^{3}(x^i)^2 \leq 1\}$ by showing $\det \rho_{\boldsymbol{x}} = \frac{1-\|\boldsymbol{x}\|^2}{4}$.

1.8. Verify that $\rho_{\boldsymbol{x}}$ is a pure state if and only if $\|\boldsymbol{x}\| = 1$.

1.9. Show that all 2×2 Hermitian matrices with trace 0 can be written as a linear combination of S^1, S^2, S^3.

1.4 Composite Systems and Tensor Products

A combined system composed of two systems $\mathcal{H}_A$ and $\mathcal{H}_B$ is called the *composite system* of $\mathcal{H}_A$ and $\mathcal{H}_B$. When the system $\mathcal{H}_A$ ($\mathcal{H}_B$) has an orthonormal basis $\{u_1^A, \ldots, u_d^A\}$ ($\{u_1^B, \ldots, u_{d_B}^B\}$), respectively, the representation space of the composite system is given by the Hilbert space $\mathcal{H}_A \otimes \mathcal{H}_B$ with the orthonormal basis $\{u_1^A \otimes u_1^B, \ldots, u_1^A \otimes u_{d_B}^B, u_2^A \otimes u_1^B, \ldots, u_2^A \otimes u_{d_B}^B, \ldots, u_{d_A}^A \otimes u_1^B, \ldots, u_{d_A}^A \otimes u_{d_B}^B\}$. The space $\mathcal{H}_A \otimes \mathcal{H}_B$ is called the *tensor product space* of $\mathcal{H}_A$ and $\mathcal{H}_B$; its dimension is $d_A \times d_B$. Using $d_A \times d_B$ complex numbers $(z^{i,j})$, the elements of $\mathcal{H}_A \otimes \mathcal{H}_B$ may be written as $\sum_{i,j} z^{i,j} u_i^A \otimes u_j^B$. The tensor product of two vectors $u^A = \sum_k x^k u_k^A$ and $u^B = \sum_j y^j u_j^B$ is defined as $u^A \otimes u^B \stackrel{\text{def}}{=} \sum_k \sum_j x^k y^j u_k^A \otimes u_j^B$. We simplify this notation by writing $|u^A \otimes u^B\rangle$ as $|u^A, u^B\rangle$.

The tensor product $X_A \otimes X_B$ of a matrix X_A on $\mathcal{H}_A$ and a matrix X_B on $\mathcal{H}_B$ is defined as

$$X_A \otimes X_B(u_i \otimes v_j) \stackrel{\text{def}}{=} X_A(u_i) \otimes X_B(v_j).$$

The trace of this tensor product satisfies the relation

$$\operatorname{Tr} X_A \otimes X_B = \operatorname{Tr} X_A \cdot \operatorname{Tr} X_B.$$

For two matrices X_A and Y_A on $\mathcal{H}_A$ and two matrices X_B and Y_B on $\mathcal{H}_B$,

$$(X_A \otimes X_B)(Y_A \otimes Y_B) = (X_A Y_A) \otimes (X_B Y_B) .$$

Hence it follows that

$$\mathrm{Tr}(X_A \otimes X_B)(Y_A \otimes Y_B) = \mathrm{Tr}(X_A Y_A) \cdot \mathrm{Tr}(X_B Y_B) .$$

If the states of systems $\mathcal{H}_A$ and $\mathcal{H}_B$ are independent and represented by the densities ρ_A and ρ_B, respectively, then the state of the composite system may be represented by the tensor product of the density matrices $\rho_A \otimes \rho_B$. Such a state is called a *tensor product state.*

When the density matrix ρ on the composite system $\mathcal{H}_A \otimes \mathcal{H}_B$ can be written as a probabilistic mixture of tensor product states, it is called *separable*:

$$\rho = \sum_i p_i \rho_A^i \otimes \rho_B^i, \quad p_i \geq 0, \quad \sum_i p_i = 1, \quad \rho_A^i \in \mathcal{S}(\mathcal{H}_A), \rho_B^i \in \mathcal{S}(\mathcal{H}_B). \tag{1.17}$$

Such separable states do not have any quantum-mechanical correlation (entanglement). When a state ρ does not have the form (1.17), it is called an *entangled state.*

When all the n systems are identical to $\mathcal{H}$, their composite system is written as $\underbrace{\mathcal{H} \otimes \cdots \otimes \mathcal{H}}_{n}$; this will be denoted by $\mathcal{H}^{\otimes n}$ for brevity. In particular, if all the quantum systems are independent, and the state in each system is given by ρ, the composite state on $\mathcal{H}^{\otimes n}$ is $\underbrace{\rho \otimes \cdots \otimes \rho}_{n}$, which is denoted by $\rho^{\otimes n}$. Such states can be regarded as quantum versions of independent and identical distributions (discussed later).

Let us now focus on the composite state of the quantum two-level systems $\mathcal{H}_A$ and $\mathcal{H}_B$. By defining $e_0^{A,B} \stackrel{\mathrm{def}}{=} \frac{1}{\sqrt{2}} \left(e_0^A \otimes e_0^B + e_1^A \otimes e_1^B \right)$, we see that $|e_0^{A,B}\rangle\langle e_0^{A,B}|$ is not separable, i.e., it is an entangled state. Other entangled states include

$$\begin{aligned}
e_1^{A,B} &\stackrel{\mathrm{def}}{=} (S_1^A \otimes I_B) e_0^{A,B} = \frac{1}{\sqrt{2}} \left(e_1^A \otimes e_0^B + e_0^A \otimes e_1^B \right) , \\
e_2^{A,B} &\stackrel{\mathrm{def}}{=} (S_2^A \otimes I_B) e_0^{A,B} = \frac{i}{\sqrt{2}} \left(e_1^A \otimes e_0^B - e_0^A \otimes e_1^B \right) , \\
e_3^{A,B} &\stackrel{\mathrm{def}}{=} (S_3^A \otimes I_B) e_0^{A,B} = \frac{1}{\sqrt{2}} \left(e_0^A \otimes e_0^B - e_1^A \otimes e_1^B \right) .
\end{aligned}$$

They are mutually orthogonal, i.e.,

$$\langle e_k^{A,B} | e_l^{A,B} \rangle = \delta_{k,l} . \tag{1.18}$$

In general, any vector $|x\rangle$ on $\mathcal{H}_A \otimes \mathcal{H}_B$ can be expressed as $|x\rangle = |\sum_{i,j} x^{i,j} u_i^A \otimes u_j^B\rangle = (X \otimes I_B)|\sum_{i=1}^{d_B} u_i^B \otimes u_i^B\rangle$ by a linear map $X = \sum_{i,j} x^{i,j} |u_i^A\rangle\langle u_j^B|$ from $\mathcal{H}_B$ to $\mathcal{H}_A$. When we describe this vector by $|X\rangle$,[3] we obtain the following properties:

$$(Y \otimes Z^T)|X\rangle = |YXZ\rangle, \tag{1.19}$$

$$\langle Y|X\rangle = \operatorname{Tr} Y^* X\,. \tag{1.20}$$

In particular, when $\sqrt{d}X$ is a unitary matrix, $|X\rangle\langle X|$ is called a *maximally entangled state* of size d. In this book, we denote the vector $|\frac{1}{\sqrt{d}} I\rangle$ by $|\Phi_d\rangle$.

Next, let us consider the independent applications of the measurements $\boldsymbol{M}_A = \{M_{A,\omega_A}\}_{\omega_A \in \Omega_A}$ and $\boldsymbol{M}_B = \{M_{B,\omega_B}\}_{\omega_B \in \Omega_B}$ on systems $\mathcal{H}_A$ and $\mathcal{H}_B$, respectively. This is equivalent to performing a measurement $\boldsymbol{M}_A \otimes \boldsymbol{M}_B \stackrel{\text{def}}{=} \{M_{A,\omega_A} \otimes M_{B,\omega_B}\}_{(\omega_A,\omega_B) \in \Omega_A \times \Omega_B}$ on the composite system. Such a measurement is called an *independent* measurement. If a measurement $\boldsymbol{M} = \{M_\omega\}_{\omega \in \Omega}$ on the composite system $\mathcal{H}_A \otimes \mathcal{H}_B$ has the form

$$M_\omega = M_{A,\omega} \otimes M_{B,\omega}, \quad M_{A,\omega} \geq 0, M_{B,\omega} \geq 0 \tag{1.21}$$

or the form

$$M_\omega = \sum_i M_{A,\omega,i} \otimes M_{B,\omega,i}, \quad M_{A,\omega,i} \geq 0, M_{B,\omega,i} \geq 0\,,$$

the measurement $\boldsymbol{M}$ is said to be separable. Otherwise, it is called collective. Of course, independent measurements are always separable, but the converse is not always true.

Since the vectors $e_0^{A,B}, \ldots, e_3^{A,B}$ defined previously form an orthonormal basis in the composite system $\mathbb{C}^2 \otimes \mathbb{C}^2$, the set $\{|e_0^{A,B}\rangle\langle e_0^{A,B}|, \ldots, |e_3^{A,B}\rangle\langle e_3^{A,B}|\}$ is a PVM. This measurement is a collective measurement because it does not have the separable form (1.21).

On the other hand, *adaptive* measurements are known as a class of separable POVMs, and their definition is given as follows.[4] Consider a case in which we perform a measurement $\boldsymbol{M}_A = \{M_{A,\omega_A}\}_{\omega_A \in \Omega_A}$ on system $\mathcal{H}_A$ and then another measurement $\boldsymbol{M}_B^{\omega_A} = \{M_{B,\omega_B}^{\omega_A}\}_{\omega_B \in \Omega_B}$ on system $\mathcal{H}_B$ according to the measurement value ω_A. The POVM of this measurement on the composite system $\mathcal{H}_A \otimes \mathcal{H}_B$ is given as

$$\{M_{A,\omega_A} \otimes M_{B,\omega_B}^{\omega_A}\}_{(\omega_A,\omega_B) \in \Omega_A \times \Omega_B}\,. \tag{1.22}$$

Such a measurement is called *adaptive*, and it satisfies the separable condition (1.21).

---

[3] A notation similar to $|X\rangle$ was introduced in [88]. However, the relations (1.19) and (1.20) were essentially pointed out in [230].

[4] Adaptive measurements are often called one-way LOCC measurements in entanglement theory. See Sect. 8.1.

Presently, there is not enough information on how different the adaptive condition (1.22) is from the separable condition (1.21). In Chaps. 3 and 4, we focus on the restriction of our measurements to separable or adaptive measurements and discuss the extent of its effects on the performance of information processing.

Similarly, a separable measurement $\boldsymbol{M} = \{M_\omega\}_{\omega\in\Omega}$ in a composite system $\mathcal{H}_1 \otimes \ldots \otimes \mathcal{H}_n$ of n systems $\mathcal{H}_1, \ldots, \mathcal{H}_n$ is given by

$$M_\omega = M_{1,\omega} \otimes \cdots \otimes M_{n,\omega}, \quad M_{1,\omega} \geq 0, \ldots, M_{n,\omega} \geq 0 .$$

An adaptive measurement may be written in terms of a POVM as

$$\{M_{1,\omega_1} \otimes \cdots \otimes M_{n,\omega_n}^{\omega_1,\ldots,\omega_{n-1}}\}_{(\omega_1,\ldots,\omega_n)\in\Omega_1\times\cdots\times\Omega_n} .$$

We also denote n applications of the POVM $\boldsymbol{M}$ on the composite system $\mathcal{H}^{\otimes n}$ by $\boldsymbol{M}^{\otimes n}$.

Consider a composite system $\mathcal{H}_A\otimes\mathcal{H}_B$ in a state $\rho \in \mathcal{S}(\mathcal{H}_A\otimes\mathcal{H}_B)$. Assume that we can directly access only system $\mathcal{H}_A$ for performing measurements. In this case, we would only be interested in the state of system $\mathcal{H}_A$, and the density matrix on $\mathcal{H}_A$ is given by the *reduced density* matrix $\mathrm{Tr}_{\mathcal{H}_B} \rho \in \mathcal{S}(\mathcal{H}_A)$, which is defined to satisfy

$$\mathrm{Tr}(\mathrm{Tr}_{\mathcal{H}_B} \rho) X = \mathrm{Tr}(X \otimes I_{\mathcal{H}_B})\rho . \tag{1.23}$$

Then, $\mathrm{Tr}_{\mathcal{H}_B}$ can be regarded as a map from the density on the composite system to the reduced density matrix and called a *partial trace*, often abbreviated to Tr_B. To specify the space on which the trace is acting, we denote the trace by Tr_A even if it is a full trace. The partial trace can be calculated according to

$$\rho^{i,j} = \sum_{k=1}^{d'} \langle u_i^A \otimes u_k^B|\rho|u_j^A \otimes u_k^B\rangle, \quad \mathrm{Tr}_B \rho = \sum_{i,j} \rho^{i,j} |u_i^A\rangle\langle u_j^A| , \tag{1.24}$$

where the orthonormal basis of $\mathcal{H}_A$ ($\mathcal{H}_B$) is $u_1^A, \ldots, u_d^A$ ($u_1^B, \ldots, u_{d'}^B$). This may also be written as

$$\mathrm{Tr}_B \rho = \sum_{i,i'} \Big(\sum_j \rho^{(i,j),(i'j)}\Big) |u_i^A\rangle\langle u_{i'}^A| , \tag{1.25}$$

where $\rho = \sum_{i,j,i',j'} \rho^{(i,j),(i'j')} |u_i^A, u_j^A\rangle\langle u_{i'}^A, u_{j'}^B|$. We may also write

$$\mathrm{Tr}_B \rho = \sum_{k=1}^{d'} P_k \rho P_k , \tag{1.26}$$

where P_k is a projection from $\mathcal{H}_A \otimes \mathcal{H}_B$ to $\mathcal{H}_A \otimes u_k^B$.

Exercises

1.10. Suppose that the spaces $\mathcal{H}_A$ and $\mathcal{H}_B$ also have other bases $\{v_1^A, \dots, v_{d_A}^A\}$ and $\{v_1^B, \dots, v_{d_B}^B\}$ and that the unitary matrices $V_A = (v_A^{ij})$ and $V_B = (v_B^{ij})$ satisfy $v_j^A = \sum_i v_A^{ij} u_i^A$ and $v_j^B = \sum_i v_B^{ij} u_i^B$. Show that $v_j^A \otimes v_k^B = \sum_{i,l} v_A^{ij} v_B^{lk} u_i^A \otimes u_l^B$. Hence, the definition of the tensor product is independent of the choice of the bases on $\mathcal{H}_A$ and $\mathcal{H}_B$.

1.11. Prove (1.18).

1.12. Prove formulas (1.24)–(1.26), which calculate the partial trace.

1.13. Show that the following two conditions are equivalent, following the steps below, for two Hermitian matrices $\rho_A \geq 0$ and $\sigma_A \geq 0$ on $\mathcal{H}_A$ and another two Hermitian matrices $\rho_B \geq 0$ and $\sigma_B \geq 0$ on $\mathcal{H}_B$.

- $[\rho_A \otimes \rho_B, \sigma_A \otimes \sigma_B] = 0.$ (1.27)
- $(\mathrm{Tr}\, \sigma_B \rho_B)[\rho_A, \sigma_A] = 0$ and $(\mathrm{Tr}\, \sigma_A \rho_A)[\rho_B, \sigma_B] = 0.$ (1.28)

a Show that (1.27) holds when $[\rho_A, \sigma_A] = [\rho_B, \sigma_B] = 0$.
b Show that (1.27) holds when $\mathrm{Tr}\, \rho_A \sigma_A = 0$.
c Show that (1.28) $\Rightarrow$ (1.27).
d Show that (1.27) $\Rightarrow$ (1.28).

1.14. Show that $\mathrm{Tr}_B X(I_A \otimes Y) = \mathrm{Tr}_B(I_A \otimes Y)X$, where X is a matrix on $\mathcal{H}_A \otimes \mathcal{H}_B$ and Y is a matrix on $\mathcal{H}_B$.

1.15. Further, show that the following formula is satisfied when ρ and ρ_0 are states on $\mathcal{H}_A$ and $\mathcal{H}_B$:

$$\begin{aligned} &\mathrm{Tr}_B \sqrt{\rho \otimes \rho_0}[X, Y \otimes I_B]\sqrt{\rho \otimes \rho_0} \\ =&\sqrt{\rho}[\mathrm{Tr}_B \left(I_A \otimes \sqrt{\rho_0}\right) X \left(I_A \otimes \sqrt{\rho_0}\right), Y]\sqrt{\rho}\,. \end{aligned}$$

1.16. Let P be a projection from $\mathcal{H}_A \otimes \mathcal{H}_B$ to the subspace $\{u^A \otimes u^B | u^B \in \mathcal{H}_B\}$ for any elemenet $u^A \in \mathcal{H}_A$. Show that

$$\mathrm{Tr}_A(|u^A\rangle\langle u^A| \otimes I_B)X = \mathrm{Tr}_A\, PXP\,.$$

1.5 Matrix Inequalities and Matrix Monotone Functions

In later chapters, we will encounter quantities such as error probabilities that require us to handle inequalities in various situations. Of course, probabilities such as error probabilities are real numbers. However, in quantum systems these probabilities are expressed in terms of matrices, as we show in (1.8). Therefore, it is often helpful to use inequalities involving matrices when evaluating probabilities. By using the definition of positive semidefiniteness defined in Sect. 1.1, we may define the order (matrix inequality)

$$X \geq Y \stackrel{\text{def}}{\Longleftrightarrow} X - Y \geq 0$$

for two Hermitian matrices X and Y. Such an order requires some care as it may involve some unexpected pitfalls arising from the noncommutativity of X and Y. In order to examine this order in greater detail, let us first analyze the properties of positive semidefiniteness again. Let X be a $d \times d$ positive semidefinite (≥ 0) Hermitian matrix and Y be a $d \times d'$ matrix. It follows that $Y^* XY$ is a $d' \times d'$ positive semidefinite Hermitian matrix. This can be verified from

$$\langle v|Y^* XY|v\rangle = \langle Yv|X|Yv\rangle \geq 0\,,$$

where v is a vector of length d'. Furthermore, if X_1 and X_2 are two $d \times d$ Hermitian matrices satisfying $X_1 \geq X_2$, it follows that

$$Y^* X_1 Y \geq Y^* X_2 Y\,. \tag{1.29}$$

If the matrices commute, then some additional types of matrix inequalities hold. For example, if $d \times d$ positive semidefinite Hermitian matrices X and Y commute, then

$$X \circ Y = \frac{1}{2}(XY + YX) \geq 0\,. \tag{1.30}$$

Inequality (1.30) does not hold unless X and Y commute. A simple counterexample exists for the noncommuting case.

Let X_1 and X_2 be two $d \times d$ Hermitian matrices satisfying $X_1 \geq X_2 \geq 0$, and Y be a $d \times d$ positive semidefinite Hermitian matrix. If all these matrices commute, we have

$$X_1 Y X_1 \geq X_2 Y X_2\,. \tag{1.31}$$

Inequality (1.31) does not hold unless all matrices commute. In general, when noncommutativity is involved, matrix inequalities are more difficult to handle and should therefore be treated with care.

Let us now define the projection $\{X \geq 0\}$ with respect to a Hermitian matrix X with a spectral decomposition $X = \sum_i x_i E_{X,i}$ [306]:

$$\{X \geq 0\} \stackrel{\text{def}}{=} \sum_{x_i \geq 0} E_{X,i}\,. \tag{1.32}$$

Consider the probability of the set $\{x_i \geq 0\}$ containing the measurement value for a measurement corresponding to the spectral decomposition $\{E_{X,i}\}$ of X. This probability is $\sum_{x_i \geq 0} \operatorname{Tr} \rho E_{X,i} = \operatorname{Tr} \rho \{X \geq 0\}$ when the state is given as a density ρ. Therefore, this notation generalizes the concept of the subset to the noncommuting case. In other words, the probability $\operatorname{Tr} \rho\{X \geq 0\}$ can be regarded as a generalization of the probability $p\{\omega \in \Omega | X(\omega) \geq 0\}$, where p is a probability distribution and X is a random variable. Similarly, we may also define $\{X > 0\}$, $\{X \leq 0\}$, $\{X < 0\}$, and $\{X \neq 0\}$.

If two Hermitian matrices X and Y commute, we obtain the matrix inequality

$$\{X \geq 0\} + \{Y \geq 0\} \geq \{X + Y \geq 0\} \tag{1.33}$$

in the sense defined above. The range of the projection $\{X \neq 0\}$ is called the *support* of X. If the support of $\{X \neq 0\}$ is not equal to I, then no inverse exists. In this case, the Hermitian matrix Y satisfying $XY = YX = \{X \neq 0\}$ is called the *generalized inverse matrix* of a Hermitian matrix X.

We may verify (1.33) by noting that the LHS − RHS is equal to $\sum_{x_i, y_i \geq 0} |u_i\rangle\langle u_i|$, for the matrices $X = \sum_i x^i |u_i\rangle\langle u_i|$ and $Y = \sum_i y_i |u_i\rangle\langle u_i|$ and orthonormal basis $\{u_1, \ldots, u_d\}$. It should be noted that this is not generally true unless X and Y commute. It is known that two noncummutative Hermitian matrices X and Y cannot be diagonalized simultaneously. This fact often causes many technical difficulities in the above method.

We now examinine *matrix monotone functions*, which are useful for dealing with matrix inequalities. Given a function f, which maps a real number to a real number, we denote the Hermitian matrix $\sum_i f(x_i) E_{X,i}$ by $f(X)$ with respect to a Hermitian matrix $X = \sum_i x_i E_{X,i}$.

f is called a matrix monotone function in $[0, \infty)$ if $f(X) \geq f(Y)$ for two Hermitian matrices X and Y satisfying $X \geq Y$ with eigenvalues $[0, \infty)$. Some known matrix monotone functions in $[0, \infty)$ are, for example, $f(x) = x^s$ $(0 < s \leq 1)$, $f(x) = \log x$, and $f(x) = -1/x$ [49]. See Exercise A.7 for the $s = 1/2$ case. Since the function $f(x) = -x^{-s}$ $(0 < s \leq 1)$ is the composite function of $-1/x$ and x^s, it is also a matrix monotone function. Note that the function $f(x) = x^s$ $(s > 1)$ $(f(x) = x^2$, etc.) is not matrix monotone [Ex. 1.21].

Exercises

1.17. Suppose that X and Y commute. Show inequalities (1.30) and (1.33) using (1.10).

1.18. Verify inequality (1.31) when X_1, X_2, and Y commute.

1.19. Show that $X = \begin{pmatrix} 1 & 0 \\ 0 & 0 \end{pmatrix}$, $Y = \frac{1}{2}\begin{pmatrix} 1 & 1 \\ 1 & 1 \end{pmatrix}$ form a counterexample to (1.30).

1.20. Show that $X_1 = I$, $X_2 = \begin{pmatrix} 1 & 0 \\ 0 & 0 \end{pmatrix}$, $Y = \frac{1}{2}\begin{pmatrix} 1 & 1 \\ 1 & 1 \end{pmatrix}$ form a counterexample to (1.31).

1.21. Verify that the following X and Y provide a counterexample to $f(x) = x^2$ as a matrix monotone function:

$$X = \begin{pmatrix} 1 & 1 \\ 1 & 1 \end{pmatrix}, \quad Y = \begin{pmatrix} 2 & 1 \\ 1 & 1 \end{pmatrix}.$$

1.22. Show that $\operatorname{rank}\{X - xI \geq 0\} \geq \operatorname{rank} P$ when a Hermitian matrix X, a projection P, and real number x satisfy $X \geq xP$.

1.23. Show that $\operatorname{Tr} X \geq \operatorname{Tr} |Y|$ for Hermitian matrices X and Y when $X \geq Y$ and $X \geq -Y$.

2

Information Quantities and Parameter Estimation in Classical Systems

Summary. For the study of quantum information theory, we require fundamental knowledge of information theory, mathematical statistics, and information geometry, which are mainly examined in a nonquantum context. This chapter briefly summarizes the fundamentals of these topics from a unified viewpoint. Since these topics are usually treated individually, this chapter will be useful even for a nonquantum applications.

Table 2.1. Denotations used in Chap. 2

Information quantities in classical system	
$H(p)$	Entropy of distribution p (2.1)
$H(X)$	Entropy of random variable X
$h(x)$	Binary entropy
$H(X\|Y)$	Conditional entropy
$D(p\|\|q)$	Relative entropy (2.8)
$D_f(p\|\|q)$	f-relative entropy
$d_2(p,q)$	Hellinger distance (2.13)
$d_1(p,q)$	Variational distance (2.16)
$\psi(s\|p)$	Rényi entropy (2.29)
$\phi(s\|p\|\|q)$	Relative Rényi entropy
$I(X:Y)$	Mutual information (2.21)
$I(X:Y\|Z)$	Conditional mutual information (2.22)
$I(p,Q)$	Transmission information (2.25)
p_{mix}	Uniform distribution
$\mathrm{E}_p(X)$	Expectation of X under distribution p
$\mathrm{V}_p(X)$	Variance of X under distribution p
$\mathrm{Cov}_p(X,Y)$	Covariance between X and Y under distribution p
J_θ	Fisher information (2.75)
$\boldsymbol{J}_\theta$	Fisher information matrix
$l_\theta(\omega)$	Logarithmic derivative
$\mu(\theta)$	Moment function (potential function) (2.83), (2.84)

Table 2.1. Continued

Error criterion	
$\hat{\mathrm{V}}_\theta(\hat{\theta})$	Mean square error of estimator $\hat{\theta}$ (2.94)
$\beta(\{\hat{\theta}_n\},\theta,\epsilon)$	Rate function of error probability (2.126)
$\alpha(\{\hat{\theta}_n\},\theta)$	First-order coefficient of rate function (2.127)

Information quantities in quantum systems	
$H(\rho)$	von Neumann entropy
$\psi(s\|\rho)$	Rényi entropy
$D(\rho\\|\sigma)$	Quantum relative entropy (2.53)
$F(\rho,\sigma)$	Fidelity $\mathrm{Tr}\,\|\sqrt{\rho}\sqrt{\sigma}\|$
$b(\rho\\|\sigma)$	Bures distance (2.55)
$d_1(\rho,\sigma)$	Trace norm distance (2.58)
$\phi(s\|\rho\\|\sigma)$	Relative Rényi entropy

2.1 Information Quantities in Classical Systems

When all the given density matrices $\rho_1,\ldots,\rho_n$ commute, they may be simultaneously diagonalized using a common orthonormal basis $\{u^1,\ldots,u^d\}$ according to $\rho_1=\sum_i p_{1,i}|u^i\rangle\langle u^i|,\ldots,\rho_n=\sum_i p_{n,i}|u^i\rangle\langle u^i|$. In this case, it is sufficient to treat only the diagonal elements, i.e., we discuss only the probability distributions $p_1,\ldots,p_n$. Henceforth we will refer to such cases as *classical* because they do not exhibit any quantum properties. Let us now examine various information quantities with respect to probability distributions.

2.1.1 Entropy

Shannon entropy is defined as

$$H(p)\stackrel{\mathrm{def}}{=}\sum_{i=1}^{k}-p_i\log p_i \tag{2.1}$$

with respect to a probability distribution $p=\{p_i\}_{i=1}^k$.[1] It is often simply called *entropy*. By denoting the probability distribution of a random variable X by P_X, we write the entropy of P_X as $H(X)$. For $k=2$, the entropy of the probability distribution $(x,1-x)$ is called a *binary entropy*, and it is given by $h(x)\stackrel{\mathrm{def}}{=}-x\log x-(1-x)\log(1-x)$. As shown later, we may use the number of elements in the probability distribution p (i.e., k for the current case) to prove that

$$H(p)\leq\log k\,. \tag{2.2}$$

[1] In this case, we consider $0\log 0$ to be 0.

The entropy $H(\mathrm{P}_{X,Y}(x,y))$ of the joint distribution $\mathrm{P}_{X,Y}(x,y)$ for two random variables X and Y is denoted by $H(X,Y)$. In particular, if Y can be expressed as $f(X)$, where f is some function, then$^{\text{Ex. 2.1}}$

$$H(X,Y) = H(X). \tag{2.3}$$

Given a conditional probability $\mathrm{P}_{X|Y=y} = \{\mathrm{P}_{X|Y}(x|y)\}_x$, the entropy of X is given by $H(X|Y=y) \stackrel{\text{def}}{=} H(\mathrm{P}_{X|Y=y})$ when the random variable Y is known to be y. The expectation value of this entropy with respect to the probability distribution of Y is called the *conditional entropy* denoted by $H(X|Y)$. We may write it as

$$\begin{aligned} H(X|Y) \stackrel{\text{def}}{=} & \sum_y \sum_x -\mathrm{P}_Y(y)\mathrm{P}_{X|Y}(x|y) \log \mathrm{P}_{X|Y}(x|y) \\ = & -\sum_{x,y} \mathrm{P}_{X,Y}(x,y) \log \frac{\mathrm{P}_{X,Y}(x,y)}{\mathrm{P}_Y(y)} \\ = & -\sum \mathrm{P}_{X,Y}(x,y) \log \mathrm{P}_{X,Y}(x,y) + \sum_y \mathrm{P}_Y(y) \log \mathrm{P}_Y(y) \\ = & H(X,Y) - H(Y). \end{aligned} \tag{2.4}$$

Since (as will be shown later)

$$H(X) + H(Y) - H(X,Y) \geq 0, \tag{2.5}$$

we have

$$H(X) \geq H(X|Y). \tag{2.6}$$

If Y takes values in $\{0,1\}$, (2.6) is equivalent to the concavity of the entropy $^{\text{Ex. 2.2}}$:

$$\lambda H(p) + (1-\lambda)H(p') \leq H(\lambda p + (1-\lambda)p'), \quad 0 < \forall \lambda < 1. \tag{2.7}$$

2.1.2 Relative Entropy

We now consider a quantity that expresses the closeness between two probability distributions $p = \{p_x\}_{x=1}^k$ and $q = \{q_x\}_{x=1}^k$. It is called an information quantity because our access to information is closely related to the difference between the distributions reflecting the information we are interested in. A typical example is the *relative entropy*[2] $D(p\|q)$, which is defined as

[2] The term relative entropy is commonly used in statistical physics. In information theory, it is generally known as the *Kullback–Leibler divergence*, while in statistics it is known as the *Kullback–Leibler information.*

$$D(p\|q) \stackrel{\text{def}}{=} \sum_{i=1}^{k} p_i \log \frac{p_i}{q_i} . \tag{2.8}$$

This quantity is always no less than 0, and it is equal to 0 if and only if $p = q$. This can be shown by applying the *logarithmic inequality*$^{\text{Ex. 2.5}}$ "$\log x \leq x - 1$ for $x > 0$" to (2.8):

$$0 - D(p\|q) = \sum_{i=1}^{k} p_i \left(-\frac{q_i}{p_i} + 1 + \log \frac{q_i}{p_i} \right) \leq \sum_{i=1}^{k} p_i \; 0 = 0 .$$

Note that the equality of $\log x \leq x - 1$ holds only when $x = 1$. We may obtain (2.2) by using the positivity of the relative entropy for the case $q = \{1/k\}$.

Let us now consider possible information processes. When an information process converts a set of data $\mathbb{N}_k \stackrel{\text{def}}{=} \{1, \dots, k\}$ to another set of data $\mathbb{N}_l$ deterministically, we may denote the information processing by a function from $\mathbb{N}_k$ to $\mathbb{N}_l$. If it converts probabilistically, it is denoted by a real-valued matrix $\{Q^i_j\}$ in which every element Q^i_j represents the probability of the output data $j \in \mathbb{N}_l$ when the input data are $i \in \mathbb{N}_k$. This matrix $Q = (Q^i_j)$ satisfies $\sum_{j=1}^{l} Q^i_j = 1$ for each i. Such a matrix Q is called a *stochastic transition matrix.* When the input signal is generated according to the probability distribution p, the output signal is generated according to the probability distribution $Q(p)_j \stackrel{\text{def}}{=} \sum_{i=1}^{k} Q^i_j p_i$. The stochastic transition matrix Q represents not only such probabilistic information processes but also probabilistic fluctuations in the data due to noise. Furthermore, since it expresses the probability distribution of the output system for each input signal, we can also use it to model a channel transmitting information.

A fundamental property of a stochastic transition matrix Q is the inequality

$$D(p\|q) \geq D(Q(p)\|Q(q)) , \tag{2.9}$$

which is called an *information-processing inequality.* This property is often called *monotonicity.*[3] The inequality implies that the amount of information should not increase via any information processing. This inequality will be proved for the general case in Theorem 2.1. It may also be shown using a logarithmic inequality.

For example, consider the stochastic transition matrix $Q = (Q^i_j)$ from $\mathbb{N}_{2k}$ to $\mathbb{N}_k$, where Q^i_j is 1 when $i = j, j + k$ and 0 otherwise. We define the probability distribution $\tilde{p}$ for $\mathbb{N}_{2k}$ as

$$\tilde{p}_i = \lambda p_i, \; \tilde{p}_{i+k} = (1 - \lambda) p'_i, \quad 1 \leq \forall i \leq k$$

[3] In this book, monotonicity refers to only the monotonicity regarding the change in probability distributions or density matrices.

for the probability distributions p, p' in $\mathbb{N}_k$, where $0 < \lambda < 1$. Similarly, we define $\tilde{q}$ for the probability distributions q, q' in $\mathbb{N}_k$. Then, it follows that

$$D(\tilde{p}\|\tilde{q}) = \lambda D(p\|q) + (1-\lambda)D(p'\|q').$$

Since $Q(\tilde{p}) = \lambda p + (1-\lambda)p'$ and $Q(\tilde{q}) = \lambda q + (1-\lambda)q'$, the information-processing inequality (2.9) yields the *joint convexity* of the relative entropy

$$\lambda D(p\|q) + (1-\lambda)D(p'\|q') \geq D(\lambda p + (1-\lambda)p' \| \lambda q + (1-\lambda)q'). \tag{2.10}$$

Next, let us consider other information quantities that express the difference between the two probability distributions p and q. In order to express the amount of information, these quantities should satisfy the property given by (2.9). This property can be satisfied by constructing the information quantity in the following manner. First, we define convex functions. When a function f satisfies

$$f(\lambda x_1 + (1-\lambda)x_2) \leq \lambda f(x_1) + (1-\lambda)f(x_2), \quad 0 \leq \forall\lambda \leq 1, \forall x_1, x_2 \in \mathbb{R},$$

it is called a *convex function*. For a probability distribution $p = \{p_i\}$, a convex function f satisfies *Jensen's inequality*:

$$\sum_i p_i f(x_i) \geq f\left(\sum_i p_i x_i\right). \tag{2.11}$$

Theorem 2.1 *(Csiszár [84]) Let f be a convex function. The information quantity $D_f(p\|q) \stackrel{\text{def}}{=} \sum_i p_i f\left(\frac{q_i}{p_i}\right)$ then satisfies the monotonicity condition*

$$D_f(p\|q) \geq D_f(Q(p)\|Q(q)). \tag{2.12}$$

Henceforth, $D_f(p\|q)$ will be called an f-relative entropy.[4]

For example, for $f(x) = -\log x$ we obtain the relative entropy. For $f(x) = 1 - \sqrt{x}$,

$$D_f(p\|q) = 1 - \sum_i \sqrt{p_i}\sqrt{q_i} = \frac{1}{2}\sum_i \left(\sqrt{p_i} - \sqrt{q_i}\right)^2. \tag{2.13}$$

Its square root is called the *Hellinger distance* and is denoted by $d_2(p,q)$. This satisfies the axioms of a distance$^{\text{Ex. 2.12}}$. When $f(x) = \frac{4}{1-\alpha^2}(1 - x^{(1+\alpha)/2})(-1 < \alpha < 1)$, $D_f(p\|q)$ is equal to the α-divergence $\frac{4}{1-\alpha^2}\left(1 - \sum_i p_i^{(1-\alpha)/2} q_i^{(1+\alpha)/2}\right)$ according to Amari and Nagaoka [11]. By

[4] This quantity is more commonly used in information theory, where it is called f-divergence. In this text, we prefer to use the term "relative entropy" for all relative-entropy-like quantities.

applying inequality (2.12) to the concave function $x \to x^s$ $(0 \le s \le 1)$ and the convex function $x \to x^s$ $(s \le 0)$, we obtain the inequalities

$$\sum_i p_i^{1-s} q_i^s \le \sum_j Q(p)_j^{1-s} Q(q)_j^s \quad \text{for } 0 \le s \le 1\,,$$
$$\sum_i p_i^{1-s} q_i^s \ge \sum_j Q(p)_j^{1-s} Q(q)_j^s \quad \text{for } s \le 0\,.$$

Hence, the *relative Rényi entropy* $\phi(s|p\|q) \stackrel{\text{def}}{=} \log(\sum_i p_i^{1-s} q_i^s)$ satisfies the *monotonicity*

$$\phi(s|p\|q) \le \phi(s|Q(p)\|Q(q)) \quad \text{for } 0 \le s \le 1\,,$$
$$\phi(s|p\|q) \ge \phi(s|Q(p)\|Q(q)) \quad \text{for } s \le 0\,.$$

The relative entropy can be expressed as

$$\phi'(0|p\|q) = -D(p\|q), \quad \phi'(1|p\|q) = D(q\|p). \tag{2.14}$$

Since $\phi(s|p\|q)$ is a convex function of $s^{\text{Ex. 2.14}}$,

$$\frac{\phi(s|p\|q)}{s} = \frac{\phi(s|p\|q) - \phi(0|p\|q)}{s} \tag{2.15}$$

is monotone increasing for s. A more precise analysis is given in Exercise 3.11 **b**. We will abbreviate it to $\phi(s)$ if it is not necessary to specify p and q explicitly.

Proof of Theorem 2.1. Since f is a convex function, Jensen's inequality ensures that

$$\sum_i \frac{Q_j^i p_i}{\sum_{i'} Q_j^{i'} p_{i'}} f\left(\frac{q_i}{p_i}\right) \ge f\left(\sum_i \frac{Q_j^i p_i}{\sum_{i'} Q_j^{i'} p_{i'}} \frac{q_i}{p_i}\right) = f\left(\frac{\sum_i Q_j^i q_i}{\sum_{i'} Q_j^{i'} p_{i'}}\right).$$

Therefore,

$$\begin{aligned} D_f(Q(p)\|Q(q)) &= \sum_j \sum_{i''} Q_j^{i''} p_{i''} f\left(\frac{(\sum_i Q_j^i q_i)}{(\sum_{i'} Q_j^{i'} p_{i'})}\right) \\ &\le \sum_j \sum_{i''} Q_j^{i''} p_{i''} \sum_i \frac{Q_j^i p_i}{\sum_{i'} Q_j^{i'} p_{i'}} f\left(\frac{q_i}{p_i}\right) \\ &= \sum_j \sum_i Q_j^i p_i f\left(\frac{q_i}{p_i}\right) = \sum_i p_i f\left(\frac{q_i}{p_i}\right) = D_f(p\|q). \end{aligned}$$

■

We consider the *variational distance* as another information quantity. It is defined as

$$d_1(p,q) \stackrel{\text{def}}{=} \frac{1}{2}\sum_i |p_i - q_i| \,. \tag{2.16}$$

It is not an f-relative entropy. However, it satisfies the *monotonicity* property
Ex. 2.8

$$d_1(Q(p),Q(q)) \le d_1(p,q) \,. \tag{2.17}$$

The variational distance, Hellinger distance, and relative entropy are related by the following formulas:

$$d_1(p,q) \ge d_2^2(p,q) \ge \frac{1}{2} d_1^2(p,q) \,, \tag{2.18}$$

$$D(p\|q) \ge -2\log\left(\sum_i \sqrt{p_i}\sqrt{q_i}\right) \ge 2d_2^2(p,q) \,. \tag{2.19}$$

The last inequality may be deduced from the logarithmic inequality.

When a stochastic transition matrix $Q = (Q^i_j)$ satisfies $\sum_i Q^i_j = 1$, i.e., its transpose is also a stochastic transition matrix, the stochastic transition matrix $Q = (Q^i_j)$ is called a *double stochastic transition matrix*. When all probabilities p_i have the same value, the probability distribution $p = (p_i)$ is called a *uniform distribution* and is denoted by p_{mix}. If it is necessary to denote the number of supports k explicitly, we will write $p_{\text{mix},k}$. Then, a stochastic transition matrix Q is a double stochastic transition matrix if and only if the output distribution $Q(p_{\text{mix}})$ is a uniform distribution. As will be shown in Sect. 8.4 by using majorization, the double stochastic transition matrix Q and the probability distribution p satisfy

$$H(Q(p)) \ge H(p) \,. \tag{2.20}$$

2.1.3 Mutual Information

Let us say that we are given the joint probability distribution $\mathrm{P}_{X,Y}$ of two random variables X and Y. Then, the *marginal distributions* P_X and P_Y of $\mathrm{P}_{X,Y}$ are given as

$$\mathrm{P}_X(x) \stackrel{\text{def}}{=} \sum_y \mathrm{P}_{X,Y}(x,y) \quad \text{and} \quad \mathrm{P}_Y(y) \stackrel{\text{def}}{=} \sum_x \mathrm{P}_{X,Y}(x,y) \,.$$

Then, the conditional distribution is calculated as

$$\mathrm{P}_{X|Y}(x|y) = \frac{\mathrm{P}_{X,Y}(x,y)}{\mathrm{P}_Y(y)} \,.$$

When $P_X(x) = P_{X|Y}(x|y)$, two random variables X and Y are *independent.* In this case, the joint distribution $P_{X,Y}(x,y)$ is equal to the product of marginal distributions $P_X(x)P_Y(y)$. That is, the relative entropy $D(P_{X,Y}\|P_X P_Y)$ is equal to zero. We now introduce *mutual information* $I(X:Y)$, which expresses how different the joint distribution $P_{X,Y}(x,y)$ is from the product of marginal distributions $P_X(x)P_Y(y)$. This quantity satisfies the following relation:

$$
\begin{aligned}
I(X:Y) &\stackrel{\text{def}}{=} D(P_{X,Y}\|P_X P_Y) = \sum_{x,y} P_{X,Y}(x,y)\log\frac{P_{X,Y}(x,y)}{P_X(x)P_Y(y)} \\
&= H(X)-H(X|Y) = H(Y)-H(Y|X) = H(X)+H(Y)-H(X,Y). \qquad (2.21)
\end{aligned}
$$

Hence, inequality (2.5) may be obtained from the above formula and the positivity of $I(X:Y)$. Further, we can define a *conditional mutual information* in a manner similar to that of the entropy. This quantity involves another random variable Z (in addition to X and Y) and is defined as

$$
\begin{aligned}
I(X:Y|Z) &\stackrel{\text{def}}{=} \sum_z P_Z(z) I(X:Y|Z=z) \qquad (2.22) \\
&= \sum_{x,y,z} P_{X,Y,Z}(x,y,z)\log\frac{P_{XY|Z}(x,y|z)}{P_{X|Z}(x|z)P_{Y|Z}(y|z)} \ge 0,
\end{aligned}
$$

where $I(X:Y|Z=z)$ is the mutual information of X and Y assuming that $Z=z$ is known. By applying (2.4) and (2.21) to the case $Z=z$, we obtain

$$
\begin{aligned}
I(X:Y|Z) &= H(X|Z) + H(Y|Z) - H(XY|Z) = H(X|Z) - H(X|YZ) \\
&= -(H(X) - H(X|Z)) + (H(X) - H(X|YZ)) \\
&= -I(X:Z) + I(X:YZ).
\end{aligned}
$$

This equation is called the *chain rule* of mutual information, which may also be written as

$$
I(X:YZ) = I(X:Z) + I(X:Y|Z). \qquad (2.23)
$$

Hence, it follows that

$$
I(X:YZ) \ge I(X:Z).
$$

Note that (2.23) can be generalized as

$$
I(X:YZ|U) = I(X:Z|U) + I(X:Y|ZU). \qquad (2.24)
$$

Next, we apply the above argument to the case where the information channel is given by a stochastic transition matrix Q and the input distribution is given by p. Let X and Y be, respectively, the random variables of an input system and output system. Then, the mutual information $I(X:Y)$ can be regarded as the amount of information transmitted via channel Q when the input signal is generated with the distribution p. This is called *transmission information*, and it is denoted by $I(p, Q)$. Therefore, we can define the transmission information by

$$I(p, Q) \stackrel{\text{def}}{=} H(Q(p)) - \sum_x p_x H(Q^x) . \tag{2.25}$$

We will now discuss *Fano's inequality*, which is given by the following theorem.

Theorem 2.2 *(Fano [113]) Let X and Y be random variables that take values in the same data set $\mathbb{N}_k = \{1, \ldots, k\}$. Then, the following inequality holds:*

$$\begin{aligned} H(X|Y) &\le \mathrm{P}\{X \neq Y\} \log(k-1) + h(\mathrm{P}\{X \neq Y\}) \\ &\le \mathrm{P}\{X \neq Y\} \log k + \log 2 . \end{aligned} \tag{2.26}$$

Proof. We define the random variable $Z \stackrel{\text{def}}{=} \begin{cases} 0 & X = Y \\ 1 & X \neq Y \end{cases}$. Applying (2.4) to X and Z under the condition $Y = y$, we obtain

$$\begin{aligned} H(X|Y = y) &= H(X, Z|Y = y) \\ &= \sum_z \mathrm{P}_Z(z) H(X|Z = z, Y = y) + H(Z|Y = y). \end{aligned}$$

The first equality follows from the fact that the random variable Z can be uniquely obtained from X. Taking the expectation value with respect to y, we get

$$\begin{aligned} H(X|Y) =& H(X|Z, Y) + H(Z|Y) \le H(X|Z, Y) + H(Z) \\ =& H(X|Z, Y) + h(\mathrm{P}\{X \neq Y\}). \end{aligned} \tag{2.27}$$

Applying (2.2), we have

$$H(X|Y = y, Z = 0) = 0, \quad H(X|Y = y, Z = 1) \le \log(k-1).$$

Therefore,

$$H(X|Y, Z) \le \mathrm{P}\{X \neq Y\} \log(k-1). \tag{2.28}$$

Finally, combining (2.27) and (2.28), we obtain (2.26). ■

2.1.4 The Independent and Identical Condition and Rényi Entropy

Given a probability distribution $p = \{p_i\}_{i=1}^k$, we can also construct the *Rényi entropy*

$$\psi(s|p) \stackrel{\text{def}}{=} \log \sum_i p_i^{1-s} \tag{2.29}$$

for a real number s in addition to the entropy $H(p)$. We will abbreviate the Rényi entropy to $\psi(s)$ when there is no risk of ambiguity. When $0 < s < 1$, the Rényi entropy $\psi(s)$ is a positive quantity that is larger when the probability distribution is closer to the uniform distribution. In particular, the Rényi entropy $\psi(s)$ is equal to $s \log k$ when the distribution is the uniform distribution $p_{\mathrm{mix},k}$. When $s < 0$, the Rényi entropy $\psi(s)$ is a negative quantity that is smaller when the probability distribution is closer to the uniform distribution. Finally, when $s = 0$, the Rényi entropy $\psi(s)$ is equal to 0. The derivative $\psi'(0)$ of $\psi(s)$ at $s = 0$ is equal to $H(p)$. Further, similarly to (2.15), the function $\frac{\psi(s|p)}{s}$ is monotone increasing for s.

Now consider n data $i_1, \dots, i_n$ that are generated independently with the same probability distribution $p = \{p_i\}_{i=1}^k$. The probability of obtaining a particular data sequence $i^n = (i_1, \dots, i_n)$ is given by $p_{i_1}, \cdots, p_{i_n}$. This probability distribution is called an n-fold *independent and identical distribution* (abbreviated as n-i.i.d.) and denoted by p^n. When a sufficiently large number n of data are generated according to the independent and identical condition, the behavior of the distribution may be characterized by the entropy and the Rényi entropy. Let us now discuss this in greater detail.

The probability of the likelihood being less than $a \geq 0$ under the probability distribution p, i.e., the probability that $\{p_i \leq a\}$, is

$$p\{p_i \leq a\} = \sum_{i:p_i \leq a} p_i \leq \sum_{i:1 \leq \frac{a}{p_i}} \left(\frac{a}{p_i}\right)^s p_i \leq \sum_{i=1}^k p_i^{1-s} a^s = e^{\psi(s)+s\log a} \tag{2.30}$$

if $0 \leq s$. Accordingly,

$$p^n\{p_{i^n}^n \leq e^{-nR}\} \leq e^{n \min_{0 \leq s \leq 1}(\psi(s)-sR)}. \tag{2.31}$$

Conversely, the probability of the likelihood being greater than a, i.e., the probability that $\{p_i > a\}$, is

$$p\{p_i > a\} \leq \sum_{i:1 > \frac{a}{p_i}} \left(\frac{a}{p_i}\right)^s p_i \leq \sum_{i=1}^k p_i^{1-s} a^s = e^{\psi(s)+s\log a} \tag{2.32}$$

if $s \leq 0$. Similarly, we obtain

$$p^n\{p_{i^n}^n > e^{-nR}\} \leq e^{n \min_{s \leq 0}(\psi(s)-sR)}. \tag{2.33}$$

Ship To:

AMZ APTP ORDER
800 AVONDALE AVE
COLUMBUS, OH 43212-3473

Order ID: 111-3141751-4810630

Thank you for buying from Burlingame Library Foundation on Amazon Marketplace.

Shipping Address:	Order Date:	Thu, Oct 22, 2026
AMZ APTP ORDER	Shipping Service:	Standard
800 AVONDALE AVE	Buyer Name:	Yvanns
COLUMBUS, OH 43212-3473	Seller Name:	Burlingame Library Foundation

Quantity	Product Details	Unit price	Order Totals
1	**Quantum Information: An Introduction [Hardcover] [2006]** **Hayashi, Masahito** **SKU:** A210191A **ASIN:** 3540302654 **Condition:** Used - Very Good **Order Item ID:** 173097934301481 **Condition note:** This hardbound copy is in very good condition: clean,	$20.00	**Item subtotal** $20.00 **Shipping total** $5.81 **Item total** **$25.81**

Grand total: $25.81

Returning your item:

Go to "Your Account" on Amazon.com, click "Your Orders" and then click the "seller profile" link for this order to get information about the return and refund policies that apply.

Visit https://www.amazon.com/returns to print a return shipping label. Please have your order ID ready.

Thanks for buying on Amazon Marketplace. To provide feedback for the seller please visit www.amazon.com/feedback. To contact the seller, go to Your Orders in Your Account. Click the seller's name under the appropriate product. Then, in the "Further Information" section, click "Contact the Seller."

The exponent on the right-hand side (RHS) of (2.31) is negative when $R > H(p)$. Hence, the probability $p^n\{p^n_{i^n} \le e^{-nR}\}$ approaches 0 exponentially. In order to investigate this further, let us consider the following limit by noting that $\psi(0) = 0$ and $\psi'(0) = H(p)$:

$$\lim_{s\to 0} \frac{\psi(s)}{s} = \lim_{s\to 0} \frac{\psi(s) - \psi(0)}{s} = \psi'(0) = H(p) < R. \tag{2.34}$$

By choosing a sufficiently small number $0 < s_1$, we have

$$\psi(s_1) < s_1 R, \tag{2.35}$$

and therefore

$$\min_{0\le s}(\psi(s) - sR) \le \psi(s_1) - s_1 R < 0.$$

We see that the exponent on the RHS of (2.31) is negative. Conversely, the exponent on the RHS of (2.33) is negative when $R < H(p)$, and the probability $p^n\{p^n_{i^n} \le e^{-nR}\}$ approaches 0 exponentially. This can be verified from (2.34) by choosing $s_2 < 0$ with a sufficiently small absolute value.

We may generalize this argument for the likelihood $q^n_{i^n}$ of a different probability distribution q as follows. Defining $\tilde{\psi}(s) \stackrel{\text{def}}{=} \log \sum_i p_i q_i^{-s}$, we can show that

$$p^n\{q^n_{i^n} \le e^{-nR}\} \le e^{n \min_{0\le s}(\tilde{\psi}(s) - sR)}, \tag{2.36}$$

$$p^n\{q^n_{i^n} > e^{-nR}\} \le e^{n \min_{s\le 0}(\tilde{\psi}(s) - sR)}. \tag{2.37}$$

The Rényi entropy $\psi(s)$ and the entropy $H(p)$ express the concentration of probability under independent and identical distributions with a sufficiently large number of data. To investigate this, let us consider the probability $P(p, L)$ of the most frequent L outcomes for a given probability distribution $p = (p_i)$.[5] This can be written as

$$P(p, L) = \sum_{i=1}^{L} p_i^{\downarrow},$$

where $p_i^{\downarrow}$ are the elements of p_i that are reordered according to size. Let us analyze this by reexamining the set $\{p_i > a\}$. The number of elements of the set $|\{p_i > a\}|$ is evaluated as

$$|\{p_i > a\}| \le \sum_{i:p_i > a} \left(\frac{p_i}{a}\right)^{1-s} \le \sum_{i=1}^{k} p_i^{1-s} a^{-1+s} = e^{\psi(s) - (1-s)\log a} \tag{2.38}$$

[5] If L is not an integer, we consider the largest integer that does not exceed L.

when $0 < s < 1$. By using (2.30) and defining $b(s,R) \stackrel{\rm def}{=} \frac{\psi(s)-R}{1-s}$ for R and $0 \le s < 1$, we have

$$|\{p_i > e^{b(s,R)}\}| \le e^R, \quad p\{p_i \le e^{b(s,R)}\} \le e^{\frac{\psi(s)-sR}{1-s}}.$$

We choose $s_0 \stackrel{\rm def}{=} \mathrm{argmin}_{0\le s\le 1} \frac{\psi(s)-sR}{1-s}$ [6] and define $P^c(p,e^R) \stackrel{\rm def}{=} 1-P(p,e^R)$; hence,

$$P^c(p,e^R) \le e^{\frac{\psi(s_0)-s_0R}{1-s_0}} = e^{\min_{0\le s\le 1}\frac{\psi(s)-sR}{1-s}}. \tag{2.39}$$

Applying this argument to the n-i.i.d p^n, we have

$$P^c(p^n,e^{nR}) \le e^{n\frac{\psi(s_0)-s_0R}{1-s_0}} = e^{n\min_{0\le s\le 1}\frac{\psi(s)-sR}{1-s}}. \tag{2.40}$$

Now, we let $R > H(p)$ and choose a sufficiently small number $0 < s_1 < 1$. Then, inequality (2.35) yields

$$\min_{0\le s<1} \frac{\psi(s)-sR}{1-s} \le \frac{\psi(s_1)-s_1R}{1-s_1} < 0\,.$$

Hence, the probability $P^c(p^n,e^{nR})$ approaches 0 exponentially. That implies that the probabilities are almost concentrated on the most frequent e^{nR} elements because $1-P^c(p^n,e^{nR})$ equals the probability on the most frequent e^{nR} elements. Since this holds when $R > H(p)$, most of the probabilities are concentrated on $e^{nH(p)}$ elements. Therefore, this can be interpreted as meaning that the entropy $H(p)$ asymptotically expresses the degree of concentration. This will play an important role in problems such as source coding, which will be discussed later.

On the other hand, when $H(p) > R$, $P(p^n,e^{nR})$ approaches 0. To prove this, let us consider the following inequality for an arbitrary subset A:

$$pA \le a|A| + p\{p_i > a\}\,. \tag{2.41}$$

We can prove this inequality by considering the set $A = (A\cap\{p_i \le a\})\cup(A\cap\{p_i > a\})$. Defining $R \stackrel{\rm def}{=} \log|A|$ and $a \stackrel{\rm def}{=} e^{b(s,R)}$ and using (2.32), we obtain $pA \le 2e^{\frac{\psi(s)-sR}{1-s}}$. Therefore, $P(p,e^R) \le 2e^{\min_{s\le 0}\frac{\psi(s)-sR}{1-s}}$, and we obtain

$$P(p^n,e^{nR}) \le 2e^{n\min_{s\le 0}\frac{\psi(s)-sR}{1-s}}. \tag{2.42}$$

We also note that in order to avoid $P(p^n,e^{nR}) \to 0$, we require $R \ge H(p)$ according to the condition $\min_{s\le 0}\frac{\psi(s)-sR}{1-s} < 0$.

[6] $\mathrm{argmin}_{0\le s\le 1} f(s)$ returns the value of s that yields $\min_{0\le s\le 1} f(s)$. argmax is similarly defined.

Exercises

2.1. Verify (2.3) if the variable Y can be written $f(X)$ for a function f.

2.2. Verify that (2.6) and (2.7) are equivalent.

2.3. Show that

$$H(p_A) + H(p_B) = H(p_A \times p_B) \tag{2.43}$$

for probability distributions p_A in Ω_A, p_B in Ω_B, and $p_A \times p_B(\omega_A, \omega_B) = p_A(\omega_A)p_B(\omega_B)$ in $\Omega_A \times \Omega_B$.

2.4. Show that

$$D(p_A\|q_A) + D(p_B\|q_B) = D(p_A \times p_B\|q_A \times q_B) \tag{2.44}$$

for probability distributions p_A, q_A in Ω_A and p_B, q_B in Ω_B.

2.5. Show the logarithmic inequality, i.e., the ineqaulity $\log x \le x - 1$, holds for $x > 0$ and the equality holds only for $x = 1$.

2.6. Show that the f-relative entropy $D_f(p\|q)$ of a convex function f satisfies $D_f(p\|q) \ge f(1)$.

2.7. Prove (2.13).

2.8. Show that the variational distance satisfies the monotonicity condition (2.17).

2.9. Show that $d_1(p,q) \ge d_2^2(p,q)$ by first proving the inequality $|x - y| \ge (\sqrt{x} - \sqrt{y})^2$.

2.10. Show that $d_2^2(p,q) \ge \frac{1}{2}d_1^2(p,q)$ following the steps below.

a Prove

$$\left(\sum_i |p_i - q_i|\right)^2 \le \left(\sum_i |\sqrt{p_i} - \sqrt{q_i}|^2\right)\left(\sum_i |\sqrt{p_i} + \sqrt{q_i}|^2\right)$$

using the Schwarz inequality.

b Show that $\sum_i |\sqrt{p_i} + \sqrt{q_i}|^2 \le 4$.

c Show that $d_2^2(p,q) \ge \frac{1}{2}d_1^2(p,q)$ using the above results.

2.11. Show that $D(p\|q) \ge -2\log\left(\sum_i \sqrt{p_i}\sqrt{q_i}\right)$.

2.12. Verify that the Hellinger distance satisfies the axioms of a distance by following the steps below.

a Prove the following for arbitrary vectors x and y

$$(\|x\| + \|y\|)^2 \geq \|x\|^2 + \langle x, y \rangle + \langle y, x \rangle + \|y\|^2 .$$

b Prove the following for arbitrary vectors x and y:

$$\|x\| + \|y\| \geq \|x + y\| .$$

c Show the following for the three probability distributions p, q, and r:

$$\sqrt{\sum_i (\sqrt{p_i} - \sqrt{q_i})^2} \leq \sqrt{\sum_i (\sqrt{p_i} - \sqrt{r_i})^2} + \sqrt{\sum_i (\sqrt{r_i} - \sqrt{q_i})^2} .$$

Note that this formula is equivalent to the axiom of a distance $d_2(p, q) \leq d_2(p, r) + d_2(r, q)$ for the Hellinger distance.

2.13. Show (2.14).

2.14. Show that $\phi(s|p\|q)$ is convex for s.

2.15. Show the chain rule of conditional mutual information (2.24) based on (2.23).

2.2 Extensions to Quantum Systems

The above-mentioned discussion on probability distributions may be extended to the density matrix ρ in quantum systems as follows. Let us first consider the *von Neumann entropy* of the density matrix ρ with the spectral decomposition $\rho = \sum_{i=1}^d p_i |u^i\rangle\langle u^i|$ as its quantum extension of the entropy.[7] The von Neumann entropy is defined as the entropy of the probability distribution $p = \{p_i\}$ of the eigenvalues of the density ρ, and it is denoted by $H(\rho)$. Applying the arguments of Sect. 1.5 to $f(x) = \log(x)$, we have $\log \rho \stackrel{\text{def}}{=} \sum_{i=1}^d (\log p_i) |u^i\rangle\langle u^i|$, and we can write $H(\rho)$ as

$$H(\rho) = -\operatorname{Tr} \rho \log \rho .$$

The von Neumann entropy also satisfies the concavity, as proved in Sect. 5.5. Similarly, the *Rényi entropy* is defined as $\psi(s|\rho) \stackrel{\text{def}}{=} \log \operatorname{Tr} \rho^{1-s}$. Henceforth, we will use its abbreviation $\psi(s)$ as mentioned previously. Regarding the diagonal elements of the diagonalized matrix to be a probability distribution, we can therefore interpret the tensor product $\rho^{\otimes n}$ as the quantum-mechanical

[7] Historically, the von Neumann entropy for a density matrix ρ was first defined by von Neumann [408]. Following this definition, Shannon [366] defined the entropy for a probability distribution.

analog of the independent and identical distribution. In other words, the eigenvalues of $\rho^{\otimes n}$ are equal to the n-i.i.d. of the probability distribution resulting from the eigenvalues of ρ.

Since $\{\rho^s a^{-s} > 1\}(\rho^s a^{-s} - I) \geq 0$ for $s \geq 0$, the inequalities

$$\{\rho > a\} = \{\rho^s a^{-s} > 1\} \leq \{\rho^s a^{-s} > 1\}\rho^s a^{-s} \leq \rho^s a^{-s} \tag{2.45}$$

hold. Similarly,

$$\{\rho \leq a\} \leq \rho^{-s} a^s .$$

Hence, we obtain

$$\operatorname{Tr}\{\rho > a\} \leq \operatorname{Tr}\rho^s a^{-s} , \tag{2.46}$$

$$\operatorname{Tr}\rho\{\rho \leq a\} \leq \operatorname{Tr}\rho^{1-s} a^s , \tag{2.47}$$

for $a > 0$ and $0 \leq s$. Treating the independent and identical distribution in a manner similar to (2.31) and (2.33), we obtain

$$\operatorname{Tr}\rho^{\otimes n}\left\{\rho^{\otimes n} \leq e^{-nR}\right\} \leq e^{n \min_{0\leq s}(\psi(s)-sR)} \tag{2.48}$$

$$\operatorname{Tr}\rho^{\otimes n}\left\{\rho^{\otimes n} > e^{-nR}\right\} \leq e^{n \min_{s\leq 0}(\psi(s)-sR)} . \tag{2.49}$$

Certainly, the relationship similar to the classical system holds concerning $\frac{\psi(s)-sR}{1-s}$ and $H(\rho)$.

Replacing ρ by a different state σ in (2.45), and defining $\tilde{\psi}(s) \stackrel{\text{def}}{=} \log \operatorname{Tr}\rho\sigma^{-s}$, we have for $s \geq 0$

$$\operatorname{Tr}\rho\{\sigma > a\} \leq \operatorname{Tr}\rho\sigma^s a^{-s} . \tag{2.50}$$

Based on an argument similar to that in Sect. 2.1.4, we can show that

$$\operatorname{Tr}\rho^{\otimes n}\left\{\sigma^{\otimes n} \leq e^{-nR}\right\} \leq e^{n \min_{0\leq s}(\tilde{\psi}(s)-sR)} , \tag{2.51}$$

$$\operatorname{Tr}\rho^{\otimes n}\left\{\sigma^{\otimes n} > e^{-nR}\right\} \leq e^{n \min_{s\leq 0}(\tilde{\psi}(s)-sR)} . \tag{2.52}$$

Therefore, the exponent on the RHS of (2.51) is negative when $R > -\operatorname{Tr}\rho\log\sigma$, and that on the RHS of (2.52) is negative when $R < -\operatorname{Tr}\rho\log\sigma$. This kind of argument can be applied to inequalities (2.48) and (2.49).

As an extension of the relative entropy, we define the *quantum relative entropy* $D(\rho\|\sigma)$ for two density matrices ρ and σ as

$$D(\rho\|\sigma) \stackrel{\text{def}}{=} \operatorname{Tr}\rho(\log\rho - \log\sigma) . \tag{2.53}$$

The quantum relative entropy satisfies an inequality similar to (2.9), which will be discussed in greater detail in Sect. 5.4. For the two quantum states ρ and σ, we can also define a quantum extension of the *relative Rényi entropy* $\phi(s|\rho\|\sigma) \stackrel{\text{def}}{=} \log(\operatorname{Tr}\rho^{1-s}\sigma^s)$ and obtain [Ex. 2.19]

$$\phi'(0|\rho\|\sigma) = -D(\rho\|\sigma), \quad \phi'(1|\rho\|\sigma) = D(\sigma\|\rho). \tag{2.54}$$

When it is not necessary to explicitly specify ρ and σ, we will abbreviate this value to $\phi(s)$. If ρ commutes with σ, $\phi(s)$ is equal to the classical relative Rényi entropy $\phi(s)$ with probability distributions that consist of eigenvalues.

The quantum version of the Hellinger distance $d_2(p,q)$ is the *Bures distance* $b(\rho,\sigma)$ defined as

$$b^2(\rho,\sigma) \stackrel{\text{def}}{=} \min_{U:\text{unitary}} \frac{1}{2} \operatorname{Tr}(\sqrt{\rho} - \sqrt{\sigma}U)(\sqrt{\rho} - \sqrt{\sigma}U)^* . \tag{2.55}$$

The Bures distance $b(\rho,\sigma)$ also satisfies the axioms of a distance in a similar way to the Hellinger distance $^{\text{Ex. 2.20}}$. Using (A.13), this quantity may be rewritten as

$$\begin{aligned} b^2(\rho,\sigma) &= 1 - \frac{1}{2} \max_{U:\text{unitary}} \operatorname{Tr}\left(U\sqrt{\rho}\sqrt{\sigma} + U^*(\sqrt{\rho}\sqrt{\sigma})^*\right) \\ &= 1 - \operatorname{Tr}|\sqrt{\rho}\sqrt{\sigma}| = 1 - \operatorname{Tr}\sqrt{\sqrt{\rho}\sigma\sqrt{\rho}}. \end{aligned}$$

Therefore, this value does not change when ρ and σ are interchanged. Later, we will also see that this quantity also satisfies similar information inequalities (Corollary 8.4). The quantity $\operatorname{Tr}|\sqrt{\rho}\sqrt{\sigma}|$ is called *fidelity* and is denoted by $F(\rho,\sigma)$. Then, it follows that

$$b^2(\rho,\sigma) = 1 - F(\rho,\sigma). \tag{2.56}$$

If one of the states is a pure state, then

$$F(|u\rangle\langle u|,\rho) = \sqrt{\langle u|\rho|u\rangle}. \tag{2.57}$$

The square of this value corresponds to a probability. If both ρ and σ are pure states $|u\rangle\langle u|$ and $|v\rangle\langle v|$, respectively, then $\operatorname{Tr}\sqrt{\sqrt{\rho}\sigma\sqrt{\rho}} = |\langle u|v\rangle|$ and the Bures distance is given by

$$b^2(|u\rangle\langle u|, |v\rangle\langle v|) = 1 - |\langle u|v\rangle|.$$

We also define the *trace norm distance* $d_1(\rho,\sigma)$ as a quantum version of the variational distance by

$$d_1(\rho,\sigma) \stackrel{\text{def}}{=} \frac{1}{2}\|\rho - \sigma\|_1, \tag{2.58}$$

where $\|\cdot\|_1$ denotes the trace norm (Sect. A.3). This also satisfies the monotonicity [see (5.39) and Exercise 5.22].

If the states involved in the above quantities are pure states such as $|u\rangle\langle u|$ and $|v\rangle\langle v|$, we shall abbreviate the notation to label the states, i.e., $b(\rho,\sigma)$ will be written as $b(u,v)$, and so on. The above information quantities satisfy the *monotonicity*[8] with respect to the measurement $\boldsymbol{M}$ as follows.

[8] Here, the monotonicity concerns only the state evolution, not parameter s.

$$D(\rho\|\sigma) \geq D(\mathrm{P}_\rho^{\boldsymbol{M}}\|\mathrm{P}_\sigma^{\boldsymbol{M}})\,, \tag{2.59}$$

$$b(\rho,\sigma) \geq d_2(\mathrm{P}_\rho^{\boldsymbol{M}},\mathrm{P}_\sigma^{\boldsymbol{M}})\,, \tag{2.60}$$

$$d_1(\rho,\sigma) \geq d_1(\mathrm{P}_\rho^{\boldsymbol{M}},\mathrm{P}_\sigma^{\boldsymbol{M}})\,, \tag{2.61}$$

$$\phi(s|\rho\|\sigma) \leq \phi(s|\mathrm{P}_\rho^{\boldsymbol{M}}\|\mathrm{P}_\sigma^{\boldsymbol{M}}) \text{ for } 0 \leq s \leq 1\,, \tag{2.62}$$

$$\phi(s|\rho\|\sigma) \geq \phi(s|\mathrm{P}_\rho^{\boldsymbol{M}}\|\mathrm{P}_\sigma^{\boldsymbol{M}}) \text{ for } s \leq 0\,. \tag{2.63}$$

Proofs of (2.59) and (2.63) are given in Sect. 3.7, and (2.60), (2.61), and (2.62) are proved in Exercise 2.25, Sect. 3.3, and Sect. A.4, respectively. As is discussed in Exercises 2.26–2.28, the equalities of (2.60) and (2.61) hold when the POVM $\boldsymbol{M}$ is chosen appropriately. In contrast, there exists a POVM M satisfying the equality in (2.59) only when ρ and σ commute, as shown in Theorem 3.5; however, there exists a sequence of POVMs attaining the equality in (2.59) in an asymptotic sense, as mentioned in Exercise 5.44. However, the equality in (2.63) for $s \leq -1$ does not necessarily hold even in an asymptotic sense, as verified from Exercises 3.18 and 5.25.

Exercises

2.16. Show that the information quantities $D(p\|q), d_2(p,q)$, and $d_1(p,q)$ between q and p are equal to their quantum versions $D(\rho\|\sigma)$, $b(\rho,\sigma)$, and $d_1(\rho,\sigma)$ for commuting ρ and σ with diagonalizations $\rho = \sum_i p_i |u_i\rangle\langle u_i|$ and $\sigma = \sum_i q_i |u_i\rangle\langle u_i|$.

2.17. Show for density matrices ρ_A, σ_A in $\mathcal{H}_A$ and density matrices ρ_B, σ_B in $\mathcal{H}_B$

$$H(\rho_A) + H(\rho_B) = H(\rho_A \otimes \rho_B)\,,$$
$$D(\rho_A\|\sigma_A) + D(\rho_B\|\sigma_B) = D(\rho_A \otimes \rho_B\|\sigma_A \otimes \sigma_B)\,.$$

2.18. Show that

$$\operatorname{Tr} \rho f(X) \geq f(\operatorname{Tr} \rho X) \tag{2.64}$$

for a convex function f, a Hermitian matrix X, and a state ρ. This is a quantum version of Jensen's inequality.

2.19. Show (2.54) using Exercise 1.4.

2.20. Show that the Bures distance satisfies the axioms of a distance by following the steps below.

a Show the following for arbitrary matrices X and Y:

$$\sqrt{\operatorname{Tr} XX^*} + \sqrt{\operatorname{Tr} YY^*} \geq \sqrt{\operatorname{Tr}(X-Y)(X-Y)^*}\,.$$

b Show the following for density matrices ρ_1, ρ_2, ρ_3 and unitary matrices U_1, U_2:

$$\sqrt{\mathrm{Tr}\,(\sqrt{\rho_1} - \sqrt{\rho_2}U_1)\,(\sqrt{\rho_1} - \sqrt{\rho_2}U_1)^*}$$
$$\leq \sqrt{\mathrm{Tr}(\sqrt{\rho_1} - \sqrt{\rho_3}U_2)(\sqrt{\rho_1} - \sqrt{\rho_3}U_2)^*} + \sqrt{\mathrm{Tr}(\sqrt{\rho_3} - \sqrt{\rho_2}U_1U_2^*)(\sqrt{\rho_3} - \sqrt{\rho_2}U_1U_2^*)^*}.$$

c Show that $b(\rho_1, \rho_2) \leq b(\rho_1, \rho_3) + b(\rho_3, \rho_2)$ for density matrices ρ_1, ρ_2, and ρ_3.

2.21. Show that the square of the Bures distance satisfies

$$b^2(\rho_1, \rho_2) \leq 2b^2(\rho_1, \rho_3) + 2b^2(\rho_3, \rho_2)\,.$$

2.22. Show the following regarding the Bures distance for two different orthogonal bases $\{u_k\}$ and $\{v_k\}$.

a Show that the vectors $u = \sum_{k=1}^d \sqrt{p_k} e^{i\theta_k} u_k$ and $v = \sum_{k=1}^d \sqrt{p_k} e^{i\theta_k} v_k$ satisfy

$$\langle u|v\rangle = \sum_i p_i \langle u_i|v_i\rangle \tag{2.65}$$

for $\langle u_k|v_j\rangle = 0, k \neq j$, an arbitrary real number θ_i, and a probability distribution p_i.

b Show that (2.65) still holds if θ_i is chosen appropriately, even if the above conditions do not hold.

2.23. Show that $d_1(u,v) = \sqrt{1 - |\langle u|v\rangle|^2}$ using the result of Exercise A.3.

2.24. Show that $\phi(s|\rho\|\sigma)$ is convex by following the steps below [320].

a Show that $\phi'(s|\rho\|\sigma) = \dfrac{\mathrm{Tr}\,\rho^{1-s}\sigma^s(\log\sigma - \log\rho)}{\mathrm{Tr}\,\rho^{1-s}\sigma^s}$ by using Exercise 1.4.

b Show that $\dfrac{d\,\mathrm{Tr}\,\rho^{1-s}\sigma^s(\log\sigma - \log\rho)}{ds} = \mathrm{Tr}\rho^{1-s}(\log\sigma - \log\rho)\sigma^s(\log\sigma - \log\rho)$.

c Show that $\phi''(s|\rho\|\sigma) = \dfrac{\mathrm{Tr}\,\rho^{1-s}(\log\sigma - \log\rho)\sigma^s(\log\sigma - \log\rho)}{\mathrm{Tr}\,\rho^{1-s}\sigma^s}$ $- \dfrac{(\mathrm{Tr}\,\rho^{1-s}\sigma^s(\log\sigma - \log\rho))^2}{(\mathrm{Tr}\,\rho^{1-s}\sigma^s)^2}$.

d Show the convexity of $\phi(s|\rho\|\sigma)$ using Schwarz's inequality.

2.25. Show that

$$\sum_i \sqrt{\mathrm{P}_\rho^M(i)}\sqrt{\mathrm{P}_\sigma^M(i)} \geq \mathrm{Tr}\,|\sqrt{\rho}\sqrt{\sigma}| \tag{2.66}$$

for a POVM $\boldsymbol{M}$ and ρ, σ following the steps below [118]. This is equivalent to (2.60).

a Show that $\sqrt{\operatorname{Tr} X^* X}\sqrt{\operatorname{Tr} Y^* Y} \geq |\operatorname{Tr} X^* Y|$ for two matrices X and Y.
b Show that $\sqrt{\operatorname{Tr} U\rho^{1/2} M_i \rho^{1/2} U^*}\sqrt{\operatorname{Tr} \sigma^{1/2} M_i \sigma^{1/2}} \geq |\operatorname{Tr} U\rho^{1/2} M_i \sigma^{1/2}|$ for a unitary matrix U.
c Show (2.66).

2.26. Suppose that the density matrix σ possesses the inverse. Show that the equality in (2.66) holds if $\boldsymbol{M} = \{M_i\}$ is chosen to be the spectral decomposition of $\rho^{1/2} U^* \sigma^{-1/2} = \sigma^{-1/2}(\sigma^{1/2}\rho\sigma^{1/2})^{1/2}\sigma^{-1/2}$, for U satisfying $|\rho^{1/2}\sigma^{1/2}| = U\rho^{1/2}\sigma^{1/2} = \sigma^{1/2}\rho^{1/2}U^*$ [118].

2.27. Suppose that the density matrix σ does not possess the inverse. Show that there exists a POVM satisfying the equality in (2.66) by following the steps below.

a Show that the support of matrix U is included in the support $\mathcal{H}_1$ of σ when $|\rho^{1/2}\sigma^{1/2}| = U\rho^{1/2}\sigma^{1/2} = \sigma^{1/2}\rho^{1/2}U^*$.
b Let $\boldsymbol{M} = \{M_i\}$ be the spectral decomposition of the matrix $\rho^{1/2}U^*\sigma^{-1/2}$ on $\mathcal{H}_1$ and let P be a projection onto $\mathcal{H}_1$. Show that the POVM $\{M_i\} \cup \{I - P\}$ in $\mathcal{H}$ satisfies the equality in (2.66).

2.28. Show that the equality in (2.61) holds when the POVM $\boldsymbol{M}$ is the diagonalization of $\rho - \sigma$.

2.3 Geometry of Probability Distribution Family

2.3.1 Inner Product for Random Variables and Fisher Information

In Sect. 2.1, we introduced the mutual information $I(X : Y)$ as a quantity that expresses the correlation between two random variables X and Y. However, for calculating this quantity, one must calculate the logarithm of each probability, which is a rather tedious process. We now introduce the *covariance* $\operatorname{Cov}_p(X, Y)$ as a quantity that expresses the correlation between two real random variables X and Y. Generally, calculations involving the covariance are less tedious than those of mutual information. Given a probability distribution p in a space Ω, the covariance is defined as

$$\operatorname{Cov}_p(X, Y) \stackrel{\text{def}}{=} \sum_{\omega \in \Omega} (X(\omega) - \mathrm{E}_p(X))(Y(\omega) - \mathrm{E}_p(Y))p(\omega),$$

$$\mathrm{E}_p(X) \stackrel{\text{def}}{=} \sum_{\omega \in \Omega} X(\omega)p(\omega).$$

If X and Y are independent, the covariance $\operatorname{Cov}_p(X, Y)$ is equal to 0 $^{\text{Ex. 2.3.2}}$. Thus far it has not been necessary to specify the probability distribution, and therefore we had no difficulties in using notations such as $H(X)$ and

$I(X:Y)$. However, it is important to emphasize the probability distribution treated in our discussion; hence, we will use the above notation without their abbreviation. If X and Y are the same, the covariance $\mathrm{Cov}_p(X,Y)$ coincides with the *variance* $\mathrm{V}_p(X)$ of X. Given a set of random variables $X_1,\ldots,X_d$, the matrix $\mathrm{Cov}_p(X_i,X_j)$ is called a *covariance matrix.* Now, starting from a given probability distribution p, we define the inner product in the space of random variables as[9]

$$\langle A,B\rangle_p^{(e)} \stackrel{\mathrm{def}}{=} \sum_\omega A(\omega)B(\omega)p(\omega)\,. \tag{2.67}$$

Then, the covariance $\mathrm{Cov}_p(X,Y)$ is equal to the above inner product between the two random variables $(X(\omega)-\mathrm{E}_p(X))$ and $(Y(\omega)-\mathrm{E}_p(Y))$ with a zero expectation value. That is, the inner product (2.67) implies the correlation between the two random variables with zero expectation value in classical systems. This inner product is also deeply related to statistical inference in another sense, as discussed below.

When we observe n independent random variables $X_1,\ldots,X_n$ identical to random variable X, the average value

$$\boldsymbol{X}^n \stackrel{\mathrm{def}}{=} \sum_{i=1}^n \frac{X^1+\ldots+X^n}{n} \tag{2.68}$$

converges to the expectation $\mathrm{E}_p(X)$ in probability. That is,

$$p^n\{|\boldsymbol{X}^n-\mathrm{E}_p(X)|>\epsilon\}\to 0,\quad \forall\epsilon>0\,, \tag{2.69}$$

which is called the *law of large numbers.* Further, the distribution of the random variable

$$\sqrt{n}(\boldsymbol{X}^n-\mathrm{E}_p(X)) \tag{2.70}$$

goes to the Gaussian distribution with the variance $V=\mathrm{V}_p(X)$:

$$P_{G,V}(x)=\frac{1}{\sqrt{2\pi V}}e^{-\frac{x^2}{2V}}\,, \tag{2.71}$$

i.e.,

$$p^n\{a\le\sqrt{n}(\boldsymbol{X}^n-\mathrm{E}_p(X))\le b\}\to\int_a^b P_{G,V}(x)dx\,, \tag{2.72}$$

which is called the *central limit theorem.* Hence, the asymptotic behavior is almost characterized by the expectation $\mathrm{E}_(X)$ and the variance $\mathrm{V}(X)$.

[9] The superscript (e) means "exponential." This is because A corresponds to the exponential representation, as discussed later.

For n random variables $X_1, \ldots, X_k$, we can similarly define the random variables $\boldsymbol{X}_1^n, \ldots, \boldsymbol{X}_k^n$. These converge to their expectation in probability. The distribution of the random variables

$$(\sqrt{n}(\boldsymbol{X}_1^n - \mathrm{E}_p(X)), \ldots, \sqrt{n}(\boldsymbol{X}_k^n - \mathrm{E}_p(X))) \tag{2.73}$$

converges the k-multirate Gaussian distribution and the covariance matrix $\boldsymbol{V} = \mathrm{Cov}_p(X_i, X_j)$:

$$P_{G,\boldsymbol{V}}(x) \stackrel{\text{def}}{=} \frac{1}{\sqrt{(2\pi)^k \det \mathrm{V}}} e^{\langle x|\boldsymbol{V}^{-1}|x\rangle} . \tag{2.74}$$

Therefore, the asymptotic behavior is almost described by the expectation and the covariance matrix.

Consider the set of probability distributions p_θ parameterized by a single real number θ. For example, we can parameterize a binomial distribution with the probability space $\{0, 1\}$ by $p_\theta(0) = \theta$, $p_\theta(1) = 1 - \theta$. When the set of probability distributions is parameterized by a single parameter, it is called a *probability distribution family* and is represented by $\{p_\theta | \theta \in \Theta \subset \mathbb{R}\}$. Based on a probability distribution family, we can define the *logarithmic derivative* as $l_{\theta_0}(\omega) \stackrel{\text{def}}{=} \frac{d \log p_\theta(\omega)}{d\theta} = \left.\frac{dp_\theta(\omega)}{d\theta}\right|_{\theta=\theta_0} / p_{\theta_0}(\omega)$. Since it is a real-valued function of the probability space, it can be regarded as a random variable. We can consider that this quantity expresses the sensitivity of the probability distribution to the variations in the parameter θ around θ_0. The *Fisher metric* (*Fisher information*) is defined as the variance of the logarithmic derivative l_{θ_0}. Since the expectation of l_{θ_0} with respect to p_{θ_0} is 0, the Fisher information can also be defined as

$$J_\theta \stackrel{\text{def}}{=} \langle l_\theta, l_\theta \rangle_{p_\theta}^{(e)} . \tag{2.75}$$

Therefore, this quantity represents the amount of variation in the probability distribution due to the variations in the parameter. Alternatively, it can indicate how much the probability distribution family represents the information related to the parameter. As discussed later, these ideas will be further refined from the viewpoint of statistical inference. The Fisher information J_θ may also be expressed as the limits of relative entropy and Hellinger distance [Ex. 2.31,2.32]:

$$\frac{J_\theta}{2} = 4 \lim_{\epsilon \to 0} \frac{d_2^2(p_\theta, p_{\theta+\epsilon})}{\epsilon^2} \tag{2.76}$$

$$= \lim_{\epsilon \to 0} \frac{D(p_\theta \| p_{\theta+\epsilon})}{\epsilon^2} = \lim_{\epsilon \to 0} \frac{D(p_{\theta+\epsilon} \| p_\theta)}{\epsilon^2} . \tag{2.77}$$

The Fisher information J_θ is also characterized by the limit of relative Rényi entropy [Ex. 2.33]:

$$\frac{J_\theta}{2} = \lim_{\epsilon \to 0} \frac{-\phi(s|p_\theta \| p_{\theta+\epsilon})}{\epsilon^2 s(1-s)} . \tag{2.78}$$

Next, let us consider the probability distribution family $\{p_\theta|\theta \in \Theta \subset \mathbb{R}^d\}$ with multiple parameters. For each parameter, we define the logarithmic derivative $l_{\theta_0:i}(\omega)$ as

$$l_{\theta_0:i}(\omega) \stackrel{\text{def}}{=} \left.\frac{\partial \log p_\theta(\omega)}{\partial \theta^i}\right|_{\theta=\theta_0} = \left.\frac{\partial p_\theta(\omega)}{\partial \theta^i}\right|_{\theta=\theta_0} \Big/ p_{\theta_0}(\omega).$$

We use the covariance matrix $\langle l_{\theta:i}, l_{\theta:j}\rangle^{(e)}_{p_\theta}$ for the logarithmic derivatives $l_{\theta_0:1}, \ldots, l_{\theta_0:d}$ instead of the Fisher information. This matrix is called the *Fisher information matrix* and will be denoted by $\boldsymbol{J}_\theta = (J_{\theta:i,j})$. This matrix takes the role of the Fisher information when there are multiple parameters; we discuss this in greater detail below.

This inner product is useful for treating the conditional expectation as follows. Suppose that we observe only the subsystem Ω_1, although the total system is given as $\Omega_1 \times \Omega_2$. Let us consider the behavior of the random variable X of the total system in the subsystem. This behavior depends on the distribution p of the total system. It is described by the conditional expectation $\kappa_p(X)$, which is defined to satisfy

$$\langle Y, \kappa_p(X)\rangle^{(e)}_p = \langle Y, X\rangle^{(e)}_p \tag{2.79}$$

for any random variable Y on Ω_1. In this case, we identify the random variable Y on Ω_1 with the random variable $\tilde{Y}$ on $\Omega_1 \times \Omega_2$ by

$$\tilde{Y}(\omega_1, \omega_2) = Y(\omega_1), \quad \forall(\omega_1, \omega_2) \in \Omega_1 \times \Omega_2\,. \tag{2.80}$$

Generally, when we focus on a subspace $\mathfrak{U}$ of random variables for an arbitrary random variable X, we can define the conditional expectation $\kappa_{\mathfrak{U},p}(X) \in \mathfrak{U}$ as

$$\langle Y, \kappa_{\mathfrak{U},p}(X)\rangle^{(e)}_p = \langle Y, X\rangle^{(e)}_p, \quad \forall Y \in \mathfrak{U}\,. \tag{2.81}$$

This implies that the map $\kappa_{\mathfrak{U},p}(\ )$ is the projection from the space of all random variables to the subspace $\mathfrak{U}$ with respect to the inner product $\langle\ ,\ \rangle_p$.

2.3.2 Exponential Family and Divergence

In Sect. 2.1, relative entropy $D(p\|q)$ is defined. In this subsection, we characterize it by a geometrical viewpoint.

Let $p(\omega)$ be a probability distribution and $X(\omega)$ be a random variable. When the family $\{p_\theta|\theta \in \Theta\}$ has the form

$$p_\theta(\omega) = p(\omega)e^{\theta X(\omega) - \mu(\theta)}\,, \tag{2.82}$$

$$\mu(\theta) \stackrel{\text{def}}{=} \log \sum_\omega p(\omega)e^{\theta X(\omega)}\,, \tag{2.83}$$

the logarithmic derivative at respective points equals the logarithmic derivative at a fixed point with the addition of a constant. In this case, the family is called an *exponential family*, and $\mu(\theta)$ is called the *moment function* of X. In particular, since the logarithmic derivative is closely related to exponential families, it is often called the exponential (e) representation of the derivative. Therefore, we use the superscript (e) in the inner product $\langle\ ,\ \rangle_p^{(e)}$. The function $\mu(\theta)$ is often called a *potential function* in the context of information geometry. Since the second derivative $\mu''(\theta)$ is the Fisher information $J_\theta \geq 0$, the moment function $\mu(\theta)$ is a convex function. Therefore, the first derivative $\mu'(\theta) = \sum_\omega p_\theta(\omega) X(\omega)$ is monotone increasing. That is, we may regard it as another parameter identifying the distribution p_θ, and denote it by η. The original parameter θ is called a *natural parameter*, and the other parameter η is an *expectation parameter*. For example, in the binomial distribution, the parameterization $p_\theta(0) = 1/(1+e^\theta)$, $p_\theta(1) = e^\theta/(1+e^\theta)$ is the natural parameter, and the parameterization $p_\eta(1) = \eta$, $p_\eta(0) = 1-\eta$ is the expectation parameter. Hence, the binomial distribution is an exponential family.

Further, let $X_1(\omega), \ldots, X_k(\omega)$ be k random variables. We can define a k-parameter exponential family

$$p_\theta(\omega) \stackrel{\text{def}}{=} p(\omega) e^{\sum_i \theta^i X_i(\omega) - \mu(\theta)}, \quad \mu(\theta) \stackrel{\text{def}}{=} \log \sum_\omega p(\omega) e^{\sum_i \theta^i X_i(\omega)}. \tag{2.84}$$

The parameters θ^i are natural parameters, and the other parameters $\eta_i \stackrel{\text{def}}{=} \frac{\partial \mu}{\partial \theta^i} = \sum_\omega p_\theta(\omega) X_i(\omega)$ are expectation parameters. Since the second derivative $\frac{\partial^2 \mu(\theta)}{\partial \theta^j \partial \theta^i}$ is equal to the Fisher information matrix $J_{\theta:i,j}$, the moment function $\mu(\theta)$ is a convex function.

Let $\mu(\theta)$ be a twice-differentiable and strictly convex function defined on a subset of the d-dimensional real vector space $\mathbb{R}^d$. The divergence concerning the convex function μ is defined by

$$D^\mu(\bar\theta\|\theta) \stackrel{\text{def}}{=} \sum_i \eta_i(\bar\theta)(\bar\theta^i - \theta^i) - \mu(\bar\theta) + \mu(\theta), \quad \eta_i(\theta) \stackrel{\text{def}}{=} \frac{\partial \mu}{\partial \theta^i}(\theta). \tag{2.85}$$

This quantity has the following two characterizations:

$$D^\mu(\bar\theta\|\theta) = \max_{\tilde\theta} \sum_i \frac{\partial \mu}{\partial \theta^i}(\bar\theta)(\tilde\theta^i - \theta^i) - \mu(\tilde\theta) + \mu(\theta) \tag{2.86}$$

$$= \int_0^1 \sum_{i,j} (\bar\theta^i - \theta^i)(\bar\theta^j - \theta^j) \frac{\partial^2 \mu}{\partial \theta^i \partial \theta^j}(\theta + (\bar\theta - \theta)t) t\, dt, \tag{2.87}$$

where $\tilde\theta = (\bar\theta - \theta)t + \theta$. In the one-parameter case, we obtain

$$\begin{aligned} D^\mu(\bar\theta\|\theta) =& \mu'(\bar\theta)(\bar\theta - \theta) - \mu(\bar\theta) + \mu(\theta) \\ =& \max_{\tilde\theta} \mu'(\bar\theta)(\tilde\theta - \theta) - \mu(\tilde\theta) + \mu(\theta) = \int_\theta^{\bar\theta} \mu''(\tilde\theta)(\tilde\theta - \theta) d\tilde\theta . \end{aligned} \tag{2.88}$$

Since the function μ is strictly convex, the correspondence $\theta^i \leftrightarrow \eta_i = \frac{\partial \mu}{\partial \theta^i}$ is one-to-one. Hence, the divergence $D^\mu(\bar{\theta}\|\theta)$ can be expressed with the parameter η. For this purpose, we define the *Legendre transform* ν of μ

$$\nu(\eta) \stackrel{\text{def}}{=} \max_{\tilde{\theta}} \sum_i \eta_i \tilde{\theta}^i - \mu(\tilde{\theta}).$$

Then, the function ν is a convex function, and we can recover the function μ of θ as

$$\mu(\theta) = \max_{\tilde{\eta}} \sum_i \theta^i \tilde{\eta}_i - \nu(\tilde{\eta}), \quad \theta^i = \frac{\partial \nu}{\partial \eta_i}.$$

The second derivative matrix $\frac{\partial^2 \nu}{\partial \eta_i \partial \eta_j}$ of ν is equal to the inverse of the matrix $\frac{\partial^2 \mu}{\partial \theta^i \partial \theta^j}$.

In particular, when $\eta_i = \frac{\partial \mu}{\partial \theta^i}(\theta)$,

$$\nu(\eta) = \sum_i \eta_i \theta^i - \mu(\theta) = D^\mu(\theta\|0) - \mu(0), \tag{2.89}$$

$$\mu(\theta) = \sum_i \theta^i \eta_i - \nu(\eta) = D^\nu(\eta\|0) - \nu(0). \tag{2.90}$$

Using these relations, we can characterize the divergence concerning the convex function μ by the divergence concerning the convex function ν as

$$D^\mu(\bar{\theta}\|\theta) = D^\nu(\eta\|\bar{\eta}) = \sum_i \theta^i(\eta_i - \bar{\eta}_i) - \nu(\eta) + \nu(\bar{\eta}). \tag{2.91}$$

Now, we apply the discussion about the divergence to a multiparametric exponential family $\{p_\theta | \theta \in \mathbb{R}\}$ defined in (2.84) [11]. Then,

$$D(p_{\bar{\theta}}\|p_\theta) = D^\mu(\bar{\theta}\|\theta) = \sum_i \eta_i(\bar{\theta})(\bar{\theta}^i - \theta^i) - \mu(\bar{\theta}) + \mu(\theta).$$

In particular, applying (2.88) to a one-parameter exponential family (2.82), we have

$$\begin{aligned} D(p_{\bar{\theta}}\|p_\theta) =& D(p_{\eta(\theta)+\epsilon}\|p_{\eta(\theta)}) = (\bar{\theta} - \theta)\eta(\bar{\theta}) - \mu(\bar{\theta}) + \mu(\theta) \\ =& \int_\theta^{\bar{\theta}} J_{\tilde{\theta}}(\tilde{\theta} - \theta) d\tilde{\theta} = \max_{\tilde{\theta}:\tilde{\theta} \ge \theta} (\tilde{\theta} - \theta)(\eta(\theta) + \epsilon) - \mu(\tilde{\theta}) + \mu(\theta). \end{aligned} \tag{2.92}$$

In what follows, we consider the case where p is the uniform distribution p_{mix}. Let the random variables $X_1(\omega), \ldots, X_k(\omega)$ be a CONS of the space of random variables with expectation 0 under the uniform distribution p_{mix}, and $Y^1(\omega), \ldots, Y^k(\omega)$ be its dual basis satisfying $\sum_\omega Y^i(\omega) X_j(\omega) = \delta^i_j$. Then, any distribution can be parameterized by the expectation parameter as

$$p_{\eta(\theta)}(\omega) = p_{\text{mix}}(\omega) + \sum_i \eta_i(\theta) Y^i(\omega) .$$

From (2.91) and (2.89),

$$D(p_{\bar{\eta}} \| p_\eta) = D^\nu(\eta \| \bar{\eta}) = \frac{\partial \nu}{\partial \eta_i}(\eta_i - \bar{\eta}_i) - \nu(\eta) + \nu(\bar{\eta}) ,$$
$$\nu(\eta) = D(p_\eta \| p_{\text{mix}}) = -H(p_\eta) + H(p_{\text{mix}})$$

because $\mu(0) = 0$. The second derivative matrix of ν is the inverse of the second derivative matrix of μ, i.e., the Fisher information matrix concerning the natural parameter θ. That is, the second derivative matrix of ν coincides with the Fisher information matrix concerning the expectation parameter η. Hence, applying (2.88) to the subspace $\{(1-t)p + tq | 0 \le t \le 1\}$, we have

$$D(p\|q) = \int_0^1 J_t t dt , \tag{2.93}$$

where J_t is the Fisher information concerning parameter t.

Exercises

2.29. Show that $\text{Cov}_p(X, Y) = 0$ for random variables X and Y if they are independent.

2.30. Let J_θ be the Fisher information of a probability distribution family $\{p_\theta | \theta \in \Theta\}$. Let p_θ^n be the n-fold independent and identical distribution of p_θ. Show that the Fisher information of the probability distribution family $\{p_\theta^n | \theta \in \Theta\}$ at p_θ is nJ_θ.

2.31. Prove (2.76) using the second equality in (2.13), and noting that $\sqrt{1+x} \cong 1 + \frac{1}{2}x - \frac{1}{8}x^2$ for small x.

2.32. Prove (2.77) following the steps below.

a Show the following approximation with the limit $\epsilon \to 0$.

$$\log p_{\theta+\epsilon}(\omega) - \log p_\theta(\omega) \cong \frac{d \log p_\theta(\omega)}{d\theta}\epsilon + \frac{1}{2}\frac{d^2 \log p_\theta(\omega)}{d^2\theta}\epsilon^2 .$$

b Prove the first equality in (2.77) using **a**.
c Show the following approximation with the limit $\epsilon \to 0$.

$$p_{\theta+\epsilon}(\omega) \cong p_\theta(\omega) + \frac{dp_\theta(\omega)}{d\theta}\epsilon + \frac{1}{2}\frac{d^2 p_\theta(\omega)}{d^2\theta}\epsilon^2 .$$

d Prove the second equality in (2.77) using **a** and **c**.

2.33. Prove (2.78) using the approximation $(1+x)^s \cong 1 + sx + \frac{s(s-1)}{2}x^2$ for small x.

2.4 Estimation in Classical Systems

An important problem in mathematical statistics is the estimation of the parameter θ from some given data $\omega \in \Omega$ for a probability distribution that generates the data. To solve this problem, a mapping $\hat{\theta}$ called an estimator from the probability space Ω to the parameter space $\Theta \subset \mathbb{R}$ is required. The accuracy of the estimator is most commonly evaluated by the *mean square error*, which is the expectation value of the square of the difference $\hat{\theta} - \theta$:

$$\hat{\mathrm{V}}_\theta(\hat{\theta}) \stackrel{\mathrm{def}}{=} \mathrm{E}_{p_\theta}((\hat{\theta} - \theta)^2), \tag{2.94}$$

where θ is the true parameter. Note that sometimes the mean square error is not the same as the variance $\mathrm{V}_{p_\theta}(X)$. The estimator

$$\mathrm{E}_\theta(\hat{\theta}) \stackrel{\mathrm{def}}{=} \mathrm{E}_{p_\theta}(\hat{\theta}) = \theta, \quad \forall \theta \in \Theta \tag{2.95}$$

is called an *unbiased estimator*, and such estimators form an important class of estimators. The mean square error of the unbiased estimator $\hat{\theta}$ satisfies the *Cramér–Rao inequality*

$$\hat{\mathrm{V}}_\theta(\hat{\theta}) \geq J_\theta^{-1} . \tag{2.96}$$

When an unbiased estimator attains the RHS of (2.96), it is called *efficient*. This inequality can be proved from the relations

$$\langle (\hat{\theta} - \theta), l_{\theta_0} \rangle_p^{(e)} = \left. \frac{d\mathrm{E}_\theta(\hat{\theta} - \theta_0)}{d\theta} \right|_{\theta = \theta_0} = 1$$

and

$$\langle (\hat{\theta} - \theta_0), (\hat{\theta} - \theta_0) \rangle_p^{(e)} \langle l_{\theta_0}, l_{\theta_0} \rangle_p^{(e)} \geq \left| \langle (\hat{\theta} - \theta), l_{\theta_0} \rangle_p^{(e)} \right|^2 = 1 ,$$

which follows from Schwarz's inequality. The equality of (2.96) holds for every value of θ if and only if the probability distribution family is a one-parameter exponential family (2.82) and the expectation parameter $\eta(\theta) = \sum_\omega X(\omega) p_\theta(\omega)$ is to be estimated. Hence, even in the estimation for an exponential family, there is necessarily no estimator for the natural parameter θ in (2.82) such that the equality of (2.96) holds for all θ. In this case, the efficient estimator for the expected parameter is given as $\hat{\eta}(\omega) = X(\omega)$ (Exercise 2.34).

Let n data $\omega^n = (\omega_1, \ldots, \omega_n) \in \Omega^n$ be generated with the n-i.i.d. of the probability distribution p_θ. The estimator may then be given by the mapping $\hat{\theta}^n$ from Ω^n to $\Theta \subset \mathbb{R}$. In this case, the Fisher information of the probability distribution family is nJ_θ, and the unbiased estimator $\hat{\theta}^n$ satisfies the Cramér–Rao inequality

$$\hat{V}_\theta(\hat{\theta}^n) \geq \frac{1}{n} J_\theta^{-1} .$$

However, in general, it is not necessary to restrict our estimator to unbiased estimators. In fact, rare estimators satisfy such conditions for finite n.

Therefore, in mathematical statistics, we often study problems in the asymptotic limit $n \to \infty$ rather than those with a finite number of data elements. For this purpose, let us apply the asymptotic unbiasedness conditions

$$\lim \mathrm{E}_\theta(\hat{\theta}_n) = \theta, \quad \lim \frac{d}{d\theta} \mathrm{E}_\theta(\hat{\theta}_n) = 1, \quad \forall \theta \in \Theta$$

to a sequence of estimators $\{\hat{\theta}^n\}$. Evaluating the accuracy with $\underline{\lim}\, n\hat{V}_\theta(\hat{\theta}_n)$, we have the *asymptotic Cramér–Rao inequality*:[10]

$$\underline{\lim}\, n\hat{V}_\theta(\hat{\theta}_n) \geq J_\theta^{-1} . \tag{2.97}$$

Then we obtain

$$n J_\theta \hat{V}_\theta(\hat{\theta}_n) \geq \left| \frac{d}{d\theta} \mathrm{E}_\theta(\hat{\theta}_n) \right|^2 ,$$

based on a derivation similar to (2.96). Inequality (2.97) may also be derived from this inequality.

Now, we consider what estimator attains the lower bound of (2.97). The *maximum likelihood estimator* $\hat{\theta}_{n,ML}(\omega^n)$

$$\hat{\theta}_{n,ML}(\omega^n) = \operatorname*{argmax}_{\theta \in \Theta} p_\theta^n(\omega^n) \tag{2.98}$$

achieves this lower bound, and the limit of its mean squared error is equal to J_θ^{-1} [396]. Indeed, in an exponential family with the expectation parameter, the maximum likelihood estimator is equal to the efficient estimator $^{\text{Ex. 2.36}}$. Hence, the maximum likelihood estimator plays an important role in statistical inference.[11]

Indeed, we choose the mean square error as the criterion of estimation error because (1) its mathematical treatment is easy and (2) in the i.i.d. case, the average random variable can be characterized by a Gaussian distribution. Hence, we can expect that a suitable estimator will also approach a Gaussian distribution asymptotically. That is, we can expect that its asymptotic

[10] This inequality still holds even if the asymptotic unbiasedness condition is replaced by another weak condition. Indeed, it is a problem to choose a suitable condition to be assumed for the inequality (2.97). For details, see van der Vaart [396].

[11] This is generally true for all probability distribution families, although some regularity conditions must be imposed. For example, consider the case in which Ω consists of finite elements. These regularity conditions are satisfied when the first and second derivatives with respect to θ are continuous. Generally, the central limit theorem is used in the proof [396].

behavior will be characterizable by the variance. In particular, the maximum likelihood estimator $\hat{\theta}_{n,ML}$ obeys the Gaussian distribution asymptotically:

$$p_\theta^n\{a \le \sqrt{n}(\hat{\theta}_{n,ML} - \theta) \le b\} \to \int_a^b P_{G,1/J_\theta}(x)dx, \quad \forall a, b\,.$$

Let us now consider the probability distribution family $\{p_\theta | \theta \in \Theta \subset \mathbb{R}^d\}$ with multiple parameters. We focus on the Fisher information matrix $\boldsymbol{J}_\theta = (J_{\theta:i,j})$, which was defined at the end of Sect. 2.3.1, instead of the Fisher information. The estimator is given by the map $\hat{\theta} = (\hat{\theta}^1, \dots, \hat{\theta}^d)$ from the probability space Ω to the parameter space Θ, similar to the one-parameter case. The unbiasedness conditions are

$$\mathrm{E}_\theta^i(\hat{\theta}) \stackrel{\text{def}}{=} \mathrm{E}_{p_\theta}(\hat{\theta}^i) = \theta^i, \quad \forall \theta \in \Theta, 1 \le \forall i \le d\,.$$

The error can be calculated using the *mean square error matrix* $\hat{\boldsymbol{V}}_\theta(\hat{\theta}) = (\hat{\mathrm{V}}_\theta^{i,j}(\hat{\theta}))$:

$$\hat{\mathrm{V}}_\theta^{i,j}(\hat{\theta}) \stackrel{\text{def}}{=} \mathrm{E}_{p_\theta}((\hat{\theta}^i - \theta^i)(\hat{\theta}^j - \theta^j)).$$

Then, we obtain the *multiparameter Cramér–Rao inequality*

$$\hat{\boldsymbol{V}}_\theta(\hat{\theta}) \ge \boldsymbol{J}_\theta^{-1}\,. \tag{2.99}$$

Proof of (2.99). For the proof, let us assume that any vectors $|b\rangle = (b_1, \dots, b_d)^T \in \mathbb{C}^d$ and $|a\rangle \in \mathbb{C}^d$ satisfy

$$\langle b|\hat{\boldsymbol{V}}_\theta(\hat{\theta})b\rangle\langle a|\boldsymbol{J}_\theta|a\rangle \ge |\langle b|a\rangle|^2\,. \tag{2.100}$$

By substituting $a = (\boldsymbol{J}_\theta)^{-1}b$, inequality (2.100) becomes

$$\langle b|\hat{\boldsymbol{V}}_\theta(\{\hat{\theta}_n\})|b\rangle \ge \langle b|(\boldsymbol{J}_\theta)^{-1}|b\rangle$$

since $(\boldsymbol{J}_\theta)^{-1}$ is a symmetric matrix. Therefore, we obtain (2.99) if (2.100) holds. We prove (2.100) as follows. Since

$$\delta_i^j = \left.\frac{d\mathrm{E}_\theta^j(\hat{\theta}) - \theta_0^j}{d\theta^i}\right|_{\theta=\theta_0} = \left\langle l_{\theta:i}, (\hat{\theta}^j - \theta_0^j)\right\rangle_\theta^{(e)},$$

similarly to the proof of (2.96), the Schwarz inequality yields

$$\begin{aligned}\langle b|\hat{\boldsymbol{V}}_\theta(\hat{\theta})b\rangle &= \left\langle \left(\sum_{i=1}^d (\hat{\theta}^i(\omega) - \theta^i)b_i\right), \left(\sum_{i=1}^d (\hat{\theta}^i(\omega) - \theta^i)b_i\right)\right\rangle_\theta^{(e)} \\ &\ge \frac{\left|\left\langle \left(\sum_{i=1}^d l_{\theta:i}a_i\right), \left(\sum_{i=1}^d (\hat{\theta}^i(\omega) - \theta^i)b_i\right)\right\rangle_\theta^{(e)}\right|^2}{\left\langle \left(\sum_{i=1}^d l_{\theta:i}a_i\right), \left(\sum_{i=1}^d l_{\theta:i}a_i\right)\right\rangle_\theta^{(e)}} = \frac{|\langle a|b\rangle|^2}{\langle a|\boldsymbol{J}_\theta|a\rangle}\,.\end{aligned}$$

■

Moreover, since the sequence of estimators $\{\hat{\theta}_n = (\hat{\theta}_n^1, \ldots, \hat{\theta}_n^d)\}$ satisfies the asymptotic unbiasedness condition

$$\lim \mathrm{E}_\theta^i(\hat{\theta}_n) = \theta^i, \quad \lim \frac{\partial}{\partial \theta^j} \mathrm{E}_\theta^i(\hat{\theta}_n) = \delta_j^i, \quad \forall \theta \in \Theta, \tag{2.101}$$

the asymptotic Cramér–Rao inequality for the multiparameter case

$$\hat{\boldsymbol{V}}_\theta(\{\hat{\theta}_n\}) \geq \boldsymbol{J}_\theta^{-1} \tag{2.102}$$

holds if the limit $\hat{\boldsymbol{V}}_\theta(\{\hat{\theta}_n\}) \stackrel{\text{def}}{=} \lim n \hat{\boldsymbol{V}}_\theta(\hat{\theta}_n)$ exists. Next, we prove (2.102). Defining $A_{n,i}^j \stackrel{\text{def}}{=} \frac{\partial}{\partial \theta^j} \mathrm{E}_\theta^i(\hat{\theta}_n)$, we have

$$n \langle a | \boldsymbol{J}_\theta | a \rangle \langle b | \hat{\boldsymbol{V}}_\theta(\hat{\theta}_n) | b \rangle \geq |\langle a | \mathbf{A}_n | b \rangle|^2$$

instead of (2.100). We then obtain

$$\langle a | \boldsymbol{J}_\theta | a \rangle \langle b | \hat{\boldsymbol{V}}_\theta(\{\hat{\theta}_n\}) | b \rangle \geq |\langle a | b \rangle|^2 ,$$

from which (2.102) may be obtained in a manner similar to (2.99).

Similarly to the one-parameter case, the equality of (2.99) holds if and only if the following conditions hold: (1) The probability distribution family is a multiparameter exponential family. (2) The expectation parameter η is to be estimated. (3) The estimator for η is given by

$$\hat{\eta}_i(\omega) = X_i(\omega) . \tag{2.103}$$

In this case, this estimator (2.103) equals the maximum likelihood estimator $\hat{\theta}_{n,ML} = (\hat{\theta}_{n,ML}^1, \ldots \hat{\theta}_{n,ML}^d)$ defined by (2.98) Ex. 2.36, i.e.,

$$\max_\eta p_\eta(\omega) = p_{\eta_i = X_i(\omega)}(\omega) . \tag{2.104}$$

A probability distribution family does not necessarily have such an estimator; however, a maximum likelihood estimator $\hat{\theta}_{n,ML}$ can be defined by (2.98). This satisfies the asymptotic unbiasedness property (2.101) in a similar way to (2.98), and it satisfies the equality of (2.102). Moreover, it is known that the maximum likelihood estimator $\hat{\theta}_{n,ML}$ satisfies [396]

$$\hat{\boldsymbol{V}}_\theta(\{\hat{\theta}_n\}) = \boldsymbol{J}_\theta^{-1} .$$

Note that this inequality holds independently of the choice of coordinate. Hence, for a large amount of data, it is best to use the maximum likelihood estimator. Its mean square error matrix is almost in inverse proportion to the number of observations n. This coefficient of the optimal case is given by the Fisher information matrix. Therefore, the Fisher information matrix can be considered to yield the best accuracy of an estimator.

Indeed, usually any statistical decision with the given probability distribution family $\{p_\theta|\theta \in \Theta\}$ is based on the likelihood ratio $\log p_\theta(\omega) - \log p_{\theta'}(\omega)$. For example, the maximum likelihood estimator depends only on the likelihood ratio. If the probability distribution family $\{p_\theta|\theta \in \Theta\}$ belongs to a larger multiparameter exponential family $\{q_\theta|\theta \in \Theta\}$, the family is called a *curved exponential family*. In this case, the likelihood ratio can be expressed by the relative entropy

$$\log p_\theta(\omega) - \log p_{\theta'}(\omega) = D(q_{\eta_i = X_i(\omega)}\|p_{\theta'}) - D(q_{\eta_i = X_i(\omega)}\|p_\theta)\,. \qquad (2.105)$$

That is, our estimation procedure can be treated from the viewpoint of the relative entropy geometry.

Exercises

2.34. Show that the following two conditions are equivalent for a probability distribution family $\{p_\theta|\theta \in \mathbb{R}\}$ and its estimator X by following the steps below.

① The estimator X is an unbiased estimator and satisfies the equality of (2.96) at all points.
② The probability distribution family $\{p_\theta|\theta \in \mathbb{R}\}$ is an exponential family, $p_\theta(\omega)$ is given by (2.82) using X, and the parameter to be estimated is the expectation parameter $\eta(\theta)$.

a Show that the estimator X is an unbiased estimator if the parameter for the exponential family (2.82) is given by the expectation parameter.
b Show that ① may be deduced from ②.
c For the exponential family (2.82), show that the natural parameter θ is given as a function of the expectation parameter η with the form $\theta = \int_0^\eta J_{\eta'}\, d\eta'$.
d Show that $\mu(\theta(\eta)) = \int_0^\eta \eta' J_{\eta'}\, d\eta'$.
e Show that $\frac{l_\theta}{J_\theta} = X - \theta$ if ① is true.
f Denote the parameter to be estimated by η. Show that $\frac{dp_\eta}{d\eta} = J_\eta(X - \eta)p_\eta$ if ① is true.
g Show that ② is true if ① is true.

2.35. Show that equation (2.105) holds.

2.36. Show equation (2.104) from (2.105).

2.37. Consider the probability distribution family $\{p_\theta|\theta \in \mathbb{R}\}$ in the probability space $\{1, \dots, l\}$ and the stochastic transition matrix $Q = (Q^i_j)$. Let the Fisher information of p_{θ_0} in the probability distribution family $\{p_\theta|\theta \in \mathbb{R}\}$ be J_{θ_0}. Let J'_{θ_0} be the Fisher information of $Q(p_{\theta_0})$ in the probability distribution family $\{Q(p_\theta)|\theta \in \mathbb{R}\}$. Show then that $J_{\theta_0} \geq J'_{\theta_0}$. This inequality is called the *monotonicity* of the Fisher information. Similarly, define $\boldsymbol{J}_{\theta_0}, \boldsymbol{J}'_{\theta_0}$ for the multiple variable case, and show that the matrix inequality $\boldsymbol{J}_{\theta_0} \geq \boldsymbol{J}'_{\theta_0}$ holds.

2.5 Type Method and Large Deviation Evaluation

In this section, we analyze the case of a sufficiently large number of data by using the following two methods. The first method involves an analysis based on empirical distributions, and it is called the *type method.* In the second method, we consider a particular random variable and examine its exponential behavior.

2.5.1 Type Method and Sanov's Theorem

Let n data be generated according to a probability distribution in a finite set of events $\mathbb{N}_d = \{1, \ldots, d\}$. Then, we can perform the following analysis by examining the empirical distribution of the data [85]. Let T_n be the set of empirical distributions obtained from n observations. We call each element of this set a type. For each type $q \in T_n$, let the subset $T_q^n \subset \mathbb{N}_d^n$ be a set of data with the empirical distribution q. Since the probability $p^n(\boldsymbol{i})$ depends only on the type q for each $\boldsymbol{i} \in T_q^n$, we can denote this probability by $p^n(q)$. Then, when the n data are generated according to the probability distribution p^n, the empirical distribution matches $q \in T_n$ with the probability $p^n(T_q^n) \left(\overset{\text{def}}{=} \sum_{\boldsymbol{i} \in T_q^n} p^n(\boldsymbol{i}) \right)$.

Theorem 2.3 *Any type $p \in T_n$ and any data $\boldsymbol{i} \in T_q^n$ satisfy the following:*

$$p^n(T_q^n) \le p^n(T_p^n), \tag{2.106}$$

$$p^n(\boldsymbol{i}) = e^{-n(H(q)+D(q\|p))}. \tag{2.107}$$

Denoting the number of elements of T_n and T_q^n by $|T_n|$ and $|T_q^n|$, respectively, we obtain the relations

$$|T_n| \le (n+1)^{d-1}, \tag{2.108}$$

$$\frac{1}{(n+1)^d} e^{nH(q)} \le |T_q^n| \le e^{nH(q)}, \tag{2.109}$$

$$\frac{1}{(n+1)^d} e^{-nD(q\|p)} \le p^n(T_q^n) \le e^{-nD(q\|p)}. \tag{2.110}$$

Proof. Let $p(i) = \frac{n_i}{n}$ and $q(i) = \frac{n_i'}{n}$. Then,

$$p^n(T_p^n) = |T_p^n| \prod_{i=1}^d p(i)^{n_i} = \frac{n!}{n_1! \cdots n_d!} \prod_{i=1}^d p(i)^{n_i},$$

$$p^n(T_q^n) = |T_q^n| \prod_{i=1}^d p(i)^{n_i'} = \frac{n!}{n_1'! \cdots n_d'!} \prod_{i=1}^d p(i)^{n_i'}.$$

Using the inequality Ex. 2.38

$$\frac{n!}{m!} \le n^{n-m}, \tag{2.111}$$

we have

$$\begin{aligned}\frac{p^n(T_q^n)}{p^n(T_p^n)} &= \prod_{i=1}^d \left(\frac{n_i!}{n_i'!} p(i)^{n_i'-n_i} \right) \le \prod_{i=1}^d \left(n_i^{n_i-n_i'} \left(\frac{n_i}{n}\right)^{n_i'-n_i} \right) \\ &= \prod_{i=1}^d \left(\frac{1}{n}\right)^{n_i'-n_i} = \left(\frac{1}{n}\right)^{\sum_{i=1}^d (n_i'-n_i)} = 1 .\end{aligned}$$

Therefore, inequality (2.106) holds. For $\boldsymbol{i} \in T_q^n$, we have

$$\begin{aligned}p(\boldsymbol{i}) &= \prod_{i=1}^d p(i)^{n_i'} = \prod_{i=1}^d p(i)^{n\left(\frac{n_i'}{n}\right)} \\ &= \prod_{i=1}^d e^{n \log p(i) \left(\frac{n_i'}{n}\right)} = e^{n \sum_{i=1}^d q(i) \log p(i)} = e^{-n(H(q)+D(q\|p))} ,\end{aligned}$$

which implies (2.107).

Each element q of T_n may be written as a d-dimensional vector. Each component of the vector then assumes one of the following $n+1$ values: $0, 1/n, \ldots, n/n$. Since $\sum_{i=1}^d q_i = 1$, the dth element is decided by the other $d-1$ elements. Therefore, inequality (2.108) follows from a combinatorial observation. Applying inequality (2.107) to the case $p = q$, we have the relation $p^n(T_q^n) = e^{-nH(p)}|T_p^n|$. Since $1 = \sum_{q \in T_n} p^n(T_q^n) \ge p^n(T_q^n)$ for $p \in T_n$, we obtain the inequality on the RHS of (2.109). Conversely, inequality (2.106) yields that $1 = \sum_{q \in T_n} p^n(T_q^n) \le \sum_{q \in T_n} p^n(T_p^n) = e^{-nH(p)}|T_p^n||T_n|$. Combining this relation with (2.108), we obtain the inequality on the LHS of (2.109). Inequality (2.110) may be obtained by combining (2.107) and (2.109). ■

We obtain *Sanov's Theorem* using these inequalities.

Theorem 2.4 *(Sanov [358]) The following holds for a subset $\mathcal{R}$ of distributions on $\mathbb{N}_d$:*

$$\begin{aligned}\frac{1}{(n+1)^d} \exp(-n \min_{q \in \mathcal{R} \cap T_n} D(q\|p)) &\le p^n(\cup_{q \in \mathcal{R} \cap T_n} T_q^n) \\ &\le (n+1)^d \exp(-n \inf_{q \in \mathcal{R}} D(q\|p)).\end{aligned}$$

In particular, when the closure of the interior of $\mathcal{R}$ coincides with the closure of $\mathcal{R}$,[12]

[12] The set is called the interior of a set X when it consists of the elements of X without its boundary. For example, for a one-dimensional set, the interior of $[0, 0.5] \cup \{0.7\}$ is $(0, 0.5)$ and the closure of the interior is $[0, 0.5]$. Therefore, the condition is not satisfied in this case.

$$\lim \frac{-1}{n} \log p^n \left(\cup_{q \in \mathcal{R} \cap T_n} T_q^n\right) = \inf_{q \in \mathcal{R}} D(q\|p)$$

in the limit $n \to \infty$.

Based on this theorem, we can analyze how different the true distribution is from the empirical distribution. More precisely, the empirical distribution belongs to the neighborhood of the true distribution with a sufficiently large probability, i.e., the probability of its complementary event approaches 0 exponentially. This exponent is then given by the relative entropy. The discussion of this exponent is called a large deviation evaluation.

However, it is difficult to consider a quantum extension of Sanov's theorem. This is because we cannot necessarily take the common eigenvectors for plural densities. That is, this problem must be treated independently of the choice of basis. One possible way to fulfill this requirement is the group representation method. If we use this method, it is possible to treat the eigenvalues of density of the system instead of the classical probabilities [252, 286]. Since eigenvalues do not identify the density matrix, they cannot be regarded as the complete quantum extension of Sanov's theorem. Indeed, a quantum extension is available if we focus only on two densities; however, it should be regarded as the quantum extension of Stein's lemma given in Sect. 3.4. Since the data are not given without our operation in the quantum case, it is impossible to directly extend Sanov's theorem to the quantum case.

In fact, the advantage of using the type method is the universality in information theory [85]. However, if we apply the type method to quantum systems independently of the basis, the universality is not available in the quantum case. A group representation method is very effective for a treatment independent of basis [169, 174, 175, 183–185, 252]. Indeed, several universal protocols have been obtained by this method.

2.5.2 Cramér Theorem and Its Application to Estimation

Next, we consider the asymptotic behavior of a random variable X in the case of independent and identical trials of the probability distribution p.

For this purpose, we first introduce two fundamental inequalities $^{\text{Ex. 2.39}}$. The *Markov inequality* states that for a random variable X where $X \geq 0$,

$$\frac{\mathrm{E}_p(X)}{c} \geq p\{X \geq c\}. \tag{2.112}$$

Applying the Markov inequality to the variable $|X - \mathrm{E}_p(X)|$, we obtain the *Chebyshev inequality*:

$$p\{|X - \mathrm{E}_p(X)| \geq a\} \leq \frac{\mathrm{V}_p(X)}{a^2}. \tag{2.113}$$

Now, consider the random variable

$$X^n \stackrel{\mathrm{def}}{=} \sum_{i=1}^{n} \frac{1}{n} X_i \,, \tag{2.114}$$

where $X_1, \ldots, X_n$ are n independent random variables that are identical to the random variable X. When the variable X^n obeys the independent and identical distribution p^n of p, the expectation of X^n coincides with the expectation $\mathrm{E}_p(X)$. Let $\mathrm{V}_p(X)$ be the variance of X. Then, its variance with n observations equals $\mathrm{V}_p(X)/n$.

Applying Chebyshev's inequality (2.113), we have

$$p^n\{|X^n - \mathrm{E}_p(X)| \geq \epsilon\} \leq \frac{\mathrm{V}_p(X)}{n\epsilon^2}$$

for arbitrary $\epsilon > 0$. This inequality yields the *(weak) law of large numbers*

$$p^n\{|X^n - \mathrm{E}_p(X)| \geq \epsilon\} \to 0, \quad \forall \epsilon > 0 \,. \tag{2.115}$$

In general, if a sequence of pairs $\{(X^n, p_n)\}$ of a random variable and a probability distribution satisfies

$$p_n\{|X^n - x| \geq \epsilon\} \to 0, \quad \forall \epsilon > 0 \tag{2.116}$$

for a real number x, then the random variable X^n is said to *converge in probability* to x.

Since the left-hand side (LHS) of (2.115) converges to 0, the next focus is the speed of this convergence. Usually, this convergence is exponential. The exponent of this convergence is characterized by *Cramér's Theorem* below.

Theorem 2.5 *(Cramér [83]) Define the moment function* $\mu(\theta) \stackrel{\mathrm{def}}{=} \log\left(\sum_\omega p(\omega) e^{\theta X(\omega)}\right)$. *Then*

$$\underline{\lim} \frac{-1}{n} \log p^n\{X^n \geq x\} \geq \max_{\theta \geq 0} (\theta x - \mu(\theta)) \,, \tag{2.117}$$

$$\overline{\lim} \frac{-1}{n} \log p^n\{X^n \geq x\} \leq \lim_{x' \to x+0} \max_{\theta \geq 0} (\theta x' - \mu(\theta)) \,, \tag{2.118}$$

$$\underline{\lim} \frac{-1}{n} \log p^n\{X^n \leq x\} \geq \max_{\theta \leq 0} (\theta x - \mu(\theta)) \tag{2.119}$$

$$\overline{\lim} \frac{-1}{n} \log p^n\{X^n \leq x\} \leq \lim_{x' \to x-0} \max_{\theta \leq 0} (\theta x' - \mu(\theta)) \,, . \tag{2.120}$$

If we replace $\{X^n \geq x\}$ *and* $\{X^n \leq x\}$ *with* $\{X^n > x\}$ *and* $\{X^n < x\}$, *respectively, the same inequalities hold.*

When the probability space consists of finite elements, the function $\max_{\theta \geq 0} (\theta x - \mu(\theta))$ is continuous, i.e., $\lim_{x' \to x+0} \max_{\theta \geq 0} (\theta x' - \mu(\theta)) =$

$\max_{\theta \geq 0} (\theta x - \mu(\theta))$. Hence, the equalities of (2.117) and (2.119) hold. Conversely, if the probability space contains an infinite number of elements as the set of real numbers $\mathbb{R}$, we should treat the difference between the RHS and LHS more carefully. Further, the inequalities of (2.117) and (2.119) hold without limit, and they are equivalent to (2.36) and (2.37) when we replace the random variable $X(\omega)$ with $-\log q(\omega)$.

Proof. Inequality (2.119) is obtained by considering $-X$ in (2.117). Therefore, we prove only (2.117). Inequality (2.120) is also obtained by considering $-X$ in (2.118). Here we prove only inequality (2.117). Inequality (2.118) will be proved at the end of this section.

For a random variable X with $X(\omega)$ for each ω,

$$\mathrm{E}_{p^n}(e^{n\theta X^n}) = \mathrm{E}_{p^n}(\prod_{i=1}^{n} e^{\theta X_i}) = (\mathrm{E}_p e^{\theta X})^n = e^{n\mu(\theta)} . \tag{2.121}$$

Using the Markov inequality (2.112), we obtain

$$p^n\{X^n \geq x\} = p^n\{e^{n\theta X^n} \geq e^{n\theta x}\} \leq \frac{e^{n\mu(\theta)}}{e^{n\theta x}} \text{ for } \theta \geq 0 .$$

Taking the logarithm of both sides, we have

$$\frac{-1}{n} \log p^n\{X^n \geq x\} \geq \theta x - \mu(\theta).$$

Let us take the maximum on the RHS with respect to $\theta \geq 0$ and then take the limit on the LHS. We obtain inequality (2.117). ■

This theorem can be extended to the non-i.i.d. case as the *Gärtner–Ellis theorem.*

Theorem 2.6 *(Gärtner [141], Ellis [106]) Let $\{p_n\}$ be a general sequence of the probabilities with the random variables X_n. Define the moment functions $\mu_n(\theta) \stackrel{\mathrm{def}}{=} \frac{1}{n} \log \left(\sum_\omega p_n(\omega) e^{\theta n X_n(\omega)}\right)$ and $\mu(\theta) \stackrel{\mathrm{def}}{=} \lim \mu_n(\theta)$ and the set $G \stackrel{\mathrm{def}}{=} \{\mu'(\theta) | \theta\}$. Then*

$$\underline{\lim} \frac{-1}{n} \log p_n\{X_n \geq x\} \geq \max_{\theta \geq 0} (\theta x - \mu(\theta)) , \tag{2.122}$$

$$\overline{\lim} \frac{-1}{n} \log p_n\{X_n \geq x\} \leq \inf_{\bar{x} \in G : \bar{x} > x} \max_{\theta \geq 0} (\theta \bar{x} - \mu(\theta)) , \tag{2.123}$$

$$\underline{\lim} \frac{-1}{n} \log p_n\{X_n \leq x\} \geq \max_{\theta \leq 0} (\theta x - \mu(\theta)) , \tag{2.124}$$

$$\overline{\lim} \frac{-1}{n} \log p_n\{X_n \leq x\} \leq \inf_{\bar{x} \in G : \bar{x} < x} \max_{\theta \leq 0} (\theta \bar{x} - \mu(\theta)) . \tag{2.125}$$

If we replace $\{X_n \geq x\}$ and $\{X_n \leq x\}$ by $\{X_n > x\}$ and $\{X_n < x\}$, respectively, the same inequalities hold.

Inequalities (2.122) and (2.124) can be proved in a similar way to Theorem 2.5.

Next, we apply large deviation arguments to estimation theory. Our arguments will focus not on the mean square error but on the decreasing rate of the probability that the estimated parameter does not belong to the ϵ-neighborhood of the true parameter. To treat the accuracy of a sequence of estimators $\{\hat{\theta}_n\}$ with a one-parameter probability distribution family $\{p_\theta|\theta \in \mathbb{R}\}$ from the viewpoint of a large deviation, we define

$$\beta(\{\hat{\theta}_n\}, \theta, \epsilon) \stackrel{\text{def}}{=} \underline{\lim} \frac{-1}{n} \log p_\theta^n\{|\hat{\theta}_n - \theta| > \epsilon\}, \tag{2.126}$$

$$\alpha(\{\hat{\theta}_n\}, \theta) \stackrel{\text{def}}{=} \underline{\lim}_{\epsilon \to 0} \frac{\beta(\{\hat{\theta}_n\}, \theta, \epsilon)}{\epsilon^2}. \tag{2.127}$$

As an approximation, we have

$$p_\theta^n\{|\hat{\theta}_n - \theta| > \epsilon\} \cong e^{-n\epsilon^2 \alpha(\{\hat{\theta}_n\}, \theta)}.$$

Hence, an estimator functions better when it has larger values of $\beta(\{\hat{\theta}_n\}, \theta, \epsilon)$ and $\alpha(\{\hat{\theta}_n\}, \theta)$.

Theorem 2.7 *(Bahadur [23–25]) Let a sequence of estimators $\{\hat{\theta}_n\}$ satisfy the weak consistency condition*

$$p_\theta^n\{|\hat{\theta}_n - \theta| > \epsilon\} \to 0, \quad \forall \epsilon > 0, \quad \forall \theta \in \mathbb{R}. \tag{2.128}$$

Then, it follows that

$$\beta(\{\hat{\theta}_n\}, \theta, \epsilon) \le \inf_{\theta' : |\theta' - \theta| > \epsilon} D(p_{\theta'} \| p_\theta). \tag{2.129}$$

Further, if

$$D(p_{\theta'} \| p_\theta) = \lim_{\bar{\theta} \to \theta'} D(p_{\bar{\theta}} \| p_\theta), \tag{2.130}$$

the following also holds:

$$\alpha(\{\hat{\theta}_n\}, \theta) \le \frac{1}{2} J_\theta. \tag{2.131}$$

If the probability space consists of finite elements, condition (2.130) holds.

Proof. Inequality (2.131) is obtained by combining (2.129) with (2.77). Inequality (2.129) may be derived from monotonicity (2.9) as follows. From the consistency condition (2.128), the sequence $a_n \stackrel{\text{def}}{=} p_\theta^n\{|\hat{\theta}_n - \theta| > \epsilon\}$ satisfies $a_n \to 0$. Assume that $|\theta - \theta'| > \epsilon$. Then, when $|\hat{\theta}_n - \theta| \le \epsilon$, we have $|\hat{\theta}_n - \theta'| > 0$. Hence, the other sequence $b_n \stackrel{\text{def}}{=} p_{\theta'}^n\{|\hat{\theta}_n - \theta| > \epsilon\}$ satisfies

$b_n \to 1$ because of the consistency condition (2.128). Thus, monotonicity (2.9) implies that

$$D(p_{\theta'}^n \| p_\theta^n) \geq b_n(\log b_n - \log a_n) + (1-b_n)(\log(1-b_n) - \log(1-a_n)).$$

Since $nD(p_{\theta'}\|p_\theta) = D(p_{\theta'}^n\|p_\theta^n)$ follows from (2.44) and $-(1-b_n)\log(1-a_n) \geq 0$, we have $nD(p_{\theta'}\|p_\theta) \geq -h(b_n) - b_n \log a_n$, and therefore

$$\frac{-1}{n}\log a_n \leq \frac{D(p_{\theta'}\|p_\theta)}{b_n} + \frac{h(b_n)}{nb_n}. \tag{2.132}$$

As the convergence $h(b_n) \to 0$ follows from the convergence $b_n \to 1$, we have

$$\beta(\{\hat{\theta}_n\}, \theta, \epsilon) \leq D(p_{\theta'}\|p_\theta).$$

Considering $\inf_{\theta':|\theta'-\theta|>\epsilon}$, we obtain (2.129). In addition, this proof is valid even if we replace $\{|\hat{\theta}_n - \theta| > \epsilon\}$ in (2.126) by $\{|\hat{\theta}_n - \theta| \geq \epsilon\}$. ■

If no estimator satisfies the equalities in inequalities (2.129) and (2.131), these inequalities cannot be considered satisfactory. The equalities of (2.129) and (2.131) are the subject of the following proposition.

Proposition 2.1 *Suppose that the probability distribution family (2.82) is exponential, and the parameter to be estimated is an expectation parameter. If a sequence of estimators is given by $X^n(\omega^n)$ (see (2.114)), then the equality of (2.129) holds. The equality of (2.131) is also satisfied.*

It is known that the maximum likelihood estimator $\hat{\theta}_{n,ML}$ satisfies (2.131) if the probability distribution family satisfies some regularity conditions [25, 117].

Now, we prove Proposition 2.1 and its related formulas ((2.117) and (2.118) in Theorem 2.5) as follows. Because of (2.92), Proposition 2.1 follows from the inequalities

$$\begin{aligned}&\underline{\lim}\, \frac{-1}{n}\log p_{\eta(\theta)}^n\{X^n(\omega^n) > \eta(\theta)+\epsilon\}) \\ &\geq \max_{\theta'\geq\theta}(\theta'-\theta)(\eta(\theta)+\epsilon) - (\mu(\theta')-\mu(\theta)),\end{aligned} \tag{2.133}$$

$$\overline{\lim}\, \frac{-1}{n}\log p_{\eta(\theta)}^n\{X^n(\omega^n) > \eta(\theta)+\epsilon\} \leq \underline{\lim}_{\,\epsilon'\to\epsilon+0} D(p_{\eta(\theta)+\epsilon'}\|p_{\eta(\theta)}) \tag{2.134}$$

for the expectation parameter η of the exponential family (2.82) and arbitrary $\epsilon > 0$. Certainly, these formulas are the same as (2.117) and (2.118) in Theorem 2.5 for the case $x = \eta(\theta) + \epsilon \geq 0$ and $\theta = 0$.

Choose arbitrary $\bar{\epsilon} > \epsilon$ and $\bar{\theta}$ such that $\mu'(\bar{\theta}) = \eta(\theta) + \bar{\epsilon}$. Based on the proof of (2.122) in Theorem 2.5, we can show that

$$\frac{-1}{n} \log p_{\theta}^{n}\{X^n(\omega^n) > \eta(\theta)+\epsilon\} \\ \geq \max_{\theta' \geq \theta} (\theta'-\theta)(\eta(\theta)+\epsilon) - (\mu(\theta')-\mu(\theta)), \quad (2.135)$$

$$\frac{-1}{n} \log p_{\bar{\theta}}^{n}\{X^n(\omega^n) \leq \eta(\theta)+\epsilon\} \\ \geq \max_{\theta' \leq \bar{\theta}} (\theta'-\bar{\theta})(\eta(\theta)+\epsilon) - (\mu(\theta')-\mu(\bar{\theta})) = D(p_{\eta(\theta)+\epsilon}\|p_{\eta(\theta)+\bar{\epsilon}}) > 0\,, \quad (2.136)$$

which implies (2.133). According to a discussion similar to the proof of (2.129) in Theorem 2.7, we have

$$\frac{-1}{n} \log p_{\eta(\theta)}^{n}\{X^n(\omega^n) > \eta(\theta)+\epsilon\} \leq \frac{D(p_{\eta(\theta)+\epsilon'}\|p_{\eta(\theta)})}{b_n} + \frac{h(b_n)}{nb_n} \quad (2.137)$$

for $\epsilon' > \epsilon$, where $b_n \stackrel{\text{def}}{=} p_{\eta(\theta)+\epsilon'}^{n}\{X^n(\omega^n) > \eta(\theta)+\epsilon\}$. From (2.136), $b_n \to 1$. Hence, we obtain the last inequality in (2.134).

Finally, we prove inequality (2.123) in Theorem 2.6, i.e., we prove that

$$\underline{\lim} \frac{-1}{n} \log p_n\{X_n(\omega) \geq x\} \leq \max_{\theta \geq 0} \left(\theta\mu'(\bar{\theta}) - \mu(\theta)\right) \quad (2.138)$$

for any $\bar{\theta}$ satisfying $\mu'(\bar{\theta}) > x$. Define the exponential family $p_{n,\theta}(\omega) \stackrel{\text{def}}{=} p_n(\omega)e^{n\theta X_n(\omega) - n\mu_n(\theta)}$. Similarly to (2.137), we have

$$\frac{-1}{n} \log p_{n,0}\{X_n(\omega) > x\} \leq \frac{D(p_{n,\bar{\theta}}\|p_{n,0})}{nb_n} + \frac{h(b_n)}{nb_n}$$

for $\epsilon' > \epsilon$, where $b_n \stackrel{\text{def}}{=} p_{n,\bar{\theta}}\{X_n(\omega) > x\}$. From (2.92), $\frac{D(p_{n,\bar{\theta}}\|p_{n,0})}{n} = \max_{\theta \geq 0} \left(\theta\mu_n'(\bar{\theta}) - \mu_n(\theta)\right)$. Hence, if we show that $b_n \to 1$, we obtain (2.138). Similarly to (2.136), the inequality

$$\frac{-1}{n} \log p_{n,\bar{\theta}}\{X_n(\omega) \leq x\} \geq \max_{\theta \leq \bar{\theta}} (\theta-\bar{\theta})x - \mu(\theta) + \mu(\bar{\theta})$$

holds. Since the set of differentiable points of μ is open and μ' is monotone increasing and continuous in this set, there exists a point θ' in this set such that

$$\theta' < \bar{\theta}, \quad x < \mu'(\theta').$$

Since μ' is monotone increasing, we obtain

$$\max_{\theta \leq \bar{\theta}} (\theta-\bar{\theta})x - \mu(\theta) + \mu(\bar{\theta}) \geq (\theta'-\bar{\theta})x - \mu(\theta') + \mu(\bar{\theta}) \\ \geq (\mu'(\theta') - x)(\bar{\theta}-\theta') > 0\,,$$

which implies that $b_n \to 1$.

Exercises

2.38. Prove (2.111) by considering the cases $n \geq m$ and $n < m$ separately.

2.39. Prove Markov's inequality by noting that it can be written $\sum_{i:x_i \geq c} p_i x_i \geq c \sum_{i:x_i \geq c} p_i$.

2.40. Using Cramér's theorem and (2.31) and (2.33), show the following equations below. Show analogous formulas for (2.36), (2.37), (2.48), (2.49), (2.51), and (2.52).

$$\lim_{n\to\infty} \frac{-1}{n} \log p^n\{p^n_{i^n} \leq e^{-nR}\} = -\min_{0\leq s}(\psi(s) - sR), \tag{2.139}$$

$$\lim_{n\to\infty} \frac{-1}{n} \log p^n\{p^n_{i^n} > e^{-nR}\} = -\min_{s\leq 0}(\psi(s) - sR). \tag{2.140}$$

2.41. Show that

$$-\min_{0\leq s\leq 1} \frac{\psi(s) - sR}{1-s} = \min_{q:H(q)\geq R} D(q\|p). \tag{2.141}$$

following the steps below. This equation will play an important role in source coding (Chap. 10).

a Show that $\psi''(s) > 0$ i.e., $\psi'(s)$ is strictly monotone increasing.

b Let $p_M \stackrel{\text{def}}{=} \operatorname{argmax}_i p_i$. Show that if $-\log p_M < S \leq \psi'(1)$, the inverse function $s(S)$ of $\psi'(s)$ can be uniquely defined.

c Let k be the number of is such that $p_M = p_i$. Show that the inverse function $S(R)$ of the function $S \mapsto (1-s(S))S + \psi(s(S))$ may be uniquely found if $\log k < R < \log d$. Show that $s(S(H(p))) = 0$.

d Show that $S(R) - R = S(R)s(S(R)) - \psi(s(S(R))) = \frac{s(S(R))R - \psi(s(S(R)))}{1-s(S(R))}$ if $\log k < R < \log d$.

e Show that $\frac{d}{ds}\frac{sR-\psi(s)}{1-s} = \frac{R-(1-s)\psi'(s)-\psi(s)}{(1-s)^2}$ and $\frac{d}{ds}(R-(1-s)\psi'(s)-\psi(s)) = -(1-s)\psi''(s)$.

f Show that $\max_{0\leq s\leq 1} \frac{sR-\psi(s)}{1-s} = \frac{s(S(R))R-\psi(s(S(R)))}{1-s(S(R))}$ if $H(p) < R < \log d$.

g Show that $H(p(s)) = (1-s)\psi'(s) + \psi(s)$, $(1-s)\sum_i (q_i - p(s)_i)\log p_i = -D(q\|p(s)) + H(p(s)) - H(q)$ and $D(q\|p) - D(p(s)\|p) = D(q\|p(s)) - s\sum_i (q_i - p(s)_i)\log p_i$ if $p(s) = \{p(s)_i \stackrel{\text{def}}{=} \frac{p_i^{1-s}}{\sum_i p_i^{1-s}}\}$.

h Show that $D(q\|p) - D(p(s)\|p) = \frac{1}{1-s}D(q\|p(s)) - \frac{s}{1-s}(H(p(s)) - H(q))$.

i Show that $H(p(s(S(R)))) = R$ and $D(q\|p) - D(p(s(S(R)))\|p) \geq 0$ if $H(q) = R$.

j Show that $\min_{q:H(q)=R} D(q\|p) = S(R) - R$.

k Show that $\frac{d}{dR}(S(R) - R) = \frac{s(S(R))}{1-s(S(R))}$.

l Show that $0 < s(S(R)) < 1$ and $S(R) - R = \min_{q:H(q)\geq R} D(q\|p)$ if $H(p) < R < \log d$.

m Show (2.141).
n Show that

$$-\min_{s\le 0}\frac{\psi(s)-sR}{1-s}=\min_{q:H(q)\le R}D(q\|p). \tag{2.142}$$

using steps similar to those above.
o Using steps similar to those above, show that

$$\max_{0\le s\le 1}\frac{-sr-\phi(s|p\|p')}{1-s}=\min_{q:D(q\|p')\le r}D(q\|p), \tag{2.143}$$

$$\max_{s\le 0}\frac{-sr-\phi(s|p\|p')}{1-s}=\min_{q:D(q\|p')\ge r}D(q\|p). \tag{2.144}$$

2.42. Show that

$$\lim_{n\to\infty}\frac{-1}{n}\log P^c(p^n,e^{nR})=-\min_{0\le s\le 1}\frac{\psi(s)-sR}{1-s} \tag{2.145}$$

by first proving (2.146) and then combining this with (2.141). The $\ge$ part may be obtained directly from (2.40)

$$\begin{aligned}P^c(p^n,e^{nR})&\ge\max_{q\in T_n:|T_q^n|>e^{nR}}(|T_q^n|-e^{nR})e^{-n(H(p)+H(p\|q))}\\&\ge\max_{q\in T_n:\frac{e^{nH(q)}}{(n+1)^d}>e^{nR}}\left(\frac{e^{nH(q)}}{(n+1)^d}-e^{nR}\right)e^{-n(H(p)+H(p\|q))}\\&=\max_{q\in T_n:\frac{e^{nH(q)}}{(n+1)^d}>e^{nR}}e^{-nD(p\|q)}\left(1-\frac{(n+1)^de^{nR}}{e^{nH(q)}}\right).\end{aligned} \tag{2.146}$$

2.43. Show that

$$\lim_{n\to\infty}\frac{-1}{n}\log P(p^n,e^{nR})=-\min_{s\le 0}\frac{\psi(s)-sR}{1-s} \tag{2.147}$$

by first proving (2.148) and then combining this with (2.142). The inequality $\ge$ may be obtained directly from (2.42)

$$P(p^n,e^{nR})\ge\max_{q\in T_n:|T_q^n|\le e^{nR}}p^n(T_q^n)\ge\max_{q\in T_n:H(q)\le R}\frac{e^{-nD(q\|p)}}{(n+1)^d}. \tag{2.148}$$

2.44. Consider the case where $\Omega_n=\{0,1\}$, $p_n(0)=e^{-na}$, $p_n(1)=1-e^{-na}$, $X_n(0)=a$, $X_n(1)=-b$. Show that $\mu(\theta)=-\min\{(1-\theta)a,\theta b\}$ and the following for $-b<x<a$:

$$\max_{\theta>0}(x\theta-\mu(\theta))=\frac{a(x+b)}{a+b}<a,\quad \lim\frac{1}{n}\log p_n\{X_n\ge x\}=a.$$

It gives a counterexample of Gärtner–Ellis Theorem in the nondifferentiable case.

2.6 Related Books

In this section, we treat several important topics in information science from the probabilistic viewpoint. In Sect. 2.1, information quantities e.g., entropy, relative entropy, mutual information, and Rényi entropy are discussed. Its discussion and its historical notes appear in Chap. 2 of Cover and Thomas [82]. Section 2.2 reviews its quantum extensions. This topic will be discussed in greater depth in Sects. 5.4 and 5.5. Ohya and Petz [323] is a good reference for this topic. Section 2.3 focuses on information geometry. Amari and Nagaoka [11] is a textbook on this topic written by the pioneers in the field. Section 2.4 briefly treats the estimation theory of probability distribution families. Lehmann and Casella [263] is a good textbook covering all of estimation theory. For a more in-depth discussion of its asymptotic aspect, see van der Vaart [396]. Section 2.5.1 reviews the type method. It has been formulated by Csiszár and Köner [85]. Section 2.5.2 treats the large deviation theory including estimation theory. Its details are given in Dembo and Zeitouni [94] and Bucklew [62]. In this book, we give a proof of Cramér's theorem and the Gártner–Ellis theorem. In fact, (2.117), (2.119), (2.122), and (2.124) follow from Markov's inequality. However, its opposite parts are not simple. Many papers and books give their proof. In this book, we prove these inequalities by combining the estimation of the exponential theory and the Legendre transform. This proof is given in this book first and seems to be the simplest of known proofs.

3

Quantum Hypothesis Testing and Discrimination of Quantum States

Summary. Various types of information processing occur in quantum systems. The most fundamental processes are state discrimination and hypothesis testing. These problems often form the basis for an analysis of other types of quantum information processes. The difficulties associated with the noncommutativity of quantum mechanics appear in the most evident way among these problems. Therefore, we examine state discrimination and hypothesis testing before examining other types of information processing in quantum systems in this text.

In two-state discrimination, we discriminate between two unknown candidate states by performing a measurement and examining the measurement data. Note that in this case, the two hypotheses for the unknown state are treated symmetrically. In contrast, if the two hypotheses are treated asymmetrically, the process is called hypothesis testing rather than state discrimination. Hypothesis testing is not only interesting in itself but is also relevant to other topics in quantum information theory. In particular, the quantum version of Stein's lemma, which is the central topic of this chapter, is closely related to quantum channel coding discussed in Chap. 4. Moreover, Stein's lemma is also connected to the distillation of maximally entangled states, as discussed in Sect. 8.5, in addition to other topics discussed in Chap. 9. The importance of Stein's lemma may not be apparent at first sight since it considers the tensor product states of identical states, which rarely appear in real communications. However, the asymptotic analysis for these tensor product states provides the key to the analysis of asymptotic problems in quantum communications. For these reasons, this topic is discussed in an earlier chapter in this text.

Table 3.1. Denotations used in Chap. 3

$\phi(s)$	Abbreviation of relative Rényi entropy $\phi(s\|\rho\|\sigma)$
$\phi(s\|\rho\|\sigma)$	Relative Rényi entropy ($\stackrel{\rm def}{=} \log \operatorname{Tr} \rho^{1-s}\sigma^s$)
$\tilde{\phi}(s\|\rho\|\sigma)$	$\stackrel{\rm def}{=} \log \operatorname{Tr} \rho\sigma^{s/2}\rho^{-s}\sigma^{s/2}$
$\bar{\phi}(s\|\rho\|\sigma)$	$\stackrel{\rm def}{=} \lim \frac{1}{n}\phi(s\|\kappa_{\sigma^{\otimes n}}(\rho^{\otimes n})\|\sigma^{\otimes n})$
$\|\boldsymbol{M}\|$	Number of POVM elements
$\beta_\epsilon^n(\rho\|\sigma)$	Minimum value of second error probability
$B(\rho\|\sigma)$	Maximun decreasing rate of second error probability when first error probability goes to 0

Table 3.1. Continued

$\tilde{B}(\rho\|\sigma)$	Maximun decreasing rate of second error probability when first error probability goes to 0 and measurment is separable
$B^{\dagger}(\rho\|\sigma)$	Maximum decreasing rate of second error probability when first error probability does not go to 1
$B(r\|\rho\|\sigma)$	Maximum decreasing rate of first error probability when second error probability goes to 0 at rate r
$B^{*}(r\|\rho\|\sigma)$	Minimum decreasing rate of first error probability when second error probability goes to 0 at rate r

3.1 Two-State Discrimination in Quantum Systems

Consider a quantum system $\mathcal{H}$ whose state is represented by the density matrix ρ or σ. Let us consider the problem of determining the density matrix that describes the true state of the quantum system by performing a measurement. This procedure may be expressed as a Hermitian matrix T satisfying $I \geq T \geq 0$ in $\mathcal{H}$, and it is called *state discrimination* for the following reason.

Consider performing a measurement corresponding to a POVM $\boldsymbol{M} = \{M_\omega\}_{\omega\in\Omega}$ to determine whether the true state is ρ or σ. For this purpose, we must first choose subsets of Ω that correspond to ρ and σ. That is, we first choose a suitable subset A of Ω, and if $\omega \in A$, we can then determine that the state is ρ, and if $\omega \in A^c$ (where A^c is the complement of A), then the state is σ. The Hermitian matrix $T \stackrel{\text{def}}{=} \sum_{\omega\in A} M_\omega$ then satisfies $I \geq T \geq 0$. When the true state is ρ, we erroneously conclude that the state is σ with the probability:

$$\sum_{\omega\in A^c} \operatorname{Tr} \rho M_\omega = \operatorname{Tr} \rho \sum_{\omega\in A^c} M_\omega = \operatorname{Tr} \rho(I - T).$$

On the other hand, when the true state is σ, we erroneously conclude that the state is ρ with the probability:

$$\sum_{\omega\in A} \operatorname{Tr} \sigma M_\omega = \operatorname{Tr} \sigma \sum_{\omega\in A} M_\omega = \operatorname{Tr} \sigma T.$$

Therefore, in order to treat state discrimination, it is sufficient to examine the Hermitian matrix T. The two-valued POVM $\{T, I - T\}$ for a Hermitian matrix T satisfying $I \geq T \geq 0$ allows us to perform the discrimination. Henceforth, T will be called a *test*.

The problem in state discrimination is to examine the tradeoff between the two error probabilities $\operatorname{Tr} \sigma T$ and $\operatorname{Tr} \rho(I - T)$. We then obtain the following lemma by assigning a weight to the two error probabilities.

Lemma 3.1 *(Holevo [211], Helstrom [203]) The following relation holds for an arbitrary real number $c > 0$:*

$$\min_{I\geq T\geq 0} (\operatorname{Tr}\rho(I-T) + c\operatorname{Tr}\sigma T) = \operatorname{Tr}\rho\{\rho-c\sigma\leq 0\}+c\operatorname{Tr}\sigma\{\rho-c\sigma > 0\}. \quad (3.1)$$

The minimum value is attained when $T = \{\rho - c\sigma \geq 0\}$. *In particular, if* $c = 1$, *relation (A.14) guarantees that*

$$\min_{I\geq T\geq 0} (\operatorname{Tr}\rho(I-T) + \operatorname{Tr}\sigma T) = \operatorname{Tr}\rho\{\rho-\sigma\leq 0\} + \operatorname{Tr}\sigma\{\rho-\sigma > 0\} \quad (3.2)$$

$$= 1 - \frac{1}{2}\|\rho-\sigma\|_1. \quad (3.3)$$

The optimal average probability of correct discrimination is

$$\frac{1}{2}\max_{I\geq T\geq 0} (\operatorname{Tr}\rho T + \operatorname{Tr}\sigma(I-T)) = \frac{1}{2}\operatorname{Tr}\rho\{\rho-\sigma\geq 0\} + \frac{1}{2}\operatorname{Tr}\sigma\{\rho-\sigma < 0\}$$

$$= \frac{1}{2} + \frac{1}{4}\|\rho-\sigma\|_1. \quad (3.4)$$

Therefore, the trace norm gives a measure for the discrimination of two states.

Hence, for examining the tradeoff between the two error probabilities $\operatorname{Tr}\sigma T$ and $\operatorname{Tr}\rho(I-T)$, it is sufficient to discuss the test $T = \{\rho - c\sigma \geq 0\}$ alone. This kind of test is called a *likelihood test.*

If ρ and σ commute, they may be simultaneously diagonalized as $\rho = \sum_i p_i|u^i\rangle\langle u^i|$ and $\sigma = \sum_i q_i|u^i\rangle\langle u^i|$ using a common orthonormal basis $\{u^1,\ldots,u^d\}$. Therefore, the problem reduces to the discrimination of the probability distributions $p = \{p_i\}$ and $q = \{q_i\}$, as discussed below. Henceforth, such cases wherein the states ρ and σ commute will be henceforth called "*classical.*"

Now, we discriminate between the two probability distributions $p = \{p_i\}$ and $q = \{q_i\}$ by the following process. When the datum i is observed, we decide the true distribution is p with the probability t_i. This discrimination may therefore be represented by a map t_i from $\{1,\ldots,d\}$ to the interval $[0,1]$. Defining the map $t_i \stackrel{\text{def}}{=} \langle u^i|T|u^i\rangle$ from $\{1,\ldots,d\}$ to the interval $[0,1]$ for an arbitrary discriminator T, we obtain

$$\sum_i (1-t_i)p_i = \operatorname{Tr}\rho(I-T), \quad \sum_i t_i q_i = \operatorname{Tr}\sigma T.$$

These are the two error probabilities for discriminating the probability distributions $p = \{p_i\}_i$ and $q = \{q_i\}_i$. If the function t_i is defined on the data set of the measurement $\{|u^i\rangle\langle u^i|\}_i$ such that it is equal to 1 on the set $\{i|q_i - cp_i < 0\}$ and 0 on the set $\{i|q_i - cp_i \geq 0\}$, then the test T is equal to $\{\sigma - c\rho < 0\}$. Therefore, if ρ and σ commute, $T = \{\sigma - c\rho < 0\}$ has a one-to-one correspondence with the subset of the data set. If these density matrices commute, the problem may be reduced to that of probability distributions, which simplifies the situation considerably.

Now, we compare two discrimination problems. One is the discrimination between two states ρ and σ, and the other is that between two probability distributions $\mathrm{P}_\rho^{\boldsymbol{M}}$ and $\mathrm{P}_\sigma^{\boldsymbol{M}}$ for a given POVM $\boldsymbol{M}$. Their optimal average correct probabilities are

$$\max_{I \ge T \ge 0} \left(\operatorname{Tr} \rho T + \operatorname{Tr} \sigma (I - T) \right), \quad \max_{t_i : 1 \ge t_i \ge 0} \sum_i \left(\mathrm{P}_\rho^{\boldsymbol{M}}(i) t_i + \mathrm{P}_\sigma^{\boldsymbol{M}}(i)(1 - t_i) \right).$$

Comparing them, we have $\|\rho - \sigma\|_1 \ge \|\mathrm{P}_\rho^{\boldsymbol{M}} - \mathrm{P}_\sigma^{\boldsymbol{M}}\|_1$ from (3.4), i.e., we obtain (2.61).

Proof of Lemma 3.1. The quantity to be minimized can be rewritten as

$$\operatorname{Tr} \rho (I - T) + c \operatorname{Tr} \sigma T = 1 + \operatorname{Tr}(c\sigma - \rho) T.$$

Now, we diagonalize $c\sigma - \rho$ as $c\sigma - \rho = \sum_i \lambda_i |u_i\rangle\langle u_i|$. Then,

$$\operatorname{Tr}(c\sigma - \rho) T = \sum_i \lambda_i \operatorname{Tr} |u_i\rangle\langle u_i| T.$$

The test T minimizing the above satisfies the following conditions: $\operatorname{Tr} |u_i\rangle\langle u_i| T = 0$ when $\lambda_i \ge 0$; $\operatorname{Tr} |u_i\rangle\langle u_i| T = 1$ when $\lambda_i < 0$. The test T satisfying these conditions is nothing other than $\{c\sigma - \rho < 0\}$. Accordingly, we have

$$\min_{I \ge T \ge 0} \operatorname{Tr}(c\sigma - \rho) T = \operatorname{Tr}(c\sigma - \rho)\{c\sigma - \rho < 0\}. \tag{3.5}$$

Equality (3.1) can be proved according to

$$\begin{aligned} \min_{I \ge T \ge 0} \operatorname{Tr} \rho (I - T) + c \operatorname{Tr} \sigma T &= 1 + \operatorname{Tr}(c\sigma - \rho)\{c\sigma - \rho < 0\} \\ &= \operatorname{Tr} \rho \{\rho - c\sigma \le 0\} + c \operatorname{Tr} \sigma \{\rho - c\sigma > 0\}. \end{aligned}$$

Then, we can also obtain (3.2). See Exercise 3.2 for the derivation of (3.3). ■

Exercises

3.1. Show (A.14) referring to the proof of (3.5).

3.2. Show (3.3) using (A.14).

3.3. Show that $\|\rho - \rho_{\mathrm{mix}}\|_1 \ge 2(1 - \frac{\operatorname{rank} \rho}{d})$.

3.2 Discrimination of Plural Quantum States

In this section, we extend the discussion of the previous section, where there were only two hypothesis states, to the case of k hypothesis states $\rho_1, \dots, \rho_k$.

The state discrimination in this case is given by a POVM $\boldsymbol{M} = \{M_i\}_{i=1}^k$ with k measurement values. For a fixed i, the quality of the discrimination is given by the error probability $1 - \operatorname{Tr} \rho_i M_i$. We are then required to determine a POVM $\boldsymbol{M} = \{M_i\}_{i=1}^k$ that minimizes this error probability. However, since it is impossible to reduce the error probability for all cases, some a priori probability distribution p_i is often assumed for the k hypotheses, and the average error probability $\sum_{i=1}^k p_i(1 - \operatorname{Tr} \rho_i M_i)$ is minimized. Therefore, we maximize the linear function

$$\Lambda(M_1, \dots, M_k) \stackrel{\text{def}}{=} \sum_{i=1}^k p_i \operatorname{Tr} \rho_i M_i = \operatorname{Tr}\left(\sum_{i=1}^k p_i \rho_i M_i\right)$$

with respect to the matrix-valued vector $(M_1, \dots, M_k)$ under the condition

$$M_i \geq 0, \quad \sum_{i=1}^k M_i = I.$$

For this maximization problem, we have the following equation[1] [431]:

$$\max\left\{\Lambda(M_1, \dots, M_k) \middle| M_i \geq 0, \sum_{i=1}^k M_i = I\right\} = \min\{\operatorname{Tr} F | F \geq p_i \rho_i\}. \tag{3.6}$$

When the matrix F and the POVM $(M_1, \dots, M_k)$ satisfy this constraint condition, they satisfy $\operatorname{Tr} F - \operatorname{Tr}\left(\sum_{i=1}^k p_i \rho_i M_i\right) = \sum_i \operatorname{Tr} M_i(F - p_i \rho_i) \geq 0$. Hence, the inequality LHS $\leq$ RHS in (3.6). The direct derivation of the reverse inequality is rather difficult; however, it can be treated by generalized linear programming. Equation (3.6) may be immediately obtained from Theorem 3.1, called the generalized duality theorem.

Theorem 3.1 *Consider the real vector spaces V_1, V_2. Let L be a closed convex cone of V_1 (Sect. A.4). Assume that arbitrary $a > 0$ and $x \in L$ satisfy $ax \in L$. If A is a linear map from V_1 to V_2, then the following relation is satisfied for $b \in V_2$ and $c \in V_1$*[2]*:*

$$\max_{x \in V_1}\{\langle c, x\rangle | x \in L, A(x) = b\} = \min_{y \in V_2}\{\langle y, b\rangle | A^T(y) - c \in L^T\}, \tag{3.7}$$

where $L^T \stackrel{\text{def}}{=} \{x \in V_1 | \langle x, x'\rangle \geq 0, \forall x' \in L\}$.

In our current problem, V_1 is the space consisting of k Hermitian matrices with an inner product $\langle (M_1, \dots, M_k), (M_1', \dots, M_k')\rangle = \sum_{i=1}^k \operatorname{Tr} M_i M_i'$, V_2 is the space of Hermitian matrices with the usual product $\langle X, Y\rangle = \operatorname{Tr} XY$, L is the subset in V_1 such that all the matrices are positive semidefinite, A is the

[1] $\max\{a|b\}$ denotes the maximum value of a satisfying condition b.

[2] A^T denotes the transpose of A.

map $(M_1, \ldots, M_k) \mapsto \sum_{i=1}^{k} M_i$, and b and c are I and $(p_1\rho_1, \ldots, p_k\rho_k)$, respectively. Applying Theorem 3.1, we obtain (3.6). Therefore, we have rewritten the multivariable maximization problem on the (left-hand side) LHS of (3.6) into a single-variable minimization problem involving only one Hermitian matrix on the (right-hand side) RHS of (3.6). In general it is difficult to further analyze such optimization problems except for problems involving special symmetries [27]. Due to the fundamental nature of this problem, it is possible to reuse our results here in the context of other problems, as will be discussed in later sections. In these problems, it is sufficient to evaluate only the upper and lower bounds. Therefore, although it is generally difficult to obtain the optimal values, their upper and lower bounds can be more readily obtained.

In this section, the problem was formulated in terms of generalized linear programming [397]. However, it is also possible to formulate the problem in terms of semidefinite programming (SDP) [238]. The semidefinite programming problem has been studied extensively, and many numerical packages are available for this problem. Therefore, for numerical calculations it is convenient to recast the given problem in terms of SDP [48].

The generalized duality theorem given here may also be applied to other problems such as the minimization problem appearing on the RHS of (6.92) in Sect. 6.6 [165–167] and the problem involving the size and accuracy of the maximally entangled state [353] that can be produced by class ② introduced in Sect. 8.11. Therefore, this theorem is also interesting from the viewpoint of several optimization problems in quantum information theory.[3]

Exercises

3.4. Show that the average correct probability $\sum_{i=1}^{k} \frac{1}{k} \operatorname{Tr} \rho_i M_i$ with the uniform distribution is less than $\frac{d}{k}$. Furthermore, show that it is less than $\frac{d}{k} \max_i \|\rho_i\|$.

3.3 Asymptotic Analysis of State Discrimination

It is generally very difficult to infer the density matrix of the state of a single quantum system. Hence, it is possible that an incorrect result will be obtained from a single measurement. To avoid this situation, one can develop many independent systems in the same state and then perform measurements on

[3] An example of a numerical solution of the maximization problem in quantum information theory is discussed in Sect. 4.1.2, where we calculate the classical capacity $C_c(W)$. Nagaoka's quantum version of the Arimoto–Blahut algorithm [13, 53], known from classical information theory [305, 311], and an application of this algorithm to the verification of the addition law discussed in Sect. 9.2 are some examples of related works [311, 325]. The connection between these quantities and linear programming has also been discussed widely [238, 375].

these. In this case, we would perform individual measurements on each system and analyze the obtained data statistically. However, it is also possible to infer the unknown state via a single quantum measurement on the composite state of these systems. There are many methods available for the second approach as compared with the first. Therefore, it would be interesting to determine if the difference between the two approaches becomes apparent in their optimized forms.

Let us consider the problem of state discrimination for unknown states given by the tensor product states such as $\rho^{\otimes n}$ and $\sigma^{\otimes n}$. This may be regarded as a quantum-mechanical extension of an independent and identical distribution. If arbitrary measurements on $\mathcal{H}^{\otimes n}$ are allowed to perform this discrimination, we can identify the Hermitian matrix T satisfying $I \geq T \geq 0$ on $\mathcal{H}^{\otimes n}$ as the discriminator. If we restrict the allowable measurements performed on $\mathcal{H}^{\otimes n}$ to be separable or adaptive, the problem becomes somewhat more complicated.

Let us consider the first case. The minimum of the sum of the two error probabilities is then given by $\min_{I \geq T \geq 0} (\operatorname{Tr} \rho^{\otimes n}(I-T) + \operatorname{Tr} \sigma^{\otimes n} T)$, and it asymptotically approaches 0 as n increases. Since this quantity approaches 0 exponentially with n, our problem is then to calculate this exponent.

Lemma 3.2 *The minimum of the sum of the two error probabilities satisfies*

$$\min_{I \geq T \geq 0} \left(\operatorname{Tr} \rho^{\otimes n}(I-T) + \operatorname{Tr} \sigma^{\otimes n} T\right) \leq e^{n\phi(1/2)},$$

where $\phi(s) = \phi(s|\rho\|\sigma)$ was defined in Sect. 2.2.

Proof. Since $\left(\sqrt{\rho^{\otimes n}} - \sqrt{\sigma^{\otimes n}}\right)\{\sqrt{\rho^{\otimes n}} - \sqrt{\sigma^{\otimes n}} < 0\} \leq 0$ and $\{\sqrt{\rho^{\otimes n}} - \sqrt{\sigma^{\otimes n}} \geq 0\}(\sqrt{\sigma^{\otimes n}} - \sqrt{\rho^{\otimes n}}) \leq 0$, we have

$$\begin{aligned}
&\operatorname{Tr} \rho^{\otimes n}\{\sqrt{\rho^{\otimes n}} - \sqrt{\sigma^{\otimes n}} < 0\} - \operatorname{Tr} \sqrt{\rho^{\otimes n}}\sqrt{\sigma^{\otimes n}}\{\sqrt{\rho^{\otimes n}} - \sqrt{\sigma^{\otimes n}} < 0\} \\
&\qquad = \operatorname{Tr} \sqrt{\rho^{\otimes n}}(\sqrt{\rho^{\otimes n}} - \sqrt{\sigma^{\otimes n}})\{\sqrt{\rho^{\otimes n}} - \sqrt{\sigma^{\otimes n}} < 0\} \leq 0
\end{aligned}$$

and

$$\begin{aligned}
&\operatorname{Tr} \sigma^{\otimes n}\{\sqrt{\rho^{\otimes n}} - \sqrt{\sigma^{\otimes n}} \geq 0\} - \operatorname{Tr} \sqrt{\rho^{\otimes n}}\sqrt{\sigma^{\otimes n}}\{\sqrt{\rho^{\otimes n}} - \sqrt{\sigma^{\otimes n}} \geq 0\} \\
&\qquad = \operatorname{Tr}(\sqrt{\sigma^{\otimes n}} - \sqrt{\rho^{\otimes n}})\sqrt{\sigma^{\otimes n}}\{\sqrt{\rho^{\otimes n}} - \sqrt{\sigma^{\otimes n}} \geq 0\} \\
&\qquad = \operatorname{Tr}\{\sqrt{\rho^{\otimes n}} - \sqrt{\sigma^{\otimes n}} \geq 0\}(\sqrt{\sigma^{\otimes n}} - \sqrt{\rho^{\otimes n}})\sqrt{\sigma^{\otimes n}} \leq 0.
\end{aligned}$$

Therefore, considering a test $T = \{\sqrt{\rho^{\otimes n}} - \sqrt{\sigma^{\otimes n}} \geq 0\}$, we obtain

$$\begin{aligned}
&\min_{I \geq T \geq 0} \left(\operatorname{Tr} \rho^{\otimes n}(I-T) + \operatorname{Tr} \sigma^{\otimes n} T\right) \\
&\leq \operatorname{Tr} \rho^{\otimes n}\{\sqrt{\rho^{\otimes n}} - \sqrt{\sigma^{\otimes n}} < 0\} + \operatorname{Tr} \sigma^{\otimes n}\{\sqrt{\rho^{\otimes n}} - \sqrt{\sigma^{\otimes n}} \geq 0\} \\
&\leq \operatorname{Tr}\sqrt{\rho^{\otimes n}}\sqrt{\sigma^{\otimes n}}\{\sqrt{\rho^{\otimes n}} - \sqrt{\sigma^{\otimes n}} < 0\} + \operatorname{Tr}\sqrt{\rho^{\otimes n}}\sqrt{\sigma^{\otimes n}}\{\sqrt{\rho^{\otimes n}} - \sqrt{\sigma^{\otimes n}} \geq 0\} \\
&= \operatorname{Tr} \sqrt{\rho^{\otimes n}}\sqrt{\sigma^{\otimes n}} = e^{n\phi(1/2)}.
\end{aligned}$$

■

If ρ and σ commute, the following lemma holds. However, it is an open problem in the noncommutative case.

Lemma 3.3 *(Chernoff [72]) Let ρ and σ commute. Then*

$$\min_{I \ge T \ge 0} \left(\operatorname{Tr} \rho^{\otimes n}(I-T) + \operatorname{Tr} \sigma^{\otimes n} T \right) \le \exp(n \inf_{1 \ge s \ge 0} \phi(s)) \tag{3.8}$$

and

$$\lim \frac{1}{n} \log \min_{I \ge T \ge 0} \left(\operatorname{Tr} \rho^{\otimes n}(I-T) + \operatorname{Tr} \sigma^{\otimes n} T \right) = \inf_{1 \ge s \ge 0} \phi(s). \tag{3.9}$$

Equation (3.9) follows from (3.2) and Theorem 2.5 (Cramér's Theorem) with $X = -\log \frac{p(\omega)}{\bar{p}(\omega)}$, $\theta = s$, $x = 0$, where p and $\bar{p}$ are probability distributions that consist of the eigenvalues of ρ and σ, respectively $^{\text{Ex. 3.5}}$. Inequality (3.8) is obtained in the same way, but we prove (3.8) here for the reader's convenience.

Proof of (3.8). If ρ and σ commute, then $\{\rho^{\otimes n} - \sigma^{\otimes n} < 0\} = \{(\rho^{\otimes n})^s - (\sigma^{\otimes n})^s < 0\}$. Therefore,

$$\begin{aligned}
&\operatorname{Tr} \rho^{\otimes n} \{\rho^{\otimes n} - \sigma^{\otimes n} < 0\} \\
&\quad = \operatorname{Tr} (\rho^{\otimes n})^{1-s} \left((\rho^{\otimes n})^s - (\sigma^{\otimes n})^s \right) \{\rho^{\otimes n} - \sigma^{\otimes n} < 0\} \\
&\qquad + \operatorname{Tr} (\rho^{\otimes n})^{1-s} (\sigma^{\otimes n})^s \{\rho^{\otimes n} - \sigma^{\otimes n} < 0\} \\
&\qquad\qquad \le \operatorname{Tr} (\rho^{\otimes n})^{1-s} (\sigma^{\otimes n})^s \{\rho^{\otimes n} - \sigma^{\otimes n} < 0\}.
\end{aligned}$$

Similarly, we obtain

$$\operatorname{Tr} \sigma^{\otimes n} \{\rho^{\otimes n} - \sigma^{\otimes n} \ge 0\} \le \operatorname{Tr} (\sigma^{\otimes n})^{1-t} (\rho^{\otimes n})^t \{\rho^{\otimes n} - \sigma^{\otimes n} \ge 0\}.$$

Substituting $t = 1 - s$,

$$\begin{aligned}
&\min_{I \ge T \ge 0} \left(\operatorname{Tr} \rho^{\otimes n}(I-T) + \operatorname{Tr} \sigma^{\otimes n} T \right) \\
&\le \operatorname{Tr} \rho^{\otimes n} \{\rho^{\otimes n} - \sigma^{\otimes n} < 0\} + \operatorname{Tr} \sigma^{\otimes n} \{\rho^{\otimes n} - \sigma^{\otimes n} \ge 0\} \\
&\le \operatorname{Tr} (\rho^{\otimes n})^{1-s} (\sigma^{\otimes n})^s \{\rho^{\otimes n} - \sigma^{\otimes n} < 0\} + \operatorname{Tr} (\sigma^{\otimes n})^s (\rho^{\otimes n})^{1-s} \{\rho^{\otimes n} - \sigma^{\otimes n} \ge 0\} \\
&= \operatorname{Tr} (\sigma^{\otimes n})^s (\rho^{\otimes n})^{1-s} = e^{n\phi(s)},
\end{aligned}$$

from which we obtain (3.8). ∎

Exercises

3.5. Show equation (3.9).

3.6. Show (3.8) when ρ is a pure state $|u\rangle\langle u|$ following the steps below.

a Show that $\inf_{1 \ge s \ge 0} \phi(s) = \log \langle u|\sigma|u\rangle$, i.e., $\inf_{1 \ge s \ge 0} \langle u|\sigma^s|u\rangle = \langle u|\sigma|u\rangle$.
b Show (3.8) when $T = |u\rangle\langle u|^{\otimes n}$.

3.7. Check that this bound can be attained by the test based on the multiple application of the POVM $\{|u\rangle\langle u|, I - |u\rangle\langle u|\}$ on the single system.

3.8. Show (3.9) when ρ and σ are pure states $|u\rangle\langle u|$ and $|v\rangle\langle v|$ following the steps below.

a Show that $\min_{I \ge T \ge 0} (\mathrm{Tr}\,\rho^{\otimes n}(I-T) + \mathrm{Tr}\,\sigma^{\otimes n}T) = 1 - \sqrt{1 - |\langle u|v\rangle|^{2n}}$.
b Show that $\lim \frac{1}{n} \log 1 - \sqrt{1 - |\langle u|v\rangle|^{2n}} = \log |\langle u|v\rangle|^2$.
c Show (3.9).

3.9. Show that $\inf_{1 \ge s \ge 0} \phi(s) = \phi(1/2)$ when σ has the form $\sigma = U\rho U^*$.

3.10. Suppose that σ has the form $\sigma = U\rho U^*$ and $\phi''(1/2) > 0$, $\phi(1/2|\rho\|\sigma) < F(\rho,\sigma)$. Show that $\phi(1/2|\rho\|\sigma) < \min_{\boldsymbol{M}} \inf_{1 \ge s \ge 0} \phi(s|\mathrm{P}_\rho^{\boldsymbol{M}}\|\mathrm{P}_\sigma^{\boldsymbol{M}})$ using (2.62) and $F(\rho,\sigma) \le \phi(1/2|\mathrm{P}_\rho^{\boldsymbol{M}}\|\mathrm{P}_\sigma^{\boldsymbol{M}})$, which is shown in Corollary 8.4 in a more general form. Also show that $\phi(1/2|\rho\|\sigma) < \underline{\lim}\, \frac{1}{n} \min_{\boldsymbol{M}^n} \inf_{1 \ge s \ge 0} \phi(s|\mathrm{P}_{\rho^{\otimes n}}^{\boldsymbol{M}^n}\|\mathrm{P}_{\sigma^{\otimes n}}^{\boldsymbol{M}^n})$.

3.4 Hypothesis Testing and Stein's Lemma

Up until now, the two hypotheses for the two unknown states have been treated equally. However, there are situations where the objective is to disprove one of the hypotheses (called the *null hypothesis*) and accept the other (called the *alternative hypothesis*). This problem in this situation is called *hypothesis testing*. In this case, our errors can be classified as follows. If the null hypothesis is rejected despite being correct, it is called the *error of the first kind*. Conversely, if the null hypothesis is accepted despite being incorrect, it is called the *error of the second kind*. Then, we make our decision only when we support the alternative hypothesis and withhold our decision when we support the null one. Hence, the probability that we make a wrong decision is equal to the *error probability of the first kind*, i.e., the probability that an error of the first kind is made (if the null hyppothesis consists of more than one element, then it is defined as the maximum probability with respect to these elements). Hence, we must guarantee that the error probability of the first kind is restricted to below a particular threshold. This threshold then represents the reliability, in a statistical sense, of our decision and is called the *level of significance*. The usual procedure in hypothesis testing is to fix the level of significance and maximize the probability of accepting the alternative hypothesis when it is true; in other words, we minimize the *error probability of the second kind*, which is defined as the probability of an error of the second kind. For simplicity, we assume that these two hypotheses consist of a single element, i.e., these are given by ρ and σ, respectively. Such hypotheses are called *simple* and are often assumed for a theoretical analysis because this assumption simplifies the mathematical treatment considerably.

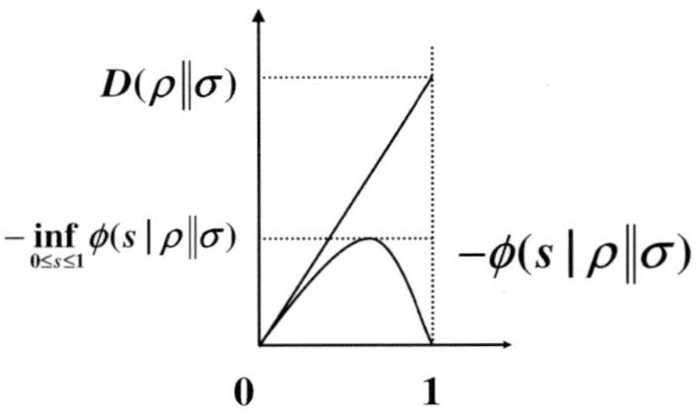

Fig. 3.1. Chernoff's bound and quantum relative entropy

As before, we denote our decision by a test T where $0 \leq T \leq I$, despite the asymmetry of the situation (the event of rejecting the null hypothesis then corresponds to $I-T$). The error probability of the first kind is $\mathrm{Tr}\, \rho(I-T)$, and the error probability of the second kind is $\mathrm{Tr}\, \sigma T$. The discussion in Sect. 3.1 comfirms that in order to optimize our test, it is sufficient to treat only the tests of the form $T = \{\sigma - c\rho < 0\}$ [see RHS of (3.1)]. However, the analysis in Sect. 3.3 cannot be reused due to the asymmetrical treatment of the problem here and therefore another kind of formalism is required. Let us first examine the asymptotic behavior of the error probability of the second kind when the null and the alternative hypotheses are given by the tensor product states $\rho^{\otimes n}$ and $\sigma^{\otimes n}$ on the tensor product space $\mathcal{H}^{\otimes n}$ with a level of significance of $\epsilon > 0$.

Theorem 3.2 *(Hiai and Petz [206], Ogawa and Nagaoka [320]) The minimum value of the error probability of the second kind $\beta_\epsilon^n(\rho\|\sigma)$ satisfies*

$$\lim \frac{-1}{n} \log \beta_\epsilon^n(\rho\|\sigma) = D(\rho\|\sigma), \quad 1 > \forall \epsilon > 0 \tag{3.10}$$

$$\beta_\epsilon^n(\rho\|\sigma) \stackrel{\mathrm{def}}{=} \min_{I \geq T \geq 0} \{\mathrm{Tr}\, \sigma^{\otimes n} T | \, \mathrm{Tr}\, \rho^{\otimes n}(I-T) \leq \epsilon\}$$

when the error probability of the first kind is below $\epsilon > 0$ (i.e., the level of significance is equal to ϵ).

This theorem is called *quantum Stein's lemma*, which is based on its classical counterpart of Stein's lemma. Of course, if ρ and σ commute, we may treat this testing problem by classical means according to the arguments given after Lemma 3.1 in the previous section. From (2.54) the relation between quantum relative entropy $D(\rho\|\sigma)$ and Chernoff's bound $\inf_{1 \geq s \geq 0} \phi(s)$ is illustrated as follows. In particular, when $\sigma = U\rho U^*$, $D(\rho\|\sigma) \geq -2\phi(1/2) = -2 \inf_{1 \geq s \geq 0} \phi(s)$.

Since the proof below also holds for commuting ρ and σ, it can be regarded as a proof of the classical Stein's lemma, although it is rather elaborate. The proof of Theorem 3.2 is obtained by first showing Lemmas 3.4 and 3.5.

Lemma 3.4 (Direct Part) *(Hiai and Petz [206]) There exists a sequence of Hermitian matrices $\{T_n\}$ on $\mathcal{H}^{\otimes n}$ with $I \geq T_n \geq 0$ such that for arbitrary $\delta > 0$*

$$\underline{\lim} \frac{-1}{n} \log \operatorname{Tr} \sigma^{\otimes n} T_n \geq D(\rho\|\sigma) - \delta, \tag{3.11}$$

$$\lim \operatorname{Tr} \rho^{\otimes n}(I - T_n) = 0. \tag{3.12}$$

Lemma 3.5 (Converse Part) *(Ogawa and Nagaoka [320]) If a sequence of Hermitian matrices $\{T_n\}$ $(I \geq T_n \geq 0)$ on $\mathcal{H}^{\otimes n}$ satisfies*

$$\underline{\lim} \frac{-1}{n} \log \operatorname{Tr} \sigma^{\otimes n} T_n > D(\rho\|\sigma), \tag{3.13}$$

then

$$\lim \operatorname{Tr} \rho^{\otimes n}(I - T_n) = 1. \tag{3.14}$$

Proof of Theorem 3.2 using Lemmas 3.4 and 3.5: For $1 > \epsilon > 0$, we take $\{T_n\}$ to satisfy (3.11) and (3.12) according to Lemma 3.4. Taking a sufficiently large N, we have $\operatorname{Tr} \rho^{\otimes n}(I - T_n) \leq \epsilon$ for $n \geq N$ from (3.12). Therefore, $\beta_\epsilon^n(\rho\|\sigma) \leq \operatorname{Tr} \sigma^{\otimes n} T_n$, and we see that $\underline{\lim} \frac{-1}{n} \log \beta_\epsilon^n(\rho\|\sigma) \geq D(\rho\|\sigma) - \delta$. Since $\delta > 0$ is arbitrary, we obtain $\underline{\lim} \frac{-1}{n} \log \beta_\epsilon^n(\rho\|\sigma) \geq D(\rho\|\sigma)$ by taking the limit $\delta \to 0$.

Now, let $\underline{\lim} \frac{-1}{n} \log \beta_\epsilon^n(\rho\|\sigma) > D(\rho\|\sigma)$ for a particular $1 > \epsilon > 0$. Then, we can take a sequence of Hermitian matrices $\{T_n\}$ on $\mathcal{H}^{\otimes n}$ with $I \geq T_n \geq 0$ that satisfies

$$\underline{\lim} \frac{-1}{n} \log \operatorname{Tr} \sigma^{\otimes n} T_n > D(\rho\|\sigma), \quad \operatorname{Tr} \rho^{\otimes n}(I - T_n) \leq \epsilon.$$

However, this contradicts Lemma 3.5, and hence it follows that $\underline{\lim} \frac{-1}{n} \log \beta_\epsilon^n(\rho\|\sigma) \leq D(\rho\|\sigma)$. This proves (3.10). ∎

It is rather difficult to prove the above two lemmas at this point. Hence, we will prove them after discussing several other lemmas forming the basis of the asymptotic theory described in Sect. 3.7. In fact, combining Lemmas 3.4 and 3.5, we obtain the following theorem (Theorem 3.3), which implies Theorem 3.2.

Theorem 3.3 *Define*

$$B(\rho\|\sigma) \stackrel{\text{def}}{=} \sup_{\{T_n\}} \left\{ \underline{\lim} \frac{-1}{n} \log \operatorname{Tr} \sigma^{\otimes n} T_n \,\middle|\, \lim \operatorname{Tr} \rho^{\otimes n}(I - T_n) = 0 \right\},$$

$$B^\dagger(\rho\|\sigma) \stackrel{\text{def}}{=} \sup_{\{T_n\}} \left\{ \underline{\lim} \frac{-1}{n} \log \operatorname{Tr} \sigma^{\otimes n} T_n \,\middle|\, \underline{\lim} \operatorname{Tr} \rho^{\otimes n}(I - T_n) < 1 \right\}.$$

Then,

$$B(\rho\|\sigma) = B^\dagger(\rho\|\sigma) = D(\rho\|\sigma).$$

As a corollary, we can show the following.

Corollary 3.1

$$D(\rho\|\sigma) = \lim_{n\to\infty} \frac{1}{n} \max_{\boldsymbol{M}} D(\mathrm{P}^{\boldsymbol{M}}_{\rho^{\otimes n}} \| \mathrm{P}^{\boldsymbol{M}}_{\sigma^{\otimes n}}). \tag{3.15}$$

Proof. Using classical Stein's lemma, we can show that $D(\mathrm{P}^{\boldsymbol{M}}_{\rho^{\otimes n}} \| \mathrm{P}^{\boldsymbol{M}}_{\sigma^{\otimes n}}) \le B(\rho^{\otimes n}\|\sigma^{\otimes n}) = nD(\rho\|\sigma)$. Then, we obtain the $\ge$ part. Let $\{T_n\}$ be a sequence of tests achieving the optimal. Then, we can prove that $D(\mathrm{P}^{\{T_n, I-T_n\}}_{\rho^{\otimes n}} \| \mathrm{P}^{\{T_n, I-T_n\}}_{\sigma^{\otimes n}}) \to B(\rho\|\sigma) = D(\rho\|\sigma)$. ∎

For a further analysis of the direct part, we focus on the decreasing exponent of the error probability of the first kind under an exponential constraint for the error probability of the second kind. For details on the converse part, we assume an exponential constraint of the error probability of the second kind and optimize the decreasing exponent of the correct probability of the first kind. In other words, we treat the following values:

$$B(r|\rho\|\sigma) \stackrel{\text{def}}{=} \sup_{\{T_n\}} \left\{ \underline{\lim} \frac{-1}{n} \log \mathrm{Tr}\, \rho^{\otimes n}(I - T_n) \middle| \underline{\lim} \frac{-1}{n} \log \mathrm{Tr}\, \sigma^{\otimes n} T_n \ge r \right\},$$

$$B^*(r|\rho\|\sigma) \stackrel{\text{def}}{=} \inf_{\{T_n\}} \left\{ \overline{\lim} \frac{-1}{n} \log \mathrm{Tr}\, \rho^{\otimes n} T_n \middle| \underline{\lim} \frac{-1}{n} \log \mathrm{Tr}\, \sigma^{\otimes n} T_n \ge r \right\}.$$

In the commutative case, i.e., in the classical case, the inequalities

$$B(r|p\|\bar{p}) = \sup_{0\le s\le 1} \frac{-sr - \phi(s|p\|\bar{p})}{1-s} = \min_{q:D(q\|\bar{p})\le r} D(q\|p), \tag{3.16}$$

$$B^*(r|p\|\bar{p}) = \sup_{s\le 0} \frac{-sr - \phi(s|p\|\bar{p})}{1-s} = \min_{q:D(q\|\bar{p})\ge r} D(q\|p) \tag{3.17}$$

hold, where p and $\bar{p}$ are the probability distributions. The former and the latter are shown by Hoeffding [210] and Han and Kobayashi [159], respectively. Note that the second equations of (3.16) and (3.17) follow from (2.143) and (2.144). In the quantum case, the relations

$$\sup_{0\le s\le 1} \frac{-sr - \tilde{\phi}(s|\rho\|\sigma)}{1-s} \le B(r|\rho\|\sigma) \le \min_{\tau:D(\tau\|\sigma)\le r} D(\tau\|\rho), \tag{3.18}$$

$$\begin{aligned} \sup_{s\le 0} \frac{-sr - \phi(s|\rho\|\sigma)}{1-s} &\le B^*(r|\rho\|\sigma) = \sup_{s\le 0} \frac{-sr - \bar{\phi}(s|\rho\|\sigma)}{1-s} \\ &\le \min_{\tau:D(\tau\|\sigma)\ge r} D(\tau\|\rho) \end{aligned} \tag{3.19}$$

hold, where $\tilde{\phi}(s|\rho\|\sigma)$ and $\bar{\phi}(s|\rho\|\sigma)$ are the other quantum extensions of relative Rényi entropy defined as $\tilde{\phi}(s|\rho\|\sigma) \stackrel{\text{def}}{=} \log \mathrm{Tr}\, \rho\sigma^{s/2}\rho^{-s}\sigma^{s/2}$ and $\bar{\phi}(s|\rho\|\sigma) \stackrel{\text{def}}{=}$

$\lim \frac{1}{n} \log \phi(s|\kappa_{\sigma^{\otimes n}}(\rho^{\otimes n})\|\sigma^{\otimes n})$, respectively. See (3.40) and (3.43) and Exercises 3.13, 3.14, 3.15, and 3.17. The equality on the first inequality in (3.19) does not necessarily hold (Exercise 5.25).

We can also characterize the asymptotic optimal performance of quantum simple hypothesis testing with a general sequence of two quantum states [306, 310]. In this general setting, the main problem is to determine the behavior of $\mathrm{Tr}\, \rho\{\rho - e^a\sigma \geq 0\}$ as a function of a [308].

Exercises

3.11. Show the equations (3.16) and (3.17) following the steps below.

a Show the following equations for $\phi(s) = \phi(s|p\|\bar{p})$ using Cramér's theorem.

$$\lim \frac{-1}{n} \log p^n \left\{ \frac{-1}{n} \log \left(\frac{p^n(x^n)}{\bar{p}^n(x^n)} \right) \geq R \right\} = \max_{s \geq 0} sR - \phi(s), \tag{3.20}$$

$$\lim \frac{-1}{n} \log p^n \left\{ \frac{-1}{n} \log \left(\frac{p^n(x^n)}{\bar{p}^n(x^n)} \right) \leq R \right\} = \max_{s \leq 0} sR - \phi(s), \tag{3.21}$$

$$\lim \frac{-1}{n} \log \bar{p}^n \left\{ \frac{-1}{n} \log \left(\frac{p^n(x^n)}{\bar{p}^n(x^n)} \right) \leq R \right\} = \max_{s \leq 1} -(1-s)R - \phi(s). \tag{3.22}$$

b Define $a \overset{\mathrm{def}}{=} \max_x \frac{p(x)}{\bar{p}(x)}$. Show the following table except ^{*1} and ^{*2}.

s	$-\infty$		0		1
$\phi(s)$	$+\infty$	$\searrow$	0	$\circlearrowleft$	0
$\phi'(s)$	$-\log a$	$\nearrow$	$-D(p\|\bar{p})$	$\nearrow$	$D(\bar{p}\|p)$
$s\phi'(s) - \phi(s)$	$-\log \bar{p} \left\{ \frac{p(x)}{\bar{p}(x)} = a \right\}^{*1}$	$\searrow$	0	$\nearrow$	$D(\bar{p}\|p)$
$(s-1)\phi'(s) - \phi(s)$	$-\log \bar{p} \left\{ \frac{p(x)}{\bar{p}(x)} = a \right\}^{*2}$	$\searrow$	$D(p\|\bar{p})$	$\searrow$	0

c Show that

$$s\phi'(s) - \phi(s) = \begin{cases} \max_{s_0 \geq 0}(s_0\phi'(s) - \phi(s_0)) & s \geq 0, \\ \max_{s_0 \leq 0}(s_0\phi'(s) - \phi(s_0)) & s \leq 0, \end{cases}$$

$$(s-1)\phi'(s) - \phi(s) = \max_{s_0 \leq 1}((s_0 - 1)\phi'(s) - \phi(s_0)).$$

d Show ^{*1} and ^{*2} in the above table using **a** and **c**.

e Assume that $0 \leq r \leq D(p\|\bar{p})$. Show that there exists a real number $1 \geq s_r \geq 0$ such that

$$r = (s_r - 1)\phi'(s_r) - \phi(s_r), \quad \max_{s' \geq 0} \frac{-s'r - \phi(s')}{1 - s'} = s_r\phi'(s_r) - \phi(s_r).$$

Show equation (3.16).

f Assume that $D(p\|\bar{p}) < r < -\log \bar{p}\{\frac{p(x)}{\bar{p}(x)} = a\}$. Show that there exists a real number $s_r^* \le 0$ such that

$$r = (s_r^* - 1)\phi'(s_r^*) - \phi(s_r^*), \quad \max_{s' \le 0} \frac{-s'r - \phi(s')}{1-s'} = s_r^*\phi'(s_r^*) - \phi(s_r^*).$$

Show equation (3.17) in this case.

g Assume that $r \ge -\log \bar{p}\{\frac{p(x)}{\bar{p}(x)} = a\}$. Show that

$$\sup_{s \le 0} \frac{-sr - \phi(s)}{1-s} = \lim_{s \to -\infty} \frac{-sr - \phi(s)}{1-s} = \lim_{s \to -\infty} \frac{-sr + s\log a}{1-s} = r - \log a.$$

Show equation (3.17) in this case.

h Show that

$$\frac{dB(r|p\|q)}{dr} = \frac{s_r}{s_r - 1}, \quad \frac{d^2B(r|p\|q)}{dr^2} = \frac{-1}{(s_r-1)^3\phi''(s_r)} \ge 0, \tag{3.23}$$

$$\frac{dB^*(r|p\|q)}{dr} = \frac{s_r^*}{s_r^* - 1}, \quad \frac{d^2B^*(r|p\|q)}{dr^2} = \frac{-1}{(s_r^*-1)^3\phi''(s_r^*)} \ge 0, \tag{3.24}$$

which implies the convexities of $B(r|p\|q)$ and $B^*(r|p\|q)$.

3.5 Hypothesis Testing by Separable Measurements

In the previous section, we performed the optimization with no restriction on the measurements on $\mathcal{H}^{\otimes n}$. In this section, we will restrict the possible measurements to separable measurements. In other words, our test T is assumed to have the separable form:

$$T = M^n_{1,\omega_n} \otimes \cdots \otimes M^n_{n,\omega_n}, \quad M^n_{1,\omega_n} \ge 0, \ldots, M^n_{n,\omega_n} \ge 0 \text{ on } \mathcal{H}^{\otimes n},$$

which is called a *separable test.* This class of tests includes cases such as making identical measurements on every system $\mathcal{H}$ and analyzing measurement data statistically. It also includes other methods such as adaptive improvement of the measurements and statistical analysis of measurement data. The following theorem evaluates the asymptotic performance of the tests based on these measurements.

Theorem 3.4 *Defining* $\tilde{B}(\rho\|\sigma)$ *as*

$$\tilde{B}(\rho\|\sigma) \stackrel{\text{def}}{=} \sup_{\{T_n\}:\text{separable}} \left\{ \underline{\lim} \frac{-1}{n} \log \mathrm{Tr}\, \sigma^{\otimes n} T_n \middle| \lim \mathrm{Tr}\, \rho^{\otimes n}(I - T_n) = 0 \right\},$$

we have

$$\tilde{B}(\rho\|\sigma) = \max_{\boldsymbol{M}} D\left(\mathrm{P}_\rho^{\boldsymbol{M}}\|\mathrm{P}_\sigma^{\boldsymbol{M}}\right). \tag{3.25}$$

When the measurement $\boldsymbol{M}_{\max} \stackrel{\mathrm{def}}{=} \operatorname*{argmax}_{\boldsymbol{M}} D\left(\mathrm{P}_\rho^{\boldsymbol{M}}\|\mathrm{P}_\sigma^{\boldsymbol{M}}\right)$ *is performed* n *times, the bound* $\max_{\boldsymbol{M}} D\left(\mathrm{P}_\rho^{\boldsymbol{M}}\|\mathrm{P}_\sigma^{\boldsymbol{M}}\right)$ *can be asymptotically attained by suitable statistical processing of the* n *data.*

This theorem shows that in terms of quantities such as $\tilde{B}(\rho\|\sigma)$, there is no asymptotic difference between the optimal classical data processing according to identical measurements $\boldsymbol{M}_{\max}$ on each system and the optimal separable test across systems.

Therefore, at least for this problem, we cannot take advantage of the correlation between quantum systems unless a nonseparable measurement is used. Since $\tilde{B}(\rho\|\sigma) \le B(\rho\|\sigma)$, we have the *monotonicity* of quantum relative entropy for a measurement

$$D\left(\mathrm{P}_\rho^{\boldsymbol{M}}\|\mathrm{P}_\sigma^{\boldsymbol{M}}\right) \le D(\rho\|\sigma). \tag{3.26}$$

The following theorem discusses the equality condition of the above inequality.

Theorem 3.5 *(Ohya and Petz [323], Nagaoka [302], Fujiwara [121]) The following conditions are equivalent for two states* ρ *and* σ *and a PVM* $\boldsymbol{M} = \{M_i\}_{i=1}^d$ *of* $\operatorname{rank} M_i = 1$.

① *The equality in (3.26) is satisfied.*
② $[\sigma, \rho] = 0$ *and there exists a set of real numbers* $\{a_i\}_{i=1}^d$ *satisfying*

$$\rho = \sigma \left(\sum_{i=1}^d a_i M_i\right) = \left(\sum_{i=1}^d a_i M_i\right) \sigma. \tag{3.27}$$

See Exercise 6.27 for the proof of this theorem.

Proof of Theorem 3.4. The fact that $\max_{\boldsymbol{M}} D\left(\mathrm{P}_\rho^{\boldsymbol{M}}\|\mathrm{P}_\sigma^{\boldsymbol{M}}\right)$ can be attained, i.e., the "$\ge$" sign in (3.25), follows from the classical Stein's lemma. Therefore, we show that $\tilde{B}(\rho\|\sigma)$ does not exceed this value, i.e., the "$\le$" sign in (3.25). It is sufficient to treat $\underline{\lim} \frac{-1}{n} \log \mathrm{P}_{\sigma^{\otimes n}}^{\boldsymbol{M}^n}(A_n)$ for the pair of separable measurements $\boldsymbol{M}^n = \{M_{\omega_n}^n\}_{\omega_n \in \Omega_n}$:

$$M_{\omega_n}^n = M_{1,\omega_n}^n \otimes \cdots \otimes M_{n,\omega_n}^n, \quad M_{1,\omega_n}^n \ge 0, \ldots, M_{n,\omega_n}^n \ge 0 \text{ on } \mathcal{H}^{\otimes n}$$

and a subset A_n of Ω_n with $\mathrm{P}_{\rho^{\otimes n}}^{\boldsymbol{M}^n}(A_n^c) \to 0$. First, we show that

$$\frac{1}{n} D\left(\mathrm{P}_{\rho^{\otimes n}}^{\boldsymbol{M}^n}\middle\|\mathrm{P}_{\sigma^{\otimes n}}^{\boldsymbol{M}^n}\right) \le \max_{\boldsymbol{M}} D\left(\mathrm{P}_\rho^{\boldsymbol{M}}\|\mathrm{P}_\sigma^{\boldsymbol{M}}\right). \tag{3.28}$$

For this purpose, we define $a_{k,\omega_n} \stackrel{\mathrm{def}}{=} \prod_{j\neq k} \operatorname{Tr} M_{j,\omega_n}^n \rho$ and $\boldsymbol{M}_{\omega_n}^{n,k} \stackrel{\mathrm{def}}{=} a_{k,\omega_n} M_{k,\omega_n}^n$. Since an arbitrary state ρ' on $\mathcal{H}$ satisfies

$$\begin{aligned}\operatorname{Tr} \rho' \sum_{\omega_n} M_{\omega_n}^{n,k} &= \operatorname{Tr} \sum_{\omega_n} \rho^{\otimes(k-1)} \otimes \rho' \otimes \rho^{\otimes(n-k)} M_{\omega_n}^{n,k} \\ &= \operatorname{Tr} \rho^{\otimes(k-1)} \otimes \rho' \otimes \rho^{\otimes(n-k)} = 1,\end{aligned}$$

we see that $\sum_{\omega_n} M_{\omega_n}^{n,k} = I$; hence, we can verify that $\{M_{\omega_n}^{n,k}\}$ is a POVM. Moreover, we can show $^{\text{Ex. 3.12}}$ that

$$D\left(\mathrm{P}_{\rho^{\otimes n}}^{\boldsymbol{M}^n} \middle\| \mathrm{P}_{\sigma^{\otimes n}}^{\boldsymbol{M}^n}\right) = \sum_{k=1}^{n} D\left(\mathrm{P}_{\rho}^{\boldsymbol{M}^{n,k}} \middle\| \mathrm{P}_{\sigma}^{\boldsymbol{M}^{n,k}}\right), \tag{3.29}$$

and thus verify (3.28). Since the monotonicity of the relative entropy for a probability distribution yields

$$\begin{aligned}&\mathrm{P}_{\rho^{\otimes n}}^{\boldsymbol{M}^n}(A_n)\left(\log \mathrm{P}_{\rho^{\otimes n}}^{\boldsymbol{M}^n}(A_n) - \log \mathrm{P}_{\sigma^{\otimes n}}^{\boldsymbol{M}^n}(A_n)\right) \\ &\qquad + \mathrm{P}_{\rho^{\otimes n}}^{\boldsymbol{M}^n}(A_n^c)\left(\log \mathrm{P}_{\rho^{\otimes n}}^{\boldsymbol{M}^n}(A_n^c) - \log \mathrm{P}_{\sigma^{\otimes n}}^{\boldsymbol{M}^n}(A_n^c)\right) \\ &\qquad\qquad \le D\left(\mathrm{P}_{\rho^{\otimes n}}^{\boldsymbol{M}^n} \middle\| \mathrm{P}_{\sigma^{\otimes n}}^{\boldsymbol{M}^n}\right) \le n \max_{\boldsymbol{M}} D\left(\mathrm{P}_{\rho}^{\boldsymbol{M}} \middle\| \mathrm{P}_{\sigma}^{\boldsymbol{M}}\right),\end{aligned}$$

we obtain

$$-\frac{1}{n} \log \mathrm{P}_{\sigma^{\otimes n}}^{\boldsymbol{M}^n}(A_n) \le \frac{\max_{\boldsymbol{M}} D\left(\mathrm{P}_{\rho}^{\boldsymbol{M}} \| \mathrm{P}_{\sigma}^{\boldsymbol{M}}\right) + \frac{1}{n} h(\mathrm{P}_{\rho^{\otimes n}}^{\boldsymbol{M}^n}(A_n))}{\mathrm{P}_{\rho^{\otimes n}}^{\boldsymbol{M}^n}(A_n)}, \tag{3.30}$$

where we used the fact that $-\mathrm{P}_{\rho^{\otimes n}}^{\boldsymbol{M}^n}(A_n^c) \log \mathrm{P}_{\sigma^{\otimes n}}^{\boldsymbol{M}^n}(A_n^c) \ge 0$, and $h(x)$ is a binary entropy that is expressed as $h(x) = -x \log x - (1-x)\log(1-x)$. Noting that $h(x) \le 2$ and $\mathrm{P}_{\rho^{\otimes n}}^{\boldsymbol{M}^n}(A_n) \to 1$, we have

$$\overline{\lim} - \frac{1}{n} \log \mathrm{P}_{\sigma^{\otimes n}}^{\boldsymbol{M}^n}(A_n) \le \max_{\boldsymbol{M}} D\left(\mathrm{P}_{\rho}^{\boldsymbol{M}} \| \mathrm{P}_{\sigma}^{\boldsymbol{M}}\right),$$

from which we obtain (3.25). ∎

Exercises

3.12. Prove (3.29).

3.6 Proof of Direct Part of Stein's Lemma

In order to prove the direct part of Lemma 3.4, we must first prove Lemma 3.6 given below.

Lemma 3.6 *Let ρ be a density matrix, and S and T be two Hermitian matrices such that S and ρ commute and $I \ge T \ge 0$ and $I \ge S \ge 0$. The following relation then holds:*

$$\operatorname{Tr} \rho(I - TST) \le \operatorname{Tr} \rho(I - S) + 2 \operatorname{Tr} \rho(I - T).$$

Proof. Since S and ρ commute, $(I-S)\rho = \rho(I-S) \geq 0$, which implies $S\rho + \rho S \leq 2\rho$. Therefore,

$$\begin{aligned}
&\operatorname{Tr}\rho(I-S) + 2\operatorname{Tr}\rho(I-T) - \operatorname{Tr}\rho(I-TST) \\
&\quad = \operatorname{Tr}\rho TST - \operatorname{Tr}\rho S + \operatorname{Tr}2\rho(I-T) \\
&\quad \geq \operatorname{Tr}\rho TST - \operatorname{Tr}\rho S + \operatorname{Tr}(S\rho+\rho S)(I-T) = \operatorname{Tr}\rho(I-T)S(I-T) \geq 0,
\end{aligned}$$

completing the proof. In the last equality, we used the relation that $(I-T)S(I-T) = TST - S + (I-T)S + S(I-T)$. ■

For proving Lemma 3.4, we define the matrices τ_n and ν_n for arbitrary $\delta > 0$:

$$\tau_n \stackrel{\text{def}}{=} \left\{\sigma^{\otimes n} < e^{n(\operatorname{Tr}\rho\log\sigma+\delta)}\right\}, \quad \nu_n \stackrel{\text{def}}{=} \left\{\rho^{\otimes n} > e^{n(\operatorname{Tr}\rho\log\rho-\delta)}\right\}.$$

The matrix $T_n \stackrel{\text{def}}{=} \tau_n\nu_n\tau_n$ satisfies $0 \leq T_n = \tau_n\nu_n\tau_n \leq \tau_n^2 = \tau_n \leq I$. Hence, (2.48) and (2.52) yield

$$1 - \lim \operatorname{Tr}\rho^{\otimes n}\nu_n = \lim \operatorname{Tr}\rho^{\otimes n}\left\{\rho^{\otimes n} \leq e^{n(\operatorname{Tr}\rho\log\rho-\delta)}\right\} = 0, \tag{3.31}$$

$$1 - \lim \operatorname{Tr}\rho^{\otimes n}\tau_n = \lim \operatorname{Tr}\rho^{\otimes n}\left\{\sigma^{\otimes n} \geq e^{n(\operatorname{Tr}\rho\log\sigma+\delta)}\right\} = 0. \tag{3.32}$$

Lemma 3.6 guarantees that

$$\lim \operatorname{Tr}\rho^{\otimes n}\tau_n\nu_n\tau_n = 1, \tag{3.33}$$

from which we obtain (3.12). Since $\tau_n\sigma^{\otimes n}\tau_n \leq e^{n(\operatorname{Tr}\rho\log\sigma+\delta)}$ according to (2.47) with $s=1$ and $\nu_n \leq \rho^{\otimes n}e^{-n(\operatorname{Tr}\rho\log\rho-\delta)}$, we have

$$\begin{aligned}
&\operatorname{Tr}\sigma^{\otimes n}\tau_n\nu_n\tau_n = \operatorname{Tr}\tau_n\sigma^{\otimes n}\tau_n\nu_n \leq \operatorname{Tr}e^{n(\operatorname{Tr}\rho\log\sigma+\delta)}\nu_n \\
&\quad \leq \operatorname{Tr}e^{n(\operatorname{Tr}\rho\log\sigma+\delta)}\rho^{\otimes n}e^{-n(\operatorname{Tr}\rho\log\rho-\delta)} = e^{n(\operatorname{Tr}\rho\log\sigma-\log\rho)+2\delta)} \\
&\quad = e^{n(2\delta-D(\rho\|\sigma)},
\end{aligned} \tag{3.34}$$

giving us (3.11). ■

Exercises

3.13. Show the following fact regarding $B(r|\rho\|\sigma)$.

a Define $T_n \stackrel{\text{def}}{=} \{\sigma^{\otimes n} \leq e^{-na}\}\rho^{\otimes n}\{\sigma^{\otimes n} \leq e^{-na}\}$ for a pure state ρ. Show that

$$\operatorname{Tr}\rho^{\otimes n}(I-T_n) \leq 2e^{n(\phi(s)+sa)}, \tag{3.35}$$

$$\operatorname{Tr}\sigma^{\otimes n}T_n \leq e^{n(\phi(s)-(1-s)a)} \tag{3.36}$$

for $1 > s > 0$.

b Show that $B(r|\rho\|\sigma) \geq \max_{0\leq s\leq 1} \frac{-sr-\phi(s)}{1-s}$ for a pure state ρ.

c Define $T_n(a) \stackrel{\text{def}}{=} \{\sigma^{\otimes n} - e^{-na}\kappa_{\sigma^{\otimes n}}(\rho^{\otimes n}) \leq 0\}$. Show that

$$\operatorname{Tr}\rho^{\otimes n}(I-T_n(a)) \leq e^{nsa}\operatorname{Tr}(\kappa_{\sigma^{\otimes n}}(\rho^{\otimes n}))^{1-s}(\sigma^{\otimes n})^s$$
$$\operatorname{Tr}\sigma^{\otimes n}T_n(a) \leq e^{n(-1+s)a}\operatorname{Tr}(\kappa_{\sigma^{\otimes n}}(\rho^{\otimes n}))^{1-s}(\sigma^{\otimes n})^s$$

for $1 > s > 0$.

d Define $\tilde{\phi}(s) \stackrel{\text{def}}{=} \log\operatorname{Tr}\rho\sigma^{s/2}\rho^{-s}\sigma^{s/2}$ for when ρ has an inverse. Show that

$$\operatorname{Tr}(\kappa_{\sigma^{\otimes n}}(\rho^{\otimes n}))^{1-s}(\sigma^{\otimes n})^s \leq (n+1)^d e^{n\tilde{\phi}(s)}$$

if $0 \leq s \leq 1$ [322]. To show this it is necessary to use Lemmas 3.7 and 3.8 given in the next section and the fact that $x \mapsto -x^{-s}$ is a matrix monotone function.

e Show that $B(r|\rho\|\sigma) \geq \max_{0\leq s\leq 1} \frac{-sr-\tilde{\phi}(s)}{1-s}$ if ρ has an inverse [322].

3.7 Information Inequalities and Proof of Converse Part of Stein's Lemma

In this section, we first prove the converse part of Stein's lemma based on inequality (2.63). After this proof, we show the information inequalities (2.59) and (2.63).

Proof of Theorem 3.5. Applying inequality (2.63) to the two-valued POVM $\{T_n, I-T_n\}$, we have

$$\begin{aligned}
&(\operatorname{Tr}\rho^{\otimes n}T_n)^{1-s}(\operatorname{Tr}\sigma^{\otimes n}T_n)^s \\
&\leq (\operatorname{Tr}\rho^{\otimes n}T_n)^{1-s}(\operatorname{Tr}\sigma^{\otimes n}T_n)^s + (\operatorname{Tr}\rho^{\otimes n}(I-T_n))^{1-s}(\operatorname{Tr}\sigma^{\otimes n}(I-T_n))^s \\
&\leq e^{n\phi(s|\rho\|\sigma)} \qquad (3.37)
\end{aligned}$$

for $s \leq 0$. Hence,

$$\frac{1-s}{n}\log(\operatorname{Tr}\rho^{\otimes n}T_n) + \frac{s}{n}\log(\operatorname{Tr}\sigma^{\otimes n}T_n) \leq \phi(s|\rho\|\sigma). \qquad (3.38)$$

Assume that the second error probability $\operatorname{Tr}\sigma^{\otimes n}T_n$ tends to 0 with the exponent exceeding the relative entropy $D(\rho\|\sigma)$, i.e.,

$$r \stackrel{\text{def}}{=} \underline{\lim}\frac{-1}{n}\log\operatorname{Tr}\sigma^{\otimes n}T_n > D(\rho\|\sigma).$$

Since $s \leq 0$, $-(1-s)\underline{\lim}\frac{-1}{n}\log(\operatorname{Tr}\rho^{\otimes n}T_n) \leq \phi(s|\rho\|\sigma) + sr$, i.e.,

$$\underline{\lim} \frac{-1}{n} \log(\operatorname{Tr} \rho^{\otimes n} T_n) \geq \frac{-\phi(s|\rho\|\sigma) - sr}{1-s}. \tag{3.39}$$

Therefore,

$$B^*(r|\rho\|\sigma) \geq \sup_{s \leq 0} \frac{-sr - \phi(s)}{1-s}. \tag{3.40}$$

The equation $-\phi'(0) = D(\rho\|\sigma)$ implies $\frac{\phi(s_0)}{-s_0} = \frac{\phi(s_0)-\phi(0)}{-s_0} < r$ for an appropriate $s_0 < 0$. Therefore, $\frac{-\phi(s_0)-s_0 r}{1-s_0} = \frac{s_0}{1-s_0}\left(\frac{\phi(s_0)}{-s_0} - r\right) > 0$, and we can say that $\operatorname{Tr} \rho^{\otimes n} T_n$ approaches 0 exponentially. ■

Next, we prove information inequalities (2.59), (2.62), and (2.63). For this purpose, we require two lemmas (Lemmas 3.7 and 3.8).

Lemma 3.7 *Let X be a Hermitian matrix in a d-dimensional space. The Hermitian matrix $X^{\otimes n}$ given by the tensor product of X has at most $(n+1)^d$ distinct eigenvalues, i.e., $|\boldsymbol{E}_{X^{\otimes n}}| \leq (n+1)^d$.*

Proof. Let $X = \sum_{i=1}^d x^i |u_i\rangle\langle u_i|$. Then, the eigenvalues of $X^{\otimes n}$ may be written as $(x_1)^{j_1} \cdots (x_d)^{j_d}$ $(n \geq j_i \geq 0)$. The possible values of $(j_1, \ldots, j_d)$ are limited to at most $(n+1)^d$ values. ■

Lemma 3.8 *(Hayashi [174]) For a PVM $\boldsymbol{M}$, any positive matrix ρ and the pinching map $\kappa_{\boldsymbol{M}}$ defined in (1.11) satisfy*

$$|\boldsymbol{M}|\kappa_{\boldsymbol{M}}(\rho) \geq \rho. \tag{3.41}$$

Proof. We first show (3.41) for when ρ is a pure state. Let us consider the case where $|\boldsymbol{M}| = k$, and its data set is $\{1, \ldots, k\}$. Then, the Schwarz inequality yields

$$\begin{aligned} k\langle v|\kappa_{\boldsymbol{M}}(|u\rangle\langle u|)|v\rangle &= \left(\sum_{i=1}^k 1\right)\left(\sum_{i=1}^k \langle v|M_i|u\rangle\langle u|M_i|v\rangle\right) \\ &\geq \left|\sum_{i=1}^k \langle v|M_i|u\rangle\right|^2 = \left|\left\langle v \left| \sum_{i=1}^k M_i \right| u\right\rangle\right|^2 = |\langle v\,|I|\,u\rangle|^2 = \langle v|u\rangle\langle u|v\rangle. \end{aligned}$$

Therefore, we obtain $|\boldsymbol{M}|\kappa_{\boldsymbol{M}}(|u\rangle\langle u|) \geq |u\rangle\langle u|$. Next, consider the case where $\rho = \sum_j \rho^j |u_j\rangle\langle u_j|$. Then,

$$\begin{aligned} |\boldsymbol{M}|\kappa_{\boldsymbol{M}}(\rho) - \rho =& |\boldsymbol{M}|\kappa_{\boldsymbol{M}}(\sum_j \rho^j |u_j\rangle\langle u_j|) - \sum_j \rho^j |u_j\rangle\langle u_j| \\ =& \sum_j \rho^j \left(|\boldsymbol{M}|\kappa_{\boldsymbol{M}}(|u_j\rangle\langle u_j|) - |u_j\rangle\langle u_j|\right) \geq 0, \end{aligned}$$

from which we obtain (3.41). ■

We are now ready to prove information inequalities (2.59), (2.62), and (2.63). In what follows, these inequalities are proved only for densities $\rho > 0$ and $\sigma > 0$. This is sufficient for the general case due to the following reason. First, we apply these inequalities to two densities $\rho_\epsilon \stackrel{\text{def}}{=} (\rho+\epsilon I)(1+d\epsilon)^{-1}$ and $\sigma_\epsilon \stackrel{\text{def}}{=} (\sigma+\epsilon I)(1+d\epsilon)^{-1}$. Taking the limit $\epsilon \to 0$, we obtain these inequalities in the general case.

Proof of (2.59) and (2.63). Inequality (2.59) follows from (2.63) by taking the limits $\lim_{s\to 0} \frac{\phi(s|\rho\|\sigma)}{-s}$ and $\lim_{s\to 0} \frac{\phi(s|\mathrm{P}_\rho^{M}\|\mathrm{P}_\sigma^{M})}{-s}$. Hence, we prove (2.63). Let $\sigma = \sum_j \sigma_j E_{\sigma,j}$ be the spectral decomposition of σ and $E_{\sigma,j}\rho E_{\sigma,j} = \sum_k \rho_{k,j} E_{k,j}$ be that of $E_{\sigma,j}\rho E_{\sigma,j}$. Hence, $\kappa_\sigma(\rho) = \sum_{k,j} \rho_{k,j} E_{k,j}$. Since $E_{k,j}\rho E_{k,j} = \rho_{k,j} E_{k,j}$, it follows that $\mathrm{Tr}\, \rho \frac{E_{k,j}}{\mathrm{Tr}\, \rho E_{k,j}} = \rho_{k,j}$. For $0 \le s$, we have

$$
\begin{aligned}
&\mathrm{Tr}\, \sigma^s \kappa_\sigma(\rho)^{1-s} = \mathrm{Tr} \sum_{k,j} \sigma_j^s \rho_{k,j}^{1-s} E_{k,j} = \sum_{k,j} \sigma_j^s \,\mathrm{Tr}\, E_{k,j} \rho_{k,j}^{1-s} \\
&= \sum_{k,j} \sigma_j^s \,\mathrm{Tr}\, E_{k,j} \left(\mathrm{Tr}\, \rho \frac{E_{k,j}}{\mathrm{Tr}\, \rho E_{k,j}} \right)^{1-s} \le \sum_{k,j} \sigma_j^s \,\mathrm{Tr}\, E_{k,j} \left(\mathrm{Tr}\, \rho^{1-s} \frac{E_{k,j}}{\mathrm{Tr}\, \rho E_{k,j}} \right) \\
&= \sum_{k,j} \sigma_j^s \,\mathrm{Tr}\, \rho^{1-s} E_{k,j} = \mathrm{Tr}\, \sigma^s \rho^{1-s},
\end{aligned}
$$

where we applied inequality (2.64) (the quantum version of Jensen inequality) with the Hermite matrix ρ and the density $\frac{E_{k,j}}{\mathrm{Tr}\, \rho E_{k,j}}$. For any POVM $\boldsymbol{M} = \{M_i\}$, we define the POVMs $\boldsymbol{M}' = \{M'_{i,j,k}\}$ and $\boldsymbol{M}'' = \{M''_i\}$ by $M'_{i,j,k} \stackrel{\text{def}}{=} E_{k,j} M_i E_{k,j}$ and $M''_i \stackrel{\text{def}}{=} \sum_{k,j} M'_{i,j,k}$, respectively. Then, $\mathrm{Tr}\, \sigma M'_{i,j,k} = \sigma_j \,\mathrm{Tr}\, E_{k,j} M_i E_{k,j}$ and $\mathrm{Tr}\, \rho M'_{i,j,k} = \rho_{k,j} \,\mathrm{Tr}\, E_{k,j} M_i E_{k,j}$. Thus,

$$
\begin{aligned}
\mathrm{Tr}\, \sigma^s \rho^{1-s} \ge \mathrm{Tr}\, \sigma^s \kappa_\sigma(\rho)^{1-s} &= \sum_{i,j,k} (\mathrm{Tr}\, \sigma M'_{i,j,k})^s (\mathrm{Tr}\, \rho M'_{i,j,k})^{1-s} \\
&\ge \sum_i (\mathrm{Tr}\, \sigma M''_i)^s (\mathrm{Tr}\, \rho M''_i)^{1-s},
\end{aligned}
$$

where the last inequality follows from the monotonicity in the classical case. In addition, $\mathrm{Tr}\, \sigma M''_i = \sum_{k,j}, \mathrm{Tr}\, \sigma_j E_{k,j} M_i E_{k,j} = \mathrm{Tr}\, \sigma M_i$, and $\mathrm{Tr}\, \rho M''_i = \sum_{k,j} \mathrm{Tr}\, \rho_{k,j} E_{k,j} M_i E_{k,j} = \mathrm{Tr}\, \kappa_\sigma(\rho) M_i$. Lemma 3.8 ensures that

$$
|\boldsymbol{E}_\sigma|^{1-s} (\mathrm{Tr}\, \kappa_\sigma(\rho) M_i)^{1-s} \ge (\mathrm{Tr}\, \rho M_i)^{1-s}.
$$

Hence,

$$
\mathrm{Tr}\, \sigma^s \rho^{1-s} \ge \mathrm{Tr}\, \sigma^s \kappa_\sigma(\rho)^{1-s} \ge \frac{1}{|\boldsymbol{E}_\sigma|^{1-s}} \sum_i (\mathrm{Tr}\, \sigma M_i)^s (\mathrm{Tr}\, \rho M_i)^{1-s},
$$

i.e.,

$$\phi(s|\rho\|\sigma) \ge \phi(s|\kappa_\sigma(\rho)\|\sigma) \ge \phi(s|\mathrm{P}_\rho^{\boldsymbol{M}}\|\mathrm{P}_\sigma^{\boldsymbol{M}}) - (1-s)\log|\boldsymbol{E}_\sigma|.$$

Next, we consider the tensor product case. Then,

$$\begin{aligned}\phi(s|\rho^{\otimes n}\|\sigma^{\otimes n}) \ge& \phi(s|\kappa_{\sigma^{\otimes n}}(\rho^{\otimes n})\|\sigma^{\otimes n}) \\ \ge& \phi(s|\mathrm{P}_{\rho^{\otimes n}}^{\boldsymbol{M}^{\otimes n}}\|\mathrm{P}_{\sigma^{\otimes n}}^{\boldsymbol{M}^{\otimes n}}) - (1-s)\log|\boldsymbol{E}_{\sigma^{\otimes n}}|.\end{aligned}$$

Since $\phi(s|\rho^{\otimes n}\|\sigma^{\otimes n}) = n\phi(s|\rho\|\sigma)$ and $\phi(s|\mathrm{P}_{\rho^{\otimes n}}^{\boldsymbol{M}^{\otimes n}}\|\mathrm{P}_{\sigma^{\otimes n}}^{\boldsymbol{M}^{\otimes n}}) = n\phi(s|\mathrm{P}_\rho^{\boldsymbol{M}}\|\mathrm{P}_\sigma^{\boldsymbol{M}})$, the inequality

$$\phi(s|\rho\|\sigma) \ge \frac{\phi(s|\kappa_{\sigma^{\otimes n}}(\rho^{\otimes n})\|\sigma^{\otimes n})}{n} \ge \phi(s|\mathrm{P}_\rho^{\boldsymbol{M}}\|\mathrm{P}_\sigma^{\boldsymbol{M}}) - \frac{\log|\boldsymbol{E}_{\sigma^{\otimes n}}|}{n}$$

holds. The convergence $\frac{\log|\boldsymbol{E}_{\sigma^{\otimes n}}|}{n} \to 0$ follows from Lemma 3.8. Thus, we obtain

$$\phi(s|\rho\|\sigma) \ge \bar{\phi}(s|\rho\|\sigma) \left(\stackrel{\text{def}}{=} \lim \frac{\phi(s|\kappa_{\sigma^{\otimes n}}(\rho^{\otimes n})\|\sigma^{\otimes n})}{n}\right) \ge \phi(s|\mathrm{P}_\rho^{\boldsymbol{M}}\|\mathrm{P}_\sigma^{\boldsymbol{M}}). \tag{3.42}$$

Note that the convergence $\lim \frac{\phi(s|\kappa_{\sigma^{\otimes n}}(\rho^{\otimes n})\|\sigma^{\otimes n})}{n}$ is guaranteed by Lemma A.1 because $\phi(s|\kappa_{\sigma^{\otimes(n+m)}}(\rho^{\otimes(n+m)})\|\sigma^{\otimes(n+m)}) \ge \phi(s|\kappa_{\sigma^{\otimes m}}(\rho^{\otimes m})\|\sigma^{\otimes m}) + \phi(s|\kappa_{\sigma^{\otimes n}}(\rho^{\otimes n})\|\sigma^{\otimes n}) - d\log|\boldsymbol{E}_\sigma|$. In addition, as is discussed in Exercise 5.25, the equality in $\phi(s|\rho\|\sigma) \ge \bar{\phi}(s|\rho\|\sigma)$ does not necessarily hold for $s \le -1$. ∎

Furthermore, applying the discussion at the beginning of this section to $\bar{\phi}(s|\rho\|\sigma)$, we obtain

$$B^*(r|\rho\|\sigma) \ge \sup_{s\le 0} \frac{-sr - \bar{\phi}(s|\rho\|\sigma)}{1-s}. \tag{3.43}$$

Exercises

3.14. Show the following inequality by following the steps below.

$$B^*(r|\rho\|\sigma) \le \inf_{\tau:D(\tau\|\sigma)>r} D(\tau\|\rho) = \min_{\tau:D(\tau\|\sigma)\ge r} D(\tau\|\rho) \tag{3.44}$$

a For any state τ, show that there exists a sequence $\{T_n\}$ such that $\lim \mathrm{Tr}\,\tau^{\otimes n}(I - T_n) = 1$ and $\lim \frac{-1}{n}\log \mathrm{Tr}\,\sigma^{\otimes n}T_n = r$.
b Show that the above sequence $\{T_n\}$ satisfies $\lim \frac{-1}{n}\log \mathrm{Tr}\,\sigma^{\otimes n}T_n \le D(\tau\|\rho)$.
c Show (3.44).

3.15. Let a sequence of tests $\{T_n\}$ satisfy $R = \underline{\lim} \frac{-1}{n}\log \mathrm{Tr}\,\rho^{\otimes n}T_n$ and $r \le \underline{\lim} \frac{-1}{n}\log \mathrm{Tr}\,\sigma^{\otimes n}(I - T_n)$. Show that $R \le D(\tau\|\rho)$ when $D(\tau\|\sigma) \le r$ using Lemma 3.5 twice. That is, show that

$$B(r|\rho\|\sigma) \le \inf_{\tau:D(\tau\|\sigma)<r} D(\tau\|\rho) = \min_{\tau:D(\tau\|\sigma)\le r} D(\tau\|\rho).$$

3.16. Show (2.63) from (3.40) by following the steps below.

a Show that

$$\sup_{s\le 0}\frac{-sr-\phi(s|\mathrm{P}^{M}_{\rho}\|\mathrm{P}^{M}_{\sigma})}{1-s}\ge \sup_{s\le 0}\frac{-sr-\phi(s|\rho\|\sigma)}{1-s}$$

by considering the relationship between $B^*(r|\rho\|\sigma)$ and $B^*(r|\mathrm{P}^{M}_{\rho}\|\mathrm{P}^{M}_{\sigma})$.

b Show that

$$\sup_{s'\le 0}\frac{-s'r_s-\phi(s'|\mathrm{P}^{M}_{\rho}\|\mathrm{P}^{M}_{\sigma})}{1-s'}=\frac{-sr_s-\phi(s|\mathrm{P}^{M}_{\rho}\|\mathrm{P}^{M}_{\sigma})}{1-s},$$

where $r_s \stackrel{\mathrm{def}}{=} (s-1)\phi'(s|\rho\|\sigma)-\phi(s|\rho\|\sigma)$.

c Show (2.63) using **a** and **b**.

3.17. Show the equation $B^*(r|\rho\|\sigma)\ge \sup_{s\le 0}\frac{-sr-\bar{\phi}(s|\rho\|\sigma)}{1-s}$ following the steps below.

a Show $B^*(r|\rho\|\sigma)\ge \sup_{s\le 0}\frac{-sr-\frac{1}{n}\phi(s|\kappa_{\sigma^{\otimes n}}\rho^{\otimes n}\|\sigma^{\otimes n})}{1-s}$.

b Show $\limsup_{s\le 0}\frac{-sr-\frac{1}{n}\phi(s|\kappa_{\sigma^{\otimes n}}\rho^{\otimes n}\|\sigma^{\otimes n})}{1-s}=\sup_{s\le 0}\frac{-sr-\bar{\phi}(s|\rho\|\sigma)}{1-s}$.

c Show the desired equation.

3.18. Show $\bar{\phi}(s|\rho\|\sigma)=\lim\frac{1}{n}\max_{M}\phi(s|\mathrm{P}^{M}_{\rho^{\otimes n}}\|\mathrm{P}^{M}_{\sigma^{\otimes n}})$.

3.8 Historical Note

The problem of discriminating two states was treated by Holevo [211] and Helstrom [203]. Its extension to multiple states was discussed by Yuen et al. [431]. If we allowed any POVM, the possibility of perfect discrimination is trivial. That is, it is possible only when the hypothesis states are orthogonal to each other. However, if our measurement is restricted to LOCC, its possibility is not trivial. This problem is called local discrimination and has been studied by many researchers recently [70, 71, 107, 147, 191, 326, 405, 406, 409].

On the nonperfect discrimination, Chernoff's lemma is essential in the asymptotic setting with two commutative states. However, no results were obtained concerning the quantum case of Chernoff's lemma. Hence, Theorem 3.2 is the first attempt to obtain its quantum extension. Regarding the quantum case of Stein's lemma, many results were obtained, the first by Hiai and Petz [206]. They proved that $B(\rho\|\sigma)=D(\rho\|\sigma)$. The part $B(\rho\|\sigma)\le D(\rho\|\sigma)$ essentially follows from the same discussion as (3.30). They proved the other part $B(\rho\|\sigma)\ge D(\rho\|\sigma)$ by showing the existence of the POVMs $\{M^n\}$ such that

$$\lim\frac{1}{n}D^{M^n}(\rho^{\otimes n}\|\sigma^{\otimes n})=D(\rho\|\sigma). \tag{3.45}$$

An impetus for this work was the first meeting between Hiai and Nagaoka in 1990 when they were at the same university (but in different departments). During their discussion, Nagaoka asked about the possibility of extending Stein's lemma to the quantum case. After their achievement, Hayashi [169] proved that there exists a sequence of POVMs $\{\boldsymbol{M}^n\}$ that satisfies (3.15) and depends only on σ. Hayashi [174] also proved that the asymptotically optimal condition for a measurement in terms of quantum hypothesis testing depends only on σ. Moreover, Ogawa and Hayashi [322] also derived a lower bound of the exponent of the second error probability (Exercise 3.13 **e**). Hayashi and Nagaoka [186] also proved $B(\rho\|\sigma) \geq D(\rho\|\sigma)$ in a simple form. Since their proof does not require any advanced knowledge, we use their approach in Sect. 3.6.

Regarding the strong converse part $B^\dagger(\rho\|\sigma) \leq D(\rho\|\sigma)$, Ogawa and Nagaoka [320] proved it by deriving the lower bound of the exponent $\sup_{-1\leq s\leq 0} \frac{-sr-\phi(s)}{1-s}$, which is equal to the RHS of (3.40) when $s \leq 0$ is replaced by $-1 \leq s \leq 0$. Its behavior is slightly worse for a large value r. After this, the same exponent was obtained by Nagaoka [307] in a more simple way. However, these two approaches are based on the monotonicity of the relative Rényi entropy $\phi(s|\rho\|\sigma)$ $(-1 \leq s \leq 0)$, and this kind of monotonicity has been proved by Petz [343] based on advanced knowledge, i.e., matrix convex functions. In this text, we prove the monotonicity of the relative Rényi entropy $\phi(s|\rho\|\sigma)$ $(s \leq 0)$ for a measurement based only on elementary knowledge, and we apply this monotonicity to Nagaoka's proof. Hence, we derive the better bound $\sup_{s\leq 0} \frac{-sr-\phi(s)}{1-s}$, which was derived by Hayashi [173] using a different method. In addition, the second inequality in (3.18) was first proved in the Japanese version of this book. All relations in (3.19) were first proved in the English version.

Furthermore, Nagaoka invented a quantum version of the information spectrum method, and Nagaoka and Hayashi [310] applied it to the simple hypothesis testing of a general sequence of quantum states.

Finally, we should remark that the formulation of hypothesis testing is based on industrial demands. In particular, in order to guarantee product quality, we usually use test based on random sampling and statistically evaluate the quality. It is natural to apply this method to check the quality of produced maximally entangled states because maximally entangled states are used as resources of quantum information processing. Tsuda et al. formulated this problem using statistical hypothesis testing [389] and demonstrated its usefulness by applying it to maximally entangled states that produced spontaneous parametric down conversion [390]. Further, Hayashi [195] analyzed this problem more extensively from a theoretical viewpoint. However, concerning quantum hypothesis testing, the research on the applied side is not sufficient. Hence, such a study is strongly desired.

4

Classical-Quantum Channel Coding (Message Transmission)

Summary. Communication systems such as the Internet have become part of our daily lives. In any data-transmission system, data are always exposed to noise, and therefore it might be expected that information will be transmitted incorrectly. In practice, however, such problems can be avoided entirely. How is this possible? For explaining this, let us say that we send some information that is either 0 or 1. Now, let us say that the sender and receiver agree that the former will send "000" instead of "0" and "111" instead of "1." If the receiver receives "010" or "100," he or she can deduce that the sender in fact sent 0. On the other hand, if the receiver receives a "110" or "101," he or she can deduce that a 1 was sent. Therefore, we can reduce the chance of error by introducing redundancies into the transmission. However, in order to further reduce the chance of an error in this method, it is necessary to indefinitely increase the redundancy. Therefore, it had been commonly believed that in order to reduce the error probability, one had to increase the redundancy indefinitely. However, in 1948, Shannon [366] showed that by using a certain type of encoding scheme,[1] it is possible to reduce the error probability indefinitely without increasing the redundancy beyond a fixed rate. This was a very surprising result since it was contrary to naive expectations at that time. The distinctive part of Shannon's method was to treat communication in the symbolic form of 0s and 1s and then to approach the problem of noise using encoding. In practical communication systems such as optical fibers and electrical wires, codes such as 0 and 1 are sent by transforming them into a physical medium. In particular, in order to achieve the theoretical optimal communication speed, we have to treat the physical medium of the communication as a microscopic object, i.e., quantum-mechanical object. In this quantum-mechanical scenario, it is most effective to treat the encoding process not as a transformation of the classical bits, e.g., 0s, 1s, and so on, but as a transformation of the message into a quantum state. Furthermore, the measurement and decoding process can be thought of as a single step wherein the outcome of the quantum-mechanical measurement directly becomes the recovered message.

[1] This is the transformation rule for the message. More precisely, the transformation rule during message transmission is called encoding and the transformation rule in the recovery process is called decoding.

Table 4.1. Denotations used in Chap. 4

Φ	Code (N, φ, Y)
$\tilde{\Phi}^{(n)}$	Feedback-allowing coding
$\lvert\Phi\rvert$	Size of code (4.7)
$\varepsilon[\Phi]$	Average error probability of code (4.7)
$I(p, W)$	Transmission informaion (4.1)
$I(\boldsymbol{M}, p, W)$	Classical transmission informaion when measurement is $\boldsymbol{M}$
$W^{(n)}$	n-fold stationary memoryless channel
W_p	Average state (4.2)
$C_c(W)$	C-q channel capacity (4.8)
$C_c^\dagger(W)$	Strong converse channel capacity (4.9)
$\tilde{C}_c(W)$	Channel capacity with adaptive decoding and feedback (4.16)
$C_{c;c\le K}(W)$	Channel capacity with a cost function
$B(R\lvert W)$	Reliability function (4.40)
$J(p, \sigma, W)$	$\sum_{x\in\mathcal{X}} p(x)D(W_x\\|\sigma)$

4.1 Formulation of the Channel Coding Process in Quantum Systems

There are two main processes involved in the transmission of classical information through a quantum channel. The first is the conversion of the classical message into a quantum state, which is called *encoding*. The second is the decoding of the message via a quantum measurement on the output system. For a reliable and economical communication, we should optimize these processes. However, it is impossible to reduce the error probability below a certain level in the single use of a quantum channel even with the optimal encoding and decoding. This is similar to a single use of a classical channel with a nonnegligible bit-flip probability. However, when we use a given channel repeatedly, it is possible in theory to reduce the error probability to almost 0 by encoding and decoding. In this case, this reduction requires that the transmission rate from the transmission bit size to the original message bit size should be less than a fixed rate. This fixed rate is the bound of the transmission rate of a reliable communication and is called the *channel capacity*. This argument has been mathematically proved by Shannon [366] and is called the *channel coding theorem*. Hence, it is possible to reduce the error probability without reducing the transmission rate; however, complex encoding and decoding processes are required. This implies that it is possible to reduce the error probability to almost 0 while keeping a fixed communication speed if we group an n-bit transmission and then perform the encoding and decoding on this group. More precisely, the logarithm of the decoding error can then be decreased in proportion to the number n of transmissions, and the number n can be considered as the level of complexity required by the encoding and decoding processes. These facts are known in the quantum case as well as in the classical case.

In this chapter, we first give a mathematical formulation for the single use of a quantum channel. Regarding n uses of the quantum channel as a single quantum channel, we treat the asymptotic theory in which the number n of the uses of the given channel is large.

In the transmission of classical information via a quantum channel, we may denote the channel as a map from the alphabet (set of letters) $\mathcal{X}$ to the set $\mathcal{S}(\mathcal{H})$ of quantum states on the output system $\mathcal{H}$, i.e., a *classical-quantum channel* (*c-q channel*) $W : \mathcal{X} \to \mathcal{S}(\mathcal{H})$.[2] The relevance of this formulation may be verified as follows. Let us consider the state transmission channel from the input system to the output system described by the map $\Gamma : \mathcal{S}(\mathcal{H}') \to \mathcal{S}(\mathcal{H})$, where $\mathcal{H}'$ denotes the Hilbert space of the input system. When the states to be produced in the input system are given by the set $\{\rho_x\}_{x\in\mathcal{X}}$, the above map W is given by $W_x = \Gamma(\rho_x)$. That is, sending classical information via the above channel reduces to the same problem as that with the c-q channel W.

When all the densities W_x are simultaneously diagonalizable, the problem is reduced to a channel given by a stochastic transition matrix. As in hypothesis testing, we may call such cases "classical." Then, Theorem 4.1 (to be discussed later) also gives the channel capacity for the classical channels given by a stochastic transition matrix.

4.1.1 Transmission Information in C-Q Channels and Its Properties

As in the classical case (2.25), the *transmission information* $I(p, W)$ and the average state W_p for the c-q channel W are defined as[3]

$$I(p, W) \stackrel{\text{def}}{=} \sum_{x\in\mathcal{X}} p(x) D(W_x \| W_p) = H(W_p) - \sum_{x\in\mathcal{X}} p(x) H(W_x), \tag{4.1}$$

$$W_p \stackrel{\text{def}}{=} \sum_{x\in\mathcal{X}} p(x) W_x\,. \tag{4.2}$$

The transmission information $I(p, W)$ satisfies the following two properties:

① (*Convexity*) Any two distributions p^1 and p^2 satisfy

$$I(\lambda p^1 + (1-\lambda)p^2, W) \le \lambda I(p^1, W) + (1-\lambda) I(p^2, W). \tag{4.3}$$

See Exercise 5.28.

[2] As discussed later, these types of channels are called c-q channels to distinguish them from channels with quantum inputs and outputs.

[3] In many papers, the quantity $I(p, W)$ is called the quantum mutual information. In this text, it will be called the transmission information of the c-q channel, for reasons given in Sect. 5.4. Occasionally we will denote this as $I(p_x, W_x)$.

② (*Subadditivity*) Given two c-q channels W^A from $\mathcal{X}_A$ to $\mathcal{H}_A$ and W^B from $\mathcal{X}_B$ to $\mathcal{H}_B$, we can naturally define the c-q channel $W^A \otimes W^B$ from $\mathcal{X}_A \times \mathcal{X}_B$ to $\mathcal{H}_A \otimes \mathcal{H}_B$ as

$$\left(W^A \otimes W^B\right)_{x_A, x_B} \stackrel{\text{def}}{=} W^A_{x_A} \otimes W^B_{x_B}.$$

Let p_A, p_B be the marginal distributions in $\mathcal{X}_A, \mathcal{X}_B$ for a probability distribution p in $\mathcal{X}_A \times \mathcal{X}_B$, respectively. Then, we have the *subadditivity* for

$$I(p, W^A \otimes W^B) \leq I(p_A, W^A) + I(p_B, W^B). \tag{4.4}$$

This inequality can be shown as follows (Exercise 4.2):

$$\begin{aligned} I(p_A, W^A) + I(p_B, W^B) - I(p, W^A \otimes W^B) \\ = D\left(W^A_{p_A} \otimes W^B_{p_B} \middle\| \left(W^A \otimes W^B\right)_p\right) \geq 0. \end{aligned} \tag{4.5}$$

From this property (4.4), we can show that

$$\max_p I(p, W^A \otimes W^B) = \max_{p_A} I(p_A, W^A) + \max_{p_B} I(p_B, W^B),$$

which is closely connected to the additivity discussed later. Another property of the transmission information is the inequality

$$I(p, W) \stackrel{\text{def}}{=} \sum_{x \in \mathcal{X}} p(x) D(W_x \| W_p) \leq \sum_{x \in \mathcal{X}} p(x) D(W_x \| \sigma), \ \forall \sigma \in \mathcal{S}(\mathcal{H}). \tag{4.6}$$

This inequality can be verified by noting that the LHS minus the RHS equals $D(W_p \| \sigma)$.

4.1.2 C-Q Channel Coding Theorem

Next, we consider the problem of sending a classical message using a c-q channel $W : \mathcal{X} \to \mathcal{S}(\mathcal{H})$. For this purpose, we must mathematically define a *code*, which is the combination of an *encoder* and a *decoder*. These are given by the triplet (N, φ, Y). The number N is a natural number corresponding to the *size* of the encoder. φ is a map, $\varphi : \{1, \ldots, N\} \to \mathcal{X}$, corresponding to the encoder. The decoder is a quantum measurement taking values in the set $\{1, \ldots, N\}$. Mathematically, it is given by the set of N positive Hermitian matrices $Y = \{Y_i\}_{i=1}^N$ with $\sum_i Y_i \leq I$. In this case, $I - \sum_i Y_i$ corresponds to the undecodable decision.

For an arbitrary code $\Phi(N, \varphi, Y)$, we define the size $|\Phi|$ and the *average error probability* $\varepsilon[\Phi]$ as

$$|\Phi| \stackrel{\text{def}}{=} N, \quad \varepsilon[\Phi] \stackrel{\text{def}}{=} \frac{1}{N} \sum_{i=1}^N (1 - \mathrm{Tr}[W_{\varphi(i)} Y_i]). \tag{4.7}$$

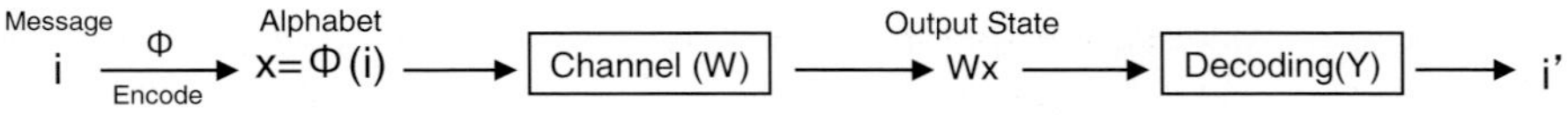

Fig. 4.1. Encoding and decoding

We then consider the encoding and decoding for n communications grouped into one. For simplicity, let us assume that each communication is independent and identical. That is, we discuss the case where the Hilbert space of the output system is $\mathcal{H}^{\otimes n}$, and the c-q channel is given by the map $W^{(n)} : x^n \stackrel{\text{def}}{=} (x_1, \ldots, x_n) \mapsto W^{(n)}_{x^n} \stackrel{\text{def}}{=} W_{x_1} \otimes \cdots \otimes W_{x_n}$ from the alphabet $\mathcal{X}^n$ to $\mathcal{S}(\mathcal{H}^{\otimes n})$. Such a channel is called *stationary memoryless*. An encoder of size N_n is given by the map $\varphi^{(n)}$ from $\{1, \ldots, N_n\}$ to $\mathcal{X}^n$, and it is written as $\varphi^{(n)}(i) = (\varphi^{(n)}_1(i), \ldots, \varphi^{(n)}_n(i))$. The decoder is also given by the POVM $Y^{(n)}$ on $\mathcal{H}^{\otimes n}$. Let us see how much information can be sent per transmission if the error probability asymptotically approaches 0. For this purpose, we look at the limit of the transmission rate $R = \underline{\lim}_{n\to\infty} \frac{1}{n} \log |\Phi^{(n)}|$ ($|\Phi^{(n)}| \cong e^{nR}$) for the sequence of reliable codes $\{\Phi^{(n)} = (N_n, \varphi^{(n)}, Y^{(n)})\}$ and discuss its bound, i.e., the c-q channel capacity $C_c(W)$:[4]

$$C_c(W) \stackrel{\text{def}}{=} \sup_{\{\Phi^{(n)}\}} \left\{ \underline{\lim} \frac{1}{n} \log |\Phi^{(n)}| \middle| \lim \varepsilon[\Phi^{(n)}] = 0 \right\}, \tag{4.8}$$

where $\Phi^{(n)}$ denotes a code for the quantum channel $W^{(n)}$. We may also define the dual capacity

$$C_c^\dagger(W) \stackrel{\text{def}}{=} \sup_{\{\Phi^{(n)}\}} \left\{ \underline{\lim} \frac{1}{n} \log |\Phi^{(n)}| \middle| \underline{\lim}\, \varepsilon[\Phi^{(n)}] < 1 \right\}, \tag{4.9}$$

which clearly satisfies the inequality $C_c(W) \le C_c^\dagger(W)$. We then have the following theorem.

Theorem 4.1 *(Holevo [212,214,217], Schumacher and Westmoreland [363], Ogawa and Nagaoka [319]) Let $\mathcal{P}(\mathcal{X})$ be the set of probability distributions with a finite support in $\mathcal{X}$. Then,*

$$C_c^\dagger(W) = C_c(W) = \sup_{p\in\mathcal{P}(\mathcal{X})} I(p, W) = \min_{\sigma\in\mathcal{S}(\mathcal{H})} \sup_{x\in\mathcal{X}} D(W_x\|\sigma) \tag{4.10}$$

holds.

Thus, this theorem connects the channel capacity $C_c(W)$ to the transmission information $I(p, W)$. Here, note that the former is operationally defined, while the latter is formally defined. The *additivity* of the c-q channel capacity

$$C_c(W^A) + C_c(W^B) = C_c(W^A \otimes W^B)$$

also follows from inequality (4.4).

[4] The subscript c of C_c indicates the sending of "classical" informaion.

This theorem may be proved using the two lemmas given below in a similar way to that of hypothesis testing.

Lemma 4.1 (Direct Part) *(Holevo [217], Schumacher and Westmoreland [363]) For an arbitrary real number $\delta > 0$ and an arbitrary distribution $p \in \mathcal{P}(\mathcal{X})$, there exists a sequence of codes $\{\Phi^{(n)}\}$ for the stationary memoryless quantum channel $W^{(n)}$ such that*

$$\underline{\lim} \frac{1}{n} \log |\Phi^{(n)}| \geq I(p, W) - \delta \,, \tag{4.11}$$

$$\lim \varepsilon[\Phi^{(n)}] = 0 \,. \tag{4.12}$$

Lemma 4.2 (Converse Part) *(Ogawa and Nagaoka [319]) If a sequence of codes $\{\Phi^{(n)}\}$ for the stationary memoryless quantum channel $W^{(n)}$ satisfies*

$$\underline{\lim} \frac{1}{n} \log |\Phi^{(n)}| > \sup_{p \in \mathcal{P}(\mathcal{X})} I(p, W) = \min_{\sigma \in \mathcal{S}(\mathcal{H})} \sup_{x \in \mathcal{X}} D(W_x \| \sigma), \tag{4.13}$$

then the following holds:

$$\lim \varepsilon[\Phi^{(n)}] = 1 \,. \tag{4.14}$$

Indeed, this theorem can be generalized to the case of M-output channel $W^1, \ldots, W^M$, with M output systems $\mathcal{H}_1, \ldots, \mathcal{H}_M$ and a single input system, where $W^i = (W^i_x)$. In this case, the encoder is defined in the same way, i.e., $\varphi : \{1, \ldots, N\} \to \mathcal{X}$. However, the decoder is defined by M POVMs $Y^1, \ldots, Y^M$. That is, the code is described by $\Phi \stackrel{\text{def}}{=} (N, \varphi, Y^1, \ldots, Y^M)$. In this case, the error of Φ is given as $\varepsilon[\Phi] \stackrel{\text{def}}{=} \frac{1}{N} \max_{1 \leq i \leq M} \frac{1}{N} \sum_{j=1}^N (1 - \operatorname{Tr} W^i_{\varphi(i)} Y^i_j)$. Further, the capacity $C_c(W^1, \ldots, W^M)$ is defined in the same way as (4.8).

Corollary 4.1

$$C_c(W^1, \ldots, W^M) = \max_p \min_{1 \leq i \leq M} I(p, W^i). \tag{4.15}$$

For a proof, see Exercises 4.13 and 4.16.

Exercises

4.1. Show that $C_c(W) = h\left(\frac{1+|\langle v|u\rangle|}{2}\right)$ if $\{W_x\}$ is composed of the two pure states $|u\rangle\langle u|$ and $|v\rangle\langle v|$.

4.2. Show (4.5).

4.2 Coding Protocols with Adaptive Decoding and Feedback

In the previous section there was no restriction on the measurements for decoding. Now, we shall restrict these measurements to adaptive decoding, and they have the following form:

$$\boldsymbol{M}^n = \{M_{1,y_1} \otimes \cdots \otimes M_{n,y_n}^{y_1,\dots,y_{n-1}}\}_{(y_1,\dots,y_n)\in\mathcal{Y}_1\times\cdots\times\mathcal{Y}_n}.$$

Therefore, the decoder may be written as the POVM $\boldsymbol{M}^n$ and the mapping $\tau^{(n)} : \mathcal{Y}_1 \times \cdots \times \mathcal{Y}_n \to \{1, \dots, N_n\}$.

We also allow feedback during the encoding process. That is, the receiver is allowed to send his or her measurement values back to the sender, who then performs the encoding based on these values. In the previous section, we considered the encoder to be a map $\varphi^{(n)}(i) = (\varphi_1^{(n)}(i), \dots, \varphi_n^{(n)}(i))$ from $\{1, \dots, N_n\}$ to $\mathcal{X}^n$. If we allow feedback, the kth encoding element will be given by a map $\tilde{\varphi}_k^{(n)}$ from $\{1, \dots, N_n\} \times \mathcal{Y}_1 \times \cdots \times \mathcal{Y}_{k-1}$ to $\mathcal{X}$. Therefore, in this case, we denote the encoder as $\tilde{\varphi}^{(n)} \stackrel{\text{def}}{=} (\tilde{\varphi}_1^{(n)}, \dots, \tilde{\varphi}_n^{(n)})$.

Henceforth, we call $\tilde{\Phi}^{(n)} = (N_n, \tilde{\varphi}^{(n)}, \boldsymbol{M}^n, \tau^{(n)})$ the code with adaptive decoding and feedback and denote its size N_n by $|\tilde{\Phi}^{(n)}|$. The average error probability of the code is denoted by $\varepsilon[\tilde{\Phi}^{(n)}]$. If the code has no feedback, it belongs to the restricted subclass of codes given in the previous section. However, if it has feedback, it does not belong to this subclass. That is, the class of codes given in this section is not a subclass of codes given in the previous section. This class of codes is the subject of the following theorem.

Theorem 4.2 *(Fujiwara and Nagaoka [128]) Define the channel capacity with adaptive decoding and feedback $\tilde{C}_c(W)$ as*

$$\tilde{C}_c(W) \stackrel{\text{def}}{=} \sup_{\{\tilde{\Phi}^{(n)}\}} \left\{ \underline{\lim} \frac{1}{n} \log |\tilde{\Phi}^{(n)}| \,\middle|\, \lim \varepsilon[\tilde{\Phi}^{(n)}] = 0 \right\}, \tag{4.16}$$

where $\tilde{\Phi}^{(n)}$ is a code with adaptive decoding and feedback. Then,

$$\tilde{C}_c(W) = \max_{\boldsymbol{M}} \sup_{p\in\mathcal{P}(\mathcal{X})} I(\boldsymbol{M}, p, W), \tag{4.17}$$

where $I(\boldsymbol{M}, p, W) \stackrel{\text{def}}{=} \sum_{x\in\mathcal{X}} p(x) D(\mathrm{P}_{W_x}^{\boldsymbol{M}} \| \mathrm{P}_{W_p}^{\boldsymbol{M}})$.[5]

In this case, the capacity $\tilde{C}_c(W)$ can be attained by performing the optimal measurement $\boldsymbol{M}_M \stackrel{\text{def}}{=} \mathrm{argmax}_{\boldsymbol{M}} \sup_{p\in\mathcal{P}(\mathcal{X})} I(\boldsymbol{M}, p, W)$ on each output system. Thus, there is no improvement if we use adaptive decoding and encoding with feedback.

[5] Since $\mathcal{H}$ is finite-dimensional, the existence of the maximum follows from the compactness of the space of the POVM.

Proof. When the receiver performs a measurement corresponding to the POVM $\boldsymbol{M}_M \stackrel{\text{def}}{=} \operatorname{argmax}_{\boldsymbol{M}} \sup_p I(\boldsymbol{M}, p, W)$, the relation between the input letter and the output measurement data is described by the stochastic transition matrix $x \mapsto \mathrm{P}_{W_x}^{\boldsymbol{M}_M}$. Applying Theorem 4.1 to the classical channel $x \mapsto \mathrm{P}_{W_x}^{\boldsymbol{M}_M}$, we see that a code attaining $\sup_p I(\boldsymbol{M}_M, p, W) = \max_{\boldsymbol{M}} \sup_p I(\boldsymbol{M}, p, W)$ must exist.

Next, we show that there is no code with a rate exceeding the RHS of (4.17). Consider a sequence of codes $\{\tilde{\Phi}^{(n)} = (N_n, \tilde{\varphi}^{(n)}, \boldsymbol{M}^n, \tau^{(n)})\}$ satisfying $\lim \varepsilon[\tilde{\Phi}^{(n)}] = 0$. Let X be a uniformly distributed random variable taking values in the input messages $\{1, \ldots, N_n\}$ and $Y^k = (Y_1, \ldots, Y_k)$ be the random variable corresponding to $y^k = (y_1, \ldots, y_k)$. Since $\varepsilon[\tilde{\Phi}^{(n)}] = \mathrm{P}\{X \neq \tau^{(n)}(Y^n)\}$, Fano's inequality yields

$$\log 2 + \varepsilon[\tilde{\Phi}^{(n)}] \log N_n \geq H(X) - I(X : \tau^{(n)}(Y^n)) = H(X|\tau^{(n)}(Y^n)).$$

Now we define $\mathrm{P}_{Y_k|X,Y^{k-1}}(y_k|x, y^{k-1}) \stackrel{\text{def}}{=} \mathrm{P}_{W_{\tilde{\varphi}_k^{(n)}(x,y^{k-1})}}^{\boldsymbol{M}_n^{y_1,\ldots,y_{k-1}}}(y_k) = \operatorname{Tr} W_{\tilde{\varphi}_k^{(n)}(x,y^{k-1})} M_{n,y_k}^{y_1,\ldots,y_{k-1}}$ to evaluate $I(W : \tau^{(n)}(Y^n))$. From the monotonicity of the classical relative entropy and the chain rule (2.23) for mutual information,

$$\begin{aligned}
I(X : \tau^{(n)}(Y^n)) &\leq I(X : Y^n) = \sum_{k=1}^{n} I(X : Y_k|Y^{k-1}) \\
&= \sum_{k=1}^{n}\sum_{y^{k-1}} p_{Y^{k-1}}(y^{k-1}) \sum_{x=1}^{N_n} p_{X|Y^{k-1}=y^{k-1}}(x) D\left(\mathrm{P}_{W_{\tilde{\varphi}_k^{(n)}(x,y^{k-1})}}^{\boldsymbol{M}_n^{y_1,\ldots,y_{k-1}}} \middle\| \mathrm{P}_{W_{p_{X|Y^{k-1}=y^{k-1}}}}^{\boldsymbol{M}_n^{y_1,\ldots,y_{k-1}}}\right) \\
&= \sum_{k=1}^{n}\sum_{y^{k-1}} p_{Y^{k-1}}(y^{k-1}) I(\boldsymbol{M}_n^{y_1,\ldots,y_{k-1}}, p_{X|Y^{k-1}=y^{k-1}}, W) \\
&\leq n \max_{\boldsymbol{M}} \sup_{p\in\mathcal{P}(\mathcal{X})} I(\boldsymbol{M}, p, W).
\end{aligned}$$

Noting that $H(X) \leq \log N_n$,

$$\log 2 + \varepsilon[\tilde{\Phi}^{(n)}] \log N_n \geq \log N_n - n \max_{\boldsymbol{M}} \sup_{p\in\mathcal{P}(\mathcal{X})} I(\boldsymbol{M}, p, W), \tag{4.18}$$

which can be rewritten as

$$\frac{1}{n} \log N_n \leq \frac{(\log 2)/n + \max_{\boldsymbol{M}} \sup_{p\in\mathcal{P}(\mathcal{X})} I(\boldsymbol{M}, p, W)}{1 - \varepsilon[\tilde{\Phi}^{(n)}]}.$$

Since $\varepsilon[\tilde{\Phi}^{(n)}] \to 0$,

$$\overline{\lim} \frac{1}{n} \log N_n \leq \max_{\boldsymbol{M}} \sup_{p\in\mathcal{P}(\mathcal{X})} I(\boldsymbol{M}, p, W),$$

completing the proof. ■

Therefore, if the decoder uses no correlation in the measuring apparatus, the channel capacity is given by $\tilde{C}_c(W)$. Next, we consider the channel capacity when correlations among n systems are allowed. In this assumption, we can regard the n uses of the channel W as a single channel $W^{(n)}$ and then reuse the arguments presented in this section. Therefore, the channel capacity is given by $\frac{\tilde{C}_c(W^{(n)})}{n}$. Its limiting case is

$$\lim \frac{\tilde{C}_c(W^{(n)})}{n} = C_c(W), \tag{4.19}$$

while $\tilde{C}_c(W) < C_c(W)$, except for special cases such as those given in Sect. 4.7. An interesting question is whether it is possible to experimentally realize a transmission rate exceeding $\tilde{C}_c(W)$. This is indeed possible, and a channel W and a measurement M have been experimentally constructed with $I(M, p, W^{(2)}) > 2\tilde{C}_c(W)$ by Fujiwara et al. [137].

Exercises

4.3. Show (4.19) using Fano's inequality in a similar way to the proof of Theorem 4.2.

4.3 Channel Capacities Under Cost Constraint

Thus far, there have been no constraints on the encoding, and we have examined only the size of the code and error probabilities. However, it is not unusual to impose a constraint that the cost, e.g., the energy required for communication, should be less than some fixed value. In this situation, we define a cost function and demand that the cost for each code should be less than some fixed value. More precisely, a cost $c(x)$ is defined for each state W_x used in the communication. In the stationary memoryless case, the cost for the states $W_{\boldsymbol{x}}^{(n)}$ is given by $c^{(n)}(\boldsymbol{x}) \stackrel{\text{def}}{=} \sum_{i=1}^{n} c(x_i)$. The states $W_{\boldsymbol{x}}^{(n)}$ used for communication are then restricted to those that satisfy $\sum_{\boldsymbol{x}} c(x_i) \leq Kn$. That is, any code $\Phi^{(n)} = (N_n, \varphi^{(n)}, Y^{(n)})$ must satisfy the restriction $\max_i c^{(n)}(\varphi^{(n)}(i)) \leq Kn$. The following theorem can be proved in a similar way to Theorem 4.1.

Theorem 4.3 *(Holevo [218], Hayashi and Nagaoka [186]) Define the channel capacities under the cost constraint*

$$C_{c;c\leq K}(W) \stackrel{\text{def}}{=} \sup_{\{\Phi^{(n)}\}} \left\{ \underline{\lim} \frac{\log |\Phi^{(n)}|}{n} \middle| \max_i \frac{c^{(n)}(\varphi^{(n)}(i))}{n} \leq K, \lim \varepsilon[\Phi^{(n)}] = 0 \right\},$$

$$C^{\dagger}_{c;c\leq K}(W) \stackrel{\text{def}}{=} \sup_{\{\Phi^{(n)}\}} \left\{ \underline{\lim} \frac{\log |\Phi^{(n)}|}{n} \middle| \max_i \frac{c^{(n)}(\varphi^{(n)}(i))}{n} \leq K, \underline{\lim}\, \varepsilon[\Phi^{(n)}] < 1 \right\}.$$

Then,

$$C_{c;c\le K}(W) = C^{\dagger}_{c;c\ \le K}(W)$$
$$= \sup_{p\in\mathcal{P}_{c\le K}(\mathcal{X})} I(p,W) = \min_{\sigma\in\mathcal{S}(\mathcal{H})} \sup_{p\in\mathcal{P}_{c\le K}(\mathcal{X})} \sum_x p_x D(W_x\|\sigma), \qquad (4.20)$$

where $\mathcal{P}_{c\le K}(\mathcal{X}) \stackrel{\text{def}}{=} \{p\in\mathcal{P}(\mathcal{X})\,|\sum_x p(x)c(x)\le K\}$.

This theorem can be obtained from the following two lemmas in a similar way to Theorem 4.1. These lemmas will be proved later in Sects. 4.5 and 4.6.

Lemma 4.3 (Direct Part) *Given arbitrary* $\delta > 0$ *and* $p \in \mathcal{P}_{c\le K}(\mathcal{X})$, *there exists a sequence* $\{\Phi^{(n)}\}$ *of codes for quantum channels* $W^{(n)}$ *such that* $\max_i c^{(n)}(\varphi^{(n)}(i)) \le Kn$ *and*

$$\underline{\lim}\,\frac{1}{n}\log|\Phi^{(n)}| \ge I(p,W)-\delta\,, \qquad (4.21)$$
$$\lim\varepsilon[\Phi^{(n)}] = 0\,. \qquad (4.22)$$

Lemma 4.4 (Converse Part) *If a sequence* $\{\Phi^{(n)}\}$ *of codes for quantum channels* $W^{(n)}$ *satisfies*

$$\underline{\lim}\,\frac{\log|\Phi^{(n)}|}{n} > \sup_{p\in\mathcal{P}_{c\le K}(\mathcal{X})} I(p,W) \qquad (4.23)$$

and $\max_i c^{(n)}(\varphi^{(n)}(i)) \le Kn$, *then the following holds:*

$$\lim\varepsilon[\Phi^{(n)}] = 1\,. \qquad (4.24)$$

Exercises

4.4. Let the set $\{W_x\}$ consist entirely of pure states. Let the cost function c be given by $c(x) = \operatorname{Tr} W_x E$, where E is a positive semidefinite Hermitian matrix in $\mathcal{H}$. Show that $C_{c;E\le K}(W) \stackrel{\text{def}}{=} C_{c;c\le K}(W) = H(\rho_{E,K})$, where $\rho_{E,K} \stackrel{\text{def}}{=} e^{-\beta_K E}/\operatorname{Tr} e^{-\beta_K E}$ and β_K satisfies $\operatorname{Tr}(e^{-\beta_K E}/\operatorname{Tr} e^{-\beta_K E})E = K$.

4.4 A Fundamental Lemma

In this section, we will prove the lemma required for the proof of Theorem 4.1.

Lemma 4.5 *(Hayashi and Nagaoka [186])* When two arbitrary Hermitian matrices S and T satisfy $I \ge S \ge 0$ and $T \ge 0$, the following inequality holds:

$$I - \sqrt{S+T}^{-1} S \sqrt{S+T}^{-1} \le 2\,(I-S) + 4\,T\,, \qquad (4.25)$$

where $\sqrt{S+T}^{-1}$ is the generalized inverse matrix of $\sqrt{S+T}$ given in Sect. 1.5.

Proof. Let P be a projection to the range of $S+T$. Since P commutes with S and T, for proving (4.25), it is sufficient to show that

$$P\left[I - \sqrt{S+T}^{-1} S \sqrt{S+T}^{-1}\right] P \leq P\left[2(I-S)+4T\right]P,$$
$$P^{\perp}\left[I - \sqrt{S+T}^{-1} S \sqrt{S+T}^{-1}\right] P^{\perp} \leq P^{\perp}\left[2(I-S)+4T\right]P^{\perp},$$

where we defined $P^{\perp} = I - P$. The second inequality follows from $P^{\perp}S = P^{\perp}T = P^{\perp}\sqrt{S+T}^{-1} = 0$. Thus, for proving (4.25), it is sufficient to show that (4.25) holds when the range of $S+T$ is equal to $\mathcal{H}$.

Since $(A-B)^*(A-B) \geq 0$ for matrices, we obtain $A^*B + B^*A \leq A^*A + B^*B$. Applying this inequality to the case of $A = \sqrt{T}$ and $B = \sqrt{T}(\sqrt{S+T}^{-1} - I)$, we obtain

$$T(\sqrt{S+T}^{-1} - I) + (\sqrt{S+T}^{-1} - I)T \leq T + (\sqrt{S+T}^{-1} - I)T(\sqrt{S+T}^{-1} - I). \quad (4.26)$$

Furthermore,

$$\sqrt{S+T} \geq \sqrt{S} \geq S \quad (4.27)$$

since $f(x) = \sqrt{x}$ is a matrix monotone function Ex. A.7 and $0 \leq S \leq I$. Finally,

$$\begin{aligned}
&I - \sqrt{S+T}^{-1} S \sqrt{S+T}^{-1} = \sqrt{S+T}^{-1} T \sqrt{S+T}^{-1} \\
&= T + T(\sqrt{S+T}^{-1} - I) + (\sqrt{S+T}^{-1} - I)T + (\sqrt{S+T}^{-1} - I)T(\sqrt{S+T}^{-1} - I) \\
&\leq 2T + 2(\sqrt{S+T}^{-1} - I)T(\sqrt{S+T}^{-1} - I) \\
&\leq 2T + 2(\sqrt{S+T}^{-1} - I)(S+T)(\sqrt{S+T}^{-1} - I) \\
&= 2T + 2(I + S + T - 2\sqrt{S+T}) \\
&\leq 2T + 2(I + S + T - 2S) = 2(I-S) + 4T,
\end{aligned}$$

where the first inequality follows from (4.26) and the third inequality follows from (4.27). Thus, we obtain the matrix inequality (4.25). ■

Exercises

4.5. Show the generalized version of inequality (4.25) under the same conditions as Lemma 4.5 [186]:

$$I - \sqrt{S+T}^{-1} S \sqrt{S+T}^{-1} \leq (1+c)(I-S) + (2+c+c^{-1})\,T. \quad (4.28)$$

4.5 Proof of Direct Part of C-Q Channel Coding Theorem

The arguments used for hypothesis testing in Chap. 3 may be reused for the proof of the converse theorem (Lemma 4.1) using the following lemma.

Lemma 4.6 *(Hayashi and Nagaoka [186]) Given a c-q channel $x \in \mathcal{X} \mapsto W_x$, there exists a code Φ of size N such that*

$$\varepsilon[\Phi] \le \sum_{x\in\mathcal{X}} p(x)\left(2\operatorname{Tr} W_x \{W_x - 2NW_p \le 0\} + 4N \operatorname{Tr} W_p \{W_x - 2NW_p > 0\}\right), \tag{4.29}$$

where p is a probability distribution in $\mathcal{X}$.

Proof of Lemma 4.1. Applying Lemma 4.6 to the pair of channels $W^{(n)}$ and the n-fold independent and identical distribution of $p_M \stackrel{\text{def}}{=} \operatorname{argmax}_p I(p,W)$, we can take a code of size N_n satisfying

$$\begin{aligned}\varepsilon[\Phi^{(n)}] \le & \sum_{x^n\in\mathcal{X}^n} p^n(x^n)\big(2\operatorname{Tr} W^{(n)}_{x^n}\{W^{(n)}_{x^n} - 2N_n W^{(n)}_{p^n} \le 0\} \\ & + 4N_n \operatorname{Tr} W^{(n)}_{p^n}\{W^{(n)}_{x^n} - 2N_n W^{(n)}_{p^n} > 0\}\big) \\ = & \, 2\operatorname{Tr} R^{\otimes n}\{R^{\otimes n} - 2N_n S^{\otimes n} \le 0\} + 4N_n \operatorname{Tr} S^{\otimes n}\{R^{\otimes n} - 2N_n S^{\otimes n} > 0\},\end{aligned} \tag{4.30}$$

where $\mathcal{X}_p \stackrel{\text{def}}{=} \{x_1, \dots, x_k\} \subset \mathcal{X}$ denotes the support of p_M, and matrices R and S are defined as

$$R \stackrel{\text{def}}{=} \begin{pmatrix} p(x_1)W_{x_1} & & 0 \\ & \ddots & \\ 0 & & p(x_k)W_{x_k} \end{pmatrix}, \quad S \stackrel{\text{def}}{=} \begin{pmatrix} p(x_1)W_p & & 0 \\ & \ddots & \\ 0 & & p(x_k)W_p \end{pmatrix}. \tag{4.31}$$

Since $D(R\|S) = I(p,W)$, from Lemma 3.4 we have

$$\operatorname{Tr} R^{\otimes n} T_n \to 0, \quad \underline{\lim} \frac{-1}{n} \log \operatorname{Tr} S^{\otimes n}(I - T_n) \ge I(p,W) - \frac{\delta}{2}, \tag{4.32}$$

for a sequence $\{T_n\}$ of Hermitian matrices satisfying $I \ge T_n \ge 0$. From condition (4.32) we can show that

$$4e^{n(I(p,W)-\delta)} \operatorname{Tr} S^{\otimes n}(I - T_n) \to 0\,.$$

Using Lemma 3.1 we have

$$
\begin{aligned}
2 \operatorname{Tr} R^{\otimes n}\{R^{\otimes n} - 2N_n S^{\otimes n} \le 0\} + 4N_n \operatorname{Tr} S^{\otimes n}\{R^{\otimes n} - 2N_n S^{\otimes n} > 0\} \\
\le 2 \operatorname{Tr} R^{\otimes n} T_n + 4N_n \operatorname{Tr} S^{\otimes n}(I - T_n) \to 0 \,,
\end{aligned}
$$

where $N_n \stackrel{\text{def}}{=} [e^{n(I(p,W)-\delta)}]$. This implies that there exists a sequence of codes of size $[e^{n(I(p,W)-\delta)}]$ whose error probability converges to 0. This completes the proof of Lemma 4.1. ■

Proof of Lemma 4.6. We prove this lemma by employing the *random coding method* in which we randomly generate a code (N, φ, Y) of fixed size and prove that the expectation of the average error probability is less than ϵ. Based on the above strategy, we can show that there exists a code whose average error probability is less than ϵ. For this purpose, we consider N random variables $X \stackrel{\text{def}}{=} (x_1, \dots, x_N)$ independently obeying a probability distribution p in $\mathcal{X}$, define the encoder $\varphi_X(i)$ by $\varphi_X(i) = x_i$, and denote the expectation value by E_X.

For a given encoder φ of size N, a decoder $Y(\varphi)$ is defined

$$
Y(\varphi)_i \stackrel{\text{def}}{=} \left(\sum_{j=1}^{N} \pi_j \right)^{-\frac{1}{2}} \pi_i \left(\sum_{j=1}^{N} \pi_j \right)^{-\frac{1}{2}} , \tag{4.33}
$$

$$
\pi_i \stackrel{\text{def}}{=} \left\{ W_{\varphi(i)} - 2NW_p > 0 \right\} . \tag{4.34}
$$

Then, the average error probability of the code $\Phi(\varphi) \stackrel{\text{def}}{=} (N, \varphi, Y(\varphi))$ is

$$
\begin{aligned}
&\varepsilon[\Phi(\varphi)] \\
&= \frac{1}{N} \sum_{i=1}^{N} \operatorname{Tr} W_{\varphi(i)} (I - Y(\varphi)_i) \le \frac{1}{N} \sum_{i=1}^{N} \operatorname{Tr} W_{\varphi(i)} \left(2(I - \pi_i) + 4 \sum_{i \ne j} \pi_j \right) \\
&= \frac{1}{N} \sum_{i=1}^{N} \operatorname{Tr} 2W_{\varphi(i)} (I - \pi_i) + \operatorname{Tr} \left(4 \sum_{i \ne j} W_{\varphi(j)} \right) \pi_i \,.
\end{aligned} \tag{4.35}
$$

For evaluating the expectation $\mathrm{E}_X[\varepsilon[\Phi(\varphi_X)]]$, let us rewrite $\mathrm{E}_X[\operatorname{Tr} W_{\varphi_X(i)}(I - \pi_i)]$ and $\mathrm{E}_X[\operatorname{Tr} W_{\varphi_X(j)} \pi_i]$ as

$$
\begin{aligned}
\mathrm{E}_X \left[\operatorname{Tr} W_{\varphi_X(i)} (I - \pi_i) \right] &= \sum_{x \in \mathcal{X}} p(x) \operatorname{Tr} W_x \left\{ W_x - 2NW_p \le 0 \right\} , \\
\mathrm{E}_X \left[\operatorname{Tr} W_{\varphi_X(j)} \pi_i \right] &= \sum_{x' \in \mathcal{X}} p(x') \sum_{x \in \mathcal{X}} p(x) \operatorname{Tr} W_{x'} \left\{ W_x - 2NW_p > 0 \right\} \\
&= \sum_{x \in \mathcal{X}} p(x) \operatorname{Tr} W_p \left\{ W_x - 2NW_p > 0 \right\} .
\end{aligned}
$$

$\mathrm{E}_X\left[\varepsilon[\Phi(\varphi_X)]\right]$ then becomes

$$
\begin{aligned}
\mathrm{E}_X\left[\varepsilon[\Phi(\varphi_X)]\right] \leq & \mathrm{E}_X\left[\frac{1}{N}\sum_{i=1}^{N}\left(\operatorname{Tr} 2W_{\varphi_X(i)}(I-\pi_i)+\operatorname{Tr}\left(4\sum_{i\neq j}W_{\varphi_X(j)}\right)\pi_i\right)\right] \\
= & \frac{1}{N}\sum_{i=1}^{N}\left(2\mathrm{E}_X\left[\operatorname{Tr} W_{\varphi_X(i)}(I-\pi_i)\right]+4\sum_{i\neq j}\mathrm{E}_X\left[\operatorname{Tr} W_{\varphi_X(j)}\pi_i\right]\right) \\
= & \sum_{x\in\mathcal{X}} p(x)\Big(2\operatorname{Tr} W_x\left\{W_x-2NW_p\leq 0\right\} \\
& + 4(N-1)\operatorname{Tr} W_p\left\{W_x-2NW_p>0\right\}\Big).
\end{aligned}
\tag{4.36}
$$

Since the RHS of this inequality is less than the RHS of (4.29), we see that a code $\Phi(\varphi)$ that satisfies (4.29) exists. ■

Lemma 4.3 can be shown using Lemma 4.7 given below.

Lemma 4.7 *(Hayashi and Nagaoka [186]) Consider the c-q channel $x \mapsto W_x$ and its cost function $c : x \mapsto c(x)$. There exists a code $\Phi = (N, \varphi, Y)$ of size N satisfying*

$$
\begin{aligned}
&\varepsilon[\Phi] \\
&\leq \frac{2}{C_K^2}\sum_{x\in\mathcal{X}} p(x)\left(\operatorname{Tr}\left[W_x\left\{W_x-2NW_p\leq 0\right\}\right]+2N\operatorname{Tr}\Big[W_p\left\{W_x-2NW_p>0\right\}\Big]\right)
\end{aligned}
$$

and $\varphi(i) \leq K$, where $C_K \stackrel{\text{def}}{=} p\{c(x) \leq K\}$.

Proof of Lemma 4.3. First, choose an element p of $\mathcal{P}_{c\leq K}(\mathcal{X})$. According to the central limit theorem, the probability $p^{(n)}\{c^{(n)}(\boldsymbol{x}) \leq nK\}$ approaches 1/2. Lemma 4.3 then follows from the arguments presented in the proof of Lemma 4.1. ■

Proof of Lemma 4.7. We show this lemma using the random coding method, as in Lemma 4.6. Consider a random coding for the probability distribution $\hat{p}(x) \stackrel{\text{def}}{=} p(x)/C_K$ in the set $\hat{\mathcal{X}} \stackrel{\text{def}}{=} \{c(x) \leq K\}$. Using the same notation as that of Lemma 4.6, we obtain

$$
\begin{aligned}
\mathrm{E}_X[\varepsilon[\Phi(\varphi)]] \leq & \sum_{x\in\hat{\mathcal{X}}}\hat{p}(x)\Bigg(2\operatorname{Tr}\left[W_x\left\{W_x-2NW_p\leq 0\right\}\right] \\
& + 4N\operatorname{Tr}\Big[\Big(\sum_{x'\in\hat{\mathcal{X}}}p(x')W_{x'}\Big)\left\{W_x-2NW_p>0\right\}\Big]\Bigg).
\end{aligned}
$$

By noting that $\hat{p}(x) = p(x)/C_K$ and $\hat{\mathcal{X}} \subset \mathcal{X}$, we find

$$\begin{aligned}
& E_X[\varepsilon[\Phi(\varphi)]] \\
& \le \sum_{x \in \hat{\mathcal{X}}} \frac{2p(x)}{C_K} \left(\operatorname{Tr}[W_x \{W_x - 2NW_p \le 0\}] + 2N \operatorname{Tr}(\frac{W_p}{C_K}) \{W_x - 2NW_p > 0\} \right) \\
& \le \frac{2}{C_K^2} \sum_{x \in \mathcal{X}} p(x) \left(\operatorname{Tr} W_x \{W_x - 2NW_p \le 0\} + 2N \operatorname{Tr} \left[W_p \{W_x - 2NW_p > 0\} \right] \right).
\end{aligned}$$

By using the same arguments as those in the random coding method, the proof is completed. ■

Exercises

4.6. Let α and β be defined by the following. Show inequality (4.29) in Lemma 4.6 with its RHS replaced by $\alpha + 2\beta + \sqrt{\beta(\alpha + \beta)}$ using Exercise 4.5.

$$\alpha \stackrel{\text{def}}{=} \sum_{x \in \mathcal{X}} p(x) \operatorname{Tr} W_x \{W_x - 2NW_p \le 0\},$$

$$\beta \stackrel{\text{def}}{=} N \operatorname{Tr} W_p \{W_x - 2NW_p > 0\}.$$

4.7. Show (4.29) with its RHS replaced by (4.37). The point here is to replace the probability π_i in (4.34) by that given in (4.38).

$$2 \sum_{x \in \mathcal{X}} p(x) \operatorname{Tr} W_x \{W_x > e^{B_x}\} + 4N \operatorname{Tr} W_p \{W_p < e^A\}, \tag{4.37}$$

$$\pi_i \stackrel{\text{def}}{=} \tau \nu_i \tau, \quad \tau \stackrel{\text{def}}{=} \left\{W_p \ge e^A\right\}, \tag{4.38}$$

$$\nu_i \stackrel{\text{def}}{=} \left\{W_{\varphi(i)} > e^{B_{\varphi(i)}}\right\}. \tag{4.39}$$

4.8. Give another proof of Lemma 4.1 letting $B_x = (\operatorname{Tr} W_{x^n}^{(n)} \log W_{\varphi(i)}^{(n)} - n\delta)$, $A = n(\operatorname{Tr} W_p \log W_p + \delta)$ in (4.39).

4.9. Show that there exists a code with an average error probability less than $4N \operatorname{Tr} W_p \{W_p < e^A\}$ when all the states W_x are pure.

4.10. Modify the decoder $Y(\varphi)$ given in (4.33), replacing π_i by

$$\pi_i \stackrel{\text{def}}{=} \left\{ W_{\varphi(i)}^{(n)} - 2 \sum_{j \ne i} W_{\varphi(j)}^{(n)} > 0 \right\}.$$

Show that the expectation of the average error with respect to the random coding satisfies inequality (4.29) and its modified version with its RHS replaced by (4.37).

4.11. Consider the sequence of codes $\{\Phi^{(n)}\}$ for the stationary memoryless channel of the c-q channel $x \mapsto W_x$. Let us focus on the optimal decreasing exponential rate of the average error probability when the communication rate of $\{\Phi^{(n)}\}$ is greater than R:

$$B(R|W) \stackrel{\text{def}}{=} \sup_{\{\Phi^{(n)}\}} \left\{ \underline{\lim} \frac{-1}{n} \log \varepsilon[\Phi^{(n)}] \middle| \underline{\lim} \frac{1}{n} \log |\Phi^{(n)}| \geq R \right\}, \tag{4.40}$$

This optimal rate is called the *reliability function* as a function of the communication rate R and is an important quantity in quantum information theory.

a Assume that all states W_x are pure states. Show that there exists a code of size $[e^{na}]$ with an average error probability less than $4e^{n(\psi(s|W_p)+(1-s)a)}$ for an arbitrary $s \geq 0$ using Exercise 4.9 with $e^A = N$ and (2.30). Based on this fact, show the following inequality:

$$B(R|W) \geq \max_{s \geq 0} -\psi(s|W_p) - (1-s)R\,. \tag{4.41}$$

b Show that

$$B(R|W) \geq \max_p \max_{0 \leq s \leq 1} -\tilde{\phi}(s|W,p) - sR\,, \tag{4.42}$$

where $\tilde{\phi}(s|W,p) \stackrel{\text{def}}{=} \log \sum_x p(x) \operatorname{Tr} W_x (W_p)^{s/2} (W_x)^{-s} (W_p)^{s/2}$ [186].

4.12. Define another channel capacity by replacing the condition that the average error probability goes to 0 by the alternative condition that the maximum error probability goes to 0. Show that the modified channel capacity is equal to the original channel capacity.

4.13. Show the inequality $C_c(W^1, \ldots, W^M) \geq \max_p \min_{1 \leq i \leq M} I(p, W^i)$ following the steps below.

a Show that there exists a code Φ for M-output channel $W^1, \ldots, W^M$ such that

$$\varepsilon[\Phi] \leq \max_{1 \leq i \leq M} M \sum_{x \in \mathcal{X}} p(x) \Big(2 \operatorname{Tr} W_x^i \{W_x^i - 2NW_p^i \leq 0\} + 4N \operatorname{Tr} W_p^i \{W_x^i - 2NW_p^i > 0\} \Big),$$

as an extension of (4.29).

b Show the desired inequality.

4.6 Proof of Converse Part of C-Q Channel Coding Theorem

In this section, we prove the converse parts of the c-q channel coding theorem, i.e., Lemmas 4.2 and 4.4, using the information inequality (2.63) proved in Sect. 3.7.

Proof of Lemma 4.2. Defining $J(p,\sigma,W) \stackrel{\text{def}}{=} \sum_{x\in\mathcal{X}} p(x) D(W_x\|\sigma)$, we see that $\min_{\sigma\in\mathcal{S}(\mathcal{H})} J(p,\sigma,W) = I(p,W)$ since $J(p,\sigma,W) - I(p,W) = D(W_p\|\sigma) \geq 0$. From the joint convexity of the quantum relative entropy (5.31) to be given later, we see that $\sigma \mapsto D(W_x\|\sigma)$ is convex[6] and Lemma A.7 can be applied. Therefore, we obtain [324, 364]

$$\sup_{p\in\mathcal{P}(\mathcal{X})} I(p,W) = \sup_{p\in\mathcal{P}(\mathcal{X})} \min_{\sigma\in\mathcal{S}(\mathcal{H})} J(p,\sigma,W)$$
$$= \min_{\sigma\in\mathcal{S}(\mathcal{H})} \sup_{p\in\mathcal{P}(\mathcal{X})} J(p,\sigma,W) = \min_{\sigma\in\mathcal{S}(\mathcal{H})} \sup_{x\in\mathcal{X}} D(W_x\|\sigma)\,. \tag{4.43}$$

For $\Phi^{(n)} = (N_n, \varphi^{(n)}, Y^{(n)})$ satisfying (4.13), we choose $\sigma \in \mathcal{S}(\mathcal{H})$ such that

$$r \stackrel{\text{def}}{=} \underline{\lim} \frac{1}{n} \log |\Phi^{(n)}| > \sup_{x\in\mathcal{X}} D(W_x\|\sigma) \tag{4.44}$$

and define density matrices R_n and S_n on $\mathcal{H}^{\otimes n} \otimes \mathbb{C}^{N_n}$ and a matrix T_n as follows:

$$S_n \stackrel{\text{def}}{=} \frac{1}{N_n} \begin{pmatrix} \sigma^{\otimes n} & & 0 \\ & \ddots & \\ 0 & & \sigma^{\otimes n} \end{pmatrix}, \quad T_n \stackrel{\text{def}}{=} \begin{pmatrix} Y_1^{(n)} & & 0 \\ & \ddots & \\ 0 & & Y_{N_n}^{(n)} \end{pmatrix},$$

$$R_n \stackrel{\text{def}}{=} \frac{1}{N_n} \begin{pmatrix} W^{(n)}_{\varphi^{(n)}(1)} & & 0 \\ & \ddots & \\ 0 & & W^{(n)}_{\varphi^{(n)}(N_n)} \end{pmatrix}.$$

Since $I \geq T_n \geq 0$, we have

$$\operatorname{Tr} R_n T_n = \sum_{i=1}^{N_n} \frac{1}{N_n} \operatorname{Tr} W^{(n)}_{\varphi^{(n)}(i)} Y_i^{(n)} = 1 - \varepsilon[\Phi^{(n)}]. \tag{4.45}$$

On the other hand, since $I = \sum_{i=1}^{N_n} Y_i^{(n)}$, we have

[6] The joint convexity (5.31) can be obtained from Theorem 5.5, but since the proof of this theorem uses only the arguments of Chap. 3, there is no problem in using this for the proof of Lemma 4.2.

$$\operatorname{Tr} S_n T_n = \sum_{i=1}^{N_n} \frac{1}{N_n} \operatorname{Tr} \sigma^{\otimes n} Y_i^{(n)} = \frac{1}{N_n} \operatorname{Tr} \sigma^{\otimes n} \sum_{i=1}^{N_n} Y_i^{(n)} = \frac{1}{N_n} \operatorname{Tr} \sigma^{\otimes n} = \frac{1}{N_n}.$$

We define a channel version of the quantum relative Rényi entropy as $\phi(s|W\|\sigma) \stackrel{\text{def}}{=} \sup_{x\in\mathcal{X}} \phi(s|W_x\|\sigma)$. Similarly to (3.37), the monotonicity of the quantum relative Rényi entropy (2.63) yields

$$\begin{aligned}
(1-\varepsilon[\Phi^{(n)}])^{1-s} N_n^{-s} &= (\operatorname{Tr} R_n T_n)^{1-s} (\operatorname{Tr} S_n T_n)^s \le \operatorname{Tr} R_n^{1-s} S_n^s \\
&= \frac{1}{N_n} \sum_{i=1}^{N_n} \operatorname{Tr}\left[(W_{\varphi^{(n)}(i)}^{(n)})^{1-s} (\sigma^{\otimes n})^s \right] = \frac{1}{N_n} \sum_{i=1}^{N_n} \prod_{l=1}^{n} \operatorname{Tr}\left[(W_{\varphi_l^{(n)}(i)})^{1-s} \sigma^s \right] \\
&\le e^{n\phi(s|W\|\sigma)} \qquad (4.46)
\end{aligned}$$

for $s \le 0$. Thus,

$$\frac{1}{n} \log(1-\varepsilon[\Phi^{(n)}]) \le \frac{\phi(s|W\|\sigma) + \frac{s}{n}\log N_n}{1-s}.$$

Letting

$$r \stackrel{\text{def}}{=} \underline{\lim} \frac{1}{n} \log N_n \,, \qquad (4.47)$$

we obtain

$$\underline{\lim} \frac{-1}{n} \log(1-\varepsilon[\Phi^{(n)}]) \ge \frac{-sr - \phi(s|W\|\sigma)}{1-s}. \qquad (4.48)$$

Reversing the order of the $\lim_{s\to 0}$ and $\sup_{x\in\mathcal{X}}$, we obtain

$$\begin{aligned}
\phi'(0|W\|\sigma) &= \lim_{s\to 0} \sup_{x\in\mathcal{X}} \frac{\log \operatorname{Tr} W_x^{1-s}\sigma^s}{-s} = \sup_{x\in\mathcal{X}} \lim_{s\to 0} \frac{\log \operatorname{Tr} W_x^{1-s}\sigma^s}{-s} \\
&= \sup_{x\in\mathcal{X}} D(W_x\|\sigma). \qquad (4.49)
\end{aligned}$$

Since $r > \sup_{x\in\mathcal{X}} D(W_x\|\sigma)$, we can choose a parameter $s_0 < 0$ such that $\frac{\phi(s_0|W\|\sigma) - \phi(0|W\|\sigma)}{s_0} < r$. Hence, we can show that

$$\frac{-s_0 r - \phi(s_0|W\|\sigma)}{1-s_0} = \frac{-s_0}{1-s_0}\left(r - \frac{\phi(s_0|W\|\sigma)}{-s_0} \right) > 0\,. \qquad (4.50)$$

Therefore, $1-\varepsilon[\Phi^{(n)}] \to 0$, and we obtain (4.14) from (4.45).

One may worry about the validity of reversing the order of $\lim_{s\to 0}$ and $\sup_{x\in\mathcal{X}}$ in (4.49). The validity of this step can be confirmed by showing that the convergence is uniform with respect to x. Since the dimension of our space is finite, $\{W_x\}_{x\in\mathcal{X}}$ is included in a compact set. The convergence with $s \to 0$, i.e., $\frac{\log \operatorname{Tr} W_x^{1+s}\sigma^{-s}}{s} \to D(W_x\|\sigma)$, is uniform in any compact set, which shows the uniformity of the convergence. Therefore, we obtain (4.49). ∎

Proof of Lemma 4.4. Similarly to (4.43), Lemma A.7 and the convexity of $\sigma \mapsto D(W_x\|\sigma)$ guarantee that

$$\begin{aligned}\sup_{p\in\mathcal{P}_{c\le K}(\mathcal{X})} I(p,W) &= \sup_{p\in\mathcal{P}_{c\le K}(\mathcal{X})} \min_{\sigma\in\mathcal{S}(\mathcal{H})} J(p,\sigma,W) \\ &= \min_{\sigma\in\mathcal{S}(\mathcal{H})} \sup_{p\in\mathcal{P}_{c\le K}(\mathcal{X})} J(p,\sigma,W) = \min_{\sigma\in\mathcal{S}(\mathcal{H})} \sup_{p\in\mathcal{P}_{c\le K}(\mathcal{X})} \sum_x p_x D(W_x\|\sigma).\end{aligned} \tag{4.51}$$

Let a code $\Phi^{(n)} = (N_n, \varphi^{(n)}, Y^{(n)})$ satisfy (4.23) and $\max_i \frac{c^{(n)}(\varphi^{(n)}(i))}{n} \le K$. Now, we choose a density $\sigma \in \mathcal{S}(\mathcal{H})$ such that

$$\underline{\lim} \frac{1}{n} \log |\Phi^{(n)}| \ge \sup_{p\in\mathcal{P}_{c\le K}(\mathcal{X})} J(p,\sigma,W). \tag{4.52}$$

We define the function $\phi_c(s|W\|\sigma) \stackrel{\text{def}}{=} \sup_{p\in\mathcal{P}_{c\le K}(\mathcal{X})} \sum_{x\in\mathcal{X}} p(x) \log \operatorname{Tr} W_x^{1-s}\sigma^s$. Since $\sum_{l=1}^n \frac{1}{n} c(\varphi_l^{(n)}(i)) = \frac{c^{(n)}(\varphi^{(n)}(i))}{n} \le K$, it follows from the definition of $\mathcal{P}_{c\le K}(\mathcal{X})$ that

$$\begin{aligned}\log \prod_{l=1}^n \operatorname{Tr}\left[(W^{(n)}_{\varphi_l^{(n)}(i)})^{1-s}\sigma^s\right] =& n\sum_{l=1}^n \frac{1}{n} \log \operatorname{Tr}\left[(W^{(n)}_{\varphi_l^{(n)}(i)})^{1-s}\sigma^s\right] \\ \le& n\phi_c(s|W\|\sigma)\end{aligned}$$

for arbitrary i and $s \le 0$. From (4.46) we have

$$(1-\varepsilon[\Phi^{(n)}])^{1-s} N_n^{-s} \le \frac{1}{N_n} \sum_{i=1}^{N_n} \prod_{l=1}^n \operatorname{Tr}\left[(W^{(n)}_{\varphi_l^{(n)}(i)})^{1-s}\sigma^s\right] \le e^{n\phi_c(s|W\|\sigma)}.$$

We define r according to (4.44). Similarly to (4.48), the relation

$$\underline{\lim} \frac{-1}{n} \log\left(1-\varepsilon[\Phi^{(n)}]\right) \ge \frac{-sr - \phi_c(s|W\|\sigma)}{1-s} \tag{4.53}$$

holds. Since

$$\begin{aligned}\lim_{s\to 0} \frac{\phi_c(s|W\|\sigma)}{-s} &= \sup_p \sum_{x\in\mathcal{X}} p(x) \lim_{s\to 0} \frac{\log \operatorname{Tr} W_x^{1-s}\sigma^s}{-s} \\ &= \sup_p \sum_{x\in\mathcal{X}} p(x) D(W_x\|\sigma),\end{aligned} \tag{4.54}$$

we may take a parameter $s_0 < 0$ such that

$$\frac{-s_0 r - \phi_c(s_0|W\|\sigma)}{1-s_0} = \frac{-s_0}{1-s_0}\left(r - \frac{\phi_c(s_0|W\|\sigma)}{-s_0}\right) > 0,$$

in the same way as in the case of (4.50). Therefore, we obtain (4.24) from (4.53).

One may again be concerned about the validity of exchanging the order of the limit and the supremum in (4.54). However, as in the case of (4.49), this step may be safely performed due to the uniformity of the convergence following from the compactness of $\mathcal{X}$. ■

Exercises

4.14. Given that $p_M \stackrel{\text{def}}{=} \operatorname{argmax}_{p \in \mathcal{P}(\mathcal{X})} I(p, W)$ exists, show that $\max_{p \in \mathcal{P}(\mathcal{X})} I(p, W) \geq \min_{\sigma \in \mathcal{S}(\mathcal{H})} \max_{x \in \mathcal{P}(\mathcal{X})} D(W_x \| \sigma)$ by proving that $\max_{x \in \mathcal{P}(\mathcal{X})} D(W_x \| W_{p_M}) = I(p_M, W)$ using the method of Lagrange multipliers. As the reverse inequality is trivial, this is an alternative proof of (4.43).

4.15. Define another channel capacity under cost constraint by replacing the condition that the maximum cost $\max_i \frac{c^{(n)}(\varphi^{(n)}(i))}{n}$ is less than the given cost K by the alternative condition that the average cost $\frac{1}{N} \sum_i \frac{c^{(n)}(\varphi^{(n)}(i))}{n}$ is less than the given cost K. Show that the modified channel capacity under cost constraint is equal to the original channel capacity under cost constraint following the steps below.

a First, assume that $c(x_0) = 0$ or redefine the cost as $c(x) - c(x_0)$, where $x_0 \stackrel{\text{def}}{=} \operatorname{argmin}_{x \in \mathcal{X}} c(x)$. Let a code $\Phi^{(n)} = (N_n, \varphi^{(n)}, Y^{(n)})$ satisfy $\varepsilon[\Phi^{(n)}] \to 0$ and $\frac{1}{N_n} \sum_i \frac{c^{(n)}(\varphi^{(n)}(i))}{n} \leq K$. For arbitrary $\delta > 0$, focus a code $\tilde{\Phi}^{((1+\delta)n)} = (\tilde{N}_{(1+\delta)n}, \tilde{\varphi}^{((1+\delta)n)}, \tilde{Y}^{((1+\delta)n)})$ satisfying $\tilde{N}_{(1+\delta)n} = N_n$, $\tilde{\varphi}^{((1+\delta)n)}(i) = \varphi^{(n)}(i) \otimes W_{x_0}^{\otimes \delta n}$, and $\tilde{Y}_i^{((1+\delta)n)} = Y^{(n)}$.
Show that there exist $k \stackrel{\text{def}}{=} [(1 - \frac{1}{1+\delta}) N_n]$ messages $i_1, \ldots, i_k$ such that $c^{((1+\delta)n)}(\tilde{\varphi}^{((1+\delta)n)}(i_k)) \leq K$.

b Examine the subcode of $\tilde{\Phi}^{(n)}$ consisting of $[(1 - \frac{1}{1+\delta}) N_n]$ messages, and show that the rate of this subcode is asymptotically equal to $\frac{1}{1+\delta} \lim \frac{1}{n} \log |\Phi^{(n)}|$.

c Show that the modified capacity is equal to the original capacity. (Note that this method gives the strong converse concerning the modified channel capacity by combining the strong converse of the original capacity.)

4.16. Show the inequality $C_c(W^1, \ldots, W^M) \leq \max_p \min_{1 \leq i \leq M} I(p, W^i)$ following the steps below.

a Show that there exists a distribution p for any distribution $p^{(n)}$ on $\mathcal{X}^n$ such that

$$nI(p, W^i) \geq I(p^{(n)}, (W^i)^{(n)}), \quad i = 1, \ldots, M$$

from (4.3) and (4.4).

b Show the desired inequality using inequality (4.18).

4.17. Derive the capacity of L-list decoding: Define a decoder Φ of L-list decoding with the size N by a $\binom{N}{L}$-valued POVM $M = \{M_{(i_1,\ldots,i_L)}\}$ and the average error probability

$$\varepsilon[\Phi_L] \stackrel{\text{def}}{=} \frac{1}{N}\sum_{i=1}^{N}\left(1 - \sum_{j_1,\ldots,j_{L-1}\neq i} \operatorname{Tr} W_{\varphi(i)} M_{i,j_1,\ldots,j_{L-1}}\right),$$

where $(i_1,\ldots,i_L)$ is the set of L different elements $i_1,\ldots,i_L$.

a When we replace the definition of $Y_i^{(n)}$ by $Y_i^{(n)} \stackrel{\text{def}}{=} \sum_{j_1,\ldots,j_{L_n-1}\neq i} M_{i,j_1,\ldots,j_{L_n-1}}$, show that $L_n I = \sum_{i=1}^{N_n} Y_i^{(n)}$.

b Show that $\operatorname{Tr} R_n T_n = 1 - \varepsilon[\Phi_{L_n}^{(n)}]$ in the notation of this section.

c Show that the strong converse part of L-list decoding when $\frac{1}{n}\log L_n \to 0$ using the same discussion as this section. In this case, the capacity is equal to $\sup_{p\in\mathcal{P}(\mathcal{X})} I(p,W)$.

d Show that the strong converse part of L-list decoding with cost constraint in the above condition.

4.7 Pseudoclassical Channels

Finally, we treat the capacity of a c-q channel when the quantum correlation is not allowed in the measuring apparatus, again. In Sect. 4.2 we showed that the channel capacity is not improved even when encoding with feedback and adaptive decoding is used as long as the quantum correlation is not used in the measuring apparatus. That is, the capacity can be attained when the optimal measurement with a single transmission is performed on each system. Then, we may ask, when does the channel capacity $\tilde{C}_c(W)$ with individual measurements equal the channel capacity $C_c(W)$ with the quantum correlation in the measuring apparatus? The answer to this question is the subject of the following theorem.

Theorem 4.4 *(Fujiwara and Nagaoka [128]) Suppose that* $\operatorname{Tr} W_x W_{x'} \neq 0$ *for any* $x, x' \in \mathcal{X}$*. Then, the following three conditions with respect to the c-q channel* W *are equivalent if* $\mathcal{X}$ *is compact.*

① *There exists a distribution* $p \in \mathcal{P}(\mathcal{X})$ *such that each element of* $\operatorname{supp}(p)$ *commutes and* $I(p,W) = C_c(W)$*.*
② $\tilde{C}_c(W) = C_c(W)$*.*
③ *There exists an integer* n *such that* $\frac{\tilde{C}_c(W^{(n)})}{n} = C_c(W)$*.*

A quantum channel W is called *pseudoclassical* if it satisfies the above conditions.

Proof. Since ①⇒② and ②⇒③ by inspection, we show that ③⇒①. The proof given below uses Theorems 3.5 and 4.5 (*Naĭmark extension* [315]). The proofs for these theorems will be given later.

Theorem 4.5 *(Naĭmark [315]) Given a POVM $\boldsymbol{M} = \{M_\omega\}_{\omega \in \Omega}$ on $\mathcal{H}_A$ with a finite data set Ω, there exist a space $\mathcal{H}_B$, a state ρ_0 on $\mathcal{H}_B$, and a PVM $\boldsymbol{E} = \{E_\omega\}_{\omega \in \Omega}$ in $\mathcal{H}_A \otimes \mathcal{H}_B$ such that*

$$\mathrm{Tr}_A\, \rho M_\omega = \mathrm{Tr}_{A,B}(\rho \otimes \rho_0) E_\omega, \quad \forall \rho \in \mathcal{S}(\mathcal{H}_A), \forall \omega \in \Omega.$$

For the proof of Theorem 4.5, see Exercise 5.5 or the comments regarding Theorem 7.1.

Using Condition ③, we choose a measurement $\boldsymbol{M}^{(n)}$ on $\mathcal{H}^{\otimes n}$ and a distribution $p^{(n)}$ on $\mathcal{X}^n$ such that $I(p^{(n)}, \boldsymbol{M}^{(n)}, W^{(n)}) = \tilde{C}_c(W^{(n)})$. Since

$$I(p^{(n)}, \boldsymbol{M}^{(n)}, W^{(n)}) \le I(p^{(n)}, W^{(n)}), \ \tilde{C}_c^{(n)}(W) \ge nC_c(W) \ge I(p^{(n)}, W^{(n)}),$$

we obtain $I(p^{(n)}, \boldsymbol{M}^{(n)}, W^{(n)}) = I(p^{(n)}, W^{(n)})$. This is equivalent to

$$\sum_{\boldsymbol{x} \in \mathrm{supp}(p^{(n)})} p^{(n)}(\boldsymbol{x}) \left(D^{\boldsymbol{M}^{(n)}}(W_{\boldsymbol{x}}^{(n)} \| W_{p^{(n)}}^{(n)}) - D(W_{\boldsymbol{x}}^{(n)} \| W_{p^{(n)}}^{(n)}) \right) = 0,$$

and we have $D^{\boldsymbol{M}^{(n)}}(W_{\boldsymbol{x}}^{(n)} \| W_{p^{(n)}}^{(n)}) = D(W_{\boldsymbol{x}}^{(n)} \| W_{p^{(n)}}^{(n)})$ for $\boldsymbol{x} \in \mathrm{supp}(p^{(n)})$, where $\boldsymbol{x}^n$ is simplified to $\boldsymbol{x}$. Following Theorem 4.5, we choose an additional system $\mathcal{H}_A$, a pure state ρ_A on $\mathcal{H}_A$, and a PVM $\boldsymbol{E} = \{E_k\}$ on the composite system $\mathcal{H}_A \otimes \mathcal{H}$ such that

$$D^{\boldsymbol{E}}(W_{\boldsymbol{x}}^{(n)} \otimes \rho_A \| W_{p^{(n)}}^{(n)} \otimes \rho_A) = D(W_{\boldsymbol{x}}^{(n)} \otimes \rho_A \| W_{p^{(n)}}^{(n)} \otimes \rho_A).$$

According to Theorem 3.5, we take a real number $a_k(\boldsymbol{x})$ for every $\boldsymbol{x} \in \mathrm{supp}(p^{(n)})$ such that the Hermitian matrix $X_{\boldsymbol{x}} = \sum_k a_k(\boldsymbol{x}) E_k$ satisfies $W_{\boldsymbol{x}}^{(n)} \otimes \rho_A = \left(W_{p^{(n)}}^{(n)} \otimes \rho_A \right) X_{\boldsymbol{x}}$. Since $X_{\boldsymbol{x}}$ is Hermitian, we obtain

$$\begin{aligned}
&\left(W_{\boldsymbol{x}}^{(n)} \otimes \rho_A \right) \left(W_{\boldsymbol{x}'}^{(n)} \otimes \rho_A \right) = (W_{p^{(n)}}^{(n)} \otimes \rho_A) X_{\boldsymbol{x}} X_{\boldsymbol{x}'} (W_{p^{(n)}}^{(n)} \otimes \rho_A) \\
=& (W_{p^{(n)}}^{(n)} \otimes \rho_A) X_{\boldsymbol{x}'} X_{\boldsymbol{x}} (W_{p^{(n)}}^{(n)} \otimes \rho_A) = \left(W_{\boldsymbol{x}'}^{(n)} \otimes \rho_A \right) \left(W_{\boldsymbol{x}}^{(n)} \otimes \rho_A \right)
\end{aligned}$$

for $\boldsymbol{x}, \boldsymbol{x}' \in \mathrm{supp}(p^{(n)})$. Therefore, we obtain

$$W_{\boldsymbol{x}}^{(n)} W_{\boldsymbol{x}'}^{(n)} = W_{\boldsymbol{x}'}^{(n)} W_{\boldsymbol{x}}^{(n)}.$$

Defining $p_i^{(n)}$ by $p_i^{(n)}(x) \stackrel{\mathrm{def}}{=} \sum_{\boldsymbol{x}=(x_1,\ldots,x_n):x_i=x} p^{(n)}(\boldsymbol{x})$, from Exercise 1.13 we find that W_x and W_y commute for $x, y \in \mathrm{supp}(p_i^{(n)})$ because $\mathrm{Tr}\, W_x W_{x'} \neq 0$ for any $x, x' \in \mathcal{X}$. Equation (4.4) yields

$$\sum_{i=1}^n I(p_i^{(n)}, W) \ge I(p^{(n)}, W^{(n)}) = nC_c(W).$$

Therefore, $I(p_i^{(n)}, W) = C_c(W)$, and thus we obtain ①. ■

4.8 Historical Note

Here, we briefly mention the history of the c-q channel coding theorem. Since this problem was independently formulated by several researchers, it is difficult to determine who formulated it first. The first important achievement for this theorem is the inequality

$$I(P, W) \geq I(M, P, W), \tag{4.55}$$

which was conjectured by Levitin [264] and proved by Holevo [212]. Indeed, this inequality can be proved easily from the monotonicity of the quantum relative entropy (5.30) [270, 393]; however, at that time, it had not been proved. During that period, a strong subadditivity of the von Neumann entropy (5.55) was proved by Lieb and Ruskai [268, 269]. Using the strong subadditivity, we can easily prove the above inequality $^{\text{Ex. 5.45}}$; however, this relation between the strong subadditivity and inequality (4.55) was not known at that time. Combining inequality (4.55) with Fano's inequality, Holevo [214] showed that the weaker version of the converse part, i.e., $C_c(W) \leq \sup_{p\in\mathcal{P}(\mathcal{X})} I(p, W)$, held. Twenty years later, Ogawa and Nagaoka [319] proved the strong converse part $C_c^{\dagger}(W) \leq \sup_{p\in\mathcal{P}(\mathcal{X})} I(p, W)$. Moreover, Nagaoka [307] invented a more simple proof of the strong converse. His proof is based on the monotonicity of the Rényi entropy (2.63). In this book, we prove (2.63) using elementary knowledge in Sect. 3.7 and give a proof of the strong converse part combining Nagaoka's proof and (2.63).

Regarding the direct part, in the late 1970s, Stratonovich and Vantsjan [382] treated the pure state case, i.e., the case in which all the states W_x are pure. In this case, $C_c(W)$ is equal to $\sup_p H(W_p)$, but they found the lower bound $\sup_p - \log \operatorname{Tr} W_p^2$ of $C_c(W)$, i.e., they proved that $\sup_p - \log \operatorname{Tr} W_p^2 \leq C_c(W)$. Sixteen years later, Hausladen et al. [164] proved the attainability of $\sup_p H(W_p)$ in the pure-states case. This result was presented by Jozsa, who is a coauthor of this paper, in the QCMC'96 conference held at Hakone. Holevo attended this conference and extended this proof to the mixed-state case during his stay at Tamagawa University after this conference. Later, Schumacher and Westmoreland [363] independently obtained the same result. Their method was based on the conditional typical sequence, and its classical version appeared in Cover and Thomas [82]. Therefore, we can conclude that Holevo played a central role in the formulation of the c-q channel coding theorem. Hence, some researchers call the capacity $C_c(W)$ the *Holevo capacity*, while Theorem 4.1 is called the HSW theorem.

In the classical case, Csiszár and Körner [85] have established the type method, which is a unified method in classical information theory and is partially summarized in Sect. 2.5.1. Applying it to its classical version, the researchers obtained another proof of this theorem and examined channel coding in greater detail. Winter [416, 417] tried to apply the type method to c-q channels. He obtained another proof of the c-q channel coding theorem but could not obtain an analysis of the error exponents as precise as that by Csiszár and Körner. Since there is an ambiguity regarding the orthogonal basis in the quantum case, a simple application of the type method to the c-q channel is not as powerful as the application to the classical case.

As another unified method in classical information theory, Han [158] established the method of information spectrum. Verdú and Han [400] applied it to classical channel coding and succeeded in obtaining the capacity of a general sequence of classical channels without any assumption, e.g., stationary memoryless, etc. As a byproduct, they clarified the relation between channel coding and hypothesis testing. Ogawa and Nagaoka [321] extended this method to the quantum case and obtained another proof of the direct part of the c-q channel coding theorem. Further, they clarified the relation between c-q channel coding and quantum hypothesis testing. However, they could not obtain the capacity of the general sequence in the quantum case. Hayashi and Nagaoka [186] were motivated by their proof. Based on a quantum analog of the information spectrum method, they derived the capacity for the general sequence of c-q channels without any condition. In consequence, they obtained another proof of the stationary memoryless case.

All of the above methods except that of Ogawa and Nagaoka [321] are based on the random coding method. In the classical case, Feinstein [114] showed the channel coding theorem based not on the random coding method but on a greedy coding method. Since the proof of Ogawa and Nagaoka [321] is based on the greedy coding method, their method can be regarded as its quantum version.

Moreover, we sometimes discuss the error exponent in channel coding. Burunashev and Holevo [63] first obtained the lower bound of the optimal error exponent in the pure-state case, which is equal to (4.41). Their method differs from the method of Exercise 4.11. In the mixed state case, Hayashi and Nagaoka [186] obtained the lower bound (4.42) by the same method as Exercise 4.11.

In addition, Fujiwara and Nagaoka [128] discussed coding protocols with adaptive decoding and feedback and obtained Theorem 4.2. They also introduced pseudoclassical channels (Sect. 4.7) and obtained the equivalence of Conditions ① and ② in Theorem 4.4. This textbook slightly improves their proof and proves the equivalence among the three Conditions ①, ②, and ③. Recently, Bennett et al. [44] obtained an interesting result regarding the classical capacity with feedback and quantum correlation. The channel capacity with a cost constraint was first treated by Holevo [218], and its strong converse part was shown by Hayashi and Nagaoka [186].

On the other hand, Stratonovich and Vantsjan [382] found the result of $-\log \operatorname{Tr} W_p^2$ and not $H(W_p)$ due to some weak evaluations with respect to the error probability for the pure-state case. It would be interesting to determine the difference between these two quantities. Fujiwara [123] considered an ensemble of pure states generated randomly under the asymptotic setting. He focused on two types of orthogonality relations and found that the two quantities $H(W_p)$ and $-\log \operatorname{Tr} W_p^2$ correspond to their respective orthogonality relations.

5

State Evolution and Trace-Preserving Completely Positive Maps

Summary. Until now, we have considered only quantum measurements among operations on quantum systems. In order to perform effective information processing with quantum systems, we should manipulate a wider class of state operations. This chapter examines what kinds of operations are allowed on quantum systems. The properties of these operations will also be examined.

Table 5.1. Denotations used in Chap. 5

$K(\kappa)$	Matrix representation of κ (5.2)
κ^E	TP-CP map to environment (5.5)
$\kappa_{\boldsymbol{M},W}$	Entanglement-breaking channel
τ	Transpose (5.11)
$\kappa_{d,\lambda}$	Depolarizing channel (5.9)
$\kappa_{d,\lambda}^T$	Transpose-depolarizing channel (5.15)
κ_p^{GP}	Generalized Pauli channel (5.13)
κ_D^{PD}	Phase-damping channel (5.16)
$\kappa_{d,n\to m}^{\mathrm{pns}}$	PNS channel (5.18)
$\kappa_{d,p}^{\mathrm{era}}$	Erasure channel (5.19)
$H_\rho(A\|B)$	Conditional entropy (5.60)
$I_\rho(A:B)$	Quantum mutual information (5.61)
$I_\rho(A:B\|C)$	Quantum conditional mutual information (5.62)
$\eta(x)$	$-x\log x$
$\eta_0(x)$	See (5.63)

5.1 Description of State Evolution in Quantum Systems

The time evolution over a time t of a closed quantum-mechanical system $\mathcal{H}$ is given by

$$\rho \mapsto e^{itH}\rho e^{-itH},$$

where H is a Hermitian matrix in $\mathcal{H}$ called the Hamiltonian. However, this is true only if there is no interaction between the system $\mathcal{H}$ and another system. The state evolution in the presence of an interaction cannot be written in the above way. Furthermore, the input system for information processing (i.e., state evolution) is not necessarily the same as its output system. In fact, in some processes it is crucial for the input and output systems to be different. Hence, we will denote the input and output system by $\mathcal{H}_A$ and $\mathcal{H}_B$, respectively, and investigate the map κ from the set $\mathcal{S}(\mathcal{H}_A)$ of densities on the system $\mathcal{H}_A$ to $\mathcal{S}(\mathcal{H}_B)$, which gives the relationship between the input and the output (state evolution). First, we require the map κ to satisfy the condition

$$\kappa(\lambda\rho_1 + (1-\lambda)\rho_2) = \lambda\kappa(\rho_1) + (1-\lambda)\kappa(\rho_2)$$

for $1 > \lambda > 0$ and arbitrary $\rho_1, \rho_2 \in \mathcal{S}(\mathcal{H}_A)$. Maps satisfying this property are called *affine* maps. Since the space $\mathcal{S}(\mathcal{H}_A)$ is not a linear space, we cannot claim that κ is linear; however, these two conditions are almost equivalent. In fact, we may extend the map κ into a linear map $\tilde{\kappa}$ that maps from the linear space $\mathcal{T}(\mathcal{H}_A)$ of Hermitian matrices on $\mathcal{H}_A$ to the linear space $\mathcal{T}(\mathcal{H}_B)$ of the Hermitian matrices on $\mathcal{H}_B$ as follows. Since an arbitrary matrix $X \in \mathcal{T}(\mathcal{H}_A)$ can be written as a linear sum

$$X = \sum_i a^i X_i, \quad a^i \in \mathbb{R} \tag{5.1}$$

using elements $X_1, \ldots, X_{d^2}$ of $\mathcal{S}(\mathcal{H}_A)$, $\tilde{\kappa}$ may be defined as

$$\tilde{\kappa}(X) \stackrel{\text{def}}{=} \sum_i a^i \kappa(X_i).$$

The affine property guarantees that this definition does not depend on (5.1). Henceforth, we shall identify $\tilde{\kappa}$ with κ. The linear combination of the elements in $\mathcal{T}(\mathcal{H}_A)$ multiplied by complex constants gives the space $\mathcal{M}(\mathcal{H}_A)$ of the matrices on $\mathcal{H}_A$. Therefore, κ may be extended to a map that takes us from the space $\mathcal{M}(\mathcal{H}_A)$ of matrices on $\mathcal{H}_A$ to the space $\mathcal{M}(\mathcal{H}_B)$ of matrices on $\mathcal{H}_B$. It is often more convenient to regard κ as a linear map in discussions on its properties; hence, we will often use κ as the linear map from $\mathcal{T}(\mathcal{H}_A)$ to $\mathcal{T}(\mathcal{H}_B)$. Occasionally, it is even more convenient to treat κ as a map from $\mathcal{M}(\mathcal{H}_A)$ to $\mathcal{M}(\mathcal{H}_B)$. We shall examine these cases explicitly.

In order to recover the map from $\mathcal{S}(\mathcal{H}_A)$ to $\mathcal{S}(\mathcal{H}_B)$ from the linear map κ from $\mathcal{T}(\mathcal{H}_A)$ to $\mathcal{T}(\mathcal{H}_B)$, we assume that the linear map transforms positive semidefinite matrices to positive semidefinite matrices. This map is called a *positive map*. The trace also needs to be preserved.

However, there are still more conditions that the state evolution must satisfy. In fact, we consider the state evolution κ occurring on the quantum system $\mathcal{H}_A$ whose state is entangled with another system $\mathbb{C}^n$. We also suppose that the additional system $\mathbb{C}^n$ is stationary and has no state evolution. Then, any state on the composite system of $\mathbb{C}^n$ and $\mathcal{H}_A$ obeys the state evolution from $\mathcal{H}_A \otimes \mathbb{C}^n$ to $\mathcal{H}_B \otimes \mathbb{C}^n$, which is given by the linear map $\kappa \otimes \iota_n$ from $\mathcal{T}(\mathcal{H}_A \otimes \mathbb{C}^n) = \mathcal{T}(\mathcal{H}_A) \otimes \mathcal{T}(\mathbb{C}^n)$ to $\mathcal{T}(\mathcal{H}_B \otimes \mathbb{C}^n) = \mathcal{T}(\mathcal{H}_B) \otimes \mathcal{T}(\mathbb{C}^n)$, where ι_n denotes the identity operator from $\mathcal{T}(\mathbb{C}^n)$ to itself. The map $\kappa \otimes \iota_n$ then must satisfy positivity and trace-preserving properties, as discussed above. In this case, the system $\mathbb{C}^n$ is called the *reference system*. The trace-preserving property follows from the trace-preserving property of κ:

$$\begin{aligned}\operatorname{Tr}(\kappa \otimes \iota_n)\left(\sum_i X_i \otimes Y_i\right) &= \sum_i \operatorname{Tr}\left(\kappa(X_i) \otimes \iota_n(Y_i)\right) \\ &= \sum_i \operatorname{Tr}\kappa(X_i) \cdot \operatorname{Tr}\iota_n(Y_i) = \sum_i \operatorname{Tr} X_i \cdot \operatorname{Tr} Y_i = \operatorname{Tr}\left(\sum_i X_i \otimes Y_i\right).\end{aligned}$$

However, as shown by the counterexample given in Example 5.7 of Sect. 5.2, it is not possible to deduce that $\kappa \otimes \iota_n$ is a positive map from the fact that κ is a positive map. In a composite system involving an n-dimensional reference system $\mathbb{C}^n$, the map κ is called an *n-positive map* if $\kappa \otimes \iota_n$ is a positive map. If κ is an n-positive map for arbitrary n, the map κ is called a *completely positive map*, which we abbreviate to *CP map*. Since the trace of a density matrix is always 1, the state evolution of a quantum system is given by a *trace-preserving completely positive map*, which is abbreviated to *TP-CP map*. It is currently believed that it is, in principle, possible to produce state evolutions corresponding to arbitrary TP-CP maps, as is shown by Theorem 5.1 discussed later. If a channel has a quantum input system as well as a quantum output system, it can be represented by a TP-CP map. Such channels are called quantum-quantum channels to distinguish them from classical-quantum channels. Strictly speaking, κ is a linear map from $\mathcal{T}(\mathcal{H}_A)$ to $\mathcal{T}(\mathcal{H}_B)$; however, for simplicity, we will call it a TP-CP map from the quantum system $\mathcal{H}_A$ to the quantum system $\mathcal{H}_B$. In particular, since the case with the pure input state is important, we abbreviate $\kappa(|x\rangle\langle x|)$ to $\kappa(x)$.

We now give the matrix representation of the linear map from $\mathcal{T}(\mathcal{H}_A)$ to $\mathcal{T}(\mathcal{H}_B)$ and the necessary and sufficient conditions for it to be a TP-CP map. We denote the basis of the quantum systems $\mathcal{H}_A$ and $\mathcal{H}_B$ by $e_1^A, \ldots, e_d^A$ and $e_1^B, \ldots, e_{d'}^B$, respectively. We define $K(\kappa)$ as a matrix in $\mathcal{H}_A \otimes \mathcal{H}_B$ for κ according to

$$K(\kappa)^{(j,l),(i,k)} \stackrel{\text{def}}{=} \langle e_k^B|\kappa(|e_i^A\rangle\langle e_j^A|)|e_l^B\rangle. \tag{5.2}$$

Let $X = \sum_{i,j} x^{i,j}|e_i^A\rangle\langle e_j^A|, Y = \sum_{k,l} y_{k,l}|e_k^B\rangle\langle e_l^B|$. Then, we can write

$$\operatorname{Tr}\kappa(X)Y = \sum_{i,j,k,l} x^{i,j} y_{k,l} \langle e_k^B|\kappa(|e_i^A\rangle\langle e_j^A|)|e_l^B\rangle = \operatorname{Tr}(X \otimes Y^T)K(\kappa).$$

Now, let $\mathcal{H}_R$ be the space spanned by $e_1^R, \dots, e_d^R$. Then, the maximally entangled state $|\Phi_d\rangle = \frac{1}{\sqrt{d}} \sum_i e_i^A \otimes e_i^R$ characterizes this matrix representation of κ as

$$
\begin{aligned}
&(\kappa \otimes \iota_R)(|\Phi_d\rangle\langle\Phi_d|) \\
=&\frac{1}{d}\sum_{i,j} \kappa(|e_i^A\rangle\langle e_j^A|) \otimes |e_i^R\rangle\langle e_j^R| = \frac{1}{d}\sum_{i,j,k,l} K(\kappa)^{(j,l),(i,k)} |e_k^B, e_i^R\rangle\langle e_l^B, e_j^R|. \quad (5.3)
\end{aligned}
$$

The definition of $K(\kappa)$ could be given here by naturally extending κ to a map from $\mathcal{M}(\mathcal{H}_A)$ to $\mathcal{M}(\mathcal{H}_B)$, as discussed above. Note that d (d') is the dimension of $\mathcal{H}_A$ ($\mathcal{H}_B$) above. $K(\kappa)$ may be used to characterize the TP-CP map as follows.

Theorem 5.1 *The following conditions are equivalent [77,130,243,261,379].*

① *κ is a TP-CP map.*
② *Define the linear map κ^* from $\mathcal{T}(\mathcal{H}_B)$ to $\mathcal{T}(\mathcal{H}_A)$ as the map satisfying (5.4). The map κ^* is a completely positive map and satisfies $\kappa^*(I_B) = I_A$.*

$$
\operatorname{Tr} \kappa(X)Y = \operatorname{Tr} X\kappa^*(Y), \quad \forall X \in \mathcal{T}(\mathcal{H}_A), \forall Y \in \mathcal{T}(\mathcal{H}_B). \tag{5.4}
$$

κ^* can be regarded as a dual map with respect to the inner product $\langle X, Y\rangle \stackrel{\text{def}}{=} \operatorname{Tr} XY$.
③ *κ is a trace-preserving $\min\{d, d'\}$-positive map.*
④ *The matrix $K(\kappa)$ in $\mathcal{H}_A \otimes \mathcal{H}_B$ is positive semidefinite and satisfies $\operatorname{Tr}_B K(\kappa) = I_A$.*
⑤ *(Stinespring representation) For a given map κ, it is possible to express κ as $\kappa(\rho) = \operatorname{Tr}_{A,C} U_\kappa(\rho \otimes \rho_0)U_\kappa^*$ by choosing an identical Hilbert space $\mathcal{H}_C$ to $\mathcal{H}_B$, a pure state $\rho_0 \in \mathcal{S}(\mathcal{H}_B \otimes \mathcal{H}_C)$, and a unitary matrix U_κ in $\mathcal{H}_A \otimes \mathcal{H}_B \otimes \mathcal{H}_C$. Note that the structure of $\mathcal{H}_C$ depends only on $\mathcal{H}_B$, not on κ. Only U_κ depends on κ itself.*
⑥ *(Choi–Kraus representation) It is possible to express κ as $\kappa(\rho) = \sum_i F_i \rho F_i^*$ using $\sum_i F_i^* F_i = I_{\mathcal{H}_A}$, where $F_1, \dots, F_{dd'}$ are a set of dd' linear maps from $\mathcal{H}_A$ to $\mathcal{H}_B$.*

The above conditions are also equivalent to a modified Condition ⑤ (which we call Condition ⑤′), where ρ_0 is not necessarily a pure state, and the dimension of $\mathcal{H}_C$ is arbitrary. Another equivalent condition is a modification of Condition ⑥ (which we call Condition ⑥′), where the number of linear maps $\{F_i\}$ is arbitrary.

If the input system $\mathcal{H}_A$ and the output system $\mathcal{H}_B$ are identical to $\mathbb{C}^d$, the channel is called a *d-dimensional channel*. In this case, the Stinespring representation can be rewritten as follows.

Corollary 5.1 *The following conditions are equivalent for a linear map κ from $\mathcal{T}(\mathcal{H}_A)$ to $\mathcal{T}(\mathcal{H}_A)$.*

① *κ is a TP-CP linear map.*
② *$\kappa(\rho) = \mathrm{Tr}_E V_\kappa(\rho \otimes \rho_0)V_\kappa^*$ for a quantum system $\mathcal{H}_E$, a state $\rho_0 \in \mathcal{S}(\mathcal{H}_E)$, and an appropriate unitary matrix V_κ in $\mathcal{H}_A \otimes \mathcal{H}_E$ for κ.*

It is possible to make the dimension of $\mathcal{H}_E$ less than d^2.

The triplet $(\mathcal{H}_C, \rho_0, U_\kappa)$ in ⑤ of Theorem 5.1 is called the Stinespring representation. The equivalence to ① has been proved by Stinespring for the dual map κ^* under general conditions.

The Stinespring representation is important not only as a mathematical representation theorem, but also in terms of its physical meaning. When the input system is identical to the output system, as in Corollary 5.1, we can interpret it as a time evolution under an interaction with an external system $\mathcal{H}_E$. The system $\mathcal{H}_E$ is therefore called the *environment*. If the input and output systems are different, we may regard $\mathcal{H}_A \otimes \mathcal{H}_C$ as the environment, which we again denote by $\mathcal{H}_E$. We can then define the map κ_E transforming the initial state in the input system to the final state in the environment as

$$\kappa^E \stackrel{\mathrm{def}}{=} \mathrm{Tr}_B U_\kappa(\rho \otimes \rho_0)U_\kappa^*. \tag{5.5}$$

The final state in the environment of a Stinespring representation is unitarily equivalent to that of another Stinespring representation as long as the initial state of the environment ρ_0 is chosen as a pure state. That is, the state $\kappa^E(\rho)$ essentially does not depend on the Stinespring representation.

In the next section, we will use this theorem to obtain a concrete example of a TP-CP map. In fact, the partial trace and the map $\rho \mapsto \rho \otimes \rho_0$ are TP-CP maps, as is easily verified from Theorem 5.1 above. Another representation is the output state (5.3) of the channel κ for the maximally entangled state Φ_d between the input system and the same-dimensional reference system. This representation has not only mathematical meaning but also an estimation theoretical importance because it is possible to identify the channel κ by identifying this final state (5.3) [87, 136]. Further, using this notation, we can describe the output state of any input pure state entangled with the reference system as follows. Using (1.19), any pure entangled state $|X\rangle$ can be described as $I_A \otimes \frac{X^T}{\sqrt{d}}|\Phi_d\rangle$. Hence, we have

$$\begin{aligned}(\kappa \otimes \iota_R)(|X\rangle\langle X|) =& (\kappa \otimes \iota_R)((I_A \otimes \frac{X^T}{\sqrt{d}})|\Phi_d\rangle\langle\Phi_d|(I_A \otimes \frac{\bar{X}}{\sqrt{d}})) \\ =& \frac{1}{d}(I_B \otimes X^T)(\kappa \otimes \iota_R)(|\Phi_d\rangle\langle\Phi_d|)(I_B \otimes \bar{X}). \end{aligned} \tag{5.6}$$

In Condition ⑥, another representation $\{F_i\}$ of the completely positive map is given and is called the Choi–Kraus representation. From (1.19) the state $(\kappa \otimes \iota_R)(|\Phi_d\rangle\langle\Phi_d|)$ has the form

$$(\kappa \otimes \iota_R)(|\Phi_d\rangle\langle\Phi_d|) = \frac{1}{d_A}\sum_i |F_i\rangle\langle F_i|. \tag{5.7}$$

Hence, when $\{F_i'\}$ is another Choi–Kraus representation of κ, F_i' is represented as a linear sum of $\{F_i\}$. As is shown in Appendix A.5, the TP-CP map κ^E can be characterized by a Choi–Kraus representation as follows.

Lemma 5.1 *When $\{F_i\}_{i=1}^d$ is a Choi–Kraus representation of κ, the environment system is described by $\mathbb{C}^d$ and the matrix elements of $\kappa^E(\rho)$ are given as*

$$\kappa^E(\rho)_{i,j} = \operatorname{Tr} F_j^* F_i \rho. \tag{5.8}$$

Using this representation we can characterize extremal points of TP-CP maps from $\mathcal{H}_A$ to $\mathcal{H}_B$ as follows.

Lemma 5.2 *(Choi [77]) A TP-CP map κ is an extremal point of TP-CP maps from $\mathcal{H}_A$ to $\mathcal{H}_B$ if and only if κ has Choi–Kraus representation $\{F_i\}$ such that $\{F_i^* F_j\}_{i,j}$ is a linearly independent set.*

Proof. Suppose that κ is an extremal point. Let $\{F_i\}$ be a Choi–Kraus representation of κ such that F_i is linearly independent $^{\text{Ex. 5.3}}$. Suppose $\sum_{i,j} \lambda_{i,j} F_i^* F_j = 0$ and the matrix norm $\|(\lambda_{i,j})\|$ is less than 1. Define $\kappa_\pm$ as $\kappa_\pm(\rho) \stackrel{\text{def}}{=} \sum_i F_i \rho F_i^* \pm \sum_{i,j} \lambda_{i,j} F_i \rho F_j^*$. Since $I \pm (\lambda_{i,j}) \geq 0$, $\kappa_\pm$ is a TP-CP map. It also follows that $\kappa = \frac{1}{2}\kappa_+ + \frac{1}{2}\kappa_-$. Since κ is extremal, $\kappa_+ = \kappa$. That is, $\lambda_{i,j} = 0$. Therefore, $\{F_i^* F_j\}_{i,j}$ is a linearly independent set.

Conversely, suppose that $\{F_i^* F_j\}_{i,j}$ is a linearly independent set. We choose TP-CP maps κ_1 and κ_2 and a real number $0 < \lambda < 1$ such that $\kappa = \lambda \kappa_1 + (1-\lambda)\kappa_2$. Let $\{F_i^k\}$ be a Choi–Kraus representation of κ_k. Then, κ has Choi–Kraus representation $\{\sqrt{\lambda} F_i^1\} \cup \{\sqrt{1-\lambda} F_i^2\}$. Thus, F_i^1 is written as $\sum_j \lambda_{i,j} F_j$. The condition $\sum_i (F_i^1)^* F_i^1 = \sum_j F_j^* F_j$ and the linearly independence of $\{F_i^* F_j\}_{i,j}$ imply that $\sum_{i,i'} \overline{\lambda_{i',j'}} \lambda_{i,j} = \delta_{j',j}$. That is, $\kappa = \kappa_1$ (Exercise 5.2). ∎

Corollary 5.2 *When a TP-CP map κ from $\mathcal{H}_A$ to $\mathcal{H}_B$ is extremal, its Choi–Kraus representation has only d_A elements and any image is included in the d_A^2-dimensional space of $\mathcal{H}_B$ at most. Hence, the dimension of the environment is less than d_A.*

A Stinespring representation guarantees that the state evolution corresponding to the TP-CP map κ can be implemented by the following procedure. The initial state ρ_0 is first prepared on $\mathcal{H}_B \otimes \mathcal{H}_C$; then, the unitary evolution U_κ is performed on $\mathcal{H}_A \otimes \mathcal{H}_B \otimes \mathcal{H}_C$. It is commonly believed that in principle, state evolutions corresponding to an arbitrary unitary matrix U_κ can be implemented, and hence state evolutions corresponding to arbitrary TP-CP maps can also in principle be implemented.

Let us now consider the case where we are given two TP-CP maps κ and κ' that map from the quantum systems $\mathcal{H}_A, \mathcal{H}_A'$ to $\mathcal{H}_B, \mathcal{H}_B'$, respectively. The state evolution of the composite system from $\mathcal{H}_A \otimes \mathcal{H}_A'$ to $\mathcal{H}_B \otimes \mathcal{H}_B'$

is given by $\kappa \otimes \kappa'$. One may wonder whether the map $\kappa \otimes \kappa'$ also satisfies the condition for TP-CP maps. Indeed, this condition is guaranteed by the following corollary.

Corollary 5.3 *Given a linear map κ' from $\mathcal{T}(\mathcal{H}'_A)$ to $\mathcal{T}(\mathcal{H}'_B)$, the following two conditions are equivalent.*

① *κ' is a TP-CP map.*
② *$\kappa \otimes \kappa'$ is a TP-CP map when κ is a TP-CP map from $\mathcal{H}_A$ to $\mathcal{H}_B$.*

As another condition for positive maps, we focus on the tensor product positivity. A positive map κ is called *tensor product positive* if $\kappa^{\otimes n}$ is positive for any integer n. It follows from the above corollary that any CP map is tensor product positive.

Proof of Corollary 5.3. The proof will be shown based on Condition ④ of Theorem 5.1. Condition ② is equivalent due to the fact that $K(\kappa \otimes \kappa')$ is positive semidefinite, which is equal to $K(\kappa) \otimes K(\kappa')$. Since $K(\kappa)$ is positive semidefinite, then $K(\kappa')$ is positive semidefinite. This is then equivalent to Condition ①. ∎

The fact that the dimension of the space of ρ is d'^2 is important in connection with quantum computation. One of the main issues in quantum computation theory is the classification of problems based on their computational complexity. One particularly important class is the class of problems that are solvable in polynomial time with respect to the input size. This class is called the polynomial class. The classification depends on whether operations are restricted to unitary time evolutions that use unitary gates such as C-NOT gates or if TP-CP maps are allowed. However, as confirmed by Theorem 5.1, TP-CP maps can be simulated by $d(d'^2)$-dimensional unitary evolutions. Therefore, it has been shown that the class of problems that can be solved in polynomial time is still the same [4].[1]

Remark 5.1 *The discussion presented here can be extended to the more general physical system, i.e., the case where the states are given as the duals of the general operator algebra, e.g., C^*-algebra, von Neumann algebra, and CCR algebra.*

Exercises

5.1. Show Corollary 5.1 using Theorem 5.1.

5.2. Let $\{F_i\}$ be a Choi–Kraus representation of the TP-CP map κ and $u_{i,j}$ be a unitary matrix. Show that $F'_i \stackrel{\text{def}}{=} \sum_j u_{i,j} F_j$ is also its Choi–Kraus representation.

[1] More precisely, we can implement only a finite number of unitary matrices in a finite amount of time. For a rigorous proof, we must approximate the respective TP-CP maps by a finite number of unitary matrices and evaluate the level of these approximations.

5.3. Show that we can choose a Choi–Kraus representation $\{F_i\}$ of any TP-CP map κ such that matrices F_i are linearly independent.

5.2 Examples of Trace-Preserving Completely Positive Maps

In addition to the above-mentioned partial trace, the following examples of TP-CP maps exist.

Example 5.1 (*Unitary evolutions*) Let U be a unitary matrix on $\mathcal{H}$. The state evolution $\kappa_U \colon \rho \mapsto \kappa_U(\rho) \stackrel{\text{def}}{=} U\rho U^*$ from $\mathcal{S}(\mathcal{H})$ to itself is a TP-CP map. This can be easily verified from Condition ⑥ in Theorem 5.1. Since an arbitrary unitary matrix U has the form $U = Ve^{i\theta}$, where $e^{i\theta}$ is a complex number with modulus 1 and V is a unitary matrix with determinant 1, we can write $U\rho U^* = V\rho V^*$. Therefore, we can restrict V to unitary matrices with determinant 1. Such matrices are called *special unitary matrices*. However, there are no such unitary state evolutions when the dimension of $\mathcal{H}_A$ is smaller than the dimension of $\mathcal{H}_B$. In such a case, we consider isometric matrices (i.e., matrices satisfying $U^*U = I$) from $\mathcal{H}_A$ to $\mathcal{H}_B$. The TP-CP map $\kappa_U(\rho) \stackrel{\text{def}}{=} U\rho U^*$ is then called an *isometric state evolution*.

Example 5.2 (*Partial trace*) The partial trace $\rho \mapsto \operatorname{Tr}_{\mathcal{H}} \rho$ can be regarded as a state evolution from quantum system $\mathcal{H} \otimes \mathcal{H}'$ to quantum system $\mathcal{H}'$. It is also a completely positive map because it is a special case of Condition ⑤ of Theorem 5.1.

Example 5.3 (*Depolarizing channels*) For arbitrary $1 \geq \lambda \geq 0$, a map

$$\kappa_{d,\lambda}(\rho) \stackrel{\text{def}}{=} \lambda\rho + (1-\lambda)(\operatorname{Tr}\rho)\rho_{\text{mix}} \tag{5.9}$$

is a d-dimensional TP-CP map and is called a depolarizing channel. In particular, when $d = 2$, we have

$$\kappa_{2,\lambda}(\rho) = \frac{3\lambda+1}{4}\rho + \frac{1-\lambda}{4}\sum_{i=1}^{3} S_i \rho S_i^* . \tag{5.10}$$

A depolarizing channel $\kappa_{d,\lambda}$ satisfies $\kappa_{d,\lambda}(U\rho U^*) = U\kappa_{d,\lambda}(\rho)U^*$ for all unitary matrices U. Conversely, when a d-dimensional channel satisfies this property, it is a depolarizing channel.

Example 5.4 (*Entanglement-breaking channels*) A TP-CP map from $\mathcal{H}_A$ to $\mathcal{H}_B$ satisfying the following conditions is called an entanglement-breaking channel. For an arbitrary reference system $\mathcal{H}_C$ and an arbitrary state $\rho \in \mathcal{S}(\mathcal{H}_A \otimes \mathcal{H}_C)$, the output state $(\kappa \otimes \iota_C)(\rho)$ on the space $\mathcal{H}_B \otimes \mathcal{H}_C$ is separable.

The entanglement-breaking channel is the subject of the following theorem Ex. 7.5.

Theorem 5.2 *(Horodecki et al. [235]) The following two conditions are equivalent for a TP-CP map κ from $\mathcal{H}_A$ to $\mathcal{H}_B$.*

① *κ is an entanglement-breaking channel.*
② *κ can be written as*

$$\kappa(\rho) = \kappa_{\boldsymbol{M},W}(\rho) \stackrel{\text{def}}{=} \sum_{\omega \in \Omega} (\operatorname{Tr} \rho M_\omega) W_\omega,$$

where $\boldsymbol{M} = \{M_\omega\}_{\omega \in \Omega}$ is an arbitrary POVM on $\mathcal{H}_A$ and W is a map from Ω to $\mathcal{S}(\mathcal{H}_B)$.

If the W_ω maps are mutually orthogonal pure states, then $\kappa_{\boldsymbol{M},W}(\rho)$ can be identified with the probability distribution $\mathrm{P}_\rho^{\boldsymbol{M}}$. A POVM $\boldsymbol{M}$ is not only a map that gives the probability distribution $\mathrm{P}_\rho^{\boldsymbol{M}}$ from the quantum state ρ, but it can also be regarded as an entanglement-breaking channel (and therefore a TP-CP map).

Example 5.5 (*Unital channels*) A TP-CP map κ from $\mathcal{H}_A$ to $\mathcal{H}_B$ satisfying $\kappa\left(\rho_{\text{mix}}^A\right) = \rho_{\text{mix}}^B$ is called a unital channel. The depolarizing channel defined previously is a unital channel.

Example 5.6 (*Pinching*) Recall that the pinching $\kappa_{\boldsymbol{M}}\colon \rho \mapsto \sum_{\omega \in \Omega} M_\omega \rho M_\omega$ is defined with respect to the PVM $\boldsymbol{M} = \{M_\omega\}_{\omega \in \Omega}$ in Sect. 1.2. This satisfies the conditions for a TP-CP map only when $\boldsymbol{M}$ is a PVM. For a general POVM $\boldsymbol{M}$, the map $\rho \mapsto \sum_{\omega \in \Omega} \sqrt{M_\omega} \rho \sqrt{M_\omega}$ is a TP-CP map.

If all the M_ω elements are one-dimensional, the pinching $\kappa_{\boldsymbol{M}}$ is an entanglement-breaking channel. If the POVM has a non-one-dimensional element M_ω, it is not an entanglement-breaking channel.

Example 5.7 (*Transpose*) For a quantum system $\mathcal{H}_A$, we define the transpose operator τ with respect to its orthonormal basis $u_0, \ldots, u_{d-1}$ as

$$\rho = \sum_{i,j} \rho_{i,j} |u_i\rangle\langle u_j| \mapsto \tau(\rho) \stackrel{\text{def}}{=} \rho^T = \sum_{i,j} \rho_{j,i} |u_i\rangle\langle u_j|. \tag{5.11}$$

τ is then a positive map, but not a two-positive map. Therefore, it is not a completely positive map. However, it is a tensor-product-positive map Ex. 5.11.

According to Exercise 1.3, any tensor product state $\rho^A \otimes \rho^B$ satisfies $(\tau_A \otimes \iota_B)(\rho^A \otimes \rho^B) = \tau_A(\rho^A) \otimes \rho^B \geq 0$. Hence, any separable state $\rho \in \mathcal{S}(\mathcal{H}_A \otimes \mathcal{H}_B)$ also satisfies $(\tau \otimes \iota_B)(\rho) \geq 0$. The converse is the subject of the following theorem.

Theorem 5.3 *(Horodecki [225]) Assign orthogonal bases for $\mathcal{H}_A$ and $\mathcal{H}_B$. Let τ be the transpose with respect to $\mathcal{H}_A$ under these coordinates. If either $\mathcal{H}_A$ or $\mathcal{H}_B$ is two-dimensional and the other is three-dimensional or less, then the condition $(\tau \otimes \iota_B)(\rho) \geq 0$ is the necessary and sufficient condition for the density matrix ρ on the composite system $\mathcal{H}_A \otimes \mathcal{H}_B$ to be separable.*

Counterexamples are available for $\mathbb{C}^2 \otimes \mathbb{C}^4$ and $\mathbb{C}^3 \otimes \mathbb{C}^3$ [236]. If the input and output systems of κ are both quantum two-level systems, we have the following corollary.

Corollary 5.4 *Let τ be a transpose under some set of coordinates. If κ is a channel for a quantum two-level system (i.e., a TP-CP map), the following two conditions are equivalent.*

1. *$\tau \circ \kappa$ is a CP map.*
2. *κ is an entanglement-breaking channel.*

Example 5.8 (*Generalized Pauli channel*) Define unitary matrices $\mathbf{X}_d$ and $\mathbf{Z}_d$ using the same basis as that in Example 5.7 for the quantum system $\mathcal{H}_A$ as follows:

$$\mathbf{X}_d|u_j\rangle = |u_{j-1 \bmod d}\rangle, \quad \mathbf{Z}_d|u_j\rangle = w^j|u_j\rangle, \tag{5.12}$$

where w is the dth root of 1. The generalized Pauli channel κ_p^{GP} is given by

$$\kappa_p^{\mathrm{GP}}(\rho) \stackrel{\mathrm{def}}{=} \sum_{i=0}^{d-1}\sum_{j=0}^{d-1} p(i,j)(\mathbf{X}_d^i\mathbf{Z}_d^j)^*\rho(\mathbf{X}_d^i\mathbf{Z}_d^j) \tag{5.13}$$

for the probability distribution p in $\{0,\dots,d-1\}^{\times 2}$. We often denote $\mathbf{X}_d$ and $\mathbf{Z}_d$ by $\mathbf{X}_A$ and $\mathbf{Z}_A$ respectively for indicating the space $\mathcal{H}_A$ that these act on. The above channel is also unital. For a quantum two-level system, we can write this channel as

$$\kappa(\rho) = \sum_{i=0}^{3} p_i S_i \rho S_i^*, \tag{5.14}$$

where p is a probability distribution in $\{0,1,2,3\}$, and the Pauli matrices S_i were defined in Sect. 1.3. This is called a *Pauli channel* and will be denoted by κ_p^{GP}.

Example 5.9 (*Antisymmetric channels*(*Werner–Holevo channels*), *Transpose depolarizing channels*) If and only if a real number λ belongs to $[\frac{-1}{d-1}, \frac{1}{d+1}]$, is the map

$$\kappa_{d,\lambda}^T(\rho) \stackrel{\mathrm{def}}{=} \lambda\rho^T + (1-\lambda)\rho_{\mathrm{mix}} \tag{5.15}$$

a TP-CP map, and it is called a transpose depolarizing channel, where d is the dimension of the system [90] (see Exercise 8.74 and Theorem 5.1). In particular, when $\lambda = \frac{-1}{d-1}$, the channel $\kappa^T_{d,\frac{-1}{d-1}}$ is called an antisymmetric channel [291] or a Werner–Holevo channel [413]. This channel satisfies the anticovariance $\kappa^T_{d,\lambda}(U\rho U^*) = \overline{U}\kappa^T_{d,\lambda}(\rho)U^T$.

Example 5.10 (*Phase-damping channels*) Let $D = (d_{i,j})$ be a positive semidefinite matrix satisfying $d_{i,i} = 1$. The following channel is called a phase-damping channel:

$$\kappa^{\mathrm{PD}}_D(\rho) \stackrel{\mathrm{def}}{=} \sum_{i,j} d_{i,j}\rho_{i,j}|e_i\rangle\langle e_j|, \tag{5.16}$$

where $\rho = \sum_{i,j} \rho_{i,j}|e_i\rangle\langle e_j|$.

For example, any pinching $\kappa_{\boldsymbol{M}}$ with a PVM $\boldsymbol{M}$ is a phase-damping channel. Since

$$\kappa^{\mathrm{PD}}_D(\frac{1}{d}\sum_{k,l}|e_k, e^R_k\rangle\langle e_l, e^R_l|) = \frac{1}{d}\sum_{k,l} d_{k,l}|e_k, e^R_k\rangle\langle e_l, e^R_l|,$$

Condition ④ of Theorem 5.1 guarantees that any phase-damping channel κ^{PD}_D is a TP-CP map.

Lemma 5.3 *A phase-damping channel κ^{PD}_D is a generalized Pauli channel κ^{GP}_p satisfying that the support of p belongs to the set $\{(0,0),\dots,(0,d-1)\}$ if and only if*

$$d_{i,j} = d_{i',0} \text{ for } i - j = i' \bmod d. \tag{5.17}$$

See Exercise 5.12.

Example 5.11 (*PNS channels*) Define the n-fold *symmetric space* $\mathcal{H}^n_{s,d}$ of $\mathbb{C}^d$ as the space spanned by $\{v^{\otimes n}|v \in \mathbb{C}^d\} \subset (\mathbb{C}^d)^{\otimes n}$. Let the input system be the n-fold symmetric space $\mathcal{H}^n_{s,d}$ and the output system be the m-fold symmetric space $\mathcal{H}^m_{s,d}$ $(n \ge m)$. The PNS (photon number splitting) channel $\kappa^{\mathrm{pns}}_{d,n\to m}$ is given by

$$\kappa^{\mathrm{pns}}_{d,n\to m}(\rho) \stackrel{\mathrm{def}}{=} \mathrm{Tr}_{(\mathbb{C}^d)^{\otimes n-m}}\,\rho, \tag{5.18}$$

where we regard ρ as a state on the n-fold tensor product space. In this case, the support of the output state is contained by the m-fold symmetric space $\mathcal{H}^m_{s,d}$. Hence, we can check that it is a TP-CP map from the n-fold symmetric space $\mathcal{H}^n_{s,d}$ to the m-fold symmetric space $\mathcal{H}^m_{s,d}$. Indeed, this channel corresponds to photon number splitting in the quantum key distribution.

Example 5.12 (*Erasure channels*) Let the input system be $\mathbb{C}^d$ with the basis $u_0,\dots,u_{d-1}$ and the input system be $\mathbb{C}^d$ with the basis $u_0,\dots,u_{d-1},u_d$. The erasure channel $\kappa^{\mathrm{era}}_{d,p}$ with the probability is given as

$$\kappa^{\mathrm{era}}_{d,p}(\rho) \stackrel{\mathrm{def}}{=} (1-p)\rho + p|u_d\rangle\langle u_d|. \tag{5.19}$$

Exercises

5.4. Show formula (5.10) for the depolarizing channel on the quantum two-level system.

5.5. Prove Theorem 4.5 from Condition ⑤ of Theorem 5.1.

5.6. Show that $(\mathbf{X}_d^{j_1}\mathbf{Z}_d^{k_1})(\mathbf{X}_d^{j_2}\mathbf{Z}_d^{k_2}) = (-1)^{j_1 k_2 - k_1 j_2}(\mathbf{X}_d^{j_2}\mathbf{Z}_d^{k_2})(\mathbf{X}_d^{j_1}\mathbf{Z}_d^{k_1})$ for symbols defined as in Example 5.8.

5.7. Show that

$$\frac{1}{d^2}\sum_{j=0}^{d-1}\sum_{k=0}^{d-1}(\mathbf{X}_d^j\mathbf{Z}_d^k)^* X(\mathbf{X}_d^j\mathbf{Z}_d^k) = (\operatorname{Tr} X)\rho_{\text{mix}} \tag{5.20}$$

for an arbitrary matrix X by following the steps below.

a Show that

$$(\mathbf{X}_d^{j'}\mathbf{Z}_d^{k'})^*\left(\frac{1}{d^2}\sum_{j=0}^{d-1}\sum_{k=0}^{d-1}(\mathbf{X}_d^j\mathbf{Z}_d^k)^* X(\mathbf{X}_d^j\mathbf{Z}_d^k)\right)(\mathbf{X}_d^{j'}\mathbf{Z}_d^{k'})$$
$$=\frac{1}{d^2}\sum_{j=0}^{d-1}\sum_{k=0}^{d-1}(\mathbf{X}_d^j\mathbf{Z}_d^k)^* X(\mathbf{X}_d^j\mathbf{Z}_d^k)$$

for arbitrary j', k'.

b Show that the matrix $A = \sum_{j,k} a^{j,k}|u_j\rangle\langle u_k|$ is diagonal if $\mathbf{Z}_d A = A\mathbf{Z}_d$. Show that its diagonal elements are all identical if $\mathbf{X}_d A = A\mathbf{X}_d$.

c Show (5.20) using the above.

5.8. Show that

$$\frac{1}{d^2}\sum_{j=0}^{d-1}\sum_{k=0}^{d-1}(\mathbf{X}_A^j\mathbf{Z}_A^k \otimes I_B)^*\rho(\mathbf{X}_A^j\mathbf{Z}_A^k \otimes I_B) = \rho_{\text{mix}}^A \otimes \operatorname{Tr}_A \rho$$

for a state ρ on $\mathcal{H}_A \otimes \mathcal{H}_B$ using formula (1.25).

5.9. Let $\mathcal{H}_A, \mathcal{H}_B$ be the spaces spanned by $u_0^A, \ldots, u_{d-1}^A$ and $u_0^B, \ldots, u_{d-1}^B$. Define

$$u_{0,0}^{A,B} \stackrel{\text{def}}{=} \frac{1}{\sqrt{d}}\sum_{i=1}^{d} u_i^A \otimes u_i^B, \quad u_{i,j}^{A,B} \stackrel{\text{def}}{=} (\mathbf{X}_A^i\mathbf{Z}_A^j \otimes I_B)u_{0,0}^{A,B},$$

and show that these vectors form a CONS of $\mathcal{H}_A \otimes \mathcal{H}_B$.

5.10. Suppose that the classical-quantum channel W_ρ is given by a depolarizing channel $\kappa_{d,\lambda}$ according to $W_\rho \stackrel{\text{def}}{=} \kappa_{d,\lambda}(\rho)$. In this case, the set of states $\mathcal{S}(\mathcal{H}_A)$ of the input system is regarded as the set of input alphabets $\mathcal{X}$. Show that the depolarizing channel $\kappa_{d,\lambda}$ is pseudoclassical and its capacity is given by

$$\begin{aligned} C &= c(\kappa_{d,\lambda}) = \tilde{C} = c(\kappa_{d,\lambda}) \\ &= \frac{1+(d-1)\lambda}{d}\log(1+(d-1)\lambda) + \frac{(d-1)(1-\lambda)}{d}\log(1-\lambda). \end{aligned}$$

5.11. Show that the transpose τ is tensor product positive.

5.12. Show Lemma 5.3 following the steps below.

a Show that if a generalized Pauli channel κ_p^{GP} satisfies that the support of p belongs to the set $\{(0,0),\ldots,(0,d-1)\}$, κ_p^{GP} has the form (5.16) and satisfies (5.17).

b Assume that a phase-damping channel κ_D^{PD} satisfies (5.17). Define $p(0,m) \stackrel{\text{def}}{=} \frac{1}{d}\operatorname{Tr} D\mathbf{Z}_d^m$. Show that $\kappa_p^{\mathrm{GP}} = \kappa_D^{\mathrm{PD}}$.

5.13. Show that $(\kappa_{d,p}^{\mathrm{era}})^E = \kappa_{d,1-p}^{\mathrm{era}}$.

5.14. Show that $(\kappa_{d,n\to m}^{\mathrm{pns}})^E = \kappa_{d,n\to n-m}^{\mathrm{pns}}$.

5.3 State Evolutions in Quantum Two-Level Systems

As mentioned in Sect. 1.3, the states in a quantum two-level system may be parameterized by a three-dimensional vector $\boldsymbol{x}$:

$$\rho_{\boldsymbol{x}} = \frac{1}{2}\left(S_0 + \sum_{i=1}^{3} x^i S_i\right). \tag{5.21}$$

Let κ be an arbitrary TP-CP map from a quantum two-level system to another quantum two-level system. We shall now investigate how this map κ can be characterized under the parameterization (5.21). As discussed in Sect. 4.1, this kind of map is characterized by a linear map from a Hermitian matrix on $\mathbb{C}^2$ to itself. Consider a state evolution of the unitary type given in Example 5.1. The special unitary matrix V may then be diagonalized by a unitary matrix. The two eigenvalues are complex numbers with an absolute value 1; their product yields 1. Therefore, we may represent the two eigenvalues by $e^{i\theta}$ and $e^{-i\theta}$ and the two eigenvectors by u_1 and u_2. We can then write $V = e^{i\theta}|u_1\rangle\langle u_1| + e^{-i\theta}|u_2\rangle\langle u_2| = \exp(i(\theta|u_1\rangle\langle u_1| - \theta|u_2\rangle\langle u_2|))$. V may therefore be written asexp(iX), where X is a Hermitian matrix with trace 0. We will use this to examine the state evolution corresponding to a special unitary matrix operating on both sides of the density matrix of a quantum two-level system.

For this purpose, let us examine some algebraic properties of the Pauli matrices. Define $\epsilon_{j,k,l}$ to be 0 if any of j, k, and l are the same, $\epsilon_{1,2,3} = \epsilon_{3,1,2} = \epsilon_{2,3,1} = 1$, and $\epsilon_{3,2,1} = \epsilon_{1,3,2} = \epsilon_{2,1,3} = -1$. Then, $\left[S^j, S^k\right] = -2i\sum_{l=1}^3 \epsilon_{j,l,k} S^l$. This is equivalent to

$$\left[\sum_{j=1}^3 x_j S^j, \sum_{k=1}^3 y_k S^k\right] = -2i \sum_{j=1}^3 \sum_{k=1}^3 \sum_{l=1}^3 x_j y_k \epsilon_{j,l,k} S^l.$$

Defining $R_j \stackrel{\text{def}}{=} [\epsilon_{j,l,k}]_{l,k}$, $S_{\boldsymbol{x}} \stackrel{\text{def}}{=} \sum_{j=1}^3 x_j S^j$, and $R_{\boldsymbol{x}} \stackrel{\text{def}}{=} \sum_{j=1}^3 x_j R^j$, we may rewrite the above expression as

$$\left[\frac{i}{2} S_{\boldsymbol{x}}, S_{\boldsymbol{y}}\right] = S_{R_{\boldsymbol{x}}\boldsymbol{y}}.$$

As shown later, this equation implies that

$$\exp(\frac{i}{2} S_{\boldsymbol{x}}) S_{\boldsymbol{y}} \exp(-\frac{i}{2} S_{\boldsymbol{x}}) = S_{\exp(R_{\boldsymbol{x}})\boldsymbol{y}}. \tag{5.22}$$

Applying this equation to states, we obtain

$$\exp(\frac{i}{2} S_{\boldsymbol{x}}) \rho_{\boldsymbol{y}} \exp(-\frac{i}{2} S_{\boldsymbol{x}}) = \rho_{\exp(R_{\boldsymbol{x}})\boldsymbol{y}}. \tag{5.23}$$

This shows that a 2×2 unitary matrix $\exp(\frac{i}{2} S_{\boldsymbol{x}})$ of determinant 1 corresponds to a 3×3 real orthogonal matrix $\exp(R_{\boldsymbol{x}})$.

Proof of (5.22). Since $S_{\boldsymbol{x}}$ is Hermitian, $isS_{\boldsymbol{x}}$ can be diagonalized by a unitary matrix with purely imaginary eigenvalues. Therefore, $\exp(iS_{\boldsymbol{x}})$ is a unitary matrix. Note that $\exp(iS_{\boldsymbol{x}})^* = \exp(-iS_{\boldsymbol{x}})$. Since $\exp(i\frac{s}{2} S_{\boldsymbol{x}}) S_{\boldsymbol{y}} \exp(-i\frac{s}{2} S_{\boldsymbol{x}})$ is a Hermitian matrix with trace 0 like $S_{\boldsymbol{y}}$, it can be rewritten as $S_{\boldsymbol{y}(s)}$ according to Exercise 1.9. Let us write down the vector $\boldsymbol{y}(s)$. Differentiating $\exp(i\frac{s}{2} S_{\boldsymbol{x}}) S_{\boldsymbol{y}} \exp(-i\frac{s}{2} S_{\boldsymbol{x}})$ with respect to s, we obtain

$$\begin{aligned}
S_{\boldsymbol{y}'(s)} &= \left(\exp(i\frac{s}{2} S_{\boldsymbol{x}}) S_{\boldsymbol{y}} \exp(-i\frac{s}{2} S_{\boldsymbol{x}})\right)' \\
&= \left(\exp(i\frac{s}{2} S_{\boldsymbol{x}})\right)' S_{\boldsymbol{y}} \exp(-i\frac{s}{2} S_{\boldsymbol{x}}) + \exp(i\frac{s}{2} S_{\boldsymbol{x}}) S_{\boldsymbol{y}} \left(\exp(-i\frac{s}{2} S_{\boldsymbol{x}})\right)' \\
&= \frac{i}{2} S_{\boldsymbol{x}} \exp(i\frac{s}{2} S_{\boldsymbol{x}}) S_{\boldsymbol{y}} \exp(-i\frac{s}{2} S_{\boldsymbol{x}}) + \exp(i\frac{s}{2} S_{\boldsymbol{x}}) S_{\boldsymbol{y}} \exp(-i\frac{s}{2} S_{\boldsymbol{x}})(-\frac{i}{2} S_{\boldsymbol{x}}) \\
&= \left[\frac{i}{2} S_{\boldsymbol{x}}, \exp(i\frac{s}{2} S_{\boldsymbol{x}}) S_{\boldsymbol{y}} \exp(-i\frac{s}{2} S_{\boldsymbol{x}})\right] = \left[\frac{i}{2} S_{\boldsymbol{x}}, S_{\boldsymbol{y}(s)}\right] = S_{R_{\boldsymbol{x}}\boldsymbol{y}(s)},
\end{aligned}$$

and we find that $\boldsymbol{y}(t)$ satisfies the differential equation

$$y'(s) = R_x y(s). \tag{5.24}$$

It can be verified that

$$y(s) = \exp(sR_x)y \tag{5.25}$$

satisfies this differential equation. From the uniqueness of the solution of an ordinary differential equation we see that only the function $y(s)$ given in (5.25) satisfies $y(0) = y$ and (5.24). Applying this to when $s = 1$, we obtain (5.22). ∎

Next, we consider a general TP-CP map. Define $\tilde{K}(\kappa)^{i,j} \stackrel{\text{def}}{=} \frac{1}{2} \operatorname{Tr} S_i \kappa(S_j)$. The trace-preserving property guarantees that

$$\tilde{K}(\kappa)^{0,0} = \frac{1}{2} \operatorname{Tr} \kappa(I) = 1, \quad \tilde{K}(\kappa)^{0,i} = \frac{1}{2} \operatorname{Tr} \kappa(S_i) = 0$$

for $i \neq 0$. Now define t according to $t^i \stackrel{\text{def}}{=} \tilde{K}(\kappa)^{i,0}$. The vector t then gives the image of the completely mixed state ρ_{mix} due to κ. Let T be a 3×3 matrix consisting only of the first, second, and third elements of $\tilde{K}(\kappa)$. The TP-CP map κ may then be denoted by a vector t and a matrix T [130]. For example, when a unitary matrix operates on either side, then $t = 0$ and T is an orthogonal matrix given by (5.23). To give a few more examples, let us rewrite the examples given in the previous section using t and T for the quantum two-level system. For a depolarizing channel, we have $t = 0$, and therefore $T = \lambda I$. The necessary and sufficient condition for the channel to be unital is then $t = 0$. For the transpose, we have

$$t = 0, \quad T = \begin{pmatrix} 1 & 0 & 0 \\ 0 & -1 & 0 \\ 0 & 0 & 1 \end{pmatrix}.$$

Next, we consider the necessary and sufficient conditions for a map to be a positive map, a completely positive map, an entanglement-breaking channel, and a Pauli channel, respectively, by assuming that the channel is unital, i.e., $t = 0$. Recall from the discussion in Sect. A.2 that special orthogonal matrices O_1, O_2 may be chosen such that $T' \stackrel{\text{def}}{=} O_1^* T O_2$ is diagonal (i.e., a singular decomposition). Taking unitary matrices U_1, U_2 corresponding to O_1, O_2 based on the correspondence (5.23), and letting κ^T be the TP-CP map corresponding to T, we have $\kappa_{U_1} \circ \kappa^T \circ \kappa_{U_2} = \kappa^{T'}$. For the analysis of the TP-CP map κ^T, it is sufficient to analyze the TP-CP map $\kappa^{T'}$. The necessary and sufficient conditions for the various types of channels can then be labeled by the eigenvalues λ_1, λ_2, and λ_3 of T'. Let us first consider the necessary conditions for a positive map. It is positive if and only if the image by T' of the unit sphere $\{x \| \|x\| \leq 1\}$ is contained by the unit sphere. Thus, its necessary and sufficient condition is

$$|\lambda_1|, |\lambda_2|, |\lambda_3| \leq 1. \tag{5.26}$$

Next, we use Condition ④ of Theorem 5.1 to examine the completely positive map. In order to check this condition, we calculate $K(\kappa^{T'})$:

$$K(\kappa^{T'}) = \frac{1}{2} \begin{pmatrix} 1+\lambda_3 & 0 & 0 & \lambda_1+\lambda_2 \\ 0 & 1-\lambda_3 & \lambda_1-\lambda_2 & 0 \\ 0 & \lambda_1-\lambda_2 & 1-\lambda_3 & 0 \\ \lambda_1+\lambda_2 & 0 & 0 & 1+\lambda_3 \end{pmatrix}.$$

Swapping the second and fourth coordinates, we have

$$\begin{pmatrix} 1+\lambda_3 & \lambda_1+\lambda_2 & 0 & 0 \\ \lambda_1+\lambda_2 & 1+\lambda_3 & 0 & 0 \\ 0 & 0 & 1-\lambda_3 & \lambda_1-\lambda_2 \\ 0 & 0 & \lambda_1-\lambda_2 & 1-\lambda_3 \end{pmatrix}.$$

Thus, the necessary and sufficient condition for $K(\kappa^{T'}) \geq 0$ is [130, 355]

$$(1+\lambda_3)^2 \geq (\lambda_1+\lambda_2)^2, \quad (1-\lambda_3)^2 \geq (\lambda_1-\lambda_2)^2.$$

This can be rewritten as

$$1 \geq \lambda_1+\lambda_2-\lambda_3, \lambda_1-\lambda_2+\lambda_3, -\lambda_1+\lambda_2+\lambda_3 \tag{5.27}$$

from Condition (5.26). Applying Corollary 5.4 to the unital case, we obtain the necessary and sufficient condition for a channel to be an entanglement-breaking channel [356]:

$$1 \geq |\lambda_1| + |\lambda_2| + |\lambda_3|.$$

Next, we treat the necessary and sufficient condition for a channel to be a Pauli channel. Since this condition depends on the coordinates of the input and output systems, we cannot reduce T to its singular decomposition. The orthogonal matrices corresponding to S_1, S_2, S_3 in the sense of (5.23) are

$$\begin{pmatrix} 1 & 0 & 0 \\ 0 & -1 & 0 \\ 0 & 0 & -1 \end{pmatrix}, \quad \begin{pmatrix} -1 & 0 & 0 \\ 0 & 1 & 0 \\ 0 & 0 & -1 \end{pmatrix}, \quad \begin{pmatrix} -1 & 0 & 0 \\ 0 & -1 & 0 \\ 0 & 0 & 1 \end{pmatrix}.$$

Using p_i in (5.14), the matrix T is given by

$$\begin{pmatrix} p_0+p_1-p_2-p_3 & 0 & 0 \\ 0 & p_0-p_1+p_2-p_3 & 0 \\ 0 & 0 & p_0-p_1-p_2+p_3 \end{pmatrix}. \tag{5.28}$$

Finally, the following theorem holds regarding the pseudoclassical property of channels examined in Sect. 4.7.

Theorem 5.4 *(Fujiwara and Nagaoka [128]) Let $\mathcal{H}$ be two-dimensional, $\mathcal{X}$ be given by $\mathcal{S}(\mathbb{C}^2)$, and W be given by the trace-preserving positive map κ from $\mathbb{C}^2$ to $\mathbb{C}^2$. A necessary and sufficient condition for a channel to be pseudoclassical is that one of the conditions given below should be satisfied.*

1. $\boldsymbol{t} = 0$.
2. *Let* $\boldsymbol{t}$ *be an eigenvector of* TT^*. *Let* r *be one of its eigenvalues and* r_0 *be the larger of the other two eigenvalues. Then,*

$$r_0 \le r^2 - \|\boldsymbol{t}\|r + \frac{(\|\boldsymbol{t}\| - r)\left(h\left(\frac{1+\|\boldsymbol{t}\|+r}{2}\right) - h\left(\frac{1+\|\boldsymbol{t}\|-r}{2}\right)\right)}{h'\left(\frac{1+\|\boldsymbol{t}\|-r}{2}\right)}.$$

Exercises

5.15. Check Condition (5.27) in the Pauli channel case (5.28).

5.16. Show that the Pauli channel given by (5.28) is entanglement-breaking if and only if $p_i \le \frac{1}{2}$.

5.17. Show that the positive map $\mathrm{Inv}_\lambda : \begin{pmatrix} a & b \\ c & d \end{pmatrix} \mapsto \lambda \begin{pmatrix} d & -b \\ -c & a \end{pmatrix} + (1 - \lambda)\begin{pmatrix} a & b \\ c & d \end{pmatrix}$ is completely positive if and only if $\frac{2}{3} \ge \lambda \ge 0$.

5.18. Show that

$$F(\rho_{\boldsymbol{x}}, \rho_{\boldsymbol{y}}) = \frac{1 + \sqrt{1 - |\boldsymbol{x}|^2}\sqrt{1 - |\boldsymbol{y}|^2} + \langle \boldsymbol{x}, \boldsymbol{y} \rangle}{2}. \tag{5.29}$$

5.4 Information-Processing Inequalities in Quantum Systems

In this section, we will show that the quantum versions of the information quantities introduced in Sect. 2.2 satisfy the information-processing inequalities (i.e., the monotonicity) under the state evolutions given previously.

Theorem 5.5 *(Lindblad [270], Uhlmann [393]) Let* κ *be a TP-CP map from* $\mathcal{H}_A$ *to* $\mathcal{H}_B$. *Then, the monotonicity of the quantum relative entropy*

$$D(\rho\|\sigma) \ge D(\kappa(\rho)\|\kappa(\sigma)) \tag{5.30}$$

holds.

This theorem may be used to show many properties of the quantum relative entropy and the von Neumann entropy. For example, let $\rho_1, \ldots, \rho_k$ and $\sigma_1, \ldots, \sigma_k$ be density matrices on $\mathcal{H}$ and let p_i be a probability distribution in $\{1, \ldots, k\}$. Consider now the density matrix

$$R \stackrel{\text{def}}{=} \begin{pmatrix} p_1\rho_1 & & O \\ & \ddots & \\ O & & p_k\rho_k \end{pmatrix}, \quad S \stackrel{\text{def}}{=} \begin{pmatrix} p_1\sigma_1 & & O \\ & \ddots & \\ O & & p_k\sigma_k \end{pmatrix}$$

on $\mathcal{H}_1 \otimes \mathbb{C}^k$. Since the partial trace $\mathrm{Tr}_{\mathbb{C}^k}$ is a TP-CP map, the inequality

$$D\left(\sum_{i=1}^{k} p_i \rho_i \middle\| \sum_{i=1}^{k} p_i \sigma_i\right) \leq D(R\|S) = \sum_{i=1}^{k} p_i D(\rho_i\|\sigma_i) \tag{5.31}$$

holds [271]. This inequality is called the *joint convexity* of the quantum relative entropy.

Proof of Theorem 5.5. Examine the connection with hypothesis testing. Let κ be a TP-CP map from $\mathcal{H}_A$ to $\mathcal{H}_B$. If a Hermitian matrix T on $\mathcal{H}_B^{\otimes n}$ satisfies $I \geq T \geq 0$, then $(\kappa^{\otimes n})^*(T)$ must also satisfy $I \geq (\kappa^{\otimes n})^*(T) \geq 0$. Therefore, from Condition ② of Theorem 5.1 and Corollary 5.3, we deduce that $(\kappa^{\otimes n})^*(T) \geq 0$. On the other hand, we see that $I \geq (\kappa^{\otimes n})^*(T)$ from

$$I - (\kappa^{\otimes n})^*(T) = (\kappa^{\otimes n})^*(I) - (\kappa^{\otimes n})^*(T) = (\kappa^{\otimes n})^*(I - T) \geq 0.$$

Since a state $\rho \in \mathcal{S}(\mathcal{H}_A)$ satisfies

$$\mathrm{Tr}(\kappa(\rho))^{\otimes n} T = \mathrm{Tr}\, \rho^{\otimes n} (\kappa^{\otimes n})^*(T),$$

the test $(\kappa^{\otimes n})^*(T)$ with the hypotheses $\rho^{\otimes n}$ and $\sigma^{\otimes n}$ has the same accuracy as the test T with the hypotheses $\kappa(\rho)^{\otimes n}$ and $\kappa(\sigma)^{\otimes n}$. That is, any test with the hypotheses $\kappa(\rho)^{\otimes n}$ and $\kappa(\sigma)^{\otimes n}$ can be simulated by a test with the hypotheses $\rho^{\otimes n}$ and $\sigma^{\otimes n}$ with the same performance. We therefore have

$$B(\rho\|\sigma) \geq B(\kappa(\rho)\|\kappa(\sigma)).$$

Note that $B(\rho\|\sigma)$ is defined in Theorem 3.4. Hence, applying Theorem 3.4 then completes the proof. ∎

Indeed, this proof requires only the tensor product positivity. Hence, since the transpose τ is tensor product positive, inequality (5.30) holds when κ is the transpose τ. Uhlmann [393] showed this inequality only with the two-positivity. Further, the equality condition of (5.30) can be characterized as follows.

Theorem 5.6 *(Petz [349]) Assume that $\kappa(\sigma) > 0$ for a state σ and a trace-preserving two-positive map κ. Then, the equality of (5.30) holds for a state ρ if and only if*

$$\rho = \sqrt{\sigma}\kappa^*\left(\sqrt{\kappa(\sigma)}^{-1}\kappa(\rho)\sqrt{\kappa(\sigma)}^{-1}\right)\sqrt{\sigma}. \tag{5.32}$$

Indeed, there exist many quantum versions of relative entropy. A quantity $\tilde{D}(\rho\|\sigma)$ can be regarded as a quantum version of relative entropy if any commutative states ρ and σ satisfy

$$\tilde{D}(\rho\|\sigma) = D(p\|\bar{p}), \tag{5.33}$$

where p and $\bar{p}$ are the probability distribution consisting of the eigenvalues of ρ and σ. If a relative entropy $\tilde{D}(\rho\|\sigma)$ satisfies the monotonicity for a measurement $\boldsymbol{M}$

$$\tilde{D}(\rho\|\sigma) \geq D(\mathrm{P}_{\rho}^{\boldsymbol{M}}\|\mathrm{P}_{\sigma}^{\boldsymbol{M}}) \tag{5.34}$$

and the additivity

$$\tilde{D}(\rho_1 \otimes \rho_2\|\sigma_1 \otimes \sigma_2) = \tilde{D}(\rho_1\|\sigma_1) + \tilde{D}(\rho_2\|\sigma_2), \tag{5.35}$$

then the relation

$$\tilde{D}(\rho\|\sigma) \geq D(\rho\|\sigma) \tag{5.36}$$

holds. That is, the quantum relative entropy $D(\rho\|\sigma)$ is the minimum quantum analog of relative entropy with the monotonicity for measurement and the additivity. This inequality can be easily checked by Corollary 3.1. Note that Condition (5.34) is a weaker requirement than the monotonicity for TP-CP map (5.30).

As will be shown in Corollary 8.4 of Sect. 8.2, the Bures distance $b(\rho,\sigma)$ also satisfies the *monotonicity* [29, 244, 392]

$$b(\rho,\sigma) \geq b(\kappa(\rho),\kappa(\sigma)) \tag{5.37}$$

with respect to an arbitrary TP-CP map κ. This inequality may be derived from Corollary 8.4 given later. From (5.37) we may also show its *joint convexity*

$$b^2\left(\sum_{i=1}^{k} p_i\rho_i, \sum_{i=1}^{k} p_i\sigma_i\right) \leq b^2(R,S) = \sum_{i=1}^{k} p_i b^2(\rho_i,\sigma_i) \tag{5.38}$$

in a similar way to (5.31). The variational distance $d_1(\rho,\sigma)$ also satisfies the *monotonicity*

$$d_1(\rho,\sigma) \geq d_1(\kappa(\rho),\kappa(\sigma)) \tag{5.39}$$

for an arbitrary TP-CP map κ [354] Ex. 5.22. Furthermore, as extensions of (2.62) and (2.63), the monotonicities of the relative Rényi entropy

$$\phi(s|\rho\|\sigma) \leq \phi(s|\kappa(\rho)\|\kappa(\sigma)) \quad \text{for } 0 \leq s \leq 1 \tag{5.40}$$

$$\phi(s|\rho\|\sigma) \geq \phi(s|\kappa(\rho)\|\kappa(\sigma)) \quad \text{for } -1 \leq s \leq 0. \tag{5.41}$$

hold. These relations will be proved in Appendix A.4 by using matrix convex or concave functions. Inequality (5.41) does not necessarily hold when $s \leq -1$, as is shown in Appendix A.4.

Considering the entanglement-breaking channel given in Example 5.4, we can replace the probability distributions $\mathrm{P}_{\rho}^{\boldsymbol{M}}$ and $\mathrm{P}_{\sigma}^{\boldsymbol{M}}$ by the output states

$\kappa(\rho)$ and $\kappa(\sigma)$. In this case, the inequalities similar to (2.59)–(2.62) hold, and the inequality corresponding to (2.63) holds if the condition $s \le 0$ is replaced by the condition $-1 \le s \le 0$.

Further, similarly to inequalities (2.18) and (2.19), the inequalities

$$d_1(\rho,\sigma) \ge b^2(\rho,\sigma) \ge \frac{1}{2} d_1^2(\rho,\sigma) \tag{5.42}$$

$$D(\rho\|\sigma) \ge -2\log \operatorname{Tr}|\sqrt{\rho}\sqrt{\sigma}| \ge 2b^2(\rho,\sigma) \tag{5.43}$$

hold [Ex. 5.19,5.20]. From these inequalities we can see that the convergence of $d_1(\rho_n,\sigma_n)$ to 0 is equivalent to the convergence of $b(\rho_n,\sigma_n)$ to 0. In order to express the difference between the two states ρ and σ, we sometimes focus on the quantity $1-F^2(\rho,\sigma)$, which is related to the Bures distance and the variational distance in the following way [Ex. 5.21,8.2]:

$$2b^2(\rho,\sigma) \ge 1-F^2(\rho,\sigma) \ge b^2(\rho,\sigma) \tag{5.44}$$

$$1-F^2(\rho,\sigma) \ge d_1^2(\rho,\sigma). \tag{5.45}$$

Exercises

5.19. Show that $b^2(\rho,\sigma) \ge \frac{1}{2} d_1^2(\rho,\sigma)$ and (5.43).

5.20. Show that $d_1(\rho,\sigma) \ge b^2(\rho,\sigma)$ by choosing a POVM $\boldsymbol{M}$ satisfying the equality in (2.60).

5.21. Show (5.44) by writing $x = F(\rho,\sigma) = 1 - b^2(\rho,\sigma)$.

5.22. Show (5.39) using (3.3).

5.23. Show the *quantum Pinsker inequality*:

$$D(\rho\|\sigma) \ge 2d_1^2(\rho,\sigma) \tag{5.46}$$

following the steps below. Note that this is a stronger requirement between $D(\rho\|\sigma)$ and $d_1(\rho,\sigma)$ than the combinaiton of (5.42) and (5.43).

a Show that binary relative entropy $h(x,y) \stackrel{\text{def}}{=} x\log\frac{x}{y} + (1-x)\log\frac{1-x}{1-y}$ satisfies $2(y-x)^2 \le h(x,y)$ for $0 \le x \le y \le 1$.
b Show that $2(\operatorname{Tr}\sigma P - \operatorname{Tr}\rho P) = \operatorname{Tr}|\sigma-\rho| = \operatorname{Tr}(\sigma-\rho)(P-(I-P)) \ge 0$ for $P = \{\sigma-\rho > 0\}$ or $\{\sigma-\rho \ge 0\}$.
c Show (5.46).

5.24. Using Exercise 3.18, show

$$\bar{\phi}(s|\rho\|\sigma) \ge \bar{\phi}(s|\kappa(\rho)\|\kappa(\sigma)) \text{ for } s \le 0. \tag{5.47}$$

5.25. Derive inequality (5.41) for $s \le -1$ assuming $\phi(s|\rho\|\sigma) = \bar{\phi}(s|\rho\|\sigma)$ for $s \le -1$. Further, show that the equality in inequality (3.42): $\phi(s|\rho\|\sigma) \ge \bar{\phi}(s|\rho\|\sigma)$ does not necessarily hold for $s \le -1$, by contradiction.

5.26. Show the monotonicity of transmission information

$$I(p, W) \geq I(p, \kappa(W)) \tag{5.48}$$

for any TP-CP map κ and any c-q channel: $W = (W_x)$, where $\kappa(W) = (\kappa(W_x))$.

5.5 Entropy Inequalities in Quantum Systems

In this section, we will derive various inequalities related to the von Neumann entropy from the properties of the quantum relative entropy.

Substituting $\sigma = \rho_{\mathrm{mix}}$ into the joint convexity of the quantum relative entropy (5.31), we obtain the *concavity* of the von Neumann entropy as follows:

$$H\left(\sum_{i=1}^{k} p_i \rho_i\right) \geq \sum_{i=1}^{k} p_i H(\rho_i). \tag{5.49}$$

Further, as shown in Sect. 8.4, when a state $\rho^{A,B}$ on $\mathcal{H}_A \otimes W_B$ is separable, the von Neumann entropy satisfies

$$H(\rho^{A,B}) \geq H(\rho^A), H(\rho^B). \tag{5.50}$$

We apply this inequality to the separable state R defined in Sect. 5.4. Since the von Neumann entropy of R is equal to $\sum_{i=1}^{k} p_i H(\rho_i) + H(p)$, we obtain the reverse inequality of (5.49):

$$H\left(\sum_{i=1}^{k} p_i \rho_i\right) \leq \sum_{i=1}^{k} p_i H(\rho_i) + H(p) \leq \sum_{i=1}^{k} p_i H(\rho_i) + \log k. \tag{5.51}$$

In particular, if the supports for the densities ρ_i are disjoint, the first inequality satisfies the equality.

Similar types of inequalities may also be obtained by examining the pinching $\kappa_{\boldsymbol{M}}$ of the PVM $\boldsymbol{M}$. The quantum relative entropy satisfies

$$H(\kappa_{\boldsymbol{M}}(\rho)) - H(\rho) = D(\rho \| \kappa_{\boldsymbol{M}}(\rho)) \geq 0. \tag{5.52}$$

Since the inequality

$$D(\rho \| \kappa_{\boldsymbol{M}}(\rho)) \leq \log |\boldsymbol{M}| \tag{5.53}$$

holds [206] Ex. 5.29, we obtain

$$H(\rho) \leq H(\kappa_{\boldsymbol{M}}(\rho)) \leq H(\rho) + \log |\boldsymbol{M}|. \tag{5.54}$$

Let $\rho^A, \rho^B, \rho^{A,B}$, and $\rho^{A,C}$ be the reduced density matrices of the density matrix $\rho = \rho^{A,B,C}$ on $\mathcal{H}_A \otimes \mathcal{H}_B \otimes \mathcal{H}_C$. From the monotonicity of the quantum relative entropy, we obtain

$$D(\rho^{A,B,C} \| \rho^{A,C} \otimes \rho^B) \geq D(\rho^{A,B} \| \rho^A \otimes \rho^B).$$

Rewriting this inequality, we may derive the following theorem called the *strong subadditivity* of the von Neumann entropy.

Theorem 5.7 *(Lieb and Ruskai [268, 269]) The inequality*

$$H(\rho^{A,B,C}) + H(\rho^A) \leq H(\rho^{A,B}) + H(\rho^{A,C}) \tag{5.55}$$

holds.

Further, the equality condition of (5.55) is given as follows.

Theorem 5.8 *(Hayden et al. [199]) The equality in (5.55) holds if and only if there is a decomposition of the system $\mathcal{H}_A$ as*

$$\mathcal{H}_A = \bigoplus_j \mathcal{H}_{A-B,j} \otimes \mathcal{H}_{A-C,j} \tag{5.56}$$

into a direct (orthogonal) sum of tensor products such that

$$\rho^{ABC} = \bigoplus_j q_j \rho_j^{AB} \otimes \rho_j^{AC} \tag{5.57}$$

with states ρ_j^{AB} on $\mathcal{H}_B \otimes \mathcal{H}_{A-B,j}$ and ρ_j^{AC} on $\mathcal{H}_C \otimes \mathcal{H}_{A-C,j}$, and probability distribution q_j.

In particular, when $\mathcal{H}_A$ is one-dimensional,

$$H(\rho^{B,C}) \leq H(\rho^B) + H(\rho^C), \tag{5.58}$$

which is called the *subadditivity*. Let us change the notation slightly and write $H(\rho^{A,B})$ as $H_\rho(A,B)$ in order to emphasize the quantum system rather than the quantum state. The strong subadditivity is then written as

$$H_\rho(A,B,C) + H_\rho(A) \leq H_\rho(A,B) + H_\rho(A,C). \tag{5.59}$$

Now, using this notation, let us define the *conditional entropy* $H_\rho(A|B) \stackrel{\text{def}}{=} H_\rho(A,B) - H_\rho(B)$ using this notation. This quantity satisfies the following *concavity*:

$$H_\rho(A|B) \geq \sum_{i=1}^{k} p_i H_{\rho_i}(A|B), \tag{5.60}$$

where $\rho = \sum_i p_i \rho_i$.

Similarly to Sect. 2.1.1, we can define the quantum mutual information $I_\rho(A:B)$ and the quantum conditional mutual information $I_\rho(A:B|C)$ as

$$I_\rho(A:B) \stackrel{\text{def}}{=} H_\rho(A) + H_\rho(B) - H_\rho(AB) \tag{5.61}$$

$$I_\rho(A:B|C) \stackrel{\text{def}}{=} H_\rho(AC) + H_\rho(BC) - H_\rho(ABC) - H_\rho(C). \tag{5.62}$$

The positivity of quantum mutual information is equivalent to the subadditivity, and that of quantum conditional mutual information is equivalent to the strong subadditivity.

For the continuity of the above quantities, the *Fannes inequality*, defined below, is particularly useful.

Theorem 5.9 *(Fannes [111]) Define*

$$\eta_0(x) \stackrel{\text{def}}{=} \begin{cases} \eta(x) & 0 \le x \le 1/e \\ 1/e & 1/e < x, \end{cases} \tag{5.63}$$

where $\eta(x) \stackrel{\text{def}}{=} -x \log x$. *Then, for two states* ρ *and* σ *on* $\mathcal{H}$ ($\dim \mathcal{H} = d$), *the inequality*

$$|H(\rho) - H(\sigma)| \le \epsilon \log d + \eta_0(\epsilon) \tag{5.64}$$

holds for $\epsilon \stackrel{\text{def}}{=} \|\rho - \sigma\|_1$.

Let us consider the following lemma before proving this theorem.

Lemma 5.4 *Write the eigenvalues of the Hermitian matrices* A *and* B *in decreasing order (largest first) including any degeneracies, i.e.,* $a_1, \dots, a_d$, $b_1, \dots, b_d$. *Then,* $\|A - B\|_1 \ge \sum_{i=1}^d |a_i - b_i|$.

Proof. Let $P \stackrel{\text{def}}{=} \{A - B \ge 0\}$, $X \stackrel{\text{def}}{=} P(A-B)$, and $Y \stackrel{\text{def}}{=} -(I-P)(A-B)$. Then, $X \ge 0, Y \ge 0$, and $A - B = X - Y$. Let $C \stackrel{\text{def}}{=} A + Y = B + X$. Then, $C \ge A, B$. Now let c_i be the eigenvalues of C arranged in decreasing order. From Exercise A.12 we know that $c_i \ge a_i, b_i$. Therefore, if $a_i - b_i \ge 0$, then $2c_i - a_i - b_i - (a_i - b_i) = 2(c_i - a_i) \ge 0$, and we obtain $2c_i - a_i - b_i \ge |a_i - b_i|$. This also holds for $a_i - b_i \le 0$, and therefore

$$\sum_i |a_i - b_i| \le \sum_i (2c_i - a_i - b_i) = \mathrm{Tr}(2C - A - B) = \mathrm{Tr}(X + Y) = \mathrm{Tr}\,|A - B|.$$

∎

Proof of Theorem 5.9. We only provide a proof for $d_1(\rho, \sigma) \le 1/e$. See Exercise 5.36 for the reverse inequality. Let a_i, b_i be the eigenvalues of ρ, σ placed in decreasing order. Define $\epsilon_i \stackrel{\text{def}}{=} |a_i - b_i|$. Then, according to Lemma 5.4 and the assumptions of the theorem, $\epsilon_i \le 1/e \le 1/2$. From Exercise 5.35 we obtain

$$|H(\rho) - H(\sigma)| \le \sum_{i=1}^{d} |\eta(a_i) - \eta(b_i)| \le \sum_{i=1}^{d} \eta(\epsilon_i).$$

Next, define $\epsilon \stackrel{\text{def}}{=} \sum_{i=1}^d \epsilon_i$. We find that $\sum_{i=1}^d \eta(\epsilon_i) = \epsilon \sum_{i=1}^d \eta\left(\frac{\epsilon_i}{\epsilon}\right) + \eta(\epsilon)$. Since $\sum_{i=1}^d \eta\left(\frac{\epsilon_i}{\epsilon}\right)$ represents the entropy of the probability distribution $(\epsilon_1/\epsilon, \epsilon_2/\epsilon, \ldots, \epsilon_d/\epsilon)$, we see that this must be less than $\log d$. From Exercise 5.35 **b** and $\|\rho - \sigma\|_1 \le 1/e$ we obtain $\eta(\epsilon) \le \eta(\|\rho - \sigma\|_1)$. Equation (5.64) can then be obtained by combining the above results. ■

Exercises

5.27. Show (5.55) using the monotonicity of the relative entropy.

5.28. Show (4.3) from the concavity of von Neumann entropy.

5.29. Show (5.53) following the steps below.

a Show (5.53) holds for a pure state.
b Show (5.53) for the general case using the joint convexity of the quantum relative entropy.

5.30. Show (5.60) using (5.59).

5.31. Show the Araki–Lieb inequality [12,266] below using the subadditivity and the state purification introduced in Sect. 8.1

$$H(\rho^{A,B}) \ge |H(\rho^A) - H(\rho^B)|. \tag{5.65}$$

5.32. Show that the strong subadditivity is equivalent to the following inequality:

$$H_\rho(AB|C) \le H_\rho(A|C) + H_\rho(B|C). \tag{5.66}$$

5.33. Show the following inequality using the strong subadditivity (5.59):

$$H_{|u\rangle\langle u|}(A, C) + H_{|u\rangle\langle u|}(A, D) \ge H_{|u\rangle\langle u|}(A) + H_{|u\rangle\langle u|}(B). \tag{5.67}$$

5.34. Using (5.65), show that

$$|H_\rho(A|B)| \le \log d_A. \tag{5.68}$$

5.35. Show that

$$|\eta(x) - \eta(y)| \le \eta(|x - y|) \tag{5.69}$$

if x and y satisfy $|x - y| \le 1/2$ following the steps below.

a Show that $\eta(x + \epsilon) - \eta(x) \le \eta(\epsilon)$ for $x \ge 0$ and $\epsilon \ge 0$.

b Show that $\eta(x)$ is strictly concave and has its maximum value when $x = 1/e$.
c Show that $\eta(\alpha - \epsilon) - \eta(\alpha) \le \eta(1-\epsilon) - \eta(1)$ for $\epsilon < \alpha < 1$.
d Show that the function $\eta(x) - \eta(1-x)$ is strictly concave and $\eta(x) - \eta(1-x) > 0$ for $0 < x < 1/2$.
e Show that $\eta(x) - \eta(x+\epsilon) \le \eta(\epsilon)$ using **c** and **d**, and hence show (5.69).

5.36. Prove Theorem 5.9 for $d_1(\rho,\sigma) > 1/e$ following the steps below.

a Show (5.64) if $\epsilon_1 \le 1/2$, i.e., all the ϵ_i are less than 1/2.
b Show that

$$|H(\rho) - H(\sigma)| \le 1/e + \epsilon' \log d + \eta_0(\epsilon'),$$

where $\epsilon' \stackrel{\text{def}}{=} \sum_{i=2}^{d} \epsilon_i$ and if $\epsilon_1 > 1/2$.
c Show that $\epsilon_1 \log d - (\epsilon' \log d + 1/e) \ge 0$ and hence show (5.64).

5.37. Show that $I(p, W) \le \delta \log d + \eta_0(\delta)$ using Theorem 5.9, where $\delta \stackrel{\text{def}}{=} \sum_x p(x) \|W_x - W_p\|_1$.

5.38. Let ρ and $\tilde{\rho}$ be two arbitrary states. For any real $0 \le \epsilon \le 1$, show that

$$|H_\rho(A|B) - H_\gamma(A|B)| \le 2\epsilon \log d_A + h(\epsilon), \tag{5.70}$$

following the steps, where $\gamma \stackrel{\text{def}}{=} (1-\epsilon)\rho + \epsilon\tilde{\rho}$ [8].

a Using (5.60) and (5.68), show that $H_\rho(A|B) - H_\gamma(A|B) \le \epsilon(H_\rho(A|B) - H_{\tilde{\rho}}(A|B)) \le \epsilon d_A$.
b Using (5.49), show that $H(\gamma_B) \le (1-\epsilon)H(\rho^B) + \epsilon H(\tilde{\rho}_B)$.
c Using the first inequality of (5.51), show that $H(\gamma) \le (1-\epsilon)H(\rho) + \epsilon H(\tilde{\rho}) + h(\epsilon)$.
d Using (5.68), show that $H_\rho(A|B) - H_\gamma(A|B) \ge \epsilon(H_\rho(A|B) - H_{\tilde{\rho}}(A|B)) - h(\epsilon) \ge -2\epsilon \log d_A - h(\epsilon)$.

5.39. Show that

$$|H_\rho(A|B) - H_\sigma(A|B)| \le 4\epsilon \log d_A + 2h(\epsilon) \tag{5.71}$$

for states ρ and σ on $\mathcal{H}_A \otimes \mathcal{H}_B$ and $\epsilon \stackrel{\text{def}}{=} \|\rho - \sigma\|_1$ following the steps below [8].

a Define the states $\tilde{\rho} \stackrel{\text{def}}{=} \frac{1}{\epsilon}|\rho - \sigma|$, $\tilde{\sigma} \stackrel{\text{def}}{=} \frac{1-\epsilon}{\epsilon}(\rho - \sigma) + \frac{1}{\epsilon}|\rho - \sigma|$, and $\gamma \stackrel{\text{def}}{=} (1-\epsilon)\rho + \epsilon\tilde{\rho}$, where $\epsilon \stackrel{\text{def}}{=} \|\rho - \sigma\|_1$. Show that $\gamma = (1-\epsilon)\sigma + \epsilon\tilde{\sigma}$.
b Using (5.70), show that $|H_\rho(A|B) - H_\sigma(A|B)| \le |H_\rho(A|B) - H_\gamma(A|B)| + |H_\sigma(A|B) - H_\gamma(A|B)| \le 4\epsilon \log d_A + 2h(\epsilon)$.

5.40. Using the above inequality, show that

$$|I_\rho(A:B) - I_\sigma(A:B)| \le 5\epsilon \log d_A + \eta_0(\epsilon) + 2h(\epsilon) \tag{5.72}$$

for states ρ and σ on $\mathcal{H}_A \otimes \mathcal{H}_B$ and $\epsilon \stackrel{\text{def}}{=} \|\rho - \sigma\|_1$.

5.41. Show that

$$|I_\rho(A:B|C) - I_\sigma(A:B|C)| \le 8\epsilon \log d_A d_B + 6h(\epsilon) \tag{5.73}$$

for states ρ and σ on $\mathcal{H}_A \otimes \mathcal{H}_B$ and $\epsilon \stackrel{\text{def}}{=} \|\rho - \sigma\|_1$ following the steps below [78].

a Show that $|I_\rho(A:B|C) - I_\sigma(A:B|C)| \le |H_\rho(A|C) - H_\sigma(A|C)| + |H_\rho(B|C) - H_\sigma(B|C)| + |H_\rho(AB|C) - H_\sigma(AB|C)|$.

b Show (5.73) using (5.71).

5.42. Show the *chain rules* of quantum mutual information and quantum conditional mutual information:

$$H_\rho(AB|C) = H_\rho(B|C) + H_\rho(A|BC) \tag{5.74}$$
$$I_\rho(A:BC) = I_\rho(A:C) + I_\rho(A:B|C) \tag{5.75}$$
$$I_\rho(A:BC|D) = I_\rho(A:C|D) + I_\rho(A:B|CD). \tag{5.76}$$

5.43. Show that the monotonicity $I_\rho(A:B) \ge I_{\kappa_A \otimes \kappa_B \rho}(A:B)$ for local TP-CP maps κ_A and κ_B.

5.44. Show the Hiai–Petz theorem [206] for two arbitrary states ρ, σ

$$\lim \frac{1}{n} D(\kappa_{\sigma^{\otimes n}}(\rho^{\otimes n}) \| \sigma^{\otimes n}) = D(\rho\|\sigma),$$

where $\kappa_{\sigma^{\otimes n}}$ represents the pinching of the measurement corresponding to the spectral decomposition of $\sigma^{\otimes n}$. Hence, the equality in (2.59) holds in an asymptotic sense when the POVM is the simultaneous spectral decomposition of $\kappa_{\sigma^{\otimes n}}(\rho^{\otimes n})$ and $\sigma^{\otimes n}$. Combining this result with the classical Stein's lemma gives an alternate proof of Lemma 3.4.

5.45. Show Holevo's inequality $I(M, p, W) \le I(p, W)$ using the strong subadditivity of von Neumann entropy (5.55).

5.46. Given densities ρ_i^A and ρ_i^B on $\mathcal{H}_A$ and $\mathcal{H}_B$, show the strong concavity of von Neumann entropy:

$$H(\sum_i p_i \rho_i^A \otimes \rho_i^B) \ge H(\sum_i p_i \rho_i^A) + \sum_i p_i H(\rho_i^B s) \tag{5.77}$$

from the joint convexity of quantum relative entropy (5.38) for states $\rho_i^A \otimes \rho_i^B$ and $\rho_{\text{mix}}^A \otimes \rho_i^B$ [108].

5.47. Show that

$$H_{(\iota_A \otimes \kappa)(\rho)}(A|B) \ge H_\rho(A|B) \tag{5.78}$$

for any TP-CP map κ on $\mathcal{H}_B$ from the strong subadditivity (5.55).

5.6 Historical Note

A completely positive map was initially introduced in the mathematical context; Stinespring [379] gave its representation theorem in the observable form, i.e., $\kappa^*(A) = PU_\kappa^*(A \otimes I)U_\kappa P$, where P is the projection from the extended space to the original space. Holevo [212] proposed that any state evolution in the quantum system could be described by a completely positive map based on the same reason considered in this text. After this, Lindblad [270] translated Stinespring's representation theorem to the state form. Then, he clarified that any state evolution by a completely positive map could be regarded as the interaction between the target system and the environment system.

Concerning other parts of Theorem 5.1, Jamiołkowski [243] showed the one-to-one correspondence between a CP map κ and a positive matrix $K(\kappa)$, firstly. After this study, Choi [77] obtained this correspondence. He also obtained the characterization ⑥ concerning CP maps. Kraus [261] also obtained this characterization. Choi [77] also characterize the extremal points as Lemma 5.2.

In this book, we proved the monotonicity of the quantum relative entropy based on the quantum Stein's lemma. Using this property, we derived many inequalities in Sects. 5.4 and 5.5. However, historically, these were proved by completely different approaches. First, Lieb and Ruskai [268, 269] proved the strong subadditivity of the von Neumann entropy (5.55) based on Lieb's convex trace functions [267]. Using these functions, they derived the monotonicity of the quantum relative entropy only concerning the partial trace. During that period, Lindblad [271] proved the joint convexity of the quantum relative entropy (5.31). After this result, using the Stinespring's representation theorem in the state form, Lindblad [270] proved the monotonicity of the quantum relative entropy (5.30) from that concerning the partial trace. Later, Uhlmann [393] invented the interpolation theory, and proved the monotonicity of the quantum relative entropy based on this approach. As an extension of the quantum relative entropy, Petz [343] defined a quantum f-divergence for a matrix convex function f. Then, he proved the monotonicity in a more general framework. His approach can be applied to that of the relative Rényi entropy (5.40) and (5.41). As is explained in Appendix A.4, it requires knowledge of the matrix convex or concave functions, which is advanced knowledge.

In this book, we prove the monotonicities of the relative Rényi entropy regarding measurements (2.62) and (2.63) using only elementary knowledge. Moreover, the monotonicity (2.63) holds with a larger parameter $s \leq 0$ as compare with monotonicity (5.41). This improvement enhances in the strong converse exponents in both quantum hypothesis testing and classical-quantum channel coding.

6

Quantum Information Geometry and Quantum Estimation

Summary. In Chap. 3 we examined the discrimination of two unknown quantum states. This chapter will consider the estimation of a parameter θ, which labels an unknown state parameterized by a continuous variable θ. It is a remarkable property of quantum mechanics that a measurement inevitably leads to the state demolition. Therefore, when one performs a measurement for state estimation, it is necessary to choose the measurement that extracts as much information as possible. This problem is called quantum estimation, and the optimization of the measurement is an important topic in quantum information theory.

In the classical theory of estimation (of probability distributions) discussed in Sect. 2.3, we saw that the estimation is intimately related to geometrical structures such as the inner product. We can expect that such geometrical structures will also play an important role in the quantum case. The study of geometrical structures in the space of quantum states is called quantum information geometry and is an important field in quantum information theory. This chapter will examine the geometrical structure of quantum systems and discuss its applications to estimation theory.

Table 6.1. Denotations used in Chap. 6

$E_{\rho,s}(X)$	See (6.1)
$E_{\rho,b}(X)$	See (6.2)
$E_{\rho,r}(X)$	See (6.3)
$E_{\rho,p}(X)$	See (6.4)
$E_{\rho,\lambda}(X)$	See (6.5)
$\langle Y, X\rangle_{\rho,x}^{(e)}$	See (6.7)
$\|X\|_{\rho,x}^{(e)}$	See (6.8)
$\langle A, B\rangle_{\rho,x}^{(m)}$	See (6.12)
$\|A\|_{\rho,x}^{(m)}$	See (6.13)
$\kappa_{\rho,x}$	See (6.15)
$J_{\theta,s}$	SLD Fisher metric
$J_{\theta,b}$	Bogoljubov Fisher metric
$J_{\theta,r}$	RLD Fisher metric

Table 6.1. Continued

$L_{\theta,s}$	SLD e representation
$L_{\theta,b}$	Bogoljubov e representation
$L_{\theta,r}$	RLD e representation
$\Pi^{\theta}_{L,s}\rho_0$	SLD e parallel transport
$\Pi^{\theta}_{L,b}\rho_0$	Bogoljubov e parallel transport
$\Pi^{\theta}_{L,r}\rho_0$	RLD e parallel transport
$D^{(e)}_x(\rho\\|\sigma)$	x-e-divergence (6.39)
$D^{(m)}_x(\rho\\|\sigma)$	x-m-divergence (6.49)
$\boldsymbol{J}_{\theta,s}$	SLD Fisher matrix
$\boldsymbol{J}_{\theta,r}$	RLD Fisher matrix
$\mathbf{Re}\, X$	Real part of matrix X
$\mathbf{Im}\, X$	Imaginary part of matrix X
$\boldsymbol{V}_\theta(\mathbf{X})$	Matrix with elements $(\operatorname{Tr} \rho_\theta X^i X^j)$

Error criterion	
$\hat{\mathrm{V}}_\theta(\boldsymbol{M}^n, \hat{\theta}_n)$	Mean square error (MSE) of estimator $(\boldsymbol{M}^n, \hat{\theta}_n)$ (6.57)
$\hat{\boldsymbol{V}}_\theta(\boldsymbol{M}^n, \hat{\theta}_n)$	Mean square error matrix (6.80)
$\hat{\boldsymbol{V}}_\theta(\{\boldsymbol{M}^n, \hat{\theta}_n\})$	Matrix of elements $\hat{\mathrm{V}}^{i,j}_\theta(\{\boldsymbol{M}^n, \hat{\theta}_n\}) \stackrel{\mathrm{def}}{=} \lim n \hat{\mathrm{V}}^{i,j}_\theta(\boldsymbol{M}^n, \hat{\theta}_n)$
$\beta(\{(\boldsymbol{M}^n, \hat{\theta}_n)\}, \theta, \epsilon)$	Rate function of error probability (6.71)
$\alpha(\{(\boldsymbol{M}^n, \hat{\theta}_n)\}, \theta)$	First-order coefficient of rate fucntion (6.72)

6.1 Inner Products in Quantum Systems

In any discussion about the geometry of quantum states, the metric plays a central role. To start talking about the metric, we must first discuss the quantum versions of the Fisher information and its associated inner product (2.67) examined in Sect. 2.3. Let A, B, p in (2.67) be the diagonal elements of the commuting Hermitian matrices Y, X, ρ, respectively. The inner product (2.67) is then equal to $\operatorname{Tr} Y(X\rho)$. Although the trace of a product of two matrices does not depend on the order of the multiplication, the trace of the product for three or more matrices is dependent on the order. If these matrices do not commute, then the inner product depends on the order of the product $X\rho$. There exist at least three possible ways of defining the product corresponding to $X\rho$:

$$E_{\rho,s}(X) \stackrel{\mathrm{def}}{=} X \circ \rho \stackrel{\mathrm{def}}{=} \frac{1}{2}(\rho X + X\rho), \tag{6.1}$$

$$E_{\rho,b}(X) \stackrel{\mathrm{def}}{=} \int_0^1 \rho^\lambda X \rho^{1-\lambda}\, d\lambda, \tag{6.2}$$

$$E_{\rho,r}(X) \stackrel{\mathrm{def}}{=} \rho X. \tag{6.3}$$

Here, X is not necessarily Hermitian. These extensions are unified in the general form [348]

$$E_{\rho,p}(X) \stackrel{\mathrm{def}}{=} \int_0^1 E_{\rho,\lambda}(X)p(d\lambda), \tag{6.4}$$

$$E_{\rho,\lambda}(X) \stackrel{\mathrm{def}}{=} \rho^{\lambda} X \rho^{1-\lambda}, \tag{6.5}$$

where p is an arbitrary probability distribution on $[0,1]$. The case (6.1) corresponds to the case (6.4) with $p(1) = p(0) = 1/2$, and the case (6.3) corresponds to the case (6.4) with $p(1) = 1$. In particular, the map $E_{\rho,x}$ is called *symmetric* for $x = s, b, r, \lambda, p$ when $E_{\rho,x}(X)$ is Hermitian if and only if X is Hermitian. Hence, when the distribution p is symmetric, i.e., $p(\lambda) = p(1-\lambda)$, the map $E_{\rho,p}$ is symmetric. When $\rho > 0$, these maps possess inverses. Moreover, the maps $E_{\rho,x}$ satisfy

$$E_{\rho,x}(U^* X U) = U^* E_{U\rho U^*,x}(X) U \tag{6.6}$$

$$E_{\rho,x}(I) = \rho, \quad E_{\rho\otimes\rho',x}(X \otimes X') = E_{\rho,x}(X) \otimes E_{\rho',x}(X'), \quad \mathrm{Tr}\, E_{\rho,x}(X) = \mathrm{Tr}\, \rho X$$

for $x = s, b, r, \lambda, p$, and in particular,

$$E_{\rho\otimes\rho',x}(X \otimes I) = E_{\rho,x}(X) \otimes \rho'.$$

Accordingly, we may define the following types of inner products:

$$\langle Y, X \rangle^{(e)}_{\rho,x} \stackrel{\mathrm{def}}{=} \mathrm{Tr}\, Y^* E_{\rho,x}(X) \quad x = s, b, r, \lambda, p. \tag{6.7}$$

If X, Y, ρ all commute, then these coincide with definition (2.67). These are called the SLD, Bogoljubov,[1] RLD, λ, and p inner products [11, 202, 203, 216, 309, 345, 347, 348], respectively (reasons for this will be given in the next section). These inner products are positive semidefinite and Hermitian, i.e.,

$$\left(\|X\|^{(e)}_{\rho,x}\right)^2 \stackrel{\mathrm{def}}{=} \langle X, X \rangle^{(e)}_{\rho,x} \geq 0, \quad \langle Y, X \rangle^{(e)}_{\rho,x} = (\langle X, Y \rangle^{(e)}_{\rho,x})^*. \tag{6.8}$$

From property (6.6) we have

$$\langle X \otimes X', Y \otimes Y' \rangle^{(e)}_{\rho\otimes\rho',x} = \langle X, Y \rangle^{(e)}_{\rho,x} \langle X', Y' \rangle^{(e)}_{\rho',x}, \tag{6.9}$$

$$\langle U^* X U, U^* Y U \rangle^{(e)}_{\rho,x} = \langle X, Y \rangle^{(e)}_{U\rho U^*,x}, \quad \|I\|^{(e)}_{\rho,x} = 1.$$

In particular, the SLD inner product and the RLD inner product satisfy

$$\|X \otimes I_{\mathcal{H}'}\|^{(e)}_{\rho,x} = \|X\|^{(e)}_{\mathrm{Tr}_{\mathcal{H}'}\rho,x}, \quad x = s, r. \tag{6.10}$$

[1] The Bogoljubov inner product is also called the canonical correlation in statistical mechanics. In linear response theory, it is often used to give an approximate correlation between two different physical quantities.

Generally, as is shown in Appendix A.4, we have

$$\|X \otimes I_{\mathcal{H}'}\|^{(e)}_{\rho,x} \le \|X\|^{(e)}_{\mathrm{Tr}_{\mathcal{H}'}\rho,x}, \quad x = b, \lambda, p. \tag{6.11}$$

A dual inner product may be defined as

$$\langle A, B\rangle^{(m)}_{\rho,x} \stackrel{\text{def}}{=} \mathrm{Tr}(E^{-1}_{\rho,x}(A))^* B \tag{6.12}$$

with respect to the correspondence $A = E_{\rho,x}(X)$. Denote the norm of these inner products as

$$\left(\|A\|^{(m)}_{\rho,x}\right)^2 \stackrel{\text{def}}{=} \langle A, A\rangle^{(m)}_{\rho,x}. \tag{6.13}$$

Hence, the inner product $\langle A, B\rangle^{(m)}_{\rho,x}$ is positive semidefinite and Hermitian. In particular, this inner product is called *symmetric* when $\langle A, B\rangle^{(m)}_{\rho,x} = \langle B, A\rangle^{(m)}_{\rho,x}$. This is equivalent to not only the symmetry of the dual inner product $\langle X, Y\rangle^{(e)}_{\rho,x}$, but also the symmetry of map $E_{\rho,x}$. When the inner product $\langle A, B\rangle^{(m)}_{\rho,x}$ is symmetric, it can be symmetrized as $\langle A, B\rangle^{(m)}_{\rho,s(x)} \stackrel{\text{def}}{=} \frac{1}{2}\left(\langle A, B\rangle^{(m)}_{\rho,x} + \langle B, A\rangle^{(m)}_{\rho,x}\right)$, i.e., the symmetrized map $E_{\rho,s(x)}$ is defined as $E^{-1}_{\rho,s(x)}(A) = \frac{1}{2}\left(E^{-1}_{\rho,x}(A) + (E^{-1}_{\rho,x}(A))^*\right)$ for any Hermitian matrix A. Hence, we call the inner product $\langle A, B\rangle^{(m)}_{\rho,s(x)}$ the *symmetrized* inner product of $\langle A, B\rangle^{(m)}_{\rho,x}$. Note that the SLD inner product is not the symmetrized inner product of the RLD inner product.

Similarly to (6.6), we have

$$\langle X, Y\rangle^{(m)}_{\rho,x} = \langle UXU^*, UYU^*\rangle^{(m)}_{U\rho U^*,x} \tag{6.14}$$

for an arbitrary unitary matrix U. For an arbitrary TP-CP map κ, we also define the map $\kappa_{\rho,x}$ associated with κ and ρ by the relation

$$\kappa(E_{\rho,x}(X)) = E_{\kappa(\rho),x}(\kappa_{\rho,x}(X)), \quad x = s, b, r, \lambda, p, \tag{6.15}$$

where for a non-Hermitian matrix A, $\kappa(A)$ is defined as $\kappa(A) \stackrel{\text{def}}{=} \kappa((A + A^*)/2) - i\kappa(i(A - A^*)/2)$. This map satisfies the associativity

$$(\kappa_1 \circ \kappa_2)_{\rho,x}(X) = \kappa_{1\,\kappa_2(\rho),x} \circ \kappa_{2\,\rho,x}(X), \quad x = s, b, r, \lambda, p. \tag{6.16}$$

Then, we have the following theorem.

Theorem 6.1 *The inequality*

$$\|A\|^{(m)}_{\rho,x} \ge \|\kappa(A)\|^{(m)}_{\kappa(\rho),x}, \quad x = s, b, r, \lambda, p \tag{6.17}$$

holds. This inequality (6.17) is also equivalent to

$$\|X\|^{(e)}_{\rho,x} \ge \|\kappa_{\rho,x}(X)\|^{(e)}_{\kappa(\rho),x}, \quad x = s, b, r, \lambda, p. \tag{6.18}$$

When an inner product satisfies property (6.17), it is called a *monotone metric*. Monotonicity implies that the amount of information does not increase due to any operation. That is, if the inner product is to be considered as a measure of information, this property should be satisfied because information processing does not cause any increase in the amount of information. It is also known that an arbitrary inner product $\|A\|^{(m)}_{\rho,x}$ satisfying property (6.17) as well as $\|\rho^{-1}\|^{(m)}_{\rho,x} = 1$ satisfies $\|A\|^{(m)}_{\rho,s} \le \|A\|^{(m)}_{\rho,x} \le \|A\|^{(m)}_{\rho,r}$, i.e., the SLD inner product is the minimum product and the RLD inner product is the maximum product [345].

Proof. Here, we prove (6.18) for $x = s, r$. The general case of (6.18) is shown assuming inequality (6.11); this will be proven in Appendix A.4.

These inner products are invariant for the operations $\rho \mapsto \rho \otimes \rho_0$ and $\rho \mapsto U\rho U^*$. It is sufficient to show (6.18) in the case of partial trace because of the Stinespring representation and associativity (6.16). First, using property (6.10), we prove (6.18) for $x = s, r$. Letting κ be the partial trace from system $\mathcal{H} \otimes \mathcal{H}'$ to subsystem $\mathcal{H}'$, we have

$$\begin{aligned}\langle Y \otimes I, \kappa_{\rho,x}(X) \otimes I \rangle^{(e)}_{\rho,x} &= \langle Y, \kappa_{\rho,x}(X) \rangle_{\mathrm{Tr}_{\mathcal{H}'}\,\rho,x} = \mathrm{Tr}\, Y^* \kappa(E_{\rho,x}(X)) \\ &= \mathrm{Tr}(Y \otimes I)^* E_{\rho,x}(X) = \langle Y \otimes I, X \rangle^{(e)}_{\rho,x}\end{aligned}$$

for any matrix X on $\mathcal{H} \otimes \mathcal{H}'$, any matrix Y on $\mathcal{H}$, and any state ρ on $\mathcal{H} \otimes \mathcal{H}'$. Hence, the map $\kappa_{\rho,x}$ is the projection from the space of all matrices on $\mathcal{H} \otimes \mathcal{H}'$ to the subspace of matrices $\{Y \otimes I\}$ with respect to the inner product $\langle\ ,\ \rangle^{(e)}_{\rho,x}$. Therefore, $\|X\|^{(e)}_{\rho,x} \ge \|\kappa_{\rho,x}(X)\|^{(e)}_{\kappa(\rho),x}$. Hence, we obtain (6.18) for $x = s, r$.

Next, we proceed to the general case, i.e., the case of $x = p, b$. Let $\mathcal{F}$ be the positive self adjoint map on the matrix space with respect to the inner product $\langle\ ,\ \rangle^{(e)}_{\mathrm{Tr}_{\mathcal{H}'}\,\rho,x}$ satisfying

$$\langle Y \otimes I, Y' \otimes I \rangle^{(e)}_{\rho,x} = \langle Y, \mathcal{F} Y' \rangle^{(e)}_{\mathrm{Tr}_{\mathcal{H}'}\,\rho,x}. \tag{6.19}$$

Since property (6.11) implies $\|Y \otimes I\|^{(e)}_{\rho,x} = \|\mathcal{F}^{1/2} Y\|^{(e)}_{\mathrm{Tr}_{\mathcal{H}'}\,\rho,x} \le \|Y\|^{(e)}_{\mathrm{Tr}_{\mathcal{H}'}\,\rho,x}$, we have

$$\|(\mathcal{F}^{-1} Y) \otimes I\|^{(e)}_{\rho,x} = \|\mathcal{F}^{-1/2} Y\|^{(e)}_{\mathrm{Tr}_{\mathcal{H}'}\,\rho,x} \ge \|Y\|^{(e)}_{\mathrm{Tr}_{\mathcal{H}'}\,\rho,x}.$$

Hence,

$$\begin{aligned}\langle Y \otimes I, (\mathcal{F}^{-1}\kappa_{\rho,x}(X)) \otimes I \rangle^{(e)}_{\rho,x} &= \langle Y, \kappa_{\rho,x}(X) \rangle_{\mathrm{Tr}_{\mathcal{H}'}\,\rho,x} = \mathrm{Tr}\, Y^* \kappa(E_{\rho,x}(X)) \\ &= \mathrm{Tr}(Y \otimes I)^* E_{\rho,x}(X) = \langle Y \otimes I, X \rangle^{(e)}_{\rho,x}.\end{aligned}$$

Similarly, we can show that $\|(\mathcal{F}^{-1}\kappa_{\rho,x}(X)) \otimes I\|^{(e)}_{\rho,x} \le \|X\|^{(e)}_{\rho,x}$. Therefore, we obtain

$$\|\kappa_{\rho,x}(X)\|^{(e)}_{\mathrm{Tr}_{\mathcal{H}'}\,\rho,x} \le \|X\|^{(e)}_{\rho,x}.$$

∎

From the discussion in the above proof, we can regard the map $\kappa_{\rho,x}$ as the conditional expectation when κ is the partial trace for $x = s, b$.

Remark 6.1 *If $\rho > 0$ does not hold, the situation is slightly more subtle. When $\rho > 0$, then maps $E_{\rho,x}$ always possess their inverse maps, and the fact $\|X\|^{(e)}_{\rho,x} = 0$ is equivalent to the fact $X = 0$. This is not true in general if $\rho > 0$ is not satisfied, such as in the case when ρ is a pure state. Therefore, although the inner product $\langle X, Y\rangle^{(e)}_{\rho,x}$ may be defined for all matrices, one must note that $\|X\|^{(e)}_{\rho,x}$ can be 0 in cases other than $X = 0$. Hence, in this case, the inner product $\langle A, B\rangle^{(m)}_{\rho,x}$ is then defined only for matrices in the range of $E_{\rho,x}$.*

Exercises

6.1. Prove the following facts for a traceless Hermitian matrix A and a density matrix ρ of the form $\rho = \sum_{j=1}^{d} \lambda_j E_j$ and rank $E_j = 1$.

a Show that $(x+y)/2 \geq \mathrm{Lm}(x, y)$, where $\mathrm{Lm}(x, y)$ is the logarithmic average defined below.

$$\mathrm{Lm}(x, y) \stackrel{\mathrm{def}}{=} \begin{cases} \frac{x-y}{\log x - \log y} & \text{if } x \neq y, \\ x & \text{if } x = y. \end{cases}$$

Also show that the equality holds if and only if $x = y$.

b Show the following [345]:

$$\|A\|^{(m)}_{\rho,s} = \sum_{j,k=1}^{d} \frac{2}{\lambda_j + \lambda_k} \operatorname{Tr} AE_j AE_k,$$

$$\|A\|^{(m)}_{\rho,b} = \sum_{j,k=1}^{d} \frac{1}{\mathrm{Lm}(\lambda_j, \lambda_k)} \operatorname{Tr} AE_j AE_k.$$

c Show that $\operatorname{Tr}(A\rho - \rho A)(A\rho - \rho A) = \sum_{j,k=1}^{d} (\lambda_j - \lambda_k)^2 \operatorname{Tr} AE_j AE_k$.

d Show the inequality $\|A\|^{(m)}_{\rho,b} \geq \|A\|^{(m)}_{\rho,s}$. Also, show the equivalence of the following.

① $\|A\|^{(m)}_{\rho,b} = \|A\|^{(m)}_{\rho,s}$.

② $[\rho, A] = 0$.

6.2. Show the following facts when κ is a pinching of a PVM $\boldsymbol{M}$, i.e., $\kappa_{\rho,s} = \kappa_{\boldsymbol{M},\rho,s} (= (\kappa_{\boldsymbol{M}})_{\rho,s})$.

a For any matrix X, show that $\kappa_{\boldsymbol{M},\rho,s}(X)$ commutes with every element M_i.

b Show that if every M_i commutes with X, then $\kappa_{\boldsymbol{M},\rho,s}(X) = X$. Show that if $\rho > 0$, then the reverse relation also holds. Give the necessary and sufficient conditions for $\kappa_{\boldsymbol{M},\rho,s}(X) = X$ if $\rho > 0$ is not true.

c Show that $\kappa_{\boldsymbol{M},\rho,s} \circ \kappa_{\boldsymbol{M},\rho,s} = \kappa_{\boldsymbol{M},\rho,s}$, i.e., $\kappa_{\boldsymbol{M},\rho,s}$ can be regarded as a projection.

d Show that $\langle Y, X\rangle^{(e)}_{\rho,s} = \langle Y, \kappa_{\boldsymbol{M},\rho,s}(X)\rangle^{(e)}_{\rho,s}$ if every matrix M_i commutes with Y.

e Verify that the above is true for the RLD case.

6.3. Show that the following two conditions are equivalent for the Hermitian matrix A, the state ρ, and the pinching $\kappa_{\boldsymbol{M}}$ corresponding to PVM $\boldsymbol{M} = \{M_i\}$.

① $\|A\|^{(m)}_{\rho,s} = \|\kappa_{\boldsymbol{M}}(A)\|^{(m)}_{\kappa_{\boldsymbol{M}}(\rho),s}$.

② There exists a Hermitian matrix X such that it commutes with every matrix M_i and satisfies $E_{\rho,s}(X) = A$.

6.4. Show the inequality $\|A\|^{(m)}_{\rho,b} \geq \|\kappa_{\boldsymbol{M}}(A)\|^{(m)}_{\kappa_{\boldsymbol{M}}(\rho),s}$ with the same assumption as above. Also, show the equivalence of the following:

① $\|A\|^{(m)}_{\rho,b} = \|\kappa_{\boldsymbol{M}}(A)\|^{(m)}_{\kappa_{\boldsymbol{M}}(\rho),s}$.

② There exists a Hermitian matrix X such that it commutes with every M_i and satisfies $A = \rho X = X\rho$.

6.5. Show that $\|X\rho Y\|_1 \leq \sqrt{\operatorname{Tr} \rho Y Y^*}\sqrt{\operatorname{Tr} \rho X^* X}$ by the Schwarz inequality for the inner product $\langle X, Y\rangle^{(e)}_{\rho,r} \stackrel{\text{def}}{=} \operatorname{Tr} \rho Y X^*$, where X, Y are matrices and ρ is a density matrix. Note that $\|\cdot\|_1$ denotes the trace-norm (Sect. A.3).

6.6. Given a matrix X and a density matrix ρ, show that

$$\|X\|_1 \leq \sqrt{\operatorname{Tr} \rho^{-1} X X^*}\sqrt{\operatorname{Tr} \rho U^* U} = \sqrt{\operatorname{Tr} \rho^{-1} X X^*}, \tag{6.20}$$

where U is a unitary matrix satisfying $\|X\|_1 = \operatorname{Tr} XU$. Use the Schwarz inequality for the inner product $\langle X, Y\rangle^{(e)}_{\rho,r}$.

6.7. Let the distribution p has zero measure at $\lambda = 1, 0$. Let $\rho = |y\rangle\langle y|$, and show that equation (6.10) does not hold.

6.8. Let κ be the pinching $\kappa_{\boldsymbol{M}}$ of a PVM $M = \{M_i\}$. Show that the map $\kappa_{\rho,x}$ can be regarded as the conditional expectation to the matrix space $\{X | [X, M_i] = 0 \ \forall i\}$ for $x = s, r$. (In general, the conditional expectation can be defined by (2.81) when the map κ is the dual map of the inclusion of a matrix subspace $\mathfrak{U}$.)

6.2 Metric-Induced Inner Products

In this section we treat the space of quantum states in a geometrical framework. In particular, we will discuss the properties of the metric, which will be defined in terms of the inner product discussed in the previous section. Consider a set of quantum states $\{\rho_\theta|\theta \in \mathbb{R}\}$ (a state family) parameterized by a single real number θ. We also assume that $\theta \mapsto \rho_\theta$ is continuous and differentiable up to the second order. The metric then represents the distance between two quantum states $\rho_{\theta_0}, \rho_{\theta_0+\epsilon}$ separated by a small $\epsilon > 0$. The difference in this case is approximately equal to $\frac{d\rho_\theta}{d\theta}(\theta_0)\epsilon$. The *SLD Fisher metric* $J_{\theta_0,s}$ is defined as the square of the size of $\frac{d\rho_\theta}{d\theta}(\theta_0)$ based on the SLD inner product at ρ_{θ_0}, i.e., $J_{\theta_0,s} \stackrel{\text{def}}{=} \left(\left\|\frac{d\rho_\theta}{d\theta}(\theta_0)\right\|^{(m)}_{\rho_{\theta_0},s}\right)^2$. The norm of the difference between two quantum states ρ_θ and $\rho_{\theta+\epsilon}$ is then approximately $\sqrt{J_{\theta_0,s}}\epsilon$. We can obviously define quantities such as the *Bogoljubov Fisher metric* $J_{\theta_0,b}$ [11,345,347,348], the *RLD metric* $J_{\theta_0,r}$ [202, 216], and the p metric in a similar way for the Bogoljubov, RLD, and p inner products, respectively. Therefore, if $u_1, \ldots, u_k$ is an orthonormal basis in $\mathcal{H}$, the SLD, Bogoljubov, RLD, and p Fisher metrics of the state family $\{\rho_\theta \stackrel{\text{def}}{=} \sum_{i=1}^k p_\theta(i)|u_i\rangle\langle u_i||\theta \in \mathbb{R}\}$ are all equal to the Fisher metric for the probability family $\{p_\theta\}$.

Thus, we have a theorem equivalent to Theorem 6.1 as given below.

Theorem 6.2 *Let κ be a TP-CP map, and $J_{\theta_0,x,\kappa}$ be the $x = s, b, r, \lambda, p$ Fisher metric for the state family $\{\kappa(\rho_\theta)|\theta \in \mathbb{R}\}$. The following relation then holds:*

$$J_{\theta_0,x} \geq J_{\theta_0,x,\kappa}, \quad x = s, b, r, \lambda, p. \tag{6.21}$$

When a metric satisfies (6.21), it is called a monotone metric. Since the derivative $\frac{d\rho_\theta}{d\theta}(\theta_0)$ plays an important role in the definition of the metric, $\frac{d\rho_\theta}{d\theta}(\theta_0)$ will be called the *m representation* of the derivative. We shall also define an operator $L_{\theta_0,x}$ by the relation

$$E_{\rho_{\theta_0},x}(L_{\theta_0,x}) = \frac{d\rho_\theta}{d\theta}(\theta_0).$$

Such an operator is called the *e representation* of the derivative. If all the ρ_θs commute, the e representation is the same as a logarithmic derivative. On the other hand, if some of the ρ_θs do not commute, the logarithmic derivative can be defined in several ways. The matrices $L_{\theta_0,s}$ and $L_{\theta_0,r}$

$$\frac{d\rho_\theta}{d\theta}(\theta_0) = \frac{1}{2}(\rho_{\theta_0}L_{\theta_0,s} + L_{\theta_0,s}\rho_{\theta_0}), \quad \frac{d\rho_\theta}{d\theta}(\theta_0) = \rho_{\theta_0}L_{\theta_0,r}$$

are called the symmetric logarithmic derivative (SLD) and the right logarithmic derivative (RLD), respectively. These matrices coincide with the e representations of the derivative concerning the SLD Fisher metric and the RLD Fisher metric, which are abbreviated to SLD e representation and RLD e representation, respectively.

Since the equation

$$\int_0^1 \rho_{\theta_0}^{\lambda} \left.\frac{d\log\rho_\theta}{d\theta}\right|_{\theta=\theta_0} \rho_{\theta_0}^{1-\lambda}\, d\lambda = \left.\frac{d\rho_\theta}{d\theta}\right|_{\theta=\theta_0} \tag{6.22}$$

holds [11] Ex. 6.13, the e representation of the derivative of the Bogoljubov Fisher metric $L_{\theta_0,b}$ is then equal to $\frac{d\log\rho_\theta}{d\theta}(\theta_0)$. Since $\operatorname{Tr}\frac{d\rho_\theta}{d\theta} = 0$, the e representation $L_{\theta,x}$ satisfies $\operatorname{Tr}\rho L_{\theta,x} = \operatorname{Tr} E_\rho(L_{\theta,x}) = 0$.

Theorem 6.3 *For a quantum state family $\{\rho_\theta | \theta \in \mathbb{R}\}$, the following relations hold [281, 299, 303, 394]:*

$$\frac{1}{8} J_{\theta,s} = \lim_{\epsilon\to 0} \frac{b^2(\rho_\theta, \rho_{\theta+\epsilon})}{\epsilon^2}, \tag{6.23}$$

$$\frac{1}{2} J_{\theta,b} = \lim_{\epsilon\to 0} \frac{D(\rho_{\theta+\epsilon}\|\rho_\theta)}{\epsilon^2}. \tag{6.24}$$

Hence, we obtain another proof of Theorem 6.1 (Theorem 6.2) for the SLD (Bogoljubov) case by combining Theorem 6.3 and (5.37) ((5.30)).

Proof. Define U_ϵ such that it satisfies

$$b^2(\rho_\theta, \rho_{\theta+\epsilon}) = \frac{1}{2} \operatorname{Tr}(\sqrt{\rho_\theta} - \sqrt{\rho_{\theta+\epsilon}}U_\epsilon)(\sqrt{\rho_\theta} - \sqrt{\rho_{\theta+\epsilon}}U_\epsilon)^*.$$

This can be rewritten as

$$\begin{aligned} 2b^2(\rho_\theta, \rho_{\theta+\epsilon}) &= \operatorname{Tr}(W(0) - W(\epsilon))(W(0) - W(\epsilon))^* \\ &\cong \operatorname{Tr}\left(-\frac{dW}{d\epsilon}(0)\epsilon\right)\left(-\frac{dW}{d\epsilon}(0)\epsilon\right)^* \cong \operatorname{Tr}\frac{dW}{d\epsilon}(0)\frac{dW}{d\epsilon}(0)^*\epsilon^2, \end{aligned}$$

where we defined $W(\epsilon) \stackrel{\text{def}}{=} \sqrt{\rho_{\theta+\epsilon}}U_\epsilon$. As will be shown later, the SLD $L_{\theta,s}$ satisfies

$$\frac{dW}{d\epsilon}(0) = \frac{1}{2} L W(0). \tag{6.25}$$

Therefore, $b^2(\rho_\theta, \rho_{\theta+\epsilon}) \cong \operatorname{Tr}\frac{1}{8} L W(0) W(0)^* L \epsilon^2 = \frac{1}{8}\operatorname{Tr} L^2 \rho_\theta \epsilon$, and we obtain (6.23). Thus, showing (6.25) will complete the proof.

From the definition of the Bures distance, we have

$$\begin{aligned} 2b^2(\rho_\theta, \rho_{\theta+\epsilon}) =& \min_{U:\text{unitary}} \operatorname{Tr}(\sqrt{\rho_\theta} - \sqrt{\rho_{\theta+\epsilon}}U)(\sqrt{\rho_\theta} - \sqrt{\rho_{\theta+\epsilon}}U)^* \\ =& 2 - \operatorname{Tr}\left(\sqrt{\rho_\theta}\sqrt{\rho_{\theta+\epsilon}}U(\epsilon)^* + U(\epsilon)\sqrt{\rho_{\theta+\epsilon}}\sqrt{\rho_\theta}\right). \end{aligned}$$

Therefore, $\sqrt{\rho_\theta}\sqrt{\rho_{\theta+\epsilon}}U(\epsilon)^* = U(\epsilon)\sqrt{\rho_{\theta+\epsilon}}\sqrt{\rho_\theta}$. Hence, $W(0)W(\epsilon)^* = W(\epsilon) W(0)^*$. Taking the derivative, we obtain $W(0)\frac{dW}{d\epsilon}(0)^* = \frac{dW}{d\epsilon}(0)W(0)^*$. This shows that there is a Hermitian matrix L satisfying $\frac{dW}{d\epsilon}(0) = \frac{1}{2}LW(0)$. Since

$\rho_{\theta+\epsilon} = W(\epsilon)W(\epsilon)^*$, we have $\frac{d\rho}{d\theta}(\theta) = \frac{1}{2}\left(LW(0)W(0)^* + W(0)W(0)^* L\right)$. We therefore see that L is an SLD.

We now prove (6.24). Since $L_{\theta,b}$ is equal to $\frac{d \log \rho_\theta}{d\theta}(\theta)$, we have

$$\begin{aligned} D(\rho_{\theta+\epsilon}\|\rho_\theta) &= \operatorname{Tr}\left(\rho_{\theta+\epsilon}(\log \rho_{\theta+\epsilon} - \log \rho_\theta)\right) \\ &\cong \operatorname{Tr}\left(\rho_\theta + \frac{d\rho_\theta}{d\theta}\epsilon\right)\left(\frac{d\log\rho_\theta}{d\theta}\epsilon + \frac{1}{2}\frac{d^2\log\rho_\theta}{d\theta^2}\epsilon^2\right) \\ &= \operatorname{Tr}(\rho_\theta L_{\theta,b})\epsilon + \left(\operatorname{Tr}\left(\frac{d\rho_\theta}{d\theta}L_{\theta,b}\right) + \frac{1}{2}\operatorname{Tr}\left(\rho_\theta\frac{d^2\log\rho_\theta}{d\theta^2}\right)\right)\epsilon^2. \end{aligned} \tag{6.26}$$

The first term on the right-hand side (RHS) may be evaluated as

$$\operatorname{Tr}(\rho_\theta L_{\theta,b}) = \int_0^1 \operatorname{Tr}\left(\rho_\theta^t L_{\theta,b}\rho_\theta^{1-t}\right) dt = \operatorname{Tr}\left(\frac{d\rho_\theta}{d\theta}\right) = 0. \tag{6.27}$$

Using this equation, we obtain

$$\begin{aligned} \operatorname{Tr}\left(\rho_\theta\frac{d^2\log\rho_\theta}{d\theta^2}\right) &= \frac{d}{d\theta}\left(\operatorname{Tr}\left(\rho_\theta\frac{d\log\rho_\theta}{d\theta}\right)\right) - \operatorname{Tr}\left(\frac{d\rho_\theta}{d\theta}\frac{d\log\rho_\theta}{d\theta}\right) \\ &= -\operatorname{Tr}\left(\frac{d\rho_\theta}{d\theta}L_{\theta,b}\right) = -J_{\theta,b}. \end{aligned} \tag{6.28}$$

Combining (6.26)–(6.28) we obtain $D(\rho_{\theta+\epsilon}\|\rho_\theta) \cong \frac{1}{2}J_{\theta,b}\epsilon^2$. ■

Next, let us consider a quantum state family $\{\rho_\theta | \theta \in \mathbb{R}^d\}$ with more than one parameter. The derivative at the point $\theta_0 = (\theta_0^1, \dots, \theta_0^d)$ may be obtained by considering the partial derivative $\frac{\partial}{\partial\theta^1}|_{\theta=\theta_0}, \dots, \frac{\partial}{\partial\theta^d}|_{\theta=\theta_0}$ with respect to each parameter. Since each partial derivative represents the size and direction of an infinitesimal transport, it may be regarded as a vector. We then call the vector space comprising these vectors the *tangent vector space* at θ_0, and its elements *tangent vectors*. The tangent vector $\frac{\partial}{\partial\theta^j}|_{\theta=\theta_0}$ can be represented as a matrix $\frac{\partial\rho_\theta}{\partial\theta^j}(\theta_0)$. This kind of representation of a tangent vector will be called an *m representation*. The matrix $L_{\theta_0,j,x}$ satisfying $E_{\rho_{\theta_0},x}(L_{\theta_0,j,x}) = \frac{\partial\rho_\theta}{\partial\theta^j}(\theta_0)$ will be called an *e representation* of the SLD (Bogoljubov, RLD) Fisher metric of $\frac{\partial}{\partial\theta^j}|_{\theta=\theta_0}$. The matrix $\boldsymbol{J}_{\theta_0,x} = [J_{\theta_0,x;i,j}]_{i,j}$

$$J_{\theta_0,x;i,j} \stackrel{\text{def}}{=} \left\langle \frac{\partial\rho_\theta}{\partial\theta^i}(\theta_0), \frac{\partial\rho_\theta}{\partial\theta^j}(\theta_0)\right\rangle^{(m)}_{\rho_0,x}$$

is called the SLD (Bogoljubov, RLD) Fisher information matrix [11,202,216], where $x = s, b, r$ corresponds to SLD, Bogoljubov, RLD, respectively. Note that the tangent vector refers to an infinitesimal change with respect to θ_0 and is different from the matrix represented by the m representation or e representation. The m representation and the e representation are nothing more than matrix representations of the infinitesimal change.

In summary, in this section we have defined the metric from the inner product given in Sect. 6.1 and investigated the relationship of this metric to the quantum relative entropy $D(\rho\|\sigma)$ and the Bures distance $b(\rho,\sigma)$. We also defined three types of Fisher information matrices for state families with more than one parameter.

Exercises

6.9. Define $\tilde{\phi}_\theta \stackrel{\text{def}}{=} \frac{d\phi_\theta}{d\theta} - \langle \phi_\theta | \frac{d\phi_\theta}{d\theta} \rangle \phi_\theta$ with respect to the pure state family $\{\rho_\theta \stackrel{\text{def}}{=} |\phi_\theta\rangle\langle\phi_\theta|\}$. Show that the SLD Fisher information $J_{\theta,s}$ is equal to $\langle \tilde{\phi}_\theta | \tilde{\phi}_\theta \rangle$. Show that both the RLD Fisher information and the Bogoljubov Fisher information diverge.

6.10. Let $J_\theta^{\boldsymbol{M}}$ be the Fisher information of the probability family $\{\mathrm{P}_{\rho_\theta}^{\boldsymbol{M}} | \theta \in \mathbb{R}\}$ (Sect. 1.2) for a one-parameter state family $\{\rho_\theta | \theta \in \mathbb{R}\}$ and a POVM $\boldsymbol{M} = \{M_i\}$. Show that

$$J_{\theta,x} \geq J_\theta^{\boldsymbol{M}} \text{ for } x = s, r, b, \lambda, p. \tag{6.29}$$

6.11. Show that $J_\theta^{\boldsymbol{M}} = \sum_i \frac{\langle M_i, L_{\theta,s} \rangle_{\rho_\theta,s}^{(e)} \langle L_{\theta,s}, M_i \rangle_{\rho_\theta,s}^{(e)}}{\langle M_i, I \rangle_{\rho_\theta,s}^{(e)}}$, with respect to the POVM $\boldsymbol{M} = \{M_i\}$.

6.12. Show the following facts with respect to the PVM $\boldsymbol{M} = \{M_i\}$ of rank $M_i = 1$.

a Show that M_i is an eigenvecotr of the linear map $X \mapsto \kappa_{\boldsymbol{M},\rho,s}(X)$ and $\frac{\langle M_i, X \rangle_{\rho,s}^{(e)}}{\langle M_i, I \rangle_{\rho,s}^{(e)}}$ is the corresponding eigenvalue.

b Show that $J_\theta^{\boldsymbol{M}} = \left(\| \kappa_{\boldsymbol{M},\rho_\theta,s}(L_{\theta,s}) \|_{\rho_\theta,s}^{(e)} \right)^2$.

c Assume that $\rho_\theta > 0$. Show that $J_{\theta,s} = J_\theta^{\boldsymbol{M}}$ if and only if every M_i commutes with $L_{\theta,s}$.

6.13. Prove (6.22) following the steps below.

a Show that $\int_0^1 \lambda^n (1-\lambda)^m \, d\lambda = \frac{n!m!}{(n+m)!}$.

b For a matrix valued function $X(\theta)$, show that

$$\int_0^1 \exp(\lambda X(\theta)) \frac{dX(\theta)}{d\theta} \exp((1-\lambda)X(\theta)) \, d\lambda = \frac{d \exp(X(\theta))}{d\theta}.$$

This is nothing other than (6.22).

6.14. Consider the state family $\{\rho_\theta^{\otimes n} | \theta \in \mathbb{R}\}$ consisting of the n-fold tensor product state of the state ρ_θ. Show that the metric $J_{\theta,x,n}$ of this state family $\{\rho_\theta^{\otimes n} | \theta \in \mathbb{R}\}$ is equal to n times the metric $J_{\theta,x}$ of the state family $\{\rho_\theta | \theta \in \mathbb{R}\}$, i.e., $J_{\theta,x,n} = nJ_{\theta,x}$ for $x = s, r, b, \lambda, p$.

6.15. Show that the Fisher information matrix $\boldsymbol{J}_{\theta,x}$ is Hermitian.

6.16. Show that the Fisher information matrix $\boldsymbol{J}_{\theta,x}$ is real symmetric for any symmetric inner product $\langle\ ,\ \rangle_{\rho,x}^{(e)}$.

6.17. Give an example of an RLD Fisher information matrix $\boldsymbol{J}_{\theta,r}$ that is not real symmetric.

6.18. For a Hermitian matrix Y and a quantum state family $\{\rho_\theta = e^{-i\theta Y} \rho e^{i\theta Y} | \theta \in \mathbb{R}\}$, show that the derivative at $\theta = 0$ has the e representation $i[\log \rho, Y]$ with respect to the Bogoljubov metric.

6.19. Show that $i[\rho, Y] = E_{\rho,b}(i[\log \rho, Y])$ if Y is Hermitian.

6.20. Define the state family $S = \left\{ \rho_{\boldsymbol{\theta}} = \frac{1}{2}\left(I + \sum_{i=1}^3 \theta_i S^i\right) \Big| \, \|\boldsymbol{\theta}\| \le 1 \right\}$ on the two-dimensional system $\mathcal{H} = \mathbb{C}^2$. Show that the three Fisher information matrices $\boldsymbol{J}_{\boldsymbol{\theta},s}, \boldsymbol{J}_{\boldsymbol{\theta},b}, \boldsymbol{J}_{\boldsymbol{\theta},r}$ can be written as

$$\boldsymbol{J}_{\boldsymbol{\theta},s}^{-1} = I - |\boldsymbol{\theta}\rangle\langle\boldsymbol{\theta}|, \tag{6.30}$$

$$\boldsymbol{J}_{\boldsymbol{\theta},b} = \frac{1}{1 - \|\boldsymbol{\theta}\|^2} |\boldsymbol{\theta}\rangle\langle\boldsymbol{\theta}| + \frac{1}{2\|\boldsymbol{\theta}\|} \log \frac{1 + \|\boldsymbol{\theta}\|}{1 - \|\boldsymbol{\theta}\|} \left(I - \frac{1}{\|\boldsymbol{\theta}\|^2} |\boldsymbol{\theta}\rangle\langle\boldsymbol{\theta}| \right), \tag{6.31}$$

$$\boldsymbol{J}_{\boldsymbol{\theta},r}^{-1} = I - |\boldsymbol{\theta}\rangle\langle\boldsymbol{\theta}| + iR_{\boldsymbol{\theta}}, \tag{6.32}$$

where R is defined in Sect. 5.3, following the steps below.

a Show the following for $\boldsymbol{\theta} = (0, 0, \theta)$, and check (6.30)–(6.32) in this case using them.

$$L_{\boldsymbol{\theta},s,1} = S_1,\ L_{\boldsymbol{\theta},s,2} = S_2,\ L_{\boldsymbol{\theta},r,1} = \rho_{\boldsymbol{\theta}}^{-1} S_1,\ L_{\boldsymbol{\theta},r,2} = \rho_{\boldsymbol{\theta}}^{-1} S_2,$$

$$L_{\boldsymbol{\theta},b,1} = \frac{1}{2\theta} \log \frac{1+\theta}{1-\theta} S_1,\ L_{\boldsymbol{\theta},b,2} = \frac{1}{2\theta} \log \frac{1+\theta}{1-\theta} S_2,$$

$$L_{\boldsymbol{\theta},s,3} = L_{\boldsymbol{\theta},b,3} = L_{\boldsymbol{\theta},r,3} = \begin{pmatrix} \frac{1}{1+\theta} & 0 \\ 0 & -\frac{1}{1-\theta} \end{pmatrix},$$

$$\boldsymbol{J}_{\boldsymbol{\theta},s} = \begin{pmatrix} 1 & 0 & 0 \\ 0 & 1 & 0 \\ 0 & 0 & \frac{1}{1-\theta^2} \end{pmatrix}, \quad \boldsymbol{J}_{\boldsymbol{\theta},r} = \begin{pmatrix} \frac{1}{1-\theta^2} & i\frac{\theta}{1-\theta^2} & 0 \\ -i\frac{\theta}{1-\theta^2} & \frac{1}{1-\theta^2} & 0 \\ 0 & 0 & \frac{1}{1-\theta^2} \end{pmatrix},$$

$$\boldsymbol{J}_{\boldsymbol{\theta},b} = \begin{pmatrix} \frac{1}{2\theta} \log \frac{1+\theta}{1-\theta} & 0 & 0 \\ 0 & \frac{1}{2\theta} \log \frac{1+\theta}{1-\theta} & 0 \\ 0 & 0 & \frac{1}{1-\theta^2} \end{pmatrix}.$$

b Show that $O^T J_{O\boldsymbol{\theta}} O = J_{\boldsymbol{\theta}}$, where O is an orthogonal matrix.
c Show (6.30)–(6.32) for an arbitrary $\boldsymbol{\theta}$.

6.21. Let $J^1_{\theta,s}$ and $J^2_{\theta,s}$ be the SLD Fisher metric of two state families $\{\rho^1_\theta|\theta \in \mathbb{R}\}$ and $\{\rho^2_\theta|\theta \in \mathbb{R}\}$, respectively. Show that the SLD Fisher metric $J_{\theta,s}$ of the state family $\{\lambda\rho^1_\theta + (1-\lambda)\rho^2_\theta|\theta \in \mathbb{R}\}$ satisfies $J_{\theta,s} \le \lambda J^1_{\theta,s} + (1-\lambda)J^2_{\theta,s}$. Show that its equality holds when the space spanned by the supports of $L_{\theta,1} \circ \rho^1_\theta$ and ρ^1_θ is orthogonal to that of $L_{\theta,2} \circ \rho^2_\theta$ and ρ^2_θ, where $L_{\theta,i}$ is the SLD of ρ^i_θ.

6.3 Geodesics and Divergences

In the previous section, we examined the inner product in the space of the quantum state. In this section, we will examine more advanced geometrical structures such as parallel transports, exponential family, and divergence. To introduce the concept of a parallel transport, consider an infinitesimal displacement in a one-parameter quantum state family $\{\rho_\theta|\theta \in \mathbb{R}\}$. The difference between $\rho_{\theta+\epsilon}$ and ρ_θ is approximately equal to $\frac{d\rho_\theta}{d\theta}(\theta)\epsilon$. Hence, the state $\rho_{\theta+\epsilon}$ can be regarded as the state transported from the state ρ_θ in the direction $\frac{d\rho_\theta}{d\theta}(\theta)$ by an amount ϵ. However, the state $\rho_{\theta+\epsilon}$ does not coincide precisely with the state displaced from the state ρ_θ by ϵ in the direction of $\frac{d\rho_\theta}{d\theta}(\theta)$. For example, when the infinitesimal displacement at the intermediate states $\rho_{\theta+\Delta}$ $(0 < \Delta < \epsilon)$ is equal to the infinitesimal displacement $\frac{d\rho_\theta}{d\theta}(\theta)\Delta$ at θ, the state $\rho_{\theta+\epsilon}$ coincides precisely with it. Then, the problem is to ascertain which infinitesimal displacement at the point $\theta+\epsilon'$ corresponds to the given infinitesimal displacement $\frac{d\rho_\theta}{d\theta}(\theta)\Delta$ at the initial point θ. The rule for matching the infinitesimal displacement at one point to the infinitesimal displacement at another point is called parallel transport. The coefficient $\frac{d\rho_\theta}{d\theta}(\theta)$ of the infinitesimal displacement at θ is called the tangent vector, as it represents the slope of the tangent line of the state family $\{\rho_\theta|\theta \in \mathbb{R}\}$ at θ. We may therefore consider the parallel transport of a tangent vector instead of the parallel transport of an infinitesimal displacement.

Commonly used parallel transports can be classified into those based on the m representation (*m parallel translation*) and those based on the e representation (*e parallel translation*). The m parallel translation $\Pi^{(m)}_{\rho_\theta,\rho_{\theta'}}$ moves the tangent vector at one point ρ_θ to the tangent vector at another point $\rho_{\theta'}$ with the same m representation. On the other hand, the e parallel translation $\Pi^{(e)}_{x,\rho_\theta,\rho_{\theta'}}$ moves the tangent vector at one point ρ_θ with the e representation L to the tangent vector at another point $\rho_{\theta'}$ with the e representation $L - \operatorname{Tr}\rho_{\theta'} L$ [11]. Of course, this definition requires a coincidence between the set of e representations at the point θ and that at another point θ'. Hence, this type of e parallel translation is defined only for the symmetric inner product $\langle X, Y\rangle^{(e)}_{\rho,x}$, and its definition depends on the choice of the metric. Indeed, the

e parallel translation can be regarded as the dual parallel translation of the m parallel translation concerning the metric $\langle X,Y\rangle^{(e)}_{\rho,x}$ in the following sense:

$$\operatorname{Tr} X^* \Pi^{(m)}_{\rho_\theta,\rho_{\theta'}}(A) = \operatorname{Tr} \Pi^{(e)}_{x,\rho_{\theta'},\rho_\theta}(X)^* A,$$

where X is the e representation of a tangent vector at $\rho_{\theta'}$ and A is the m representation of another tangent vector at ρ_θ.

Further, a one-parameter quantum state family is called a *geodesic* or an *autoparallel curve* when the tangent vector (i.e., the derivative) at each point is given as a parallel transport of a tangent vector at a fixed point. In particular, the e geodesic is called a *one-parameter exponential family.* For example, in an e geodesic with respect to SLD $\{\rho_\theta|\theta \in \mathbb{R}\}$, any state ρ_θ coincides with the state transported from the state ρ_0 along the autoparallel curve in the direction L by an amount θ, where L denotes the SLD e representation of the derivative at ρ_0. We shall henceforth denote the state as $\Pi^\theta_{L,s}\rho_0$. Similarly, $\Pi^\theta_{L,b}\rho_0$ denotes the state transported autoparallely with respect to the Bogoljubov e representation from ρ_0 in the direction L by an amount θ.

When the given metric is not symmetric, the e parallel translation moves the tangent vector at one point θ under the e representation $\tilde{L}$ to the tangent vector at another point θ' with the e representation $\tilde{L}' - \operatorname{Tr} \rho_{\theta'}\tilde{L}'$ with the condition $\tilde{L} + \tilde{L}^* = \tilde{L}' + (\tilde{L}')^*$. That is, we require the same Hermitian part in the e representation. Hence, the e parallel translation $\Pi^{(e)}_{x,\rho_\theta,\rho_{\theta'}}$ coincides with the e parallel translation $\Pi^{(e)}_{s(x),\rho_\theta,\rho_{\theta'}}$ with regard to its symmetrized inner product. Therefore, we can define the state transported from the state ρ_0 along the autoparallel curve in the direction with the Hermitian part L by an amount θ with respect to RLD (λ, p), and we denote them by $\Pi^\theta_{L,r}\rho_0$ ($\Pi^\theta_{L,\lambda}\rho_0$, $\Pi^\theta_{L,p}\rho_0$). However, only the SLD one-parameter exponential family $\{\Pi^\theta_{L,s}\rho_0|s \in \mathbb{R}\}$ plays an important role in quantum estimation examined in the next section.

Lemma 6.1 *$\Pi^\theta_{L,s}\sigma$, $\Pi^\theta_{L,b}\sigma$, $\Pi^\theta_{L,r}\sigma$, and $\Pi^\theta_{L,\frac{1}{2}}\sigma$ can be written in the following form [11, 300, 304]:*

$$\Pi^\theta_{L,s}\sigma = e^{-\mu_s(\theta)} e^{\frac{\theta}{2}L} \sigma e^{\frac{\theta}{2}L}, \tag{6.33}$$

$$\Pi^\theta_{L,b}\sigma = e^{-\mu_b(\theta)} e^{\log\sigma + \theta L}, \tag{6.34}$$

$$\Pi^\theta_{L,r}\sigma = e^{-\mu_r(\theta)} \sqrt{\sigma} e^{\theta L_r} \sqrt{\sigma}, \tag{6.35}$$

$$\Pi^\theta_{L,\frac{1}{2}}\sigma = e^{-\mu_{\frac{1}{2}}(\theta)} \sigma^{\frac{1}{4}} e^{\frac{\theta}{2}L_{\frac{1}{2}}} \sigma^{\frac{1}{2}} e^{\frac{\theta}{2}L_{\frac{1}{2}}} \sigma^{\frac{1}{4}}, \tag{6.36}$$

where we choose Hermitian matrices L_r and $L_{\frac{1}{2}}$ as $L = \frac{1}{2}(\sigma^{-\frac{1}{2}} L_r \sigma^{\frac{1}{2}} + \sigma^{\frac{1}{2}} L_r \sigma^{-\frac{1}{2}})$ and $L = \frac{1}{2}(\sigma^{-\frac{1}{4}} L_{\frac{1}{2}} \sigma^{\frac{1}{4}} + \sigma^{\frac{1}{4}} L_{\frac{1}{2}} \sigma^{-\frac{1}{4}})$, respectively, and

$$\mu_s(\theta) \stackrel{\text{def}}{=} \log \operatorname{Tr} e^{\frac{\theta}{2}L}\sigma e^{\frac{\theta}{2}L}, \qquad \mu_b(\theta) \stackrel{\text{def}}{=} \log \operatorname{Tr} e^{\log\sigma+\theta L}, \tag{6.37}$$

$$\mu_r(\theta) \stackrel{\text{def}}{=} \log \operatorname{Tr} \sqrt{\sigma} e^{\theta L_r} \sqrt{\sigma}, \quad \mu_{1/2}(\theta) \stackrel{\text{def}}{=} \log \operatorname{Tr} \sigma^{\frac{1}{4}} e^{\frac{\theta}{2}L_{\frac{1}{2}}} \sigma^{\frac{1}{2}} e^{\frac{\theta}{2}L_{\frac{1}{2}}} \sigma^{\frac{1}{4}}.$$

Proof. By taking the derivative of the RHS of (6.33) and (6.34), we see that the SLD (or Bogoljubov) e representation of the derivative at each point is equal to the parallel transported e representation of the derivative L at σ (use (6.22)). On the RHS of (6.35), the RLD e representation of the derivative at each point is equal to the parallel transported e representation of the derivative $\sqrt{\sigma}^{-1} L_r \sqrt{\sigma}$ at σ. Further, on the RHS of (6.36), the $\frac{1}{2}$ e representation of the derivative at each point is equal to the parallel transported e representation of the derivative L_r at σ.

Conversely, from the definition of $\Pi^\theta_{L,x}\sigma$ we have

$$\frac{d\Pi^\theta_{L,x}\sigma}{d\theta} = E_{\rho_\theta,x}(L - \operatorname{Tr} L\rho_\theta), \quad x = s, r, \frac{1}{2}.$$

Since this has only one variable, this is actually an ordinary differential equation. From the uniqueness of the solution of an ordinary differential equation, the only $\Pi^\theta_{L,x}\sigma$ satisfying $\Pi^0_{L,x}\sigma = \sigma$ is the solution of the above differential equation. Since any e representation σ has the form $\sqrt{\sigma}^{-1} L\sqrt{\sigma}$ with a Hermitian matrix L, we only discuss $\rho_\theta = \Pi^\theta_{\sqrt{\sigma}^{-1}L\sqrt{\sigma},r}\sigma$. By taking its derivative, we have

$$\frac{\Pi^\theta_{\sqrt{\sigma}^{-1}L\sqrt{\sigma},r}\sigma}{d\theta} = \rho_\theta(\sqrt{\sigma}^{-1} L\sqrt{\sigma} - \operatorname{Tr}\rho_\theta\sqrt{\sigma}^{-1} L\sqrt{\sigma}).$$

Similarly, from the uniqueness of the solution of an ordinary differential equation only the state family (6.35) satisfies this condition. ∎

Now, using the concept of the exponential family, we extend the divergence based on the first equation in (2.92). For any two states ρ and σ, we choose the Hermitian matrix L such that the exponential family $\{\Pi^\theta_{L,x}\sigma\}_{\theta\in[0,1]}$ with regard to the inner product $J_{\theta,x}$ satisfies

$$\Pi^1_{L,x}\sigma = \rho. \tag{6.38}$$

Then, we define the x-e-divergences as follows:

$$D^{(e)}_x(\rho\|\sigma) = \int_0^1 J_{\theta,x}\theta d\theta, \tag{6.39}$$

where $J_{\theta,x}$ is the Fisher information concerning the exponential family $\Pi^\theta_{L,x}\sigma$. Since $\Pi^\theta_{L^1\otimes I+I\otimes L^2,x}(\sigma_1\otimes\sigma_2)$ equals $(\Pi^\theta_{L^1,x}\sigma_1)\otimes(\Pi^\theta_{L^2,x}\sigma_2)$,

$$D^{(e)}_x(\rho_1\otimes\rho_2\|\sigma_1\otimes\sigma_2) = D^{(e)}_x(\rho_1\|\sigma_1) + D^{(e)}_x(\rho_2\|\sigma_2), \tag{6.40}$$

i.e., the e-divergence satisfies the *additivity* for any inner product.

Theorem 6.4 *When*

$$L = \begin{cases} 2\log\sigma^{-\frac{1}{2}}(\sigma^{\frac{1}{2}}\rho\sigma^{\frac{1}{2}})^{\frac{1}{2}}\sigma^{-\frac{1}{2}} & \text{for } x = s, \\ \log\rho - \log\sigma & \text{for } x = b, \\ \frac{1}{2}(\sigma^{-\frac{1}{2}}\log(\sigma^{-\frac{1}{2}}\rho\sigma^{-\frac{1}{2}})\sigma^{\frac{1}{2}} + \sigma^{\frac{1}{2}}\log(\sigma^{-\frac{1}{2}}\rho\sigma^{-\frac{1}{2}})\sigma^{-\frac{1}{2}}) & \text{for } x = r, \\ \sigma^{-\frac{1}{4}}\log(\sigma^{-\frac{1}{4}}\rho^{\frac{1}{2}}\sigma^{-\frac{1}{4}})\sigma^{\frac{1}{4}} + \sigma^{\frac{1}{4}}\log(\sigma^{-\frac{1}{4}}\rho^{\frac{1}{2}}\sigma^{-\frac{1}{4}})\sigma^{-\frac{1}{4}} & \text{for } x = \frac{1}{2}, \end{cases} \tag{6.41}$$

condition (6.38) holds. Hence, we obtain

$$D_s^{(e)}(\rho\|\sigma) = 2\operatorname{Tr}\rho\log\sigma^{-\frac{1}{2}}(\sigma^{\frac{1}{2}}\rho\sigma^{\frac{1}{2}})^{\frac{1}{2}}\sigma^{-\frac{1}{2}}, \tag{6.42}$$

$$D_b^{(e)}(\rho\|\sigma) = \operatorname{Tr}\rho(\log\rho - \log\sigma) = D(\rho\|\sigma), \tag{6.43}$$

$$D_r^{(e)}(\rho\|\sigma) = \operatorname{Tr}\rho\log(\rho^{\frac{1}{2}}\sigma^{-1}\rho^{\frac{1}{2}}), \tag{6.44}$$

$$D_{\frac{1}{2}}^{(e)}(\rho\|\sigma) = 2\operatorname{Tr}(\sigma^{\frac{1}{4}}\rho^{\frac{1}{2}}\sigma^{\frac{1}{4}})(\sigma^{-\frac{1}{4}}\rho^{\frac{1}{2}}\sigma^{-\frac{1}{4}})\log(\sigma^{-\frac{1}{4}}\rho^{\frac{1}{2}}\sigma^{-\frac{1}{4}}). \tag{6.45}$$

Proof. When we substitute (6.41) into L, condition (6.38) can be checked by using Lemma 6.1. In this case, $L_r = \log(\sigma^{-\frac{1}{2}}\rho\sigma^{-\frac{1}{2}})$, $L_{\frac{1}{2}} = 2\log(\sigma^{-\frac{1}{4}}\rho^{\frac{1}{2}}\sigma^{-\frac{1}{4}})$, and we can show that

$$\frac{d^2\mu_x(\theta)}{d\theta^2} = J_{\theta,x}. \tag{6.46}$$

For $x = b$, see (6.22). Hence, from a discussion similar to (2.92), we can prove that

$$D_x^{(e)}(\rho\|\sigma) = \left.\frac{d\mu_x(\theta)}{d\theta}\right|_{\theta=1}(1-0) - \mu_x(1) + \mu_x(0) = \left.\frac{d\mu_x(\theta)}{d\theta}\right|_{\theta=1}, \tag{6.47}$$

where $\mu_x(\theta)$ is defined in Theorem 6.1. Using this relation, we can check (6.42), (6.43), and (6.45). Concering (6.44), we obtain

$$D_r^{(e)}(\rho\|\sigma) = \operatorname{Tr}\sigma\sigma^{-\frac{1}{2}}\rho\sigma^{-\frac{1}{2}}\log(\sigma^{-\frac{1}{2}}\rho\sigma^{-\frac{1}{2}}), = \operatorname{Tr}\rho\log(\rho^{\frac{1}{2}}\sigma^{-1}\rho^{\frac{1}{2}})$$

where the last equation follows from Exercise A.2. ■

Now we compare these quantum analogs of relative entropy given in (6.42)–(6.45). As is easily checked, these satisfy condition (5.33) for quantum analogs of relative entropy. From Exercise 6.25 and the monotonicity for measurement concerning the quantum relative entropy $D(\rho\|\sigma)$,

$$D(\rho\|\sigma) \geq D_s^{(e)}(\rho\|\sigma) = 2\operatorname{Tr}\rho\log\sigma^{-\frac{1}{2}}(\sigma^{\frac{1}{2}}\rho\sigma^{\frac{1}{2}})^{\frac{1}{2}}\sigma^{-\frac{1}{2}}. \tag{6.48}$$

Hence, from inequality (5.36) and additivity (6.40), $D_s^{(e)}(\rho\|\sigma)$ do not satisfy the monotonicity even for measurements because the equality in (6.48) does not always hold.

Further, we can extend the divergence based on equation (2.93). For any two states ρ and σ, the family $\{(1-t)\rho + t\sigma | 0 \leq t \leq 1\}$ is the m geodesic joining ρ and σ. Hence, as an extension of (2.93), we can define the x-m divergences as

$$D_x^{(m)}(\rho\|\sigma) \stackrel{\text{def}}{=} \int_0^1 J_{t,x} t dt\,. \tag{6.49}$$

Since the family $\{(1-t)\kappa(\rho)+t\kappa(\sigma)|0 \le t \le 1\}$ is the m geodesic joining $\kappa(\rho)$ and $\kappa(\sigma)$ for any TP-CP map κ, we have

$$D_x^{(m)}(\rho\|\sigma) \ge D_x^{(m)}(\kappa(\rho)\|\kappa(\sigma)), \tag{6.50}$$

i.e., the m divergence satisfies the *monotonicity*. Since the RLD is the largest inner product,

$$D_r^{(m)}(\rho\|\sigma) \ge D_x^{(m)}(\rho\|\sigma). \tag{6.51}$$

We can calculate the m divergences as [Ex. 6.24]

$$D_b^{(m)}(\rho\|\sigma) = \operatorname{Tr}\rho(\log\rho - \log\sigma) = D(\rho\|\sigma), \tag{6.52}$$

$$D_r^{(m)}(\rho\|\sigma) = \operatorname{Tr}\rho\log(\sqrt{\rho}\sigma^{-1}\sqrt{\rho}). \tag{6.53}$$

The Bogoljubov case follows from Theorem 6.5. Hence, $\operatorname{Tr}\rho\log(\sqrt{\rho}\sigma^{-1}\sqrt{\rho}) = D_r^{(m)}(\rho\|\sigma)$ satisfies the monotonicity for TP-CP maps. Further, from (6.51) we obtain $\operatorname{Tr}\rho\log(\sqrt{\rho}\sigma^{-1}\sqrt{\rho}) \ge D(\rho\|\sigma)$ [206].

Not all x-m divergences necessarily satisfy additivity (6.40). At least when the inner product $J_{x,\theta}$ is smaller than the Bogoljubov inner product $J_{b,\theta}$, i.e., $J_{\theta,x} \le J_{\theta,b}$, we have $D(\rho\|\sigma) \ge D_x^{(m)}(\rho\|\sigma)$. From (5.36) and monotonicity (6.50), $D_x^{(m)}(\rho\|\sigma)$ does not satisfy additivity (6.40). For example, the SLD m divergence does not satisfy additivity (6.40).

We can now verify whether it is possible in two-parameter-state families to have states that are e autoparallel transported in the direction of L_1 by θ^1 and in the direction L_2 by θ^2. In order to define such a state, we require that the state that be e autoparallel transported first in the L_1 direction by θ^1 from ρ_0, then further e autoparallel transported in the L_2 direction by θ^2 so as to coincide with the state that is e autoparallel transported in the L_2 direction by θ^2 from ρ_0, then e autoparallel transported in the L_1 direction by θ^1. That is, if such a state were defined, the relation

$$\Pi_{L_2,x}^{\theta^2}\Pi_{L_1,x}^{\theta^1}\sigma = \Pi_{L_1,x}^{\theta^1}\Pi_{L_2,x}^{\theta^2}\sigma \tag{6.54}$$

should hold. Otherwise, the torsion $T(L_1,L_2)_x$ is defined as follows:

$$T(L_1,L_2)_{\rho,x} \stackrel{\text{def}}{=} \lim_{\epsilon\to 0}\frac{\Pi_{L_2,x}^{\epsilon}\Pi_{L_1,x}^{\epsilon}\rho - \Pi_{L_1,x}^{\epsilon}\Pi_{L_2,x}^{\epsilon}\rho}{\epsilon^2}.$$

Concerning condition (6.54), we have the following theorem.

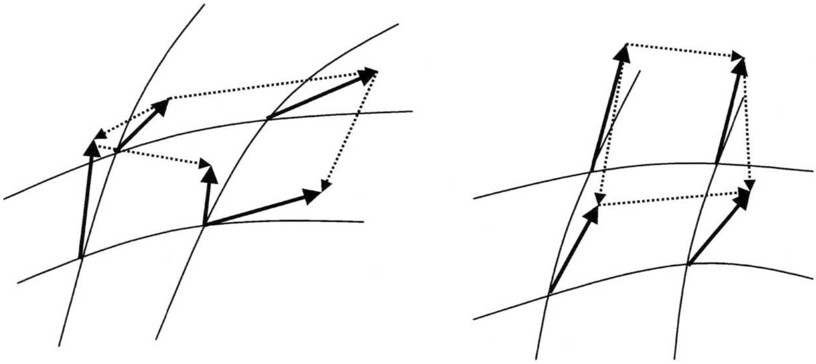

Fig. 6.1. Torsion

Theorem 6.5 *(Amari and Nagaoka [11]) The following conditions for the inner product $J_{\theta,x}$ are equivalent.*

① $J_{\theta,x}$ *is the Bogoljubov inner product, i.e.,* $x = b$.
② *Condition (6.54) holds for any two Hermitian matrices L_1 and L_2 and any state ρ_0.*
③ $D_x^{(e)}(\rho_{\bar{\theta}}\|\rho_\theta) = D^\mu(\bar{\theta}\|\theta)$.
④ $D_x^{(e)}(\rho\|\sigma) = D(\rho\|\sigma)$.
⑤ $D_x^{(m)}(\rho_{\bar{\eta}}\|\rho_\eta) = D^\nu(\eta\|\bar{\eta})$.
⑥ $D_x^{(m)}(\rho\|\sigma) = D(\rho\|\sigma)$.

Here, the convex functions $\mu(\theta)$ and $\nu(\eta)$ and the states ρ_θ and ρ_η are defined by

$$\rho_\theta \stackrel{\text{def}}{=} \exp(\sum_i \theta^i X_i - \mu(\theta)), \quad \mu(\theta) \stackrel{\text{def}}{=} \log \operatorname{Tr} \exp(\sum_i \theta^i X_i), \tag{6.55}$$

$$\rho_\eta \stackrel{\text{def}}{=} \rho_{\text{mix}} + \sum_j \eta_j Y^j, \qquad \nu(\eta) \stackrel{\text{def}}{=} D_x^{(m)}(\rho_0\|\rho_\eta) = -H(\rho_\eta) + H(\rho_{\text{mix}}),$$

where $X_1, \ldots, X_k$ is a basis of the set of traceless Hermitian matrices, and $Y^1, \ldots, Y^k$ is its dual basis.

Proof. First, we prove ① ⇒ ②. Theorem 6.1 guarantees that the Bogoljubov e autoparallel transport satisfies

$$\Pi_{L_2,b}^{\theta^2}\Pi_{L_1,b}^{\theta^1}\rho = \Pi_{L_1,b}^{\theta^1}\Pi_{L_2,b}^{\theta^2}\rho = e^{-\mu_b(\theta^1,\theta^2)}e^{\log\rho+\theta^1 L_1+\theta^2 L_2},$$

where $\mu_b(\theta) \stackrel{\text{def}}{=} \log \operatorname{Tr} e^{\log\rho+\theta^1 L_1+\theta^2 L_2}$. Hence, we obtain ②.

Next, we prove that ② ⇒ ③. We define $\tilde{\rho}_\theta \stackrel{\text{def}}{=} \Pi_{X_k,x}^{\theta^k}, \cdots, \Pi_{X_1,b}^{\theta^1}\rho_{\text{mix}}$ for $\theta = (\theta^1, \ldots, \theta^k)$. Then, condition ② guarantees that $\tilde{\rho}_{\bar{\theta}} = \Pi^1_{\sum_i(\bar{\theta}^i-\theta^i)X_i,x}\tilde{\rho}_\theta$.

In particular, when $\theta = 0$, we obtain $\tilde{\rho}_{\bar{\theta}} = \Pi^1_{\sum_i \bar{\theta}^i X_i, x} \rho_{\mathrm{mix}}$. Since $\sum_i \bar{\theta}^i X_i$ is commutative with ρ_{mix}, we can apply the classical observation to this case. Hence, state $\tilde{\rho}_{\bar{\theta}}$ coincides with state $\rho_{\bar{\theta}}$ defined in (6.55).

Let $\tilde{X}_{j,\theta}$ be the x-e representation of the partial derivative concerning θ^j at ρ_θ. It can be expressed as

$$\tilde{X}_{j,\theta} = X_j - \operatorname{Tr} \rho_\theta X_j + \bar{X}_{\theta,j},$$

where $\bar{X}_{\theta,j}$ is the skew-Hermitian part. Thus,

$$\begin{aligned}\frac{\partial \operatorname{Tr} \rho_\theta X_j}{\partial \theta^i} &= \operatorname{Tr} \left(\frac{\partial \rho_\theta}{\partial \theta^i} X_j \right) = \operatorname{Tr} \left(\frac{\partial \rho_\theta}{\partial \theta^i} (X_j - \operatorname{Tr} \rho_\theta X_j) \right) \\ &= \mathbf{Re} \operatorname{Tr} \left(\frac{\partial \rho_\theta}{\partial \theta^i} (X_j - \operatorname{Tr} \rho_\theta X_j + \bar{X}_{\theta,j}) \right) = \mathbf{Re}\, J_{\theta,x;i,j}.\end{aligned}$$

Note that the trace of the product of a Hermitian matrix and a skew-Hermitian matrix is an imaginary number. Since $\mathbf{Re}\, J_{\theta,x;i,j} = \mathbf{Re}\, J_{\theta,x;j,i}$, we have $\frac{\partial \operatorname{Tr} \rho_\theta X_j}{\partial \theta^i} = \frac{\partial \operatorname{Tr} \rho_\theta X_i}{\partial \theta^j}$. Thus, there exists a function $\bar{\mu}(\theta)$ such that $\bar{\mu}(0) = \mu(0)$ and

$$\frac{\partial \bar{\mu}(\theta)}{\partial \theta^i} = \operatorname{Tr} \rho_\theta X_i.$$

This function $\bar{\mu}$ satisfies the condition ③.

Moreover, since $\operatorname{Tr} \rho_{\mathrm{mix}} X_i = 0$, from definition (2.85), we have $\bar{\mu}(\theta) - \bar{\mu}(0) = D^{\bar{\mu}}(0\|\theta)$. Since state ρ_{mix} commutes with state ρ_θ, the relation $D^{(e)}(\rho_{\mathrm{mix}}\|\rho_\theta) = \mu(\theta) - \mu(0)$ holds. Hence, we obtain $\bar{\mu}(\theta) = \mu(\theta)$.

Further, we have $D^\mu(\bar{\theta}\|\theta) = D(\rho\|\theta)$. Thus, the equivalence between ③ and ④ is trivial since the limit of $D(\rho_{\bar{\theta}}\|\rho_\theta)$ equals the Bogoljubov inner product $J_{b,\theta}$. Hence, we obtain ④ ⇒ ①.

Now we proceed to the proof of ①+②+③+④ ⇒ ⑤. In this case, the function $\nu(\eta)$ coincides with the Legendre transform of $\mu(\theta)$, and $\eta_i = \frac{\partial \mu}{\partial \theta^i}(\theta)$. Hence, $D^\nu(\eta\|\bar{\eta}) = D^\mu(\bar{\theta}\|\theta) = D(\rho_{\bar{\eta}}\|\rho_\eta)$. The second derivative matrix $\frac{\partial^2 \nu}{\partial \eta^i \partial \eta^j}$ coincides with the inverse of the second derivative matrix $\frac{\partial^2 \mu}{\partial \theta^i \partial \theta^j}$, which equals the Bogoljubov Fisher information matrix concerning the parameter θ. Since the Bogoljubov Fisher information matrix concerning the parameter η equals the inverse of the Bogoljubov Fisher information matrix concerning the parameter θ, the Bogoljubov Fisher information matrix concerning the parameter η coincides with the second derivative matrix $\frac{\partial^2 \nu}{\partial \eta^i \partial \eta^j}$. Hence, from (2.87) we have $D^\nu(\eta\|\bar{\eta}) = D_b^{(m)}(\rho_{\bar{\theta}}\|\rho_\theta)$.

Next, we prove ⑤ ⇒ ⑥. Since $\rho_{\mathrm{mix}} = \rho_0$ commutes with ρ_η, the m divergence $D_x^{(m)}(\rho_0\|\rho_\eta)$ coincides with the Bogoljubov m divergence $D_b^{(m)}(\rho_0\|\rho_\eta)$, which equals the Legendre transform of $\mu(\theta)$ defined in (6.55). Thus, $D_x^{(m)}(\rho_{\bar{\eta}}\|\rho_\eta) = D^\nu(\eta\|\bar{\eta}) = D(\rho_{\bar{\eta}}\|\rho_\eta)$. Finally, taking the limit $\bar{\eta} \to \eta$, we obtain $J_{x,\eta} = J_{b,\eta}$, i.e., ⑥ ⇒ ①. ∎

This theorem shows that the Bogoljubov inner product is most natural from a geometrical point of view. However, from the viewpoint of estimation theory the Bogoljubov metric is rather inconvenient, as will be shown in the next section.

In summary, this section examined several geometrical structures that may be derived from the inner product. In the next section, we will discuss the connection between these structures and estimation theory.

Exercises

6.22. Define the SLD and the state as $L = S_1$ and $\rho_0 = 1/2(I + \sum_{i=1}^3 x_i S^i)$ in the two-dimensional system $\mathcal{H} = \mathbb{C}^2$. Show that the SLD e geodesic $\Pi^t_{L,s}\rho_0$ is given by

$$\Pi^t_{L,s}\rho_0 = \frac{1}{2}(I + \sum_{i=1}^{3} x^i(t)S_i), \quad x_1(t) = \frac{e^t(1+x_1) - e^{-t}(1-x_1)}{e^t(1+x_1) + e^{-t}(1-x_1)},$$

$$x_2(t) = \frac{2x_2}{e^t(1+x_1) + e^{-t}(1-x_1)}, \quad x_3(t) = \frac{2x_3}{e^t(1+x_1) + e^{-t}(1-x_1)}.$$

6.23. Show that an arbitrary SLD e geodesic on the two-dimensional system $\mathcal{H} = \mathbb{C}^2$ is unitarily equivalent to $\mathcal{S}_\alpha$ if a suitable $\alpha \in [0,1]$ is chosen [119], where $\mathcal{S}_\alpha \stackrel{\text{def}}{=} \left\{ \frac{1}{2}\begin{pmatrix} 1 + \alpha/\cosh t & \tanh t \\ \tanh t & 1 - \alpha/\cosh t \end{pmatrix} \middle| t \in \mathbb{R} \right\}$.

6.24. Show equation (6.53) following the steps below.

a Show the equation $\int_0^1 (X^2 t)(I + tX)^{-1} dt = X - \log(I + X)$ for any Hermitian matrix X.

b Show the equation $\int_0^1 \mathrm{Tr}(\sigma - \rho)^2(\rho + t(\sigma - \rho))^{-1} dt = \mathrm{Tr}\,\rho \log(\sqrt{\rho}\sigma^{-1}\sqrt{\rho})$.

6.25. Let $\boldsymbol{M}$ be a measurement corresponding to the spectral decomposition of $\sigma^{-1/2}(\sigma^{1/2}\rho\sigma^{1/2})^{1/2}\sigma^{-1/2}$. Show that $D_s^{(e)}(\rho\|\sigma) = D(\mathrm{P}_\rho^{\boldsymbol{M}}\|\mathrm{P}_\sigma^{\boldsymbol{M}})$ [304]. Show that $D_s^{(e)}(\rho\|\sigma) \geq -2\log \mathrm{Tr}\,|\sqrt{\rho}\sqrt{\sigma}|$ from Exercise 2.26 and (2.19).

6.26. Show equation (6.52) following the steps below [121, 302].

a Show that $\int_0^1 \frac{d\log\rho_t}{dt}\frac{d\rho_t}{dt} t\, dt$
$= \left[(\log\rho_t)\frac{d\rho_t}{dt}t\right]_0^1 - \int_0^1 (\log\rho_t)\frac{d^2\rho_t}{d^2t} t\, dt - [(\log\rho_t)\rho_t]_0^1 + \int_0^1 \frac{d\log\rho_t}{dt}\rho_t\, dt.$

b Show that $J_{t,b} = \mathrm{Tr}\,\frac{d\log\rho_t}{dt}\frac{d\rho_t}{dt}$ for the Bogoljubov metric.

c Show that $\mathrm{Tr}\,\frac{d\log\rho_t}{dt}\rho_t = 0$.

d Show (6.52) for the m geodesic $\rho_t = t\sigma + (1-t)\rho$ connecting two states ρ and σ.

6.27. Show that the following three conditions are equivalent for two states ρ and σ and a PVM $\boldsymbol{M} = \{M_i\}$ of rank $M_i = 1$ [121, 302]. The equivalence of ① and ③ is nothing other than Theorem 3.5.

① $D(\rho\|\sigma) = D^{\boldsymbol{M}}(\rho\|\sigma)$.
② The m geodesic $\rho_\theta = \theta\sigma + (1-\theta)\rho$ satisfies $J_{\theta,b} = J_\theta^{\boldsymbol{M}}$ for $\theta \in [0,1]$.
③ $[\sigma, \rho] = 0$, and there exists a set of real numbers $\{a_i\}_{i=1}^d$ satisfying

$$\rho = \sigma \left(\sum_{i=1}^{d} a_i M_i \right) = \left(\sum_{i=1}^{d} a_i M_i \right) \sigma. \tag{6.56}$$

6.28. Show that $\lim \frac{1}{n} D_x^{(m)}(\rho^{\otimes n}\|\sigma^{\otimes n}) = D(\rho\|\sigma)$ when $J_{\theta,x} \le J_{\theta,b}$.

6.4 Quantum State Estimation

In Chap. 2, we only considered the case of two hypotheses existing for the quantum states. In this section, we will consider the problem of efficiently estimating an unknown quantum state that is included in a state family $\{\rho_\theta | \theta \in \mathbb{R}\}$ by performing a measurement. The goal is to find θ. We will assume that a system has been prepared with n identical states in a similar manner to Chap. 2. In this case, the estimator is denoted by the pair $(\boldsymbol{M}^n, \hat{\theta}_n)$, where $\boldsymbol{M}^n$ is a POVM representing the measurement on the quantum system $\mathcal{H}^{\otimes n}$ (with the measurement data set Ω_n) and $\hat{\theta}_n$ is the map from Ω_n to the parameter space. In a similar way to the estimation of the probability distributions examined in Sect. 2.4, we assume the mean square error to be the measure of the error. If the parameter space is one-dimensional, an estimator with a smaller mean square error (MSE)

$$\hat{\mathrm{V}}_\theta(\boldsymbol{M}^n, \hat{\theta}_n) \stackrel{\mathrm{def}}{=} \sum_\omega (\hat{\theta}_n(\omega) - \theta)^2 \operatorname{Tr} \rho_\theta^{\otimes n} M^n(\omega) \tag{6.57}$$

results in a better estimation. We may then ask: what kind of estimator is then most appropriate for estimating a given state? One method is to choose a POVM $\boldsymbol{M}$, perform it n times, and decide the final estimate to be the maximum likelihood estimator $\hat{\theta}_{n,ML}$ with respect to the n trials of the probability distribution $\{\mathrm{P}_\theta^{\boldsymbol{M}} | \theta \in \mathbb{R}\}$. The mean square error is approximately $\frac{1}{n}(J_\theta^{\boldsymbol{M}})^{-1}$ in the asymptotic limit according to the discussion in Sect. 2.4, where $J_\theta^{\boldsymbol{M}}$ is the Fisher information at θ for the probability distribution $\{\mathrm{P}_\theta^{\boldsymbol{M}} | \theta \in \mathbb{R}\}$.

According to this argument there exists some arbitrariness in the choice of measurement $\boldsymbol{M}$. The essential point of quantum estimation is therefore to optimize this estimation procedure, including the choice of the POVM $\boldsymbol{M}$. It is evident that certain conditions must be imposed on the estimators. For example, consider an estimator $\hat{\theta}$ that always gives the value 0. If the true parameter is 0, the mean squared error is 0. On the other hand, if the true parameter is not 0, the mean squared error will be large. Such an estimator is clearly not useful; this indicates that the problem of the formulation of the optimization problem for our estimator must be considered more carefully. A simple example of such a condition is the *unbiasedness condition*

$$\mathrm{E}_\theta(\boldsymbol{M}^n,\hat{\theta}_n) \stackrel{\mathrm{def}}{=} \sum_\omega \hat{\theta}_n(\omega) \operatorname{Tr} \rho_\theta^{\otimes n} M^n(\omega) = \theta, \quad \forall \theta \in \Theta. \tag{6.58}$$

However, in general, the unbiasedness condition is too restrictive. In order to avoid this, we often consider the locally unbiased condition:

$$\mathrm{E}_\theta(\boldsymbol{M}^n,\hat{\theta}_n) = 0, \qquad \frac{d\mathrm{E}_\theta(\boldsymbol{M}^n,\hat{\theta}_n)}{d\theta} = 1 \tag{6.59}$$

at a fixed point $\theta \in \Theta$. However, since it depends on the true parameter, it is not so natural. As an intermediate condition, we often treat the asymptotic case, i.e., the asymptotic behavior when the number n of prepared sample spaces goes to infinity. In this case, the *asymptotic unbiasedness condition*:

$$\lim \mathrm{E}_\theta(\boldsymbol{M}^n,\hat{\theta}_n) = \theta, \quad \lim \frac{d}{d\theta}\mathrm{E}_\theta(\boldsymbol{M}^n,\hat{\theta}_n) = 1, \quad \forall \theta \in \Theta \tag{6.60}$$

is often imposed for a sequence of estimators $\{(\boldsymbol{M}^n,\hat{\theta}_n)\}_{n=1}^\infty$. The second condition guarantees a kind of uniformity of the convergence of $\mathrm{E}_\theta(\boldsymbol{M}^n,\hat{\theta}_n)$ to θ. We are now ready for the following theorem.

Theorem 6.6 *(Helstrom [202], Nagaoka [300]) If a sequence of estimators $\{(\boldsymbol{M}^n,\hat{\theta}_n)\}$ satisfies (6.60), then the following inequality holds:*

$$\underline{\lim}\, n\hat{\mathrm{V}}_\theta(\boldsymbol{M}^n,\hat{\theta}_n) \geq J_{\theta,s}^{-1}. \tag{6.61}$$

In the nonasymptotic case, when the locally unbiased condition (6.59) holds, inequality (6.61) also holds without limit [202]. Its proof is similar to the proof of Theorem 6.6.

In the above discussion, we focus on the asymptotic unbiasedness condition. Using the van Trees inequality [144,398], we can prove the same inequality almost everywhere without any assumption [143]. However, our methed has the following advantage. Indeed, our method concerns one point. Hence, by choosing a coordinate suitable for one point, we can treat the general error function in the asymptotic setting. However, the van Trees method can be applied only to an error function with a quadratic form because the van Trees method concerns a Bayes prior distribution, i.e., all points.

Proof.
Define $O(\boldsymbol{M}^n,\hat{\theta}_n) \stackrel{\mathrm{def}}{=} \sum_\omega (\hat{\theta}_n(\omega)-\theta)\boldsymbol{M}^n(\omega)$. Since $\sum_\omega \left(\hat{\theta}_n(\omega)-\theta\right)\boldsymbol{M}^n(\omega) = O(\boldsymbol{M}^n,\hat{\theta}_n)$,

$$\begin{aligned} 0 \leq & \sum_\omega \left((\hat{\theta}_n(\omega)-\theta) - O(\boldsymbol{M}^n,\hat{\theta}_n)\right) \boldsymbol{M}^n(\omega) \left((\hat{\theta}_n(\omega)-\theta) - O(\boldsymbol{M}^n,\hat{\theta}_n)\right) \\ = & \sum_\omega \left(\hat{\theta}_n(\omega)-\theta\right) \boldsymbol{M}^n(\omega) \left(\hat{\theta}_n(\omega)-\theta\right) - O(\boldsymbol{M}^n,\hat{\theta}_n)^2. \end{aligned} \tag{6.62}$$

The Schwarz inequality for the metric $\langle\ ,\ \rangle^{(e)}_{\rho_\theta^{\otimes n},s}$ yields

$$\begin{aligned}\hat{\mathrm{V}}_\theta(\boldsymbol{M}^n,\hat{\theta}_n) =& \mathrm{Tr} \sum_\omega \left(\hat{\theta}_n(\omega)-\theta\right) \boldsymbol{M}^n(\omega) \left(\hat{\theta}_n(\omega)-\theta\right) \rho_\theta^{\otimes n} \\ \geq& \mathrm{Tr}\, O(\boldsymbol{M}^n,\hat{\theta}_n)^2 \rho_\theta^{\otimes n} = \left(\left\| O(\boldsymbol{M}^n,\hat{\theta}_n) \right\|^{(e)}_{\rho_\theta^{\otimes n},s} \right)^2 \qquad (6.63) \\ \geq& \frac{\left(\langle L_{\theta,s,n}, O(\boldsymbol{M}^n,\hat{\theta}_n) \rangle^{(e)}_{\rho_\theta^{\otimes n},s} \right)^2}{\left(\| L_{\theta,s,n} \|^{(e)}_{\rho_\theta^{\otimes n},s} \right)^2}, \qquad (6.64)\end{aligned}$$

where $L_{\theta,s,n}$ denotes the SLD e representation of the derivative of the state family $\{\rho_\theta^{\otimes n} | \theta \in \mathbb{R}\}$. Since $\mathrm{Tr}\, \theta \frac{d\rho_\theta^{\otimes n}}{d\theta} = \theta\, \mathrm{Tr} \frac{d\rho_\theta^{\otimes n}}{d\theta} = 0$,

$$\begin{aligned}\frac{d}{d\theta} \mathrm{E}_\theta(\boldsymbol{M}^n,\hat{\theta}_n) =& \mathrm{Tr}\, O(\boldsymbol{M}^n,\hat{\theta}_n) \frac{d\rho_\theta^{\otimes n}}{d\theta} = \mathrm{Tr} \left(O(\boldsymbol{M}^n,\hat{\theta}_n) - \theta \right) \frac{d\rho_\theta^{\otimes n}}{d\theta} \\ =& \langle L_{\theta,s,n}, O(\boldsymbol{M}^n,\hat{\theta}_n) - \theta \rangle^{(e)}_{\rho_\theta^{\otimes n},s} \qquad (6.65)\end{aligned}$$

from the definition of $\mathrm{E}_\theta(\boldsymbol{M}^n,\hat{\theta}_n)$. Combining the above two formulas with Exercise 6.14, we obtain

$$n\hat{\mathrm{V}}_\theta(\boldsymbol{M}^n,\hat{\theta}_n) \geq n \frac{\left(\frac{d}{d\theta} \mathrm{E}_\theta(\boldsymbol{M}^n,\hat{\theta}_n) \right)^2}{nJ_{\theta,s}}.$$

Taking the limit, we obtain (6.61). ∎

According to the above proof, if equality (6.60) holds for a finite number n, then inequality (6.61) also holds for the same number n. Conversely, given a point $\theta_0 \in \mathbb{R}$, we choose the function $\hat{\theta}_{\theta_0,n}$ and projections $E_{\theta_0,n}(\omega)$ such that the following is the spectral decomposition of the matrix $\frac{L_{\theta_0,s,n}}{nJ_{\theta_0,s}} + \theta_0$:

$$\frac{L_{\theta_0,s,n}}{nJ_{\theta_0,s}} + \theta = \sum_\omega E_{\theta_0,n}(\omega) \hat{\theta}_{\theta_0,n}(\omega). \qquad (6.66)$$

Then, $(\boldsymbol{E}_{\theta_0,n} = \{E_{\theta_0,n}(\omega)\}, \hat{\theta}_{\theta_0,n})$ gives an estimator satisfying

$$\hat{\mathrm{V}}_{\theta_0}(\boldsymbol{E}_{\theta_0,n},\hat{\theta}_n) = \frac{1}{nJ_{\theta_0,s}}, \quad \mathrm{E}_{\theta_0}(\boldsymbol{E}_{\theta_0,n},\hat{\theta}_n) = \theta_0.$$

We may then show that

$$\begin{aligned}\left. \frac{d\mathrm{E}_\theta(\boldsymbol{E}_{\theta_0,n},\hat{\theta}_{\theta_0,n})}{d\theta} \right|_{\theta=\theta_0} =& \langle L_{\theta_0,s,n}, O(\boldsymbol{M}^n,\hat{\theta}_n) - \theta \rangle^{(e)}_{\rho_{\theta_0}^{\otimes n},s} \\ =& \langle L_{\theta_0,s,n}, \frac{1}{nJ_{\theta_0,s}} L_{\theta_0,s,n} \rangle^{(e)}_{\rho_{\theta_0}^{\otimes n},s} = 1, \qquad (6.67)\end{aligned}$$

in a similar way to (6.65). This guarantees the existence of an estimator satisfying (6.61) under condition (6.60). However, it is crucial that the construction of $(E_{\theta_0,n}, \hat{\theta}_n)$ depend on θ_0. We may expect from (6.67) that if θ is in the neighborhood of θ_0, $\hat{\mathrm{V}}_\theta(E_{\theta_0,n}, \hat{\theta}_n)$ will not be very different from $\hat{\mathrm{V}}_{\theta_0}(E_{\theta_0,n}, \hat{\theta}_n)$. However, if θ is far away from θ_0, it is impossible to estimate $\hat{\mathrm{V}}_\theta(E_{\theta_0,n}, \hat{\theta}_n)$. The reason is that the SLD $L_{\theta_0,s,n}$ depends on θ_0. If the SLD $L_{\theta_0,s,n}$ did not depend on θ_0, one would expect that an appropriate estimator could be constructed independently of θ_0. This is the subject of the following theorem.

Theorem 6.7 *(Nagaoka [298, 300]) Assume that a distribution $p(\theta)$ satisfying*

$$\int \rho_\theta p(\theta)\, d\theta > 0 \tag{6.68}$$

exists. Then, the following two conditions for the quantum state family ρ_θ and the estimator $(\boldsymbol{M}, \hat{\theta})$ are equivalent.

① *The estimator $(\boldsymbol{M}, \hat{\theta})$ satisfies the unbiasedness condition (6.58), and the MSE $\hat{\mathrm{V}}_\theta(\boldsymbol{M}, \hat{\theta})$ satisfies*

$$\hat{\mathrm{V}}_\theta(\boldsymbol{M}, \hat{\theta}) = J_{\theta,s}^{-1}. \tag{6.69}$$

② *The state family is an SLD e geodesic $\rho_\theta = \Pi_{L,s}^\theta \rho_0$ given by (6.33); further, the parameter to be estimated equals the expectation parameter $\eta = \operatorname{Tr} L\rho_\theta$, and the estimator $(\boldsymbol{M}, \hat{\theta})$ equals the spectral decomposition of L.*

See Exercises 6.31 and 6.32 for the proof of the above theorem.

Therefore, the bound $\frac{1}{nJ_{\theta,s}}$ is attained in the nonasymptotic case only for the case (6.33), i.e., the SLD e geodesic curve. Another example is the case when a POVM $\boldsymbol{M}$ exists such that

$$J_\theta^{\boldsymbol{M}} = J_{\theta,s} \text{ for } \forall\theta, \tag{6.70}$$

where $J_\theta^{\boldsymbol{M}}$ is the Fisher information of the probability distribution $\{\mathrm{P}_{\rho_\theta}^{\boldsymbol{M}} | \theta \in \mathbb{R}\}$. Then, if one performs the measurement $\boldsymbol{M}$ on n prepared systems and chooses the maximum likelihood estimator of the n trials of the probability distribution $\{\mathrm{P}_{\rho_\theta}^{\boldsymbol{M}} | \theta \in \mathbb{R}\}$, the equality in inequality (6.61) is ensured according to the discussion in Sect. 2.4. Therefore, $J_{\theta,s}^{-1}$ is also attainable asymptotically in this case. In general, a POVM $\boldsymbol{M}$ satisfying (6.70) rarely exists; however, such a state family is called a *quasiclassical* POVM [429].

Besides the above rare examples, the equality of (6.61) is satisfied in the limit $n \to \infty$ at all points, provided a sequence of estimators $\{(\boldsymbol{M}^n, \hat{\theta}_n)\}$ is constructed according to the following two-step estimation procedure [142, 182]. First, perform the measurement $\boldsymbol{M}$ satisfying $J_\theta^{\boldsymbol{M}} > 0$ for the first $\sqrt{n}$ systems. Next, perform the measurement $E_{\hat{\theta}_{ML,\sqrt{n}},n-\sqrt{n}}$ (defined previously) for the remaining $n-\sqrt{n}$ systems, based on the maximum likelihood estimator $\hat{\theta}_{ML,\sqrt{n}}$ for the probability distribution $\{\mathrm{P}^{\boldsymbol{M}}_{\rho_\theta}|\theta \in \mathbb{R}\}$. Finally, choose the final estimate according to $\hat{\theta}_n \stackrel{\text{def}}{=} \hat{\theta}_{\hat{\theta}_{ML,\sqrt{n}},n-\sqrt{n}}$, as given in (6.66). If n is sufficiently large, $\hat{\theta}_{ML,\sqrt{n}}$ will be in the neighborhood of the true parameter θ with a high probability. Hence, the expectation value of $(\hat{\theta}_n-\theta)^2$ is approximately equal to $\dfrac{1}{(n-\sqrt{n})J_{\theta,s}}$. Since $\lim n\dfrac{1}{(n-\sqrt{n})J_{\theta,s}} = \dfrac{1}{J_{\theta,s}}$, we can expect this estimator to satisfy the equality in (6.61). In fact, it is known that such an estimator does satisfy the equality in (6.61) [142, 182].

In summary, for the single-parameter case, it is the SLD Fisher metric and not the Bogoljubov Fisher metric that gives the bound in estimation theory. On the other hand, the Bogoljubov Fisher metric does play a role in large deviation evaluation, although it appears in a rather restricted way.

Exercises

6.29. Show that the measurement $\boldsymbol{E}_{\theta,n}$ defined in (6.66) satisfies $nJ_{\theta,s} = J_\theta^{\boldsymbol{E}_{\theta,n}}$.

6.30. Using the above result, show that an arbitrary inner product $\|A\|^{(m)}_{\rho,x}$ satisfying property (6.17) and $\|\rho^{-1}\|^{(m)}_{\rho,x} = 1$ satisfies $\|A\|^{(m)}_{\rho,s} \le \|A\|^{(m)}_{\rho,x}$.

6.31. Prove Theorem 6.7 for $\rho_\theta > 0$ following the steps below [298, 300].

a Assume that the estimator $(\boldsymbol{M},\hat{\theta})$ for the SLD e geodesic $\Pi^\theta_{L,s}\rho$ is given by the spectral decomposition of L. Show that the estimator $(\boldsymbol{M},\hat{\theta})$ satisfies the unbiasedness condition with respect to the expectation parameter.

b For an SLD e geodesic, show that $\frac{d\eta}{d\theta} = J_{\theta,s}$.

c For an SLD e geodesic, show that the SLD Fisher information $J_{\eta,s}$ for an expectation parameter η is equal to the inverse $J^{-1}_{\theta,s}$ of the SLD Fisher information for the natural parameter θ.

d Show that ① follows from ②.

e Show that $\theta(\eta) = \int_0^\eta J_{\eta',s}\,d\eta'$ for an SLD e geodesic curve.

f Show that $\mu_s(\theta(\eta)) = \int_0^\eta \eta' J_{\eta',s}\,d\eta'$ for an SLD e geodesic curve.

g Show that if ① is true, $\frac{1}{J_{\theta,s}}L_{\theta,s} = O(\boldsymbol{M},\hat{\theta}) - \theta$.

h Show that if ① is true, then $\frac{d\rho_\theta}{d\theta} = \frac{J_\eta}{2}((O(\boldsymbol{M},\hat{\theta})-\eta)\rho_\theta + \rho_\theta(O(\boldsymbol{M},\hat{\theta})-\eta))$, where η is the parameter to be estimated.

i Show that if $n=1$ and $\rho_\theta > 0$, the equality in (6.63) is satisfied only if the estimator $(\boldsymbol{M},\hat{\theta})$ is the spectral decomposition of $O(\boldsymbol{M},\hat{\theta})$.

j Show that if ① holds, then ② holds.

6.32. Show that Theorem 6.7 holds even if $\rho_\theta > 0$ is not true, following the steps below. The fact that ② ⇒ ① still follows from above.

a Show that **h** in Exercise 6.31 still holds for $\rho_\theta > 0$.
b Show that if ① holds, then the RHS of (6.62) is equal to 0.
c Show that if ① holds, then ② holds.

6.33. Similarly to (6.63), show

$$\sum_\omega \mathrm{Tr}\left(\hat\theta_n(\omega)-\theta\right)^2 \boldsymbol{M}^n(\omega)\rho_\theta^{\otimes n} \geq \left(\left\|O(\boldsymbol{M}^n,\hat\theta_n)\right\|^{(e)}_{\rho_\theta^{\otimes n},r}\right)^2.$$

6.5 Large Deviation Evaluation

In Sect. 2.5.2, we discussed the large deviation type estimation of a probability distribution for the case of a single parameter. In this section, we will examine the theory for large deviation in the case of quantum state estimation. As defined in (2.126) and (2.127), $\beta(\{\boldsymbol{M}^n,\hat\theta_n\})$ and $\alpha(\{\boldsymbol{M}^n,\hat\theta_n\})$ are defined as follows:

$$\beta(\{\boldsymbol{M}^n,\hat\theta_n\},\theta,\epsilon) \stackrel{\mathrm{def}}{=} \underline{\lim}\, \frac{-1}{n}\log \mathrm{Tr}\,\rho^{\otimes n}\boldsymbol{M}^n\{|\hat\theta_n-\theta|\geq\epsilon\}, \tag{6.71}$$

$$\alpha(\{\boldsymbol{M}^n,\hat\theta_n\},\theta) \stackrel{\mathrm{def}}{=} \underline{\lim}_{\,\epsilon\to 0}\frac{\beta(\{\boldsymbol{M}^n,\hat\theta_n\},\theta,\epsilon)}{\epsilon^2}. \tag{6.72}$$

The notation $\boldsymbol{M}^n\{|\hat\theta_n-\theta|\geq\epsilon\}$ requires some explanation. For the general POVM $\boldsymbol{M}=\{M(\omega)\}$ and the set B, let us define $\boldsymbol{M}B$ according to

$$\boldsymbol{M}B \stackrel{\mathrm{def}}{=} \sum_{\omega\in B} M(\omega). \tag{6.73}$$

Then, we have the following theorem in analogy to Theorem 2.7.

Theorem 6.8 *(Nagaoka [303]) Let the sequence of estimators* $\boldsymbol{M} = \{(\boldsymbol{M}^n,\hat\theta_n)\}$ *satisfy the weak consistency condition*

$$\mathrm{Tr}\,\rho^{\otimes n}\boldsymbol{M}^n\{|\hat\theta_n-\theta|\geq\epsilon\}\to 0, \quad \forall\epsilon>0, \forall\theta\in\mathbb{R}. \tag{6.74}$$

The following then holds:

$$\beta(\{\boldsymbol{M}^n,\hat\theta_n\},\theta,\epsilon) \leq \inf_{\theta':|\theta'-\theta|>\epsilon} D(\rho_{\theta'}\|\rho_\theta), \tag{6.75}$$

$$\alpha(\{\boldsymbol{M}^n,\hat\theta_n\},\theta) \leq \frac{1}{2}J_{\theta,b}. \tag{6.76}$$

A different inequality that evaluates the performance of the estimator may be obtained by employing a slight reformulation. That is, a relation similar to (6.75) can be obtained as given by the following lemma.

Lemma 6.2 *(Hayashi [177]) Define* $\beta'(\boldsymbol{M}, \theta, \delta) \stackrel{\rm def}{=} \lim_{\epsilon \to +0} \beta(\boldsymbol{M}, \theta, \delta - \epsilon)$ *for the sequence of estimators* $\boldsymbol{M} = \{(\boldsymbol{M}^n, \hat{\theta}_n)\}$*. The following inequality then holds:*

$$\inf_{\{s|1 \geq s \geq 0\}} (\beta'(\boldsymbol{M}, \theta, s\delta) + \beta'(\boldsymbol{M}, \theta + \delta, (1-s)\delta)) \leq -2 \log \mathrm{Tr}\, |\sqrt{\rho_\theta}\sqrt{\rho_{\theta+\delta}}|. \tag{6.77}$$

The essential part in the proof of this lemma is that the information $-\log \mathrm{Tr}\, |\sqrt{\rho}\sqrt{\sigma}|$ satisfies the information-processing inequality Ex. 6.35.

The relation corresponding to (6.76) is then given by the following theorem.

Theorem 6.9 *(Hayashi [177]) Let the sequence of estimators* $\boldsymbol{M} = \{(\boldsymbol{M}^n, \hat{\theta}_n)\}$ *satisfy the weak consistency condition and the uniformal convergence on the RHS of (6.72) with respect to* θ*. Define* $\alpha'(\boldsymbol{M}, \theta_0) \stackrel{\rm def}{=} \underline{\lim}_{\,\theta \to \theta_0} \alpha(\boldsymbol{M}, \theta_0)$*. The following inequality then holds:*

$$\alpha'(\boldsymbol{M}, \theta) \leq \frac{J_{\theta,s}}{2}. \tag{6.78}$$

Hence, the bound $\frac{J_{\theta,s}}{2}$ can be regarded as the bound under the condition for the following sequence of estimators:

$$\alpha(\boldsymbol{M}, \theta_0) = \underline{\lim}_{\,\theta \to \theta_0} \alpha(\boldsymbol{M}, \theta_0), \quad \beta(\boldsymbol{M}, \theta, \delta) = \lim_{\epsilon \to +0} \beta(\boldsymbol{M}, \theta, \delta - \epsilon). \tag{6.79}$$

So far, we have discussed the upper bound of $\alpha(\boldsymbol{M}, \theta)$ in two ways. The upper bound given here can be attained in both ways, as we now describe. Let us first focus on the upper bound $\frac{J_{\theta,s}}{2}$ given by (6.78), which is based on the SLD Fisher information. This upper bound can be attained by a sequence of estimators $\boldsymbol{M} = \{(\boldsymbol{M}^n, \hat{\theta}_n)\}$ such that $\alpha'(\boldsymbol{M}, \theta_0) = \alpha(\boldsymbol{M}, \theta_0)$ and the RHS of (6.72) converges uniformly concerning θ. This kind of estimator can be constructed according to the two-step estimator given in the previous section [177]. Let us now examine the upper bound given by (6.76) using the Bogoljubov Fisher information, which equals $\frac{J_{\theta,b}}{2}$ in this case. This bound can be attained by a sequence of estimators satisfying the weak coincidence condition but not the uniform convergence on the RHS of (6.72). However, this estimator can attain this only at a single point [177]. Although this method of construction is rather obtuse, the method is similar to the construction of the measurement that attains the bound $D(\rho\|\sigma)$ given in Stein's lemma for hypothesis testing. Of course, such an estimator is extremely unnatural and cannot be used in practice. Therefore, we see that the two bounds provide the respective answers for two completely separate problems. In a classical system, the bounds for these two problems are identical. This difference arises due to the quantum nature of the problem.

The above discussion indicates that geometrical characterization does not connect to quantum state estimation. However, there are two different approaches from the geometrical viewpoint. For example, Hayashi [179] focused on the scalar curvature of the Riemannian connection and clarified the relation between the scalar curvature and the second-order asymptotics of estimation error only for specific state families. These approaches treat the translation of the tangent bundle of state space. Matsumoto [283–285] focused on that of the line bundle and discovered the relation between the curvature and the bound of estimation error for the pure-state family. He pointed out that the difficulty of two parameters is closely related to the curvature.

Exercises

6.34. Prove Theorem 6.8 referring to the proof of Theorem 2.7.

6.35. Prove Lemma 6.2 following the steps below.

a Show that $\log \operatorname{Tr} |\sqrt{\rho^{\otimes n}}\sqrt{\sigma^{\otimes n}}| = n \log \operatorname{Tr} |\sqrt{\rho}\sqrt{\sigma}|$.
b Show that

$$
\begin{aligned}
&\log \operatorname{Tr} \left|\sqrt{\rho_\theta^{\otimes n}}\sqrt{\rho_{\theta+\delta}^{\otimes n}}\right| \\
\leq &\log\left[\left(\operatorname{Tr} \rho_{\theta+\delta}^{\otimes n} \boldsymbol{M}^n \left\{|\hat{\theta}-(\theta+\delta)| \geq \frac{\delta(m-1)}{m}\right\}\right)^{\frac{1}{2}} + \left(\operatorname{Tr} \rho_\theta^{\otimes n} \boldsymbol{M}^n \left\{|\hat{\theta}-\theta| \geq \delta\right\}\right)^{\frac{1}{2}}\right. \\
&\left. + \sum_{i=1}^m \left(\operatorname{Tr} \rho_\theta^{\otimes n} \boldsymbol{M}^n \left\{|\hat{\theta}-\theta| > \frac{\delta(i-1)}{m}\right\}\right)^{\frac{1}{2}} \left(\operatorname{Tr} \rho_{\theta+\delta}^{\otimes n} \boldsymbol{M}^n \left\{|\hat{\theta}-(\theta+\delta)| \geq \frac{\delta(m-i-1)}{m}\right\}\right)^{\frac{1}{2}}\right]
\end{aligned}
$$

for an arbitrary integer m from the fact that the amount of information $-\log \operatorname{Tr} |\sqrt{\rho}\sqrt{\sigma}|$ satisfies the information-processing inequality.
c Choosing a sufficiently large integer N for a real number $\epsilon > 0$ and an integer m, we have

$$
\frac{1}{n} \log \operatorname{Tr} \rho_\theta^{\otimes n} \boldsymbol{M}^n \left\{|\hat{\theta} - \theta| \geq \frac{\delta i}{m}\right\} \leq -\beta\left(\boldsymbol{M}, \theta, \frac{\delta i}{m}\right) + \epsilon
$$

$$
\frac{1}{n} \log \operatorname{Tr} \rho_{\theta+\delta}^{\otimes n} \boldsymbol{M}^n \left\{|\hat{\theta} - (\theta+\delta)| \geq \frac{\delta(m-i)}{m}\right\} \leq -\beta\left(\boldsymbol{M}, \theta+\delta, \frac{\delta(m-i)}{m}\right) + \epsilon,
$$

for $\forall n \geq N, 0 \leq \forall i \leq m$. Show that

$$
\begin{aligned}
&n \log \operatorname{Tr} |\sqrt{\rho_\theta}\sqrt{\rho_{\theta+\delta}}| \\
&\leq \log(m+2) - \frac{n}{2}\left(\min_{0\leq i\leq m} \beta\left(\boldsymbol{M}, \theta, \frac{\delta(i-1)}{m}\right) + \beta\left(\boldsymbol{M}, \theta+\delta, \frac{\delta(m-i-1)}{m}\right) - 2\epsilon\right).
\end{aligned}
$$

d Show the following for an arbitrary integer m:

$$
\begin{aligned}
&-\log \operatorname{Tr} |\sqrt{\rho_\theta}\sqrt{\rho_{\theta+\delta}}| \\
&\geq \frac{1}{2}\min_{0\le i\le m}\left(\beta\left(\boldsymbol{M},\theta,\frac{\delta(i-1)}{m}\right)+\beta\left(\boldsymbol{M},\theta+\delta,\frac{\delta(m-i-1)}{m}\right)-2\epsilon\right).
\end{aligned}
$$

e Prove (6.77) using **d**.

6.36. Prove Theorem 6.9 using Lemma 6.2.

6.6 Multiparameter Estimation

Let us now examine the case of a multidimensional parameter space Θ (dimension d). Assume that the unknown state lies in the multiparameter quantum state family $\{\rho_\theta|\theta \in \Theta \subset \mathbb{R}^d\}$. A typical estimation procedure is as follows. An appropriate POVM $\boldsymbol{M}$ is chosen in a manner similar to the previous one (excluding those for a Fisher matrix $\boldsymbol{J}_\theta^{\boldsymbol{M}}$ with zero eigenvalues), and a measurement corresponding to $\boldsymbol{M}$ is performed on each n quantum system in unknown but identical states. The final estimate is then given by the maximum likelihood estimator for the probability distribution $\{\mathrm{P}_\theta^{\boldsymbol{M}}|\theta \in \Theta \subset \mathbb{R}^d\}$. According to Sect. 2.4, the mean square error asymptotically approaches $\frac{1}{n}(\boldsymbol{J}_\theta^{\boldsymbol{M}})^{-1}$ in this case. The maximum likelihood estimator then approaches the true parameter θ in probability. As mentioned in the previous section, our problem is the optimization of the quantum measurement $\boldsymbol{M}$ for our estimation. To this end, we need to find an estimator minimizing the mean square error $\hat{\mathrm{V}}_\theta^{i,j}(\boldsymbol{M}^n,\hat{\theta}_n)$ for the ith parameter θ^i or the mean square error matrix $\hat{\boldsymbol{V}}_\theta(\boldsymbol{M}^n,\hat{\theta}_n) = [\hat{\mathrm{V}}_\theta^{i,j}(\boldsymbol{M}^n,\hat{\theta}_n)]$ by taking into account the correlations between the θ^i, where

$$
\hat{\mathrm{V}}_\theta^{i,j}(\boldsymbol{M}^n,\hat{\theta}_n) \stackrel{\text{def}}{=} \sum_\omega (\hat{\theta}_n^i(\omega)-\theta^i)(\hat{\theta}_n^j(\omega)-\theta^j)\operatorname{Tr}\rho_\theta^{\otimes n}M^n(\omega). \tag{6.80}
$$

The unbiasedness condition is then given by

$$
\mathrm{E}_\theta^i(\boldsymbol{M}^n,\hat{\theta}_n) \stackrel{\text{def}}{=} \sum_\omega \hat{\theta}_n^i(\omega)\operatorname{Tr}\rho_\theta^{\otimes n}M^n(\omega) = \theta^i, \quad \forall\theta\in\Theta.
$$

In the asymptotic case, for a sequence of estimators $\{(\boldsymbol{M}^n,\hat{\theta}_n)\}$, we can also write down the asymptotic unbiasedness condition

$$
\lim \mathrm{E}_\theta^i(\boldsymbol{M}^n,\hat{\theta}_n) = \theta^i, \quad \lim \frac{\partial}{\partial\theta^j}\mathrm{E}_\theta^i(\boldsymbol{M}^n,\hat{\theta}_n) = \delta_j^i, \quad \forall\theta\in\Theta. \tag{6.81}
$$

Theorem 6.10 *Let the sequence of estimators $\{(\boldsymbol{M}^n,\hat{\theta}_n)\}$ satisfy the asymptotic unbiasedness condition (6.81) and have the limit $\hat{\mathrm{V}}_\theta^{i,j}(\{\boldsymbol{M}^n,\hat{\theta}_n\}) \stackrel{\text{def}}{=} \lim n\hat{\mathrm{V}}_\theta^{i,j}(\boldsymbol{M}^n,\hat{\theta}_n)$. The following matrix inequality then holds [202, 216]:*

$$
\lim n\hat{\boldsymbol{V}}_\theta(\boldsymbol{M}^n,\hat{\theta}_n) \geq (\boldsymbol{J}_{\theta,x})^{-1}, \quad x=s,r. \tag{6.82}
$$

Proof. First, assume that any two complex vectors $|b\rangle = (b_1, \ldots, b_d)^T \in \mathbb{C}^d$ and $|a\rangle \in \mathbb{C}^d$ satisfy

$$\langle b|\hat{\boldsymbol{V}}_\theta(\{\boldsymbol{M}^n, \hat{\theta}_n\})|b\rangle\langle a|\boldsymbol{J}_{\theta,x}|a\rangle \geq |\langle b|a\rangle|^2. \tag{6.83}$$

Substituting $a = (\boldsymbol{J}_{\theta,x})^{-1}b$ into (6.83), we have $\langle b|\hat{\boldsymbol{V}}_\theta(\{\boldsymbol{M}^n, \hat{\theta}_n\})|b\rangle \geq \langle b|(\boldsymbol{J}_{\theta,x})^{-1}|b\rangle$, since $(\boldsymbol{J}_{\theta,x})^{-1}$ is Hermitian. We therefore obtain (6.82).

We next show (6.83). Define $O_n \stackrel{\text{def}}{=} \sum_\omega \left(\sum_i (\hat{\theta}^i_n(\omega) - \theta^i)b_i\right) M^n(\omega)$ and $L_n \stackrel{\text{def}}{=} \sum_j L_{\theta,j,x,n}a_j$. Using (6.63) and Exercise 6.33, we can show that

$$\begin{aligned}\langle b|\hat{\boldsymbol{V}}_\theta(\{\boldsymbol{M}^n, \hat{\theta}_n\})|b\rangle &= \lim n \sum_\omega \left|\sum_i (\hat{\theta}^i_n(\omega) - \theta^i)b_i\right|^2 \operatorname{Tr} \rho_\theta^{\otimes n} M^n(\omega) \\ &\geq \lim n \left(\|O_n\|^{(e)}_{\rho_\theta^{\otimes n},x}\right)^2,\end{aligned}$$

in a manner similar to (6.62). Then

$$\langle b|a\rangle = \lim \sum_{i,j} \overline{b}_i \frac{\partial}{\partial \theta^j} \mathrm{E}^i_\theta(\boldsymbol{M}^n, \hat{\theta}_n)a_j = \lim \langle O_n, L_n\rangle^{(e)}_{\rho_\theta^{\otimes n},x},$$

in a manner similar to (6.67). Using the Schwarz inequality, we can show that

$$\left(\|O_n\|^{(e)}_{\rho_\theta^{\otimes n},x}\right)^2 \left(\|L_n\|^{(e)}_{\rho_\theta^{\otimes n},x}\right)^2 \geq \left|\langle O_n, L_n\rangle^{(e)}_{\rho_\theta^{\otimes n},x}\right|^2.$$

Inequality (6.83) can be obtained on taking the limit because $\langle a|\boldsymbol{J}_{\theta,x}|a\rangle = n\left(\|L_n\|^{(e)}_{\rho_\theta^{\otimes n},x}\right)^2$. ∎

In general, there is no sequence of estimators that satisfies the equality in (6.82). Furthermore, as the matrix $\hat{\mathrm{V}}^{i,j}_\theta(\boldsymbol{M}^n, \hat{\theta}_n)$ is a real symmetric matrix and not a real number, there is no general minimum matrix $\hat{\mathrm{V}}^{i,j}_\theta(\boldsymbol{M}^n, \hat{\theta}_n)$ among the estimators satisfying (6.81). Instead, one can adopt the sum of MSE, i.e., the trace of $\hat{\mathrm{V}}^{i,j}_\theta(\boldsymbol{M}^n, \hat{\theta}_n)$ as our error criterion. It is therefore necessary to consider the minimum of $\operatorname{tr} \lim n\hat{\boldsymbol{V}}_\theta(\boldsymbol{M}^n, \hat{\theta}_n)$ in the asymptotic case.

From (6.82) the lower bound of the minimum value of this quantity can be evaluated as

$$\operatorname{tr} \lim n\hat{\boldsymbol{V}}_\theta(\boldsymbol{M}^n, \hat{\theta}_n) \geq \min\{\operatorname{tr} \boldsymbol{V} | \boldsymbol{V} : \text{real symmetric } \boldsymbol{V} \geq (\boldsymbol{J}_{\theta,x})^{-1}\} \tag{6.84}$$

because $\hat{\boldsymbol{V}}_\theta(\boldsymbol{M}^n, \hat{\theta}_n)$ is real symmetric. If $(\boldsymbol{J}_{\theta,x})^{-1}$ is real symmetric, the RHS is equal to $\operatorname{tr}(\boldsymbol{J}_{\theta,x})^{-1}$. If $(\boldsymbol{J}_{\theta,x})^{-1}$ is a Hermitian matrix but contains imaginary elements, the RHS will be larger than $\operatorname{tr}(\boldsymbol{J}_{\theta,x})^{-1}$. In this case, we may calculate [216]

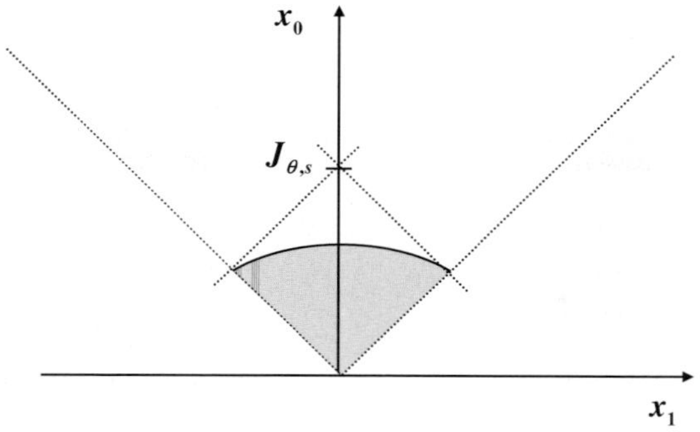

Fig. 6.2. Fisher information matrices

$$\min\{\operatorname{tr} \boldsymbol{V} | \boldsymbol{V} : \text{real symmetric} \boldsymbol{V} \geq \boldsymbol{J}_{\theta,x}^{-1}\} = \operatorname{tr} \mathbf{Re}(\boldsymbol{J}_{\theta,x}^{-1}) + \operatorname{tr} |\mathbf{Im}(\boldsymbol{J}_{\theta,x}^{-1})|. \tag{6.85}$$

For example, $(\boldsymbol{J}_{\theta,s})^{-1}$ and $(\boldsymbol{J}_{\theta,b})^{-1}$ are real symmetric matrices, as discussed in Exercise 6.17. However, since the RLD Fisher information matrix $(\boldsymbol{J}_{\theta,r})^{-1}$ possesses imaginary components, the RHS of (6.85) in the RLD case will be larger than $\operatorname{tr}(\boldsymbol{J}_{\theta,r})^{-1}$. Moreover, in order to treat the set of the limits of MSE matrices, we often minimize $\operatorname{tr} G\hat{\boldsymbol{V}}_\theta(\{\boldsymbol{M}^n, \hat{\theta}_n\})$. From a discussion similar to (6.85), it can be shown that the minimum is greater than $\operatorname{tr} \sqrt{G} \mathbf{Re}(\boldsymbol{J}_{\theta,r}^{-1})\sqrt{G} + \operatorname{tr} |\sqrt{G} \mathbf{Im}(\boldsymbol{J}_{\theta,r}^{-1})\sqrt{G}|$. Its equality holds only when $\hat{\boldsymbol{V}}_\theta(\{\boldsymbol{M}^n, \hat{\theta}_n\}) = \mathbf{Re}(\boldsymbol{J}_{\theta,r}^{-1}) + \sqrt{G}^{-1} |\sqrt{G} \mathbf{Im}(\boldsymbol{J}_{\theta,r}^{-1})\sqrt{G}| \sqrt{G}^{-1}$. When the family in the two-dimensional space has the form $\{\rho_{\boldsymbol{\theta}} | \|\boldsymbol{\theta}\| = r\}$, the set of MSE matrices is restricted by the RLD Fisher information matrix, as shown in Fig. 6.2.

In Fig. 6.2, we use the parameterization $\begin{pmatrix} x_0 + x_1 & x_2 \\ x_2 & x_0 - x_1 \end{pmatrix}$ and assume that $J_{\theta,s}$ is a constant time of the identity matrix. In addition, it was shown that these limits of MSE matrices can be attained [188]. The above figure also illustrates that the set of MSE matrices can be realized by the adaptive estimators. See Exercises 6.20 and 6.45.

The following theorem gives the asymptotic lower bound of $\operatorname{tr} \hat{\boldsymbol{V}}_\theta(\boldsymbol{M}^n, \hat{\theta}_n)$.

Theorem 6.11 *Let the sequence of estimators $\{(\boldsymbol{M}^n, \hat{\theta}_n)\}$ satisfy the same conditions as Theorem 6.10. The following inequality then holds:*

$$\operatorname{tr} \lim n\hat{\boldsymbol{V}}_\theta(\boldsymbol{M}^n, \hat{\theta}_n) \geq \overline{\lim}\, n \inf_{\boldsymbol{M}^n:\ POVM\ on\ \mathcal{H}^{\otimes n}} \operatorname{tr}(\boldsymbol{J}_\theta^{\boldsymbol{M}^n})^{-1}. \tag{6.86}$$

Conversely, we can construct the estimator attaining the bound $\min_{\boldsymbol{M}} \operatorname{tr}(\boldsymbol{J}_\theta^{\boldsymbol{M}})^{-1}$ by using the adaptive method in a manner similar to the one-parameter case. Moreover, applying this method to the n-fold tensor product system $\mathcal{H}^{\otimes n}$, we can construct an estimator attaining the bound

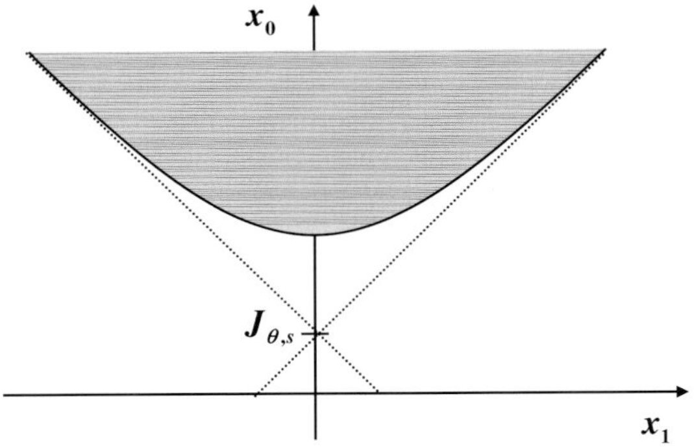

Fig. 6.3. MSE matrices

$n \min_{M^n} \operatorname{tr}(J_\theta^{M^n})^{-1}$. Hence, the set of realizable classical Fisher information J_θ^{M} and the set of $\frac{1}{n} J_\theta^{M^n}$ characterize the bound of estimation performance. When the family in the two-dimensional space has the form $\{\rho_{\boldsymbol{\theta}} | \|\boldsymbol{\theta}\| = r\}$, they are as illustrated in Fig. 6.3. In Fig. 6.3, we assume that $J_{\theta,s}$ is a constant time of the identity matrix.

Proof. Let us apply the same argument as in the proof of Theorem 6.10 to the probability distribution $\{\mathrm{P}_{\rho_\theta^{\otimes n}}^{M^n} | \theta\}$. Then

$$\langle b|\hat{V}_\theta(M^n, \hat{\theta}_n)|b\rangle \langle a|J_\theta^{M^n}|a\rangle \geq \left| \sum_{i,j} \overline{b_i} \frac{\partial}{\partial \theta^j} \mathrm{E}_\theta(M^n, \hat{\theta}_n)^i a_j \right|^2$$

for complex vectors $|b\rangle = (b_1, \ldots, b_d)^T \in \mathbb{C}^d$ and $|a\rangle \in \mathbb{C}^d$. Define $(A_n)^i_j \stackrel{\text{def}}{=} \frac{\partial}{\partial \theta^j} \mathrm{E}_\theta(M^n, \hat{\theta}_n)^i$ and substitute $a = (J_\theta^{M^n})^{-1} A_n b$. Then $\langle b|\hat{V}_\theta(M^n, \hat{\theta}_n)|b\rangle \geq \langle b|A_n^*(J_\theta^{M^n})^{-1} A_n|b\rangle$. Therefore,

$$\begin{aligned}
&\lim \operatorname{tr} n\hat{V}_\theta(M^n, \hat{\theta}_n) \geq \overline{\lim} \operatorname{tr} A_n A_n^* n(J_\theta^{M^n})^{-1} \\
&\geq \overline{\lim} \inf_{M^n\colon \text{POVM on } \mathcal{H}^{\otimes n}} \operatorname{tr} A_n A_n^* n(J_\theta^{M^n})^{-1} \\
&= \overline{\lim}\, n \inf_{M^n\colon \text{POVM on } \mathcal{H}^{\otimes n}} \operatorname{tr}(J_\theta^{M^n})^{-1},
\end{aligned}$$

which completes the proof. ∎

More generally, under the same conditions as in the previous theorem, we have [216]

$$\begin{aligned}
&\operatorname{tr} \lim n\hat{V}_\theta(M^n, \hat{\theta}_n) \\
&\geq \min_{\mathbf{X}} \left\{ \operatorname{tr} \mathbf{Re}\, V_\theta(\mathbf{X}) + \operatorname{tr} |\mathbf{Im}\, V_\theta(\mathbf{X})| \,\middle|\, \delta_i^j = \operatorname{Tr} \frac{\partial \rho_\theta}{\partial \theta^i} X^j \right\}, \qquad (6.87)
\end{aligned}$$

where X is a matrix, $\mathbf{Re}X$ is the matrix consisting of the real part of each component of X, $\mathbf{Im}X$ is the matrix consisting of the imaginary part of each component of X, and $\boldsymbol{V}_\theta(\mathbf{X}) \stackrel{\text{def}}{=} (\operatorname{Tr} \rho_\theta X^i X^j)$ for a vector of matrices $\mathbf{X} = (X^1, \cdots, X^d)$. It is known that there exists a sequence of estimators satisfying the equality in (6.87) [178,286]. In the proof of this argument, the quantum central limit theorem [145,344] plays an essential role [178].

Such an argument can be given for infinite-dimensional systems. In particular, the quantum Gaussian state family is known as the quantum analog of the Gaussian distribution family and is a typical example in an infinite-dimensional system. In the classical case, the Gaussian distribution family has been extensively investigated. Similarly, the quantum Gaussian state family has been extensively investigated in the classical case [119,170,171,178,188,213,216,432].

Another related topic to state estimation is approximate state cloning. Of course, it is impossible to completely clone a given state. However, an approximate cloning is possible by first estimating the state to be cloned, then generating this estimated state twice. Although the initial state is demolished in this case, it can be approximately recovered from the knowledge gained via the estimation. An approximate cloning is therefore possible via state estimation. In fact, it is more straightforward to treat the cloning process directly without performing the estimation. Then, the optimum cloning method is strictly better than the method via estimation [61]. In particular, the analysis for approximate state cloning is simplified for spaces having a group symmetry, e.g., sets of pure states [251,412]. An investigation has also been done in an attempt to find the interaction that realizes this [109]. The analysis is more difficult for problems with less symmetry [110].

The probabilistic framework of mathematical statistics has been applied to many fields where statistical methods are necessary. In many cases, this probabilistic framework is merely a convenience for the applied field. That is, the probabilistic description is often used to supplement the lack of knowledge of the system of interest. Therefore, it is not uncommon for statistical methods to be superceded by other methods during as the field advances due to reasons such as increasing computer speed and improvements in analysis. However, as discussed in Chap. 1, the probabilistic nature of quantum mechanics is intrinsic to the theory itself. Therefore, in fact, the framework of mathematical statistics can be naturally applied to quantum mechanics. Unfortunately, at present, it is not possible to operate a large number of quantum-mechanical particles as a single quantum system. Therefore, when we measure the order of 10^{23} particles, we often obtain only the average of the measured ensemble as the final data. The quantum-mechanical correlations cannot be observed in this situation. Furthermore, quantum-mechanical effects such as those given in this text cannot be realized. Additionally, when an observable X is measured with respect to a system in the state ρ, the measurement data coincide with $\operatorname{Tr} \rho X$ with a probability nearly equal to 1. Therefore, statistical methods are clearly not necessary in this case.

We can expect an increase in the demand for individually operating a large number of quantum-mechanical particles in proportion to experimental technology advances in microscopic systems. The measurement data will behave probabilistically in this situation, and therefore mathematical statistical methods will become more necessary. In fact, in several experiments, statistical methods have already been used to determine the generated quantum state [395]. Therefore, the theory presented here should become more important with future experimental progress.

Exercises

6.37. Show the following facts when a separable POVM $\boldsymbol{M}^n = \{M^n(\omega)\}_{\omega \in \Omega_n}$ in $\mathcal{H}^{\otimes n}$ is written as $M^n(\omega) = M_1^n(\omega) \otimes \cdots \otimes M_n^n(\omega)$.

a Show that a POVM $\boldsymbol{M}_{\theta:n,i}$ defined by (6.88) satisfies the conditions for a POVM and satisfies (6.89):

$$M_{\theta:n,i}(\omega) \stackrel{\text{def}}{=} M_i^n(\omega) \mathrm{Tr} \rho_\theta M_1^n(\omega) \cdots \mathrm{Tr} \rho_\theta M_{i-1}^n(\omega) \cdot \mathrm{Tr} \rho_\theta M_{i+1}^n(\omega) \cdots \mathrm{Tr} \rho_\theta M_n^n(\omega), \tag{6.88}$$

$$\sum_{i=1}^d J_\theta^{\boldsymbol{M}_{\theta:n,i}} = J_\theta^{\boldsymbol{M}_n}. \tag{6.89}$$

b Show that

$$\mathrm{tr} \lim n \hat{\boldsymbol{V}}_\theta(\boldsymbol{M}^n, \hat{\theta}_n) \geq \inf \left\{ \mathrm{tr}(\boldsymbol{J}_\theta^{\boldsymbol{M}})^{-1} \middle| \boldsymbol{M} \text{ POVM on } \mathcal{H} \right\}. \tag{6.90}$$

6.38. Show the following given a POVM $\boldsymbol{M} = \{M_\omega\}$ in $\mathcal{H}$ of rank $M(\omega) = 1$ [142].

a Show that $\displaystyle\sum_\omega \frac{\langle M(\omega), M(\omega)\rangle_{\theta,s}^{(e)}}{\langle M(\omega), I\rangle_{\theta,s}^{(e)}} = \dim \mathcal{H}$.

b Show that $\mathrm{tr}\, \boldsymbol{J}_{\theta,s}^{-1} \boldsymbol{J}_\theta^{\boldsymbol{M}} + 1 = \displaystyle\sum_\omega \sum_{j=0}^d \frac{\langle M(\omega), L_{\theta,s}^j\rangle_{\theta,s}^{(e)} \langle L_{\theta,j,s}, M(\omega)\rangle_{\theta,s}^{(e)}}{\langle M(\omega), I\rangle_{\theta,s}^{(e)}}$, where $L_{\theta,s}^j \stackrel{\text{def}}{=} \sum_{i=1}^d (\boldsymbol{J}_{\theta,s}^{-1})_{i,j} L_{\theta,j,s}$.

c Show that

$$\mathrm{tr}\, \boldsymbol{J}_{\theta,s}^{-1} \boldsymbol{J}_\theta^{\boldsymbol{M}} \leq \dim \mathcal{H} - 1. \tag{6.91}$$

When $\rho_\theta > 0$, show that the equality holds if and only if every element $M(\omega)$ can be written as a linear sum of $I, L_{\theta,1,s}, \ldots, L_{\theta,d,s}$.

d Give the condition for the equality in (6.91) for cases other than $\rho_\theta > 0$.

e Show that inequality (6.91) also holds if the POVM $\boldsymbol{M} = \{M_\omega\}$ is not of rank $M(\omega) = 1$.

6.39. When an estimator $(\boldsymbol{M},\hat{\theta})$ for the state family $\{\rho_\theta|\theta\in\mathbb{R}^d\}$ in $\mathcal{H}$ satisfies

$$\frac{\partial}{\partial\theta^j}\mathrm{E}^i_\theta(\boldsymbol{M}^n,\hat{\theta}_n)\bigg|_{\theta=\theta_0}=\delta^i_j,\quad \mathrm{E}^i_{\theta_0}(\boldsymbol{M}^n,\hat{\theta}_n)=\theta^i_0,$$

it is called a *locally unbiased estimator* at θ_0. Show that

$$\begin{aligned}&\inf\left\{\operatorname{tr}(\boldsymbol{J}^{\boldsymbol{M}}_\theta)^{-1}\middle|\boldsymbol{M}\text{ POVM on }\mathcal{H}\right\}\\&=\inf\left\{\operatorname{tr}\hat{\boldsymbol{V}}_\theta(\boldsymbol{M},\hat{\theta})\middle|(\boldsymbol{M},\hat{\theta}):\text{ a locally unbiased estimator}\right\}.\end{aligned}\tag{6.92}$$

6.40. Show the following equation and that $\boldsymbol{J}=\frac{(d-1)}{\operatorname{tr}\boldsymbol{J}^{-\frac{1}{2}}_{\theta,s}}\boldsymbol{J}^{\frac{1}{2}}_{\theta,s}$ gives the minimum value [142].

$$\min_{\boldsymbol{J}:\text{ symmetric matrix}}\left\{\operatorname{Tr}(\boldsymbol{J}^{-1})\middle|\operatorname{Tr}\boldsymbol{J}^{-1}_{\theta,s}\boldsymbol{J}=d-1\right\}=\frac{(\operatorname{tr}\boldsymbol{J}^{-\frac{1}{2}}_{\theta,s})^2}{d-1}.$$

6.41. Let $\boldsymbol{M}^u$ be a measurement corresponding to the spectral decomposition of $L(u)\stackrel{\text{def}}{=}\sum_{j=1}^d u^jL_{\theta,j,s}$, where $\|u=(u^1,\ldots,u^d)\|=1$. Show that the Fisher information satisfies

$$\boldsymbol{J}^{\boldsymbol{M}^u}_\theta\geq\frac{1}{\langle u|\boldsymbol{J}_{\theta,s}|u\rangle}\boldsymbol{J}_{\theta,s}|u\rangle\langle u|\boldsymbol{J}_{\theta,s}.\tag{6.93}$$

6.42. Let $\boldsymbol{M}$ and $\boldsymbol{M}'$ be POVMs with measurement sets Ω and Ω', respectively. Let $\boldsymbol{M}''$ be a POVM that performs the measurements $\boldsymbol{M},\boldsymbol{M}'$ with probability $\lambda,(1-\lambda)$. Show that $\lambda J^{\boldsymbol{M}}_\theta+(1-\lambda)J^{\boldsymbol{M}'}_\theta=J^{\boldsymbol{M}''}_\theta$.

6.43. Consider the set of vectors $u_1,\ldots,u_k\in\mathbb{R}^d$ with norm 1 in parameter space. Let $\boldsymbol{M}^p$ be the POVM corresponding to the probabilistic mixture of spectral decomposition $L(u^i)$ with probability p_i. Show that the Fisher information matrix satisfies

$$\boldsymbol{J}^{\boldsymbol{M}^p}_\theta\geq\sum_{i=1}^k p_i\frac{1}{\langle u_i|\boldsymbol{J}_{\theta,s}|u_i\rangle}\boldsymbol{J}_{\theta,s}|u_i\rangle\langle u_i|\boldsymbol{J}_{\theta,s}.\tag{6.94}$$

6.44. Using the result of the preceding exercise, show the following for $\dim\mathcal{H}=2$ regardless of the number of parameters [142, 165–167]:

$$\inf\left\{\operatorname{tr}(\boldsymbol{J}^{\boldsymbol{M}}_\theta)^{-1}\middle|\boldsymbol{M}\text{ POVM on }\mathcal{H}\right\}=\left(\operatorname{tr}\boldsymbol{J}^{-\frac{1}{2}}_{\theta,s}\right)^2.$$

6.45. Using the result of the preceding exercise, show the following under the above assumption [142, 165–167].

$$\inf\left\{\operatorname{tr}G(\boldsymbol{J}^{\boldsymbol{M}}_\theta)^{-1}\middle|\boldsymbol{M}\text{ POVM on }\mathcal{H}\right\}=\left(\operatorname{tr}\left(\sqrt{G}^{-1}\boldsymbol{J}_{\theta,s}\sqrt{G}^{-1}\right)^{-\frac{1}{2}}\right)^2.\tag{6.95}$$

6.46. Let $\{p_\theta(i)|\theta = (\theta^1, \dots, \theta^d) \in \mathbb{R}^d\}$ be a probability family and $U_i (i = 1, \dots, k)$ be unitary matrices in $\mathcal{H}$. Consider the estimation of the TP-CP map $\kappa_\theta : \rho \mapsto \sum_{i=1}^k p_\theta(i) U_i \rho U_i^*$ with the following procedure. A state ρ is input to the input system $\mathcal{H}_A$, a measurement $\boldsymbol{M}$ is performed at the output system $\mathcal{H}_B$, and the parameter θ is estimated based on the obtained data. Show that

$$\boldsymbol{J}_\theta \geq \boldsymbol{J}_{\theta,\rho,x}, \tag{6.96}$$

where $\boldsymbol{J}_{\theta,\rho,x}$ is the Fisher information matrix of the quantum state family $\{\kappa_\theta(\rho)|\theta \in \mathbb{R}^d\}$ for $x = s, r, b, \lambda, p$ and $\boldsymbol{J}_\theta$ is the Fisher information matrix of the probability distribution family $\{p_\theta(i)|\theta = (\theta^1, \dots, \theta^d) \in \mathbb{R}^d\}$.

6.47. Show that the equality in (6.96) holds if

$$\operatorname{Tr} U_i \rho U_i^* U_j \rho U_j^* = 0 \tag{6.97}$$

holds for $i \neq j$. In addition, let P_i be the projection to the range of $U_i \rho U_i^*$. Show that the output distribution of the PVM $\{P_i\}$ is equal to $p_\theta(i)$, i.e., the problem reduces to estimating this probability distribution.

6.48. Consider the problem of estimating the probability distribution θ with the generalized Pauli channel κ_{p_θ} given in Example 5.8 as the estimation of the channel $\kappa_{p_\theta} \otimes \iota_A$. Show that (6.97) holds if $\rho = |\Phi_d\rangle\langle\Phi_d|$, where $|\Phi_d\rangle = \frac{1}{d}\sum_{i=0}^d |u_i\rangle \otimes |u_i\rangle$ and $d = \dim \mathcal{H}_A$.

6.49. As in the preceding problem, show that no estimator can improve the estimation accuracy of the estimator with the input state $|\Phi_d\rangle\langle\Phi_d|^{\otimes n}$, even though any entangled state is input to a channel $(\kappa_{p_\theta} \otimes \iota_A)$ defined with respect to the generalized Pauli channel κ_{p_θ}.

6.50. Prove (6.85) following the steps below [216].

a Show that

$$\max_{X\text{:real antisymmetric matrix}} \{\operatorname{Tr} X | X - iY \geq 0\} = \operatorname{Tr} |iY|. \tag{6.98}$$

for the real 2×2 antisymmetric matrix Y.

b Show that an arbitrary antisymmetric matrix Y may be rewritten as

$$VYV^t = \begin{pmatrix} 0 & \alpha_1 & & & \\ -\alpha_1 & 0 & & & \\ & & 0 & \alpha_2 & \\ & & -\alpha_2 & 0 & \\ & & & & \ddots \end{pmatrix}$$

for a suitable real orthogonal matrix V.

c Let E_j be a projection with only nonzero elements in the $2i$th and $2j-1$th components. Show (6.98) for the general case. Note that $\operatorname{Tr} X = \sum_j \operatorname{Tr} M_j X M_j$ and that if $X - iY \geq 0$, then $M_j X M_j - i M_j Y M_j \geq 0$, where $M_j \stackrel{\text{def}}{=} V^t E_j V$.
d Show (6.85) using **c**.

6.51. Show inequality (6.87) following the steps below [119, 216].

a Show that the RHS of (6.87) has the same value as the original when the minimization is replaced by that with the condition $\operatorname{Tr} \rho_\theta X^i = 0$.
b Assume that an estimator $(\boldsymbol{M}, \hat{\theta})$ and a vector $\mathbf{X} = (X^i)$ of Hermitian matrices satisfy $X^i = \sum_i \hat{\theta}^i(\omega) M_\omega$. Show that $\hat{\boldsymbol{V}}_\theta(\boldsymbol{M}, \hat{\theta}) \geq \boldsymbol{V}_\theta(\mathbf{X})$ according to a similar argument to (6.62) by replacing $O(\boldsymbol{M}^n, \hat{\theta}_n)$ by $\sum_i a_i X^i$, where (a_i) is an arbitrary complex-valued vector.
c Show that $\operatorname{tr} \hat{\boldsymbol{V}}_\theta(\boldsymbol{M}, \hat{\theta}) \geq \operatorname{tr}(\mathbf{Re}\, \boldsymbol{V}_\theta(\mathbf{X}) + |\mathbf{Im}\, \boldsymbol{V}_\theta(\mathbf{X})|)$ in a similar way to Exercise 6.50.
d Show (6.87) using Exercise 6.39.

6.52. Show that the equality in (6.87) holds for a pure state following the steps below [126, 127, 129, 282].

a Let $\rho_\theta = |u\rangle\langle u|$ and let $\mathbf{X} = (X^i)$ be a vector of Hermitian matrices satisfying $\operatorname{Tr} \rho_\theta X^i = 0$. Show that the vectors $u^i \stackrel{\text{def}}{=} X^i u$ are orthogonal to $|u\rangle$ and satisfy $\boldsymbol{V}_\theta(\mathbf{X}) = (\langle u^i, u^j \rangle)$. Hence, when the derivative $\frac{\partial \rho_\theta}{\partial \theta^i}$ has the form $\frac{\partial \rho_\theta}{\partial \theta^i} = (|x_i\rangle\langle u| + |u\rangle\langle x_i|)/2$ with the condition $\langle u|x_i\rangle = 0$, the RHS of (6.87) can be regarded as an optimization problem for the case $\boldsymbol{V}_\theta(\mathbf{X}) = (\langle u^i, u^j \rangle)$ with respect to the vectors $u^1, \ldots, u^d$ with the condition $\langle u^i|x_j\rangle = \delta^i_j$.
b Consider the case where all $\langle u^i|u^j\rangle$ are real. Suppose that each element v_k of the POVM $\{|v_k\rangle\langle v_k|\}$ is a real linear sum of the vectors $u, u^1, \ldots, u^d$, and each $\langle v_i|u\rangle$ is nonzero. Show that the estimator $(\boldsymbol{M}, \hat{\theta}) \stackrel{\text{def}}{=} (\{|v_k\rangle\langle v_k|\}, \frac{\langle v_k|u^j\rangle}{\langle v_k|u\rangle})$ satisfies $\mathrm{E}^i_\theta(\boldsymbol{M}, \hat{\theta}) u = u^j$. Also, show that $\hat{\boldsymbol{V}}_\theta(\boldsymbol{M}, \hat{\theta}) = \boldsymbol{V}_\theta(\mathbf{X})$.
c Show that there exists a set of vectors $w^1, \ldots, w^d$ in a d-dimensional space such that $|\mathbf{Im}\, \boldsymbol{V}_\theta(\mathbf{X})| - \mathbf{Im}\, \boldsymbol{V}_\theta(\mathbf{X}) = (\langle w^i|w^j\rangle)$.
d Let $u^1, \ldots, u^d$ be a set of vectors such that $\langle u^i|u^j\rangle$ are not necessarily real. Show that $\langle y^i|y^j\rangle$ are all real, where $y^i \stackrel{\text{def}}{=} u^i \oplus w^i$ and $\oplus$ denotes the direct sum product.
e For a given set of vectors $u^1, \ldots, u^d$, show that there exists an estimator $(\boldsymbol{M}, \hat{\theta})$ such that $\mathrm{E}^i_\theta(\boldsymbol{M}, \hat{\theta}) u = u^j$ and $\hat{\boldsymbol{V}}_\theta(\boldsymbol{M}, \hat{\theta}) = \mathbf{Re}(\boldsymbol{V}_\theta(\mathbf{X})) + \mathbf{Im}(\boldsymbol{V}_\theta(\mathbf{X}))$.
f Show that the equality in (6.87) holds for a pure state.
g Examine the state family consisting of pure states with $2l$ parameters, where l is the dimension of the Hilbert space. Show that the RHS of (6.87) is equal to $\operatorname{tr}(\mathbf{Re}\, J)^{-1} + \operatorname{tr} |(\mathbf{Re}\, J)^{-1} \mathbf{Im}\, J (\mathbf{Re}\, J)^{-1}|$, where $J = (J_{i,j} \stackrel{\text{def}}{=} \langle x_i|x_j\rangle)$.

6.7 Historical Note

Research on quantum state estimation was initiated by Helstrom [202] in 1967. He derived the one-parameter Cramér–Rao inequality (6.61) for the nonasymptotic version. He also proved the multiparameter SLD Cramér–Rao inequality (6.82) for the nonasymptotic version [203]. Yuen and Lax [432] developed the RLD version of the Cramér–Rao inequality for estimation with a complex multiparameter. They applied it to the estimation of the complex amplitude of the Gaussian state. Belavkin [34] derived a neccessary and sufficient condition for the achievement of this bound. Further, Holevo [216] derived the RLD Cramér–Rao inequality (6.82) with a real multiparameter and obtained the lower bound (6.87) with the locally unbiased condition in the nonasymptotic case [216].

Young introduced the concept of quasiclassical POVM concerning the state family [429]. Nagaoka [300] focused on equation (6.92) and derived the SLD one-parameter Cramér–Rao inequality (6.61) with an asymptotic framework. He derived its lower bound based on inequality (7.32) [312]. This bound is called the Nagaoka bound. Applying it to the quantum two-level system, he obtained equation (6.95) for the two-parameter case [301]. Hayashi [165, 167] applied the duality theorem in infinite-dimensional linear programming to quantum state estimation and obtained equation (6.95) in the three-parameter case as well as in the two-parameter case. After these, Gill and Massar [142] derived the same equation by a simpler method, which is explained in Exercise 6.45. Fujiwara and Nagaoka [129] defined the coherent model as a special case of pure-state families and showed that bound (6.87) can be attained with the locally unbiased and nonasymptotic framework in this case. Following this result, Matsumoto [282, 284] extended it to the general pure-state case. Further, Hayashi and Matsumoto [188] showed that bound (6.87) can be attained with the asymptotic framework in the quantum two-level system using the Cramér–Rao approach. The achievability of bound (6.87) is discussed in Matsumoto [286] in a general framework using irreducible decompositions of group representation. It has also been examined in Hayashi [178] using the quantum central limit theorem.

As a nonasymptotic extension of the quantum Crámer–Rao inequality, Tsuda and Matsumoto [388] treated its nondifferentiable extenstion (Hammersley–Chapman–Robbins–Kshiragar bound). They also derived the lower bound of mean square errors of unbiased estimators based on higher-order derivatives (quantum Bhattacharyya bound). The quantum Bhattacharyya bound has also been obtained by Brody and Hughston [59] in the pure-state case. Using this bound, Tsuda [391] derived an interesting bound for the estimation of polynomials of complex amplitude of quantum Gaussian states. Further, nonparametric estimation has been researched by D'Ariano [86] and Artiles et al. [16].

The group covariant approach was initiated by Helstrom [204]. He treated the estimation problem of one-parameter covariant pure-state families. Holevo has established the general framework of this approach [215] and applied it to several problems. Ozawa [339] and Bogomolov [55] extended it to the case of the noncompact parameter space. Holevo applied it to the estimation of the shifted one-parameter pure-state family [223]. Holevo [216] and Massar and Popescu [278] treated the estimation of a pure qubit state with n-i.i.d. samples using the Fidelity risk function. Hayashi [168] extended it to an arbitrary dimensional case with the general invariant risk function. Brußet al. [61] discussed its relation with approximate cloning.

Further, Hayashi [192] applied this method to the estimation of the squeezed parameter with vacuum-squeezed-state families. Hayashi and Matsumoto [188] also treated the estimation of the full-parameter model in quantum two-level systems using this approach. Bagan et al. [20] treated the same problem by the covariant and Bayesian approach.

Nagaoka [303] extended Bahadur's large deviation approach to the quantum estimation and found that the estimation accuracy with condition (6.74) is bounded by the Bogoljubov Fisher information in this approach. Hayashi [177] introduced a more strict condition (6.79) and showed that the estimation accuracy with condition (6.79) is bounded by the SLD Fisher information. Fujiwara [131] started to treat the estimation of a quantum channel within the framework of quantum state estimation. Sasaki et al. [359] discussed a similar estimation problem with the Bayesian approach in a nonasymptotic setting. Fischer et al. [115] focused on the use of the maximally entangled input state for the estimation of the Pauli channel. Fujiwara and Imai [134] showed that in the estimation of the Pauli channel κ_θ, the best estimation performance is obtained if and only if the input state is the n-fold tensor product of the maximally entangled state $|\Phi_d\rangle\langle\Phi_d|^{\otimes n}$. Exercise 6.49 treats the same problem using a different approach. After this result, Fujiwara [135] and Tanaka [384] treated, independently, the estimation problem of the amplitude damping channel. Especially, Fujiwara [135] proceeded to the estimation problem of the generalized amplitude damping channel, which is the more general and difficult part. De Martini et al. [279] implemented an experiment for the estimation of an unknown unitary.

Concerning the estimation of unitary operations, Bužek et al. [65] focused on estimating an unknown one-parameter unitary action first time. They showed that the error goes to 0 with the order $\frac{1}{n^2}$, where n is the number of applications of the unknown operation. Acín et al. [2] characterized the optimal input state for the SU(d) estimation where the input state is entangled with the reference system.

On the other hand, Fujiwara [132] treats this problem using the Cramér–Rao approach in the SU(2) case. This result was extended by Ballester [26]. Bagan et al. [21] treated the estimation of the unknown n-identical SU(2) operations using entanglement with the reference system. They also showed that the optimal error goes to 0 at a rate of $\frac{\pi^2}{n^2}$ and effectively applied the Clebsch–Gordan coefficient method to this problem. Hayashi [189, 190] treated the same problem using a different method. He derived a relation between this problem and that of Bužek et al. [65] and applied the obtained relation to this problem. He also pointed out that the multiplicity of the same irreducible representations can be regarded as the reference system, i.e., the effect of "self-entanglement." Indeed, independently of Hayashi, Chiribella et al. [73] and Bagan et al. [22] also pointed out this effect of the multiplicity based on the idea of Chiribella et al. [74]. That is, these three groups proved that the error of the estimation of SU(2) goes to 0 at a rate of $\frac{\pi^2}{n^2}$. The role of this "self-entanglement" is widely discussed in Chiribella et al. [76]. Note that, as was mentioned by Hayashi [189], the Cramér–Rao approach does not necessarily provide the optimal coefficient in the estimation of unitary operations by the use of entanglement. Chiribella et al. [75] derived the optimal estimator in the Bayesian setup.

The study of monotone metric in quantum state family was initiated by Morozowa and Chentsov [296]. Following this research, Petz [345] showed that every monotone metric is constructed from the matrix monotone function or the matrix average. Nagaoka introduced an SLD one-parameter exponential family [300] and a Bogoljubov one-parameter exponential family [304], characterized them as (6.33) and (6.34), respectively, and calculated the corresponding divergences (6.42) and (6.43) [304]. He also calculated the Bogoljubov m divergence as (6.52) [302]. Other formulas (6.45), and (6.53) for divergences were first obtained by Hayashi [180]. Further, Matsumoto [289] obtained an interesting characterization of RLD (m)-divergence. Moreover, he showed that an efficient estimator exists only in the SLD one-parameter exponential family (Theorem 6.7) [298, 300]. However, before this study, Belavkin [34] introduced a complex-parameterized exponential family and showed that the RLD version of the Cramér–Rao bound with the complex multi-parameter could be attained only in special cases. Theorem 6.7 coincides with its real-one-parameter case. Following this result, Fujiwara [120] showed that any unitary SLD one-parameter exponential family is generated by an observable satisfying the canonical commutation relation.

In addition, Amari and Nagaoka [11] introduced the torsion concerning the e parallel translation as the limit of the RHS–LHS in (6.54) and showed that the torsion-free inner product is only a Bogoljubov metric. They proved that the torsions of e-connection vanish only for a Bogoljubov inner product. They also showed that the divergence can be defined by a convex function if and only if the torsions of e-connection and m-connection vanish. Combining these facts, we can derive Theorem 6.5. However, their proof is based on the calculation of Christoffel symbols. In this textbook, Theorem 6.5 is proved without any use of Christoffel symbols.

Further, Nagaoka [299, 303] showed that the Bogoljubov metric is characterized by the limit of the quantum relative entropy as (6.24). Concerning the SLD inner product, Uhlmann [394] showed that the SLD metric is the limit of the Bures distance in the mixed-state case as (6.23). Matsumoto [281] extended it to the general case. In addition, Ohya and Petz [323] characterized the existence of a measurement attaining equality in the monotonicity of the relative entropy (2.59), and Nagaoka [302] improved their discussion (Exercise 6.27). Fujiwara [121] improved these discussions further.

7

Quantum Measurements and State Reduction

Summary. In quantum mechanics, the state reduction due to a measurement is called the collapse of a wavefunction. Its study is often perceived as a somewhat mystical phenomenon because of the lack of proper understanding. As a result, the formalism for the state reduction is often somewhat inadequately presented. However, as will be explained in Sect. 7.1, the state reduction due to a measurement follows automatically from the formulation of quantum mechanics, as described in Sect. 1.2. Starting with the formulation of quantum mechanics given in Sects. 1.2 and 1.4, we give a detailed formulation of the state reduction due to a measurement. In Sect. 7.2, we discuss the relation with the uncertainty relation using these concepts. Finally, in Sect. 7.3, we propose a measurement with negligible state demolition.

Table 7.1. Denotations used in Chap. 7

$\boldsymbol{\kappa}$	instrument
$\mathcal{I}$	Indirect measurement $(\mathcal{H}_D, V, \rho_0, \boldsymbol{E})$
$O(\boldsymbol{M})$	Average matrix (7.14)
$\mathrm{Cov}_\rho(X, Y)$	Correlation between two Hermitian matrices X and Y (7.37)
$\Delta_1(X, \rho)$	Uncertainty of an observable (7.12)
$\Delta_2(\boldsymbol{M}, \rho)$	Uncertainty of a measurement (7.13)
$\Delta_3(\boldsymbol{M}, X, \rho)$	Deviation of POVM $\boldsymbol{M}$ from observable X (7.17)
$\Delta_4(\kappa, X, \rho)$	Disturbance of X caused by κ (7.23)
$\Delta_4(\boldsymbol{\kappa}, X, \rho)$	Disturbance of X caused by $\boldsymbol{\kappa}$ (7.24)
$\varepsilon(\rho, \boldsymbol{\kappa})$	Amount of state demolition (reduction) by $\boldsymbol{\kappa}$ (7.40)

7.1 State Reduction Due to Quantum Measurement

In previous chapters, we examined several issues related to quantum measurement; these issues were concerned only with the probability distribution of the measurement outcomes. However, if we want to examine multiple trials of measurements on a single system, we need to describe the state reduction due to measurement. First, we discuss the state reduction due to a typical measurement corresponding to a POVM $\boldsymbol{M} = \{M_\omega\}$. Then, we give the general conditions for state reduction from the axiomatic framework given in Sect. 1.2.

Assume that we perform a measurement corresponding to the POVM $\boldsymbol{M} = \{M_\omega\}$ leading to the observation of a measurement value ω. When the state reduction has the typical form due to the POVM $\boldsymbol{M} = \{M_\omega\}$, the final state is

$$\frac{1}{\operatorname{Tr} \rho M_\omega} \sqrt{M_\omega} \rho \sqrt{M_\omega}, \tag{7.1}$$

where $\frac{1}{\operatorname{Tr} \rho M_\omega}$ is the normalization factor.[1] In particular, if $\boldsymbol{M}$ is a PVM, then M_ω is a projection and therefore the above state is [408]

$$\frac{1}{\operatorname{Tr} \rho M_\omega} M_\omega \rho M_\omega. \tag{7.2}$$

Since (7.2) is sandwiched by projection operators, the above-mentioned state reduction is called the *projection hypothesis*. In many books on quantum mechanics, the state reduction due to a measurement is restricted only to that satisfying the projection hypothesis. However, this is in fact incorrect, and such a state reduction is merely typical. That is, it is not necessarily true that any state reduction corresponding to the POVM $\boldsymbol{M}$ satisfies the above equation (7.2). In fact, many state reductions can occur due to a single POVM $\boldsymbol{M}$, as will be described later. Let us say that we are given an initial state ρ on a composite system $\mathcal{H}_A \otimes \mathcal{H}_B$. Now, let us perform a measurement corresponding to the POVM $\boldsymbol{M}^B = \{M_\omega^B\}_\omega$ on the system $\mathcal{H}_B$ and assume that the measurement value ω is obtained. The final state of $\mathcal{H}_A$ is then

$$\frac{1}{\operatorname{Tr} \rho (I_A \otimes M_\omega^B)} \operatorname{Tr}_B (I_A \otimes \sqrt{M_\omega^B}) \rho (I_A \otimes \sqrt{M_\omega^B}), \tag{7.3}$$

regardless of the type of state reduction on $\mathcal{H}_B$, as long as the measurement value ω obeys the distribution $\operatorname{Tr}_B (\operatorname{Tr}_A \rho) M_\omega^B$.

To prove this, consider an arbitrary POVM $\boldsymbol{M}^A = \{M_x^A\}_{x \in \mathcal{X}}$ on $\mathcal{H}_A$. Since $(M_x^A \otimes M_\omega^B) = (I_A \otimes \sqrt{M_\omega^B})(M_x^A \otimes I_B)(I_A \otimes \sqrt{M_\omega^B})$, the joint distribution of (x, ω) is given by

$$\begin{aligned} \operatorname{Tr} \rho (M_x^A \otimes M_\omega^B) &= \operatorname{Tr} (I_A \otimes \sqrt{M_\omega^B}) \rho (I_A \otimes \sqrt{M_\omega^B})(M_x^A \otimes I_B) \\ &= \operatorname{Tr} [\operatorname{Tr}_B (I_A \otimes \sqrt{M_\omega^B}) \rho (I_A \otimes \sqrt{M_\omega^B})] M_x^A, \end{aligned}$$

[1] Normalization here implies the division of the matrix by its trace such that its trace is equal to 1.

according to the discussion of Sect. 1.4, e.g., (1.23). When the measurement value ω is observed, the probability distribution of the other value x is

$$\frac{1}{\operatorname{Tr}\rho(I_A \otimes M_\omega^B)} \operatorname{Tr}[\operatorname{Tr}_B(I_A \otimes \sqrt{M_\omega^B})\rho(I_A \otimes \sqrt{M_\omega^B})] M_x^A,$$

which is the conditional distribution of x when the the measurement value on $\mathcal{H}_A$ is ω. Since this condition holds for an arbitrary POVM $\boldsymbol{M}^A = \{M_x^A\}_{x\in\mathcal{X}}$ on $\mathcal{H}_A$, equation (7.3) gives the final state of $\mathcal{H}_A$ when ω is the measurement value of the measurement $\boldsymbol{M}^B$ on $\mathcal{H}_B$.

However, since this only describes the state reduction of a system that is not directly measured, we still do not know anything about the kind of state reduction that would occur on the system $\mathcal{H}_B$, e.g., (7.2). As shown from the Naĭmark–Ozawa extension [315, 327] given below, it is theoretically possible to perform a measurement such that the state reduction follows (7.1) or (7.2).

Theorem 7.1 *(Naĭmark [315], Ozawa [327]) Consider an arbitrary POVM $\boldsymbol{M} = \{M_\omega\}_{\omega\in\Omega}$ on $\mathcal{H}$ and arbitrary $\omega_0 \in \Omega$. Let $\mathcal{H}_0$ be the additional space with the orthonormal basis $\{u_\omega\}_\omega$, and let us define the PVM $E_\omega = |u_\omega\rangle\langle u_\omega|$ on $\mathcal{H}_0$. There exists a unitary matrix U such that*

$$\operatorname{Tr} M_\omega \rho = \operatorname{Tr}(I \otimes E_\omega) U(\rho \otimes \rho_0) U^*, \tag{7.4}$$

$$\sqrt{M_\omega}\rho\sqrt{M_\omega} = \operatorname{Tr}_{\mathcal{H}_0}(I \otimes E_\omega) U(\rho \otimes \rho_0) U^*(I \otimes E_\omega), \tag{7.5}$$

where $\rho_0 = |u_{\omega_0}\rangle\langle u_{\omega_0}|$.

Theorem 4.5 can be obtained from this theorem by considering a PVM $\{U^*(I \otimes E_\omega)U\}_\omega$.

In the following section, we make several observations only concerning the probabilities for the measurement values ω based on this theorem. As described by (7.4), a measurement corresponding to an arbitrary POVM $\boldsymbol{M}$ can be realized by a PVM $\boldsymbol{E} = \{E_\omega\}$ with an appropriate time evolution U between $\mathcal{H}$ and $\mathcal{H}_0$. Furthermore, according to the above arguments, when a measurement corresponding to the PVM $\boldsymbol{E}$ on the system $\mathcal{H}_0$ is performed, the following gives the final state of $\mathcal{H}$ with the measurement value ω [327]:

$$\frac{1}{\operatorname{Tr} U(\rho \otimes \rho_0) U^*(I \otimes E_\omega)} \operatorname{Tr}_{\mathcal{H}_0}(I \otimes E_\omega) U(\rho \otimes \rho_0) U^*(I \otimes E_\omega).$$

Theorem 7.1 therefore shows that the above procedure produces a measurement corresponding to the final state (7.1). This model of measurement is called an *indirect measurement* model. The additional space $\mathcal{H}_0$ is called an *ancilla*, which interacts directly with the macroscopic system. Hence, this model does not reveal anything about the process whereby the measurement value is obtained in the ancilla. However, it does give information about the final state of system $\mathcal{H}$ (the system that we want to measure) when the measurement value ω is obtained in the ancilla. Hence, there remains an undiscussed part in the process whereby the measurement value is obtained via an ancilla. This is called the measurement problem.

In almost all real experiments, the target system $\mathcal{H}$ is not directly but indirectly measured via the measurement on an ancilla. For example, consider the Stern–Gerlach experiment, which involves the measurement of the spin of silver atoms. In this case, the spin is measured indirectly by measuring the momentum of the atom after the interaction between the spin system and the system of the momentum of the atom. Therefore, in such experiments, it is natural to apply the indirect measurement model to the measurement process.

The above theorem can be regarded as the refinement of the Naĭmark extension for the measurement of real quantum systems [327]. As this was first carried out by Ozawa, the triple $(\mathcal{H}_0, \rho_0, U)$ given in Theorem 7.1 will be called the *Naĭmark–Ozawa extension.*

Proof of Theorem 7.1. The proof will consider the case where Ω is $\{1, \ldots, n\}$, the orthonormal basis of $\mathcal{H}_0$ is given by $\{u_i\}_{i=1}^n$, and ρ_0 is given by $|u_1\rangle\langle u_1|$, for simplicity. First let us check that an arbitrary matrix U on $\mathcal{H}\otimes\mathcal{H}_0$ can be written as $(U^{i,j})_{i,j}$, using a matrix $U^{i,j}$ on $\mathcal{H}$. This implies that $(I\otimes|u_i\rangle\langle u_i|)U(I\otimes|u_j\rangle\langle u_j|) = U^{i,j}\otimes|u_i\rangle\langle u_j|$. Accordingly, (7.5) is equivalent to $\sqrt{M_i}\rho\sqrt{M_i} = U^{i,1}\rho(U^{i,1})^*$ with $\omega = i$. Choosing $U^{i,1} = \sqrt{M_i}$, we have $\sum_{i=1}^n U^{i,1}(U^{i,1})^* = I$, and therefore, it is possible to choose the remaining elements such that $U = (U^{i,j})_{i,j}$ is a unitary matrix, according to Exercise 1.2. This confirms the existence of a unitary matrix U that satisfies (7.5). Taking the trace in (7.5) gives us (7.4). ■

We next consider the possible state reductions according to the framework for a quantum measurement given in Sect. 1.2. Perform the measurement corresponding to a POVM $\boldsymbol{M}$ on a quantum system in a state ρ. Then, using the map κ_ω, we describe the final state with a measurement value ω by

$$\frac{1}{\operatorname{Tr}\rho M_\omega}\kappa_\omega(\rho). \tag{7.6}$$

The map κ_ω can be restricted to a completely positive map as shown below. In order to show this, we prove that

$$\kappa_\omega(\lambda\rho_1 + (1-\lambda)\rho_2) = \lambda\kappa_\omega(\rho_1) + (1-\lambda)\kappa_\omega(\rho_2) \tag{7.7}$$

for two arbitrary states ρ_1 and ρ_2 and an arbitrary real number λ satisfying $0 \le \lambda \le 1$. Consider performing a measurement corresponding to another arbitrary POVM $\{M'_{\omega'}\}_{\omega'\in\Omega'}$ after the first measurement. This is equivalent to performing a measurement within the data set $\Omega\times\Omega'$ with respect to the initial state. The joint probability distribution of ω and ω' is then given by $\operatorname{Tr}\kappa_\omega(\rho)M'_{\omega'}$ (Exercise 7.3). Consider the convex combination of the density matrix. Then, similarly to (1.9), the equation

$$\operatorname{Tr}\kappa_\omega(\lambda\rho_1 + (1-\lambda)\rho_2)M'_{\omega'} = \lambda\operatorname{Tr}\kappa_\omega(\rho_1)M'_{\omega'} + (1-\lambda)\operatorname{Tr}\kappa_\omega(\rho_2)M'_{\omega'}$$

should hold. Since $\{M'_{\omega'}\}_{\omega' \in \Omega'}$ is an arbitrary POVM, it is also possible to choose $M'_{\omega'}$ to be an arbitrary one-dimensional projection. Therefore, we obtain (7.7). Physically, we then require κ_ω to be a completely positive map. This is achieved using arguments similar to that of Sect. 5.1. That is, it can be verified by adding a reference system. Since (7.6) is a density matrix, we obtain [330]

$$\operatorname{Tr} \kappa_\omega(\rho) = \operatorname{Tr} \rho M_\omega, \quad \forall \rho, \tag{7.8}$$

which is equivalent to $M_\omega = \kappa_\omega^*(I)$. Thus $\operatorname{Tr} \sum_\omega \kappa_\omega(\rho) = 1$. The measurement with the state reduction is represented by the set of completely positive maps $\boldsymbol{\kappa} = \{\kappa_\omega\}_{\omega \in \Omega}$, where the map $\sum_\omega \kappa_\omega$ preserves the trace [327]. Henceforth, we shall call $\boldsymbol{\kappa} = \{\kappa_\omega\}_{\omega \in \Omega}$ an *instrument*. In this framework, if ρ is the initial state, the probability of obtaining a measurement value ω is given by $\operatorname{Tr} \kappa_\omega(\rho)$. When a measurement value ω is obtained, the final state is given by $\frac{1}{\operatorname{Tr} \kappa_\omega(\rho)} \kappa_\omega(\rho)$. We can also regard the Choi–Kraus representation $\{A_i\}$ of a TP-CP map as an instrument $\{\kappa_i\}$ with the correspondence $\kappa_i(\rho) = A_i \rho A_i^*$. The notation "an instrument $\{A_i\}$" then actually implies an instrument $\{\kappa_i\}$. Therefore, the state evolution with a Choi–Kraus representation $\{A_i\}$ can be regarded as a state reduction given by the instrument $\{A_i\}$ when the measurement value is not recorded. (The measurement is performed, but the experimenter does not read the outcome.)

When the instrument is in the form of the square roots $\{\sqrt{M_\omega}\}$ of a POVM $\boldsymbol{M} = \{M_\omega\}_{\omega \in \Omega}$, the final state is given by (7.1) and will be denoted by $\boldsymbol{\kappa_M}$. If the instrument $\boldsymbol{\kappa} = \{\kappa_\omega\}_{\omega \in \Omega}$ and the POVM $\boldsymbol{M} = \{M_\omega\}_{\omega \in \Omega}$ satisfy condition (7.8), we shall call the instrument $\boldsymbol{\kappa}$ an *instrument corresponding to the POVM* $\boldsymbol{M}$. We can characterize the instrument corresponding to a POVM $\boldsymbol{M}$ as follows.

Theorem 7.2 *Let* $\boldsymbol{\kappa} = \{\kappa_\omega\}_{\omega \in \Omega}$ *be an instrument corresponding to a POVM* $\boldsymbol{M} = \{M_\omega\}$ *in a quantum system* $\mathcal{H}$. *There exists a TP-CP map* $\boldsymbol{\kappa}'_\omega$ *for each measurement value* ω *such that*

$$\kappa_\omega(\rho) = \kappa'_\omega\left(\sqrt{M_\omega}\rho\sqrt{M_\omega}\right). \tag{7.9}$$

According to this theorem, it is possible to represent any state reduction $\boldsymbol{\kappa}$ due to a measurement as the state reduction given by the joint of the typical state reduction (7.1) and the state evolution κ'_ω that depends on the measurement value ω of the POVM $\boldsymbol{M}$.[2]

[2] It can also be understood as follows: the state reduction due to any measurement by PVM can be characterized as the state reduction satisfying the projection hypothesis, followed by the state evolution κ_ω. Indeed, many texts state that the state reduction due to any measurement is given by the projection hypothesis. Therefore, it is correct in a sense.

Proof. From Condition ⑥ (the Choi–Kraus representation) of Theorem A.2, there exists a set of matrices $E_1, \ldots, E_k$ such that $\kappa_\omega(\rho) = \sum_{i=1}^k E_i \rho E_i^*$. Since $\mathrm{Tr}\, \kappa_\omega(\rho) = \mathrm{Tr}\, \rho \sum_{i=1}^k E_i^* E_i = \mathrm{Tr}\, \rho M_\omega$ for an arbitrary state ρ, then $\sum_{i=1}^k E_i^* E_i = M_\omega$. Using the generalized inverse matrix $\sqrt{M_\omega}^{-1}$ defined in Sect. 1.5, and letting P be the projection to the range of M_ω (or $\sqrt{M_\omega}$), we have

$$\sum_{i=1}^k \sqrt{M_\omega}^{-1} E_i^* E_i \sqrt{M_\omega}^{-1} = P.$$

This implies that the matrices $E_1\sqrt{M_\omega}^{-1}, \ldots, E_k\sqrt{M_\omega}^{-1}, I-P$ are the Choi–Kraus representations of the TP-CP map. Denoting this TP-CP map as κ'_ω, we have

$$\begin{aligned}
&\kappa'_\omega(\sqrt{M_\omega}\rho\sqrt{M_\omega}) \\
&= (I-P)\sqrt{M_\omega}\rho\sqrt{M_\omega}(I-P) + \sum_{i=1}^k E_i\sqrt{M_\omega}^{-1}\sqrt{M_\omega}\rho\sqrt{M_\omega}\sqrt{M_\omega}^{-1}E_i^* \\
&= \sum_{i=1}^k E_i \rho E_i^*.
\end{aligned}$$

Therefore, we see that (7.9) holds. ■

Combining Theorems 7.1 and 7.2, we can construct a model of indirect measurement in a manner similar to Theorem 7.1 for an arbitrary instrument $\boldsymbol{\kappa} = \{\kappa_\omega\}$.

Theorem 7.3 *Let $\mathcal{H}_A$ and $\mathcal{H}_B$ be two quantum systems. The following two conditions are equivalent for the set of linear maps $\boldsymbol{\kappa} = \{\kappa_\omega\}_{\omega\in\Omega}$ from $\mathcal{T}(\mathcal{H}_A)$ to $\mathcal{T}(\mathcal{H}_B)$.*

① *$\boldsymbol{\kappa}$ is an instrument.*
② *$\boldsymbol{\kappa}$ can be expressed as*

$$\kappa_\omega(\rho) = \mathrm{Tr}_{A,C}\,(I_{A,B}\otimes E_\omega)\,U\,(\rho\otimes\rho_0)\,U^*\,(I_{A,B}\otimes E_\omega)\,, \tag{7.10}$$

where $\mathcal{H}_C$ is a quantum system with the dimension $\dim \mathcal{H}_B \times$ (*Number of elements in Ω*), *ρ_0 is a pure state on $\mathcal{H}_B\otimes\mathcal{H}_C$, $\boldsymbol{E} = \{E_\omega\}_\omega$ is a PVM on $\mathcal{H}_C$, and U is a unitary matrix on $\mathcal{H}_A\otimes\mathcal{H}_B\otimes\mathcal{H}_C$.*

The above is also equivalent to Condition ② with arbitrary-dimensional space $\mathcal{H}_C$, which is called Condition ②′.

If $\mathcal{H}_A = \mathcal{H}_B$, the following corollary holds.

Corollary 7.1 *(Ozawa [327]) The following two conditions for the set of linear maps $\boldsymbol{\kappa} = \{\kappa_\omega\}_{\omega \in \Omega}$ from $\mathcal{T}(\mathcal{H}_A)$ to $\mathcal{T}(\mathcal{H}_A)$ are equivalent [327].*

① *$\boldsymbol{\kappa}$ is an instrument.*

② *$\boldsymbol{\kappa}$ can be expressed as*

$$\kappa_\omega(\rho) = \mathrm{Tr}_D \left(I_A \otimes E_\omega\right) V \left(\rho \otimes \rho_0\right) V^* \left(I_A \otimes E_\omega\right), \tag{7.11}$$

where $\mathcal{H}_D$ is a $(\dim \mathcal{H}_A)^2 \times$(Number of elements in Ω)-dimensional quantum system, ρ_0 is a pure state on $\mathcal{H}_D$, $\boldsymbol{E} = \{E_\omega\}_\omega$ is a PVM on $\mathcal{H}_D$, and V is a unitary matrix on $\mathcal{H}_A \otimes \mathcal{H}_D$.

Therefore, a model of indirect measurement exists for an arbitrary instrument $\boldsymbol{\kappa} = \{\kappa_\omega\}$. Henceforth, we shall call $(\mathcal{H}_D, V, \rho_0, \boldsymbol{E})$ and $(\mathcal{H}_C, U, \rho_0, \boldsymbol{E})$ *indirect measurements* and denote them by $\mathcal{I}$. Let us now rewrite the above relation among the three different notations for measurements $\boldsymbol{M} = \{M_\omega\}, \boldsymbol{\kappa} = \{\kappa_\omega\}$, and $\mathcal{I} = (\mathcal{H}_D, V, \rho_0, \boldsymbol{E})$. The POVM $\boldsymbol{M}$ only describes the probability distribution of the measurement values and contains the least amount of information among the three notations. The instrument $\boldsymbol{\kappa}$ refers not only to the measurement value itself, but also to the final state of the measurement. Hence, a POVM $\boldsymbol{M}$ corresponds uniquely to an instrument $\boldsymbol{\kappa}$; however, the converse is not unique. Furthermore, the indirect measurement $\mathcal{I}$ denotes the unitary evolution required to realize the measurement device as well as the final state and the probability distribution of the observed data. This is the most detailed of the three notations (i.e., it contains the most information). Hence, a POVM $\boldsymbol{M}$ and an instrument $\boldsymbol{\kappa} = \{\kappa_\omega\}$ correspond uniquely to an indirect measurement $\mathcal{I}$, although the converse is not unique.

The proof of Theorem 7.3 is as follows: ②′ ⇒ ① and ② ⇒ ②′ follows from inspection. See Exercise 7.1 for ① ⇒ ②.

Exercises

7.1. Show that Condition ② in Theorem 7.3 may be derived from the Naĭmark–Ozawa extension $(\mathcal{H}_0, \rho_1, U)$ and Condition ① using the Stinespring representation $(\mathcal{H}_C, \rho_0, U_{\kappa_{\omega'}})$ for the TP-CP map κ'_ω given in Theorem 7.2.

7.2. Prove Corollary 7.1 using Theorem 7.3.

7.3. Apply an instrument $\boldsymbol{\kappa} = \{\kappa_\omega\}$ to a quantum system $\mathcal{H}$, followed by a measurement $\boldsymbol{M} = \{M_{\omega'}\}$. Define the POVM $\boldsymbol{M}' = \{M'_{\omega,\omega'}\}$ by $M'_{\omega,\omega'} \stackrel{\mathrm{def}}{=} \kappa^*_\omega(M_{\omega'})$, where κ^*_ω is the dual map of κ_ω. Show that $\mathrm{Tr}\, \kappa_\omega(\rho) M_{\omega'} = \mathrm{Tr}\, \rho M'_{\omega,\omega'}$ for an arbitrary input state ρ.

7.4. Given an initial state given by a pure state $x = (x^{k,i}) \in \mathcal{H}_A \otimes \mathcal{H}_B$, perform a measurement given by the PVM $\{|u_i\rangle\langle u_i|\}$ (where $u_i = (u_i^j) \in \mathcal{H}_B$) on $\mathcal{H}_B$. Show that the final state on $\mathcal{H}_A$ with the measurement value i is given by $\frac{v_j}{\|v_j\|}$, assuming that $v_j \stackrel{\mathrm{def}}{=} (\sum_j u_i^j x^{k,i}) \in \mathcal{H}_A$.

7.5. Prove Theorem 5.2 following the steps below.

a Using formula (7.3) for the state reduction due to a measurement, show that ② ⇒ ①.

b Consider a quantum system $\mathcal{H}_C$. When a maximally entangled state is input to an entanglement-breaking channel κ, show that (i) the output is a separable state and (ii) ① ⇒ ② using relationship (5.3).

7.2 Uncertainty and Measurement

The concept of uncertainty is often discussed in quantum mechanics in various contexts. In fact, there are no less than four distinct implications of the word *uncertainty*. Despite this, the differences between these implications are rarely discussed, and consequently, "uncertainty" is often used in a somewhat confused manner. In particular, there appears to be some confusion regarding its implication in the context of the Heisenberg uncertainty principle. We define the four meanings of uncertainty in the following and discuss the Heisenberg uncertainty principle and related topics in some detail [see (7.27), (7.30), and (7.32)].

First, let us define the *uncertainty of an observable* $\Delta_1(X, \rho)$ for a Hermitian matrix X (this can be considered an observable) and the state ρ by

$$\Delta_1^2(X,\rho) \stackrel{\text{def}}{=} \operatorname{Tr}\rho X^2 - (\operatorname{Tr}\rho X)^2 = \operatorname{Tr}\rho(X - \operatorname{Tr}\rho X)^2. \tag{7.12}$$

Next, let us define the *uncertainty of a measurement* $\Delta_2(\boldsymbol{M}, \rho)$ for a POVM $\boldsymbol{M} = \{(M_i, x_i)\}$ with real-valued measurement outcomes and a state ρ by

$$\Delta_2^2(\boldsymbol{M},\rho) \stackrel{\text{def}}{=} \sum_i (x_i - \mathrm{E}_\rho(\boldsymbol{M}))^2 \operatorname{Tr}\rho M_i, \quad \mathrm{E}_\rho(\boldsymbol{M}) \stackrel{\text{def}}{=} \sum_i x_i \operatorname{Tr}\rho M_i. \tag{7.13}$$

Defining the *average matrix* $O(\boldsymbol{M})$ for the POVM $\boldsymbol{M}$ as below, the formula

$$\Delta_1^2(O(\boldsymbol{M}),\rho) = \Delta_2^2(\boldsymbol{M},\rho), \quad O(\boldsymbol{M}) \stackrel{\text{def}}{=} \sum_i x_i M_i \tag{7.14}$$

holds. In particular, an indirect measurement $\mathcal{I} = (\mathcal{H}_D, V, \rho_0, \boldsymbol{E})$ corresponding to $\boldsymbol{M}$ satisfies

$$\Delta_2(\boldsymbol{M},\rho) = \Delta_2(\boldsymbol{E}, \rho\otimes\rho_0) = \Delta_1(O(\boldsymbol{E}), \rho\otimes\rho_0) \tag{7.15}$$

because $\boldsymbol{E}$ is a PVM. Similarly, the Naĭmark extension $(\mathcal{H}_B, \boldsymbol{E}, \rho_0)$ of the POVM $\boldsymbol{M}$ satisfies

$$\Delta_2(\boldsymbol{M},\rho) = \Delta_2(\boldsymbol{E}, \rho\otimes\rho_0) = \Delta_1(O(\boldsymbol{E}), \rho\otimes\rho_0). \tag{7.16}$$

Let us define the deviation $\Delta_3(\boldsymbol{M}, X, \rho)$ of the POVM $\boldsymbol{M}$ from the observable X for the state ρ by

$$\Delta_3^2(\boldsymbol{M}, X, \rho) \stackrel{\text{def}}{=} \sum_i \operatorname{Tr}(x_i - X) M_i (x_i - X)\rho. \tag{7.17}$$

It then follows that the deviation of $\boldsymbol{M}$ from $O(\boldsymbol{M})$ becomes zero if $\boldsymbol{M}$ is a PVM. The square of the uncertainty $\Delta_2^2(\boldsymbol{M}, \rho)$ of the measurement $\boldsymbol{M}$ can be decomposed into the sum of the square of the uncertainty of the average matrix $O(\boldsymbol{M})$ and the square of the deviation of $\boldsymbol{M}$ from $O(\boldsymbol{M})$ as follows:

$$\Delta_2^2(\boldsymbol{M}, \rho) = \Delta_3^2(\boldsymbol{M}, O(\boldsymbol{M}), \rho) + \Delta_1^2(O(\boldsymbol{M}), \rho). \tag{7.18}$$

When the POVM $\boldsymbol{M}$ and the observable X do not necessarily satisfy $O(\boldsymbol{M}) = X$, the square of their deviation can be written as the sum of the square of the uncertainty of the observable $X - O(\boldsymbol{M})$ and the square of the deviation of the POVM $\boldsymbol{M}$ from $O(\boldsymbol{M})$ as follows:

$$\Delta_3^2(\boldsymbol{M}, X, \rho) = \Delta_3^2(\boldsymbol{M}, O(\boldsymbol{M}), \rho) + \Delta_1^2(O(\boldsymbol{M}) - X, \rho). \tag{7.19}$$

Now, consider the disturbance caused by the state evolution κ from quantum system $\mathcal{H}_A$ to quantum system $\mathcal{H}_B$. For this purpose, we examine how well the POVM $\boldsymbol{M} = \{(M_i, x_i)\}_i$ on the final system $\mathcal{H}_B$ recovers the observable X on $\mathcal{H}_A$. Its quality can be measured by the quantity $\Delta_3(\kappa^*(\boldsymbol{M}), X, \rho)$, where $\kappa^*(\boldsymbol{M})$ denotes the POVM $\{(\kappa^*(M_i), x_i)\}_i$ on the initial system $\mathcal{H}_A$. Since κ^* is the dual map of the map κ, the minimum value $\Delta_4(\kappa, X, \rho) \stackrel{\text{def}}{=} \min_{\boldsymbol{M}} \Delta_3(\kappa^*(\boldsymbol{M}), X, \rho)$ is thought to present the *disturbance* with respect to the observable X caused by the state evolution κ. Using the Stinespring representation, equation (7.18) yields

$$\Delta_3^2(\kappa^*(\boldsymbol{M}), X, \rho) = \Delta_3^2(\boldsymbol{M}, O(\boldsymbol{M}), \kappa(\rho)) + \Delta_3^2(\kappa^*(\boldsymbol{E}_{O(\boldsymbol{M})}), X, \rho). \tag{7.20}$$

Thus, our minimization can be reduced to $\min_Y \Delta_3(\kappa^*(\boldsymbol{E}_Y), X, \rho)$. Interestingly, using the Stinespring representation $(\mathcal{H}_C, \rho_0, U_\kappa)$ of κ, we can express the quantity $\Delta_3(\kappa^*(\boldsymbol{E}_Y), X, \rho)$ as [Ex. 7.10]

$$\Delta_3(\kappa^*(\boldsymbol{E}_Y), X, \rho) = \operatorname{Tr}\left(U_\kappa(X \otimes I_{B,C})U_\kappa^* - (I_{A,C} \otimes Y)\right)^2 U_\kappa(\rho \otimes \rho_0)U_\kappa^*. \tag{7.21}$$

As discussed in Sect. 6.1, the matrix $\kappa_{\rho,s}(X)$ can be regarded as the image of $U_\kappa(X \otimes I_{B,C})U_\kappa^*$ by the projection to the space $\{Y \otimes I\}$. Hence, using property (6.10), the above can be calculated as

$$\begin{aligned}
&\operatorname{Tr}\left(U_\kappa(X \otimes I_{B,C})U_\kappa^* - (I_{A,C} \otimes \kappa_{\rho,s}(X))\right)^2 U_\kappa(\rho \otimes \rho_0)U_\kappa^* \\
&\quad + \operatorname{Tr}\left((I_{A,C} \otimes \kappa_{\rho,s}(X)) - (I_{A,C} \otimes Y)\right)^2 U_\kappa(\rho \otimes \rho_0)U_\kappa^* \\
&= \operatorname{Tr} X^2\rho - \operatorname{Tr}\left((\kappa_{\rho,s}(X))^2 \kappa(\rho) + (Y - \kappa_{\rho,s}(X))^2 \kappa(\rho)\right).
\end{aligned} \tag{7.22}$$

Thus, this quantity gives the minimum when $Y = \kappa_{\rho,s}(X)$. That is, the matrix $\kappa_{\rho,s}(X)$ on the output system gives the best approximation of the matrix

X on the input system. In particular, since $\kappa_{\rho,s}$ can be regarded as the conditional expectation in the partial trace case, this argument gives the quantum version of conditional expectation an interesting meaning. Therefore, the disturbance of X caused by κ has the form

$$\Delta_4(\kappa, X, \rho) = \min_{\boldsymbol{M}} \Delta_3(\kappa^*(\boldsymbol{M}), X, \rho) = \min_{Y} \Delta_3(\kappa^*(\boldsymbol{E}_Y), X, \rho)$$
$$= \Delta_3(\kappa^*(\boldsymbol{E}_{\kappa_{\rho,s}(X)}), X, \rho) = \sqrt{\left(\|X\|_{\rho,s}^{(e)}\right)^2 - \left(\|\kappa_{\rho,x}(X)\|_{\kappa(\rho),s}^{(e)}\right)^2}. \quad (7.23)$$

Hence, if X is the SLD e representation $L_{\theta,s}$ of the derivative, this can be regarded as the loss of the SLD Fisher metric.

Furthermore, when an instrument $\boldsymbol{\kappa} = \{\kappa_\omega\}$ is used, the disturbance of the observable X is defined as

$$\Delta_4(\boldsymbol{\kappa}, X, \rho) \stackrel{\text{def}}{=} \Delta_4\left(\overline{\boldsymbol{\kappa}}, X, \rho\right), \quad \overline{\boldsymbol{\kappa}} \stackrel{\text{def}}{=} \sum_\omega \kappa_\omega. \quad (7.24)$$

Letting $\mathcal{I} = (\mathcal{H}_C, U, \rho_0, \boldsymbol{E})$ be an indirect measurement corresponding to the instrument $\boldsymbol{\kappa}$, we can describe the disturbance of the observable X by

$$\Delta_4^2(\boldsymbol{\kappa}, X, \rho) = \mathrm{Tr}((X \otimes I_{B,C}) - U^*(I_{A,C} \otimes \overline{\boldsymbol{\kappa}}_{\rho,s}(X))U)^2(\rho \otimes \rho_0) \quad (7.25)$$
$$= \left(\|X\|_{\rho,s}^{(e)}\right)^2 - \left(\|\overline{\boldsymbol{\kappa}}_{\rho,x}(X)\|_{\overline{\boldsymbol{\kappa}}(\rho),s}^{(e)}\right)^2,$$

which may be found in a manner similar to (7.22). The four uncertainties given here are often confused and are often denoted by $\Delta^2(X)$. Some care is therefore necessary to ensure these quantities.

The most famous uncertainty relation is

$$\Delta_1(X, |u\rangle\langle u|)\Delta_1(Y, |u\rangle\langle u|) \geq \frac{|\langle u|[X,Y]|u\rangle|}{2}. \quad (7.26)$$

This may be generalized to

$$\Delta_1(X, \rho)\Delta_1(Y, \rho) \geq \frac{\mathrm{Tr}\,|\sqrt{\rho}[X,Y]\sqrt{\rho}|}{2}. \quad (7.27)$$

Indeed, the above inequality still holds if the right-hand side (RHS) is replaced by $\frac{|\mathrm{Tr}\,\rho[X,Y]|}{2}$. However, if the state is not a pure state, inequality (7.27) is a stronger requirement. For the rest of this section, we assume that ρ is a density matrix, although the essential point is that $\rho \geq 0$ and not that its trace is equal to 1.

Now, let us prove (7.27). The problem is reduced to the case of $\mathrm{Tr}\,\rho X = 0$ by replacing X by $X - \mathrm{Tr}\,\rho X$. Since

$$0 \leq (X \pm iY)(X \pm iY)^* = (X \pm iY)(X \mp iY) = X^2 + Y^2 \mp i[X,Y],$$

we have $\sqrt{\rho}(X^2 + Y^2)\sqrt{\rho} \geq \pm i\sqrt{\rho}[X,Y]\sqrt{\rho}$. From Exercise 1.23 we thus obtain

$$\Delta_1^2(X,\rho) + \Delta_1^2(Y,\rho) \geq \operatorname{Tr}|i\sqrt{\rho}[X,Y]\sqrt{\rho}| = \operatorname{Tr}|\sqrt{\rho}[X,Y]\sqrt{\rho}|. \tag{7.28}$$

Replacing X by tX, we see that

$$\Delta_1^2(X,\rho)t^2 - \operatorname{Tr}|\sqrt{\rho}[X,Y]\sqrt{\rho}|t + \Delta_2^2(Y,\rho) \geq 0.$$

Equation (7.27) can then be obtained from the discriminant equation for t.

The original uncertainty relation proposed by Heisenberg [200] was obtained through a gedanken experiment, and it relates the accuracy of measurements to the disturbance of measurements. The implications of the accuracy and the disturbance due to the measurement are not neccesarily clear in (7.27). At least, it is incorrect to call (7.27) the Heisenberg uncertainty relation because it does not involve quantities related to measurement [335,336].

One may think that Heisenberg's arguments would be formulated as

$$\Delta_3(\boldsymbol{M},X,\rho)\Delta_4(\boldsymbol{\kappa},Y,\rho) \geq \frac{\operatorname{Tr}|\sqrt{\rho}[X,Y]\sqrt{\rho}|}{2}. \tag{7.29}$$

However, this is in fact incorrect, for the following reason [335, 337, 338]. Consider the POVM $\boldsymbol{M}$ that always gives the measurement value 0 without making any measurement. Then, $\Delta_3(\boldsymbol{M},X,\rho)$ is finite while $\Delta_4(\boldsymbol{\kappa},Y,\rho)$ is 0. Therefore, this inequality does not hold in general. The primary reason for this is that the RHS has two quantities having no connection to the POVM $\boldsymbol{M}$. Hence, we need to seek a more appropriate formulation. Indeed, since Heisenberg's gedanken experiment treats only the eigen state of the observable X, the measurement error in his experiment coincides with $\Delta_2(\boldsymbol{M},\rho)$ as well as $\Delta_3(\boldsymbol{M},X,\rho)$. When we apply his gedanken experiment to the noneigen state, his gendanken experiment seems to hold even with $\Delta_2(\boldsymbol{M},\rho)$. Hence, it is more appropriate to treat $\Delta_2(\boldsymbol{M},\rho)$ than $\Delta_3(\boldsymbol{M},X,\rho)$ and write[3]

$$\Delta_2(\boldsymbol{M},\rho)\Delta_4(\boldsymbol{\kappa},Y,\rho) \geq \frac{\operatorname{Tr}|\sqrt{\rho}[O(\boldsymbol{M}),Y]\sqrt{\rho}|}{2}. \tag{7.30}$$

This inequality means that the product of the standard deviation $\Delta_2(\boldsymbol{M},\rho)$ of the measurement values and the disturbance $\Delta_4(\boldsymbol{\kappa},Y,\rho)$ of the observable Y due to the measurement $\boldsymbol{\kappa}$ is lower bounded by a quantity involving $\boldsymbol{M}$ and Y. In particular, when Y is the SLD e representation $L_{\theta,s}$ of the derivative, the information loss $\Delta_4(\boldsymbol{\kappa},L_{\theta,s},\rho)$ satisfies

[3] As discussed in Sect. 6.2, $\Delta_4(\boldsymbol{\kappa},Y,\rho)$ also has the meaning of the amount of loss of the SLD Fisher information. Therefore, this inequality is interesting from the point of view of estimation theory. It indicates the naturalness of the SLD inner product. This is in contrast to the naturalness of the Bogoljubov inner product from a geometrical viewpoint.

$$\Delta_2(\boldsymbol{M},\rho)\Delta_4(\boldsymbol{\kappa},L_{\theta,s},\rho) \ge \frac{\operatorname{Tr}|\sqrt{\rho}[O(\boldsymbol{M}),L_{\theta,s}]\sqrt{\rho}|}{2}.$$

A proof of this relation will be given later. Equation (7.29) may be corrected by changing the left-hand side (LHS) of the inequality. This is the subject of the following theorem.

Theorem 7.4 *(Ozawa [335, 337, 338]) Let $\boldsymbol{\kappa}$ be an instrument corresponding to the POVM $\boldsymbol{M}$. A state ρ on $\mathcal{H}_A$ then satisfies*

$$\Delta_2(\boldsymbol{M},\rho)\Delta_4(\boldsymbol{\kappa},Y,\rho) + \Delta_1(O(\boldsymbol{M})-X,\rho)\Delta_1(Y,\rho) \ge \frac{\operatorname{Tr}|\sqrt{\rho}[X,Y]\sqrt{\rho}|}{2}. \tag{7.31}$$

Inequality (7.30) may then be shown by substituting $X = O(\boldsymbol{M})$ into this theorem.

Next consider trying to measure two observables simultaneously. For this purpose, we denote a POVM with two measurement values by $\boldsymbol{M} = (\{M_\omega\},\{x_\omega^1\},\{x_\omega^2\})$. Then, the two average matrices are given by

$$O^1(\boldsymbol{M}) \stackrel{\text{def}}{=} \sum_\omega x_\omega^1 M_\omega, \quad O^2(\boldsymbol{M}) \stackrel{\text{def}}{=} \sum_\omega x_\omega^2 M_\omega.$$

Theorem 7.5 *The POVM $\boldsymbol{M} = (\{M_\omega\},\{x_\omega^1\},\{y_\omega^2\})$ satisfies*[4]

$$\begin{aligned}&\Delta_3(\boldsymbol{M},O^1(\boldsymbol{M}),\rho)\Delta_3(\boldsymbol{M},O^2(\boldsymbol{M}),\rho)\\ &\ge \frac{\operatorname{Tr}|\sqrt{\rho}[O^1(\boldsymbol{M}),O^2(\boldsymbol{M})]\sqrt{\rho}|}{2}.\end{aligned} \tag{7.32}$$

There have also been numerous discussions relating to uncertainties, including its relation to quantum computation [337, 338].

Proof of Theorem 7.4. The theorem is proven by considering an indirect measurement $\mathcal{I} = (\mathcal{H}_C, U, \rho_0, \boldsymbol{E})$ corresponding to the instrument $\boldsymbol{\kappa}$. Let $Z = \overline{\boldsymbol{\kappa}}_{\rho,s}(X)$. Then, from (7.25),

$$\Delta_4(\boldsymbol{\kappa},Y,\rho) = \Delta_1(Y\otimes I_{B,C} - U^*(Z\otimes I_{A,C})U, \rho\otimes\rho_0). \tag{7.33}$$

Since

$$\begin{aligned}0 =& U^*[I_{A,B}\otimes O(\boldsymbol{E}), Z\otimes I_{A,C}]U = [U^*(I_{A,B}\otimes O(\boldsymbol{E}))U, U^*(Z\otimes I_{A,C})U]\\ =& [U^*(I_{A,B}\otimes O(\boldsymbol{E}))U, Y\otimes I_{B,C}]\\ &+ [U^*(I_{A,B}\otimes O(\boldsymbol{E}))U, U^*(Z\otimes I_{A,C})U - Y\otimes I_{B,C}]\\ =& [U^*(I_{A,B}\otimes O(\boldsymbol{E}))U - X\otimes I_{B,C}, Y\otimes I_{B,C}] + [X\otimes I_{B,C}, Y\otimes I_{B,C}]\\ &+ [U^*(I_{A,B}\otimes O(\boldsymbol{E}))U, U^*(Z\otimes I_{A,C})U - Y\otimes I_{B,C}],\end{aligned}$$

[4] The equality holds when an appropriate POVM $\boldsymbol{M}$ is performed in a quantum two-level system [301]. For its more general equality condition, see Exercise 7.16.

then

$$[X, Y] \otimes I_{B,C} = [X \otimes I_{B,C} - U^*(I_{A,B} \otimes O(\boldsymbol{E}))U, Y \otimes I_{B,C}] \\ + [U^*(I_{A,B} \otimes O(\boldsymbol{E}))U, Y \otimes I_{B,C} - U^*(Z \otimes I_{A,C})U].$$

Multiplying both sides by $\sqrt{\rho \otimes \rho_0}$, taking the partial trace $\mathrm{Tr}_{B,C}$, and applying the general formula $\mathrm{Tr}\,|X_1 + X_2| \le \mathrm{Tr}\,|X_1| + \mathrm{Tr}\,|X_2|$ to the matrices on $\mathcal{H}_A$, we obtain

$$\begin{aligned}&\mathrm{Tr}\,|\sqrt{\rho}[X, Y]\sqrt{\rho}| \\ &\quad \le \mathrm{Tr}_A\,|\,\mathrm{Tr}_{B,C}\sqrt{\rho \otimes \rho_0}[X \otimes I_{B,C} - U^*(I_{A,B} \otimes O(\boldsymbol{E}))U, Y \otimes I_{B,C}]\sqrt{\rho \otimes \rho_0}| \\ &\quad +\mathrm{Tr}_A|\mathrm{Tr}_{B,C}\sqrt{\rho\otimes\rho_0}[U^*(I_{A,B}\otimes O(\boldsymbol{E}))U, Y\otimes I_{B,C} - U^*(Z\otimes I_{A,C})U]\sqrt{\rho\otimes\rho_0}|.\end{aligned}$$

Since the indirect measurement $\mathcal{I} = (\mathcal{H}_C, U, \rho_0, \boldsymbol{E})$ corresponds to the POVM $\boldsymbol{M}$, we have $\mathrm{Tr}_{B,C}(I \otimes \sqrt{\rho_0})U^*(I_{A,B} \otimes O(\boldsymbol{E}))U(I \otimes \sqrt{\rho_0}) = O(\boldsymbol{M})$ [Ex. 7.18]. Referring to Exercise 1.15, the first term is

$$\begin{aligned}&\mathrm{Tr}_A\,|\,\mathrm{Tr}_{B,C}\sqrt{\rho \otimes \rho_0}[X \otimes I_{B,C} - U^*(I_{A,B} \otimes O(\boldsymbol{E}))U, Y \otimes I_{B,C}]\sqrt{\rho \otimes \rho_0}| \\ &\quad = \mathrm{Tr}_A\,|\sqrt{\rho}[X - O(\boldsymbol{M}), Y]\sqrt{\rho}| \le \Delta_1(X - O(\boldsymbol{M}), \rho)\Delta_1(Y, \rho).\end{aligned}$$

The second term is

$$\begin{aligned}&\mathrm{Tr}_A\,|\,\mathrm{Tr}_{B,C}\sqrt{\rho\otimes\rho_0}[U^*(I_{A,B}\otimes O(\boldsymbol{E}))U, Y\otimes I_{B,C} - U^*(Z\otimes I_{A,C})U]\sqrt{\rho\otimes\rho_0}| \\ &\quad \le \mathrm{Tr}\,|\sqrt{\rho \otimes \rho_0}[U^*(I_{A,B} \otimes O(\boldsymbol{E}))U, Y \otimes I_{B,C} - U^*(Z \otimes I_{A,C})U]\sqrt{\rho \otimes \rho_0}| \\ &\quad \le \Delta_1(U^*(I_{A,B} \otimes O(\boldsymbol{E}))U, \rho \otimes \rho_0)\Delta_1(Y \otimes I_{B,C} - U^*(Z \otimes I_{A,C})U, \rho \otimes \rho_0).\end{aligned}$$

Finally, from (7.15) we have $\Delta_2(\boldsymbol{M}, \rho) = \Delta_1(U^*(I_{A,B} \otimes O(\boldsymbol{E}))U, \rho \otimes \rho_0)$. Combining this with (7.33), we obtain (7.31). ∎

In particular, inequality (7.30) with the state reduction $\boldsymbol{\kappa_E}$ and the spectral decomposision $\boldsymbol{E}$ of X can be derived more simply as follows. In this case, the indirect measurement is given as $(\mathcal{H}_B, U, \rho = |u_0\rangle\langle u_0|, \boldsymbol{E})$. From the construction of indirect measurement, we have

$$U^*(I_A \otimes X_B)U|u\rangle \otimes |u_0\rangle = (X|u\rangle) \otimes |u_0\rangle. \tag{7.34}$$

Hence,

$$\begin{aligned}&\sqrt{\rho} \otimes \sqrt{\rho_0}[U^*(I_A \otimes X_B)U, Y \otimes I_B - U^*(Z \otimes I_B)U]\sqrt{\rho} \otimes \sqrt{\rho_0} \\ &\quad = \sqrt{\rho} \otimes \sqrt{\rho_0}[U^*(I_A \otimes X_B)U, Y \otimes I_B]\sqrt{\rho} \otimes \sqrt{\rho_0} = (\sqrt{\rho}[X, Y]\sqrt{\rho}) \otimes \rho_0.\end{aligned}$$

Thus, we have (7.30).

In addition, when the equality of (7.27) holds and $\mathrm{Tr}\,\rho X = \mathrm{Tr}\,\rho Y = 0$, the relation $(X \pm iY)\sqrt{\rho} = 0$ holds. From (7.34) this relation implies $(U^*(I_A \otimes X_B)U \pm iY \otimes I_B)\sqrt{\rho} \otimes \sqrt{\rho_0} = 0$. Hence, the equality of (7.30) holds and $\overline{\boldsymbol{\kappa_E}}_{\rho,s}(Y) = 0$.

Proof of Theorem 7.5. We apply Exercise 5.5. Let us choose $\mathcal{H}_B$, a PVM on $\mathcal{H} \otimes \mathcal{H}_B$, and a state ρ_0 on $\mathcal{H}_B$ such that

$$\operatorname{Tr} \rho M_\omega = \operatorname{Tr}(\rho \otimes \rho_0) E_\omega,$$

with respect to an arbitrary state ρ on $\mathcal{H}$. Let $(\mathcal{H}_B, \boldsymbol{E}, \rho_0)$ be the Naĭmark extension of $\boldsymbol{M}$. Then,

$$\Delta_3(\boldsymbol{M}, O^i(\boldsymbol{M}), \rho) = \Delta_1(O^i(\boldsymbol{E}) - O^i(\boldsymbol{M}) \otimes I_B, \rho \otimes \rho_0).$$

Since $[O^1(\boldsymbol{E}), O^2(\boldsymbol{E})] = 0$, we have

$$\begin{aligned} &[O^1(\boldsymbol{E}) - O^1(\boldsymbol{M}) \otimes I_B, O^2(\boldsymbol{E}) - O^2(\boldsymbol{M}) \otimes I_B] \\ &\quad = -[O^1(\boldsymbol{E}), O^2(\boldsymbol{M}) \otimes I_B] - [O^1(\boldsymbol{M}) \otimes I_B, O^2(\boldsymbol{E}) - O^2(\boldsymbol{M}) \otimes I_B]. \end{aligned}$$

Accordingly,

$$\begin{aligned} &\Delta_1(O^1(\boldsymbol{E}) - O^1(\boldsymbol{M}) \otimes I_B, \rho \otimes \rho_0) \Delta_1(O^2(\boldsymbol{E}) - O^2(\boldsymbol{M}) \otimes I_B, \rho \otimes \rho_0) \\ &\quad \geq \operatorname{Tr} |\sqrt{\rho \otimes \rho_0}[O^1(\boldsymbol{E}) - O^1(\boldsymbol{M}) \otimes I_B, O^2(\boldsymbol{E}) - O^2(\boldsymbol{M}) \otimes I_B]\sqrt{\rho \otimes \rho_0}| \\ &\quad \geq \operatorname{Tr}_A |\operatorname{Tr}_B \sqrt{\rho \otimes \rho_0}[O^1(\boldsymbol{E}) - O^1(\boldsymbol{M}) \otimes I_B, O^2(\boldsymbol{E}) - O^2(\boldsymbol{M}) \otimes I_B]\sqrt{\rho \otimes \rho_0}| \\ &\quad = \operatorname{Tr}_A | \operatorname{Tr}_B \sqrt{\rho \otimes \rho_0}(-[O^1(\boldsymbol{E}), O^2(\boldsymbol{M}) \otimes I_B] \\ &\qquad - [O^1(\boldsymbol{M}) \otimes I_B, O^2(\boldsymbol{E}) - O^2(\boldsymbol{M}) \otimes I_B])\sqrt{\rho \otimes \rho_0}| \\ &\quad = \operatorname{Tr}_A | \operatorname{Tr}_B \sqrt{\rho \otimes \rho_0}[O^1(\boldsymbol{E}), O^2(\boldsymbol{M}) \otimes I_B]\sqrt{\rho \otimes \rho_0}|. \end{aligned}$$

This completes the proof. In the last equality, we used the fact that $\operatorname{Tr}_B(I \otimes \sqrt{\rho_0})(O^2(\boldsymbol{E}) - O^2(\boldsymbol{M}) \otimes I_B)(I \otimes \sqrt{\rho_0}) = 0$ and Exercise 1.15. ∎

Exercises

7.6. Show (7.14) for the PVM $\boldsymbol{M}$.

7.7. Show (7.18).

7.8. Show (7.19).

7.9. Show (7.20) using a Stinespring representation.

7.10. Show (7.21) following the steps below.

a Let $(\mathcal{H}_C, \rho_0, U_\kappa)$ be a Stinespring representation of κ. Show that

$$\operatorname{Tr}(X \otimes I - xU(I \otimes E)U)^2 \rho = \operatorname{Tr}(X - x)\kappa^*(E)(X - x)\rho.$$

b Show (7.21).

7.11. Show that $\Delta_2^2(\boldsymbol{M}, \rho) \leq \Delta_3^2(\boldsymbol{M}, X, \rho) + \Delta_1^2(X, \rho)$.

7.12. Let $(\mathcal{H}_B, \boldsymbol{E}, \rho_0)$ be a Naĭmark extension $^{\text{Ex. 5.5}}$ of $\boldsymbol{M} = (\{M_\omega\}, \{x_\omega\})$. Show that $\Delta_3^2(\boldsymbol{M}, X, \rho) = \Delta_1^2(O(\boldsymbol{E}) - X \otimes I_B, \rho \otimes \rho_0)$.

7.13. Show the following using (7.31) [335].

$$\begin{aligned}&\Delta_3(\boldsymbol{M}, X, \rho)\Delta_4(\boldsymbol{\kappa}, Y, \rho) + \Delta_1(X, \rho)\Delta_4(\boldsymbol{\kappa}, Y, \rho) + \Delta_3(\boldsymbol{M}, X, \rho)\Delta_1(Y, \rho)\\ \ge &\frac{\operatorname{Tr}|\sqrt{\rho}[X, Y]\sqrt{\rho}|}{2}. \end{aligned} \tag{7.35}$$

7.14. Show that

$$\det \begin{pmatrix} \operatorname{Cov}_\rho(X, X) & \operatorname{Cov}_\rho(X, Y) \\ \operatorname{Cov}_\rho(X, Y) & \operatorname{Cov}_\rho(Y, Y) \end{pmatrix} \ge \left(\frac{\operatorname{Tr}|\sqrt{\rho}[X, Y]\sqrt{\rho}|}{2} \right)^2, \tag{7.36}$$

which is a stronger inequality than (7.27), using (7.27). Note that we define the correlation between two Hermitian matrices X and Y under the state ρ as

$$\operatorname{Cov}_\rho(X, Y) \stackrel{\text{def}}{=} \operatorname{Tr}(X - \operatorname{Tr} X\rho) \circ (Y - \operatorname{Tr} Y\rho)\rho. \tag{7.37}$$

7.15. Show that the equality in (7.36) always holds if $\mathcal{H} = \mathbb{C}^2$ [301] by following the steps below. From this fact it follows that the equality in (7.27) holds when $\operatorname{Cov}_\rho(X, Y) = 0$ for $\mathcal{H} = \mathbb{C}^2$. In the next proof, we define $X = \sum_{i=1}^3 x_i S_i, Y = \sum_{i=1}^3 y_i S_i$, $\rho = \frac{1}{2}(I + \sum_{i=1}^3 a_i S_i)$, and fist treat the case where $\operatorname{Tr} X = \operatorname{Tr} Y = 0$. After this special case, we consider the general case.

- **a** Show that $\operatorname{Cov}_\rho(X, Y) = \langle \boldsymbol{x}, \boldsymbol{y}\rangle - \langle \boldsymbol{x}, \boldsymbol{a}\rangle\langle \boldsymbol{a}, \boldsymbol{y}\rangle$.
- **b** Let $\boldsymbol{z}$ be the vector product (outer product) of $\boldsymbol{x}$ and $\boldsymbol{y}$, i.e., $\boldsymbol{z} = \boldsymbol{x} \times \boldsymbol{y} \stackrel{\text{def}}{=} (x_2 y_3 - x_3 y_2, x_3 y_1 - x_1 y_3, x_1 y_2 - x_2 y_1)$. Show that $\frac{-i}{2}[X, Y] = Z \stackrel{\text{def}}{=} \sum_{i=1}^3 z_i S_i$.
- **c** Show that $\operatorname{Tr}|\sqrt{\rho} Z \sqrt{\rho}| = \sqrt{\|\boldsymbol{z}\|^2 - \|\boldsymbol{z} \times \boldsymbol{a}\|^2}$.
- **d** Show that (7.36) is equivalent to

$$\begin{aligned}&(\|\boldsymbol{x}\|^2 - \langle \boldsymbol{x}, \boldsymbol{a}\rangle^2)(\|\boldsymbol{y}\|^2 - \langle \boldsymbol{y}, \boldsymbol{a}\rangle^2) - (\langle \boldsymbol{x}, \boldsymbol{y}\rangle - \langle \boldsymbol{x}, \boldsymbol{a}\rangle\langle \boldsymbol{a}, \boldsymbol{y}\rangle)^2\\ &\quad\ge \|\boldsymbol{x} \times \boldsymbol{y}\|^2 - \|(\boldsymbol{x} \times \boldsymbol{y}) \times \boldsymbol{a}\|^2 \end{aligned} \tag{7.38}$$

  when $\operatorname{Tr} X = \operatorname{Tr} Y = 0$ in a quantum two-level system.
- **e** Show (7.38) if $\langle \boldsymbol{x}, \boldsymbol{y}\rangle = \langle \boldsymbol{x}, \boldsymbol{a}\rangle = 0$.
- **f** Show that a 2×2 matrix $(b_{i,j})$ with determinant 0 can be taken such that $\langle \tilde{\boldsymbol{x}}, \boldsymbol{a}\rangle = 0$ and $\langle \tilde{\boldsymbol{x}}, \boldsymbol{a}\rangle = 0$, where $\tilde{\boldsymbol{x}} \stackrel{\text{def}}{=} b_{1,1}\boldsymbol{x} + b_{1,2}\boldsymbol{y}$ and $\tilde{\boldsymbol{y}} \stackrel{\text{def}}{=} b_{2,1}\boldsymbol{x} + b_{2,2}\boldsymbol{y}$.
- **g** Show (7.38) if $\operatorname{Tr} X = \operatorname{Tr} Y = 0$.
- **h** Show that (7.36) still holds even if $\operatorname{Tr} X = \operatorname{Tr} Y = 0$ is not true.

7.16. Show that the POVM $\boldsymbol{M}_{X,Y,\rho}$ below satisfies $O^1(\boldsymbol{M}) = X$, $O^2(\boldsymbol{M}) = Y$, and the equality in (7.32), for X, Y, ρ satisfying the equality in (7.27).
Construction of the POVM $\boldsymbol{M}_{X,Y,\rho}$: Let the spectral decomposition of X and Y be $X = \sum_i x_i E_{X,i}$ and $Y = \sum_j y_j E_{Y,j}$, respectively. Let p and

q be two nonnegative real numbers such that $p+q=1$. Define the POVM $\boldsymbol{M}_{X,Y,\rho}$ with the data set $\Omega = \{i\} \cup \{j\}$ as follows. Let $M_{X,Y,\rho,i} = pE_{X,i}$ and $M_{X,Y,\rho,j} = qE_{Y,j}$. Define $x_i^1 \stackrel{\text{def}}{=} \frac{1}{p}(x_i - \operatorname{Tr}\rho X) + \operatorname{Tr}\rho X$, $x_j^1 \stackrel{\text{def}}{=} \operatorname{Tr}\rho X$, $x_i^2 \stackrel{\text{def}}{=} \operatorname{Tr}\rho Y$, $x_j^2 \stackrel{\text{def}}{=} \frac{1}{q}(y_i - \operatorname{Tr}\rho Y) + \operatorname{Tr}\rho Y$. The POVM is then defined as $\boldsymbol{M}_{X,Y,\rho} \stackrel{\text{def}}{=} \{(M_{X,Y,\rho,i}, x_i^1, x_i^2)\} \cup \{(M_{X,Y,\rho,j}, x_j^1, x_j^2)\}$.

7.17. Using (7.30), show that the following two conditions are equivalent for two Hermitian matrices X and Y.

① $[X,Y]=0$.
② There exist an instrument $\boldsymbol{\kappa} = \{\kappa_\omega\}$ and a set $\{x_\omega\}$ such that the following two conditions

$$\operatorname{Tr}\rho X = \sum_\omega x_\omega \operatorname{Tr}\kappa_\omega(\rho), \quad \Delta_4(\boldsymbol{\kappa}, Y, \rho) = 0$$

hold for an arbitrary state ρ. The first equation implies that the instrument $\boldsymbol{\kappa}$ corresponds to the observable X. The second equation implies that the instrument $\boldsymbol{\kappa}$ does not disturb the observable Y.

7.18. Show that

$$\operatorname{Tr}_{B,C}\left((I_A \otimes \sqrt{\rho_0})\, U^* \,(I_{A,B} \otimes O(\boldsymbol{E}))\, U \,(I_A \otimes \sqrt{\rho_0})\right) = O(\boldsymbol{M}) \qquad (7.39)$$

for an indirect measurement $\mathcal{I} = (\mathcal{H}_C, U, \rho_0, \boldsymbol{E})$ corresponding to $\boldsymbol{M}$.

7.19. Given two state evolutions κ_1 and κ_2, show that $\Delta_4^2(\lambda\kappa_1 + (1-\lambda)\kappa_2, X, \rho) \geq \lambda\Delta_4^2(\kappa_1, X, \rho) + (1-\lambda)\Delta_4^2(\kappa_2, X, \rho)$. Show that the equality holds when the space spanned by the supports of $\kappa_1(X \circ \rho)$ and $\kappa_1(\rho)$ is orthogonal to the space spanned by the supports of $\kappa_2(X \circ \rho)$ and $\kappa_2(\rho)$.

7.20. Let two Hermitian matrices X and Y on $\mathcal{H}$ satisfy $\operatorname{Cov}_\rho(X,X)$ $\operatorname{Cov}_\rho(Y,Y) = \left(\frac{\operatorname{Tr}|\sqrt{\rho}[X,Y]\sqrt{\rho}|}{2}\right)^2$ and $\operatorname{Cov}_\rho(X,Y) = 0$. Let $X = \sum_{i=1}^k x_i E_i$, and define the POVM $\boldsymbol{M} = \{M_i, \hat{x}_i\}_{i=0}^k$ according to the condition $\hat{x}_0 = \operatorname{Tr} X\rho$, $\hat{x}_i = \operatorname{Tr} X\rho + \frac{1}{p}(x_i - \operatorname{Tr} X\rho)$, $M_0 = (1-p)I$, and $M_i = pE_i$. Now consider equivalent another space $\mathcal{H}'$ to $\mathcal{H}$, and the unitary map U from $\mathcal{H}$ to $\mathcal{H}'$. Define $\boldsymbol{\kappa} = \{\kappa_i\}$ according to $\kappa_0(\rho) = (1-p)U\rho U^*$ and $\kappa_i(\rho) = \sqrt{M_i}\rho\sqrt{M_i}$. Show that $O(\boldsymbol{M}) = X$ and that the equality of (7.30) holds.

7.3 Measurements with Negligible State Demolition

As discussed previously, any measurement inevitably demolishes the measured state. In this section, we propose a method constructing a measurement with negligible state demolition. When a measurement described by an instrument $\boldsymbol{\kappa}$ is made on a system in a state ρ, the amount of the state demolition (reduction) is given by

$$\varepsilon(\rho, \boldsymbol{\kappa}) \stackrel{\text{def}}{=} \sum_{\omega} \operatorname{Tr} \kappa_{\omega}(\rho) b^2 \left(\rho, \frac{1}{\operatorname{Tr} \kappa_{\omega}(\rho)} \kappa_{\omega}(\rho) \right), \tag{7.40}$$

where b is the Bures distance. In the following discussion, we consider the typical state reduction $\boldsymbol{\kappa}_{\boldsymbol{M}}$ of a POVM $\boldsymbol{M} = \{M_i\}$. Then, the amount of the state demolition can be found to be [185]

$$\begin{aligned}
\varepsilon(\rho, \boldsymbol{\kappa}_{\boldsymbol{M}}) &= \sum_i \operatorname{Tr} \rho M_i \left(1 - \operatorname{Tr} \sqrt{\rho^{1/2} \frac{\sqrt{M_i} \rho \sqrt{M_i}}{\operatorname{Tr} \rho M_i} \rho^{1/2}} \right) \\
&= 1 - \sum_i \sqrt{\operatorname{Tr} \rho M_i} \operatorname{Tr} \sqrt{\rho^{1/2} \sqrt{M_i} \rho \sqrt{M_i} \rho^{1/2}} \\
&= 1 - \sum_i \sqrt{\operatorname{Tr} \rho M_i} \operatorname{Tr} \rho^{1/2} \sqrt{M_i} \rho^{1/2} \\
&= 1 - \sum_i \sqrt{\operatorname{Tr} \rho M_i} \operatorname{Tr} \rho \sqrt{M_i} \le 1 - \sum_i \left(\operatorname{Tr} \rho \sqrt{M_i} \right)^2,
\end{aligned} \tag{7.41}$$

where we used the quantum version of Jensen's inequality Ex. 2.18 in the last formula. Conversely, since $M_i \le I$, we have $M_i \le \sqrt{M_i}$, and therefore $\varepsilon(\rho, \boldsymbol{\kappa}_{\boldsymbol{M}}) \ge 1 - \sum_i \left(\operatorname{Tr} \rho \sqrt{M_i} \right)^{3/2}$. When there is no scope of ambiguity, we will use the abbreviation $\varepsilon(\rho, \boldsymbol{\kappa}_{\boldsymbol{M}})$ for $\varepsilon(\rho, \boldsymbol{M})$. In particular, if ρ_j is generated with a probability distribution $p = \{p_j\}$, the average of $\varepsilon(\rho_j, \boldsymbol{\kappa}_{\boldsymbol{M}})$ can be evaluated as

$$\begin{aligned}
\varepsilon(p, \boldsymbol{\kappa}_{\boldsymbol{M}}) &\stackrel{\text{def}}{=} \sum_j p_j \varepsilon(\rho_j, \boldsymbol{\kappa}_{\boldsymbol{M}}) \le 1 - \sum_j p_j \sum_i \left(\operatorname{Tr} \rho_j \sqrt{M_i} \right)^2 \\
&\le 1 - \sum_i \left(\sum_j p_j \operatorname{Tr} \rho_j \sqrt{M_i} \right)^2 = 1 - \sum_i \left(\operatorname{Tr} \bar{\rho}_p \sqrt{M_i} \right)^2,
\end{aligned} \tag{7.42}$$

where $\bar{\rho}_p \stackrel{\text{def}}{=} \sum_j p_j \rho_j$. Hence, the analysis of $\varepsilon(p, \boldsymbol{\kappa}_{\boldsymbol{M}})$ is reduced to that of $1 - \sum_i \left(\operatorname{Tr} \bar{\rho}_p \sqrt{M_i} \right)^2$.

Let us now consider which POVM M has a negligible state demolition. For this analysis, we focus on the number $i_m \stackrel{\text{def}}{=} \operatorname{argmax}_i \operatorname{Tr} M_i \rho$ and the probability $\mathrm{P}_{\rho}^{\boldsymbol{M},\max} \stackrel{\text{def}}{=} \operatorname{Tr} M_{i_m} \rho$. Then, we obtain

$$\begin{aligned}
&(1 - \mathrm{P}_{\rho}^{\boldsymbol{M},\max})(1 + \mathrm{P}_{\rho}^{\boldsymbol{M},\max}) = 1 - (\mathrm{P}_{\rho}^{\boldsymbol{M},\max})^2 \ge 1 - \left(\operatorname{Tr} \rho \sqrt{M_{i_m}} \right)^2 \\
&\quad \ge 1 - \sum_i \left(\operatorname{Tr} \rho \sqrt{M_i} \right)^2 \ge \varepsilon(\rho, \boldsymbol{M}).
\end{aligned}$$

Therefore, we see that $\varepsilon(\rho, \boldsymbol{M})$ will approach 0 if $\mathrm{P}_{\rho}^{\boldsymbol{M},\max}$ approaches 1.

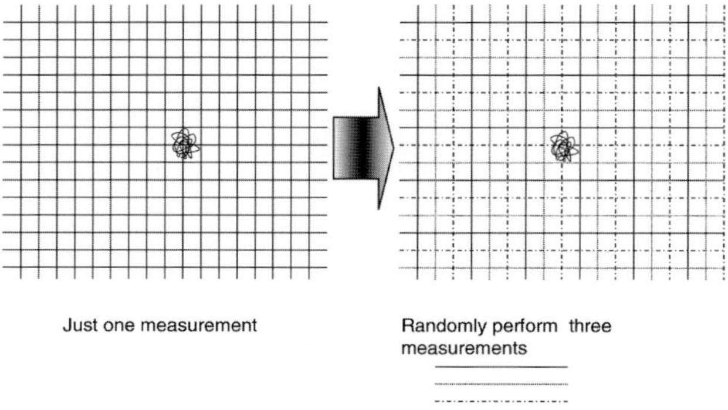

Fig. 7.1. Measurement with negligible state demolition

However, a meaningful POVM does not necessarily have the above property, but usually has the following property in the asymptotic case, i.e., in the case of n-fold tensor product state $\rho^{\otimes n}$ on $\mathcal{H}^{\otimes n}$. Let $(\boldsymbol{M}^{(n)}, \boldsymbol{x}^n) = \{(\boldsymbol{M}^{(n)}, \boldsymbol{x}^n)\}$ be a sequence of pairs of POVMs and functions to $\mathbb{R}^d$. Hence, the vector $\boldsymbol{x}^n(i) = (x^{n,k}(i))$ is the measurement value following the probability distribution $\mathrm{P}_{\rho^{\otimes n}}^{\boldsymbol{M}^{(n)}}$. Suppose that the measurement value $\boldsymbol{x}^n(i) = (x^{n,k}(i))$ satisfies the weak law of large numbers as a random variable. That is, for a given density ρ, there exists a vector $\boldsymbol{a} \in \mathbb{R}^d$ such that

$$\mathrm{Tr}\,\rho^{\otimes n} \boldsymbol{M}^{(n)}\{\|\boldsymbol{x}^n(i) - \boldsymbol{a}\| \geq \epsilon'\} \to 0, \quad \forall \epsilon' > 0. \tag{7.43}$$

For the definition of the notation $\boldsymbol{M}^{(n)}\{\|\boldsymbol{x}^n(i) - \boldsymbol{a}\| \geq \epsilon'\}$, see (6.73). Therefore, we propose a method producing a POVM with negligible state demolition from a POVM satisfying (7.43) as follows.

Theorem 7.6 *For a given positive real number δ and a given positive integer l, we define the modified POVM $\boldsymbol{M}^{(n),\delta,l}$ taking values in $\mathbb{Z}^d$ in the following way. We also define the function $\boldsymbol{x}_\delta^n$ from the set $\mathbb{Z}^d$ to $\mathbb{R}^d$ as $\boldsymbol{x}_\delta^n(\boldsymbol{j}) = \delta \boldsymbol{j}$. If a sequence $\{\delta_n\}$ of real numbers and another sequence $\{l_n\}$ of integers satisfy $\delta_n \to 0$, $l_n \to \infty$, and*

$$\mathrm{Tr}\,\rho^{\otimes n} \boldsymbol{M}^{(n)}\{\|\boldsymbol{x}^n(i) - \boldsymbol{a}\| \geq \delta_n\} \to 0, \tag{7.44}$$

$$l_n \delta_n \to 0, \tag{7.45}$$

we then have

$$\varepsilon(\rho^{\otimes n}, \boldsymbol{M}^{(n),\delta_n,l_n}) \to 0, \tag{7.46}$$

$$\mathrm{Tr}\,\rho^{\otimes n} \boldsymbol{M}^{(n),\delta_n,l_n}\{\|\boldsymbol{x}_{\delta_n}^n(\boldsymbol{j}) - \boldsymbol{a}\| \geq \epsilon'\} \to 0, \quad \forall \epsilon' > 0. \tag{7.47}$$

Construction of $\boldsymbol{M}^{(n),\delta,l}$: *Define*

$$U_{\boldsymbol{y},\epsilon} \stackrel{\text{def}}{=} \left\{ \boldsymbol{x} \in \mathbb{R}^d \,\middle|\, \forall k, y^k - \frac{1}{2}\epsilon \le x^k < y^k + \frac{1}{2}\epsilon \right\},$$

$$\tilde{U}_{\boldsymbol{y},\epsilon} \stackrel{\text{def}}{=} \left\{ \boldsymbol{x} \in \mathbb{R}^d \,\middle|\, \forall k, y^k - \frac{1}{2}\epsilon < x^k \le y^k + \frac{1}{2}\epsilon \right\}$$

for $\boldsymbol{y} = (y^k) \in \mathbb{R}^d$. *Define* $M^{(n)}_{\boldsymbol{y},\delta} \stackrel{\text{def}}{=} \sum_{\boldsymbol{x}^n_i \in U_{\boldsymbol{y},\delta}} M^{(n)}_i$. *Then,* $\{M^{(n)}_{\delta\boldsymbol{j},\delta}\}_{\boldsymbol{j}\in\mathbb{Z}^d}$ *is a POVM since* $\sum_{\boldsymbol{j}\in\mathbb{Z}^d} M^{(n)}_{\delta\boldsymbol{j},\delta} = I$ *for arbitrary* $\delta > 0$. *Moreover, we define* $M^{(n),\delta,l}_{\boldsymbol{j}} \stackrel{\text{def}}{=} \frac{1}{l^d} M^n_{\delta\boldsymbol{j},l\delta}$. *Using a similar discussion, we can check that* $\sum_{\boldsymbol{j}\in\mathbb{Z}^d} M^{(n)}_{\delta\boldsymbol{j},l\delta} = l^d I$, *and then verify that* $\boldsymbol{M}^{(n),\delta,l} = \{M^{(n),\delta,l}_{\boldsymbol{j}}\}_{\boldsymbol{j}\in\mathbb{Z}^d}$ *is a POVM with measurement values in* $\mathbb{Z}^d$.

The existence of a sequence $\{\delta_n\}$ that satisfies condition (7.44) and $\delta_n \to 0$ can be verified from Lemma A.3 in the appendix. Note that the choice of the POVM $\boldsymbol{M}^{(n),\delta,l}$ depends only on the choice of δ and not $\rho^{\otimes n}$. If the convergence of (7.43) is uniform for every $\epsilon > 0$, then the convergences of (7.46) and (7.47) also are not dependent on $\rho^{\otimes n}$.

Proof of (7.46): Let $\delta\boldsymbol{j} \in \tilde{U}_{\boldsymbol{a},(l-1)\delta} \cap \delta\mathbb{Z}^d$. Since $\{\boldsymbol{x}^n_i | \, \|\boldsymbol{x}^n(i) - \boldsymbol{a}\| < \delta\} \subset U_{\delta\boldsymbol{j},l\delta}$, we obtain

$$\frac{1}{l^d}\boldsymbol{M}^{(n)}\{\|\boldsymbol{x}^n(i) - \boldsymbol{a}\| < \delta\} \le M^{(n),\delta,l}_{\boldsymbol{j}}. \tag{7.48}$$

From the matrix monotonicity of $x \mapsto \sqrt{x}$ and the fact that $0 \le \boldsymbol{M}^{(n)}\{\|\boldsymbol{x}^n(i) - \boldsymbol{a}\| < \delta\} \le I$, we obtain

$$\frac{1}{l^{d/2}}\boldsymbol{M}^{(n)}\{\|\boldsymbol{x}^n(i) - \boldsymbol{a}\| < \delta\} \le \frac{1}{l^{d/2}}\sqrt{\boldsymbol{M}^{(n)}\{\|\boldsymbol{x}^n(i) - \boldsymbol{a}\| < \delta\}} \le \sqrt{M^{(n),\delta,l}_{\boldsymbol{j}}}.$$

Meanwhile, since $\#(\tilde{U}_{\boldsymbol{a},(l-1)\delta} \cap \delta\mathbb{Z}^d) = (l-1)^d$, we have

$$\begin{aligned}
\varepsilon(\rho^{\otimes n}, \boldsymbol{M}^{(n),\delta,l}) &\le 1 - \sum_{\boldsymbol{j}\in\mathbb{Z}^d} \left(\operatorname{Tr} \rho^{\otimes n} \sqrt{M^{(n),\delta,l}_{\boldsymbol{j}}}\right)^2 \\
&\le 1 - \sum_{\delta\boldsymbol{j}\in\tilde{U}_{\boldsymbol{a},(l-1)\delta}} \left(\operatorname{Tr} \rho^{\otimes n} \sqrt{M^{(n),\delta,l}_{\boldsymbol{j}}}\right)^2 \\
&\le 1 - \sum_{\delta\boldsymbol{j}\in\tilde{U}_{\boldsymbol{a},(l-1)\delta}} \left(\frac{1}{l^{d/2}} \operatorname{Tr} \rho^{\otimes n} \boldsymbol{M}^{(n)}\{\|\boldsymbol{x}^n(i) - \boldsymbol{a}\| < \delta\}\right)^2 \\
&= 1 - \frac{(l-1)^d}{l^d}\left(\operatorname{Tr} \rho^{\otimes n} \boldsymbol{M}^{(n)}\{\|\boldsymbol{x}^n(i) - \boldsymbol{a}\| < \delta\}\right)^2.
\end{aligned}$$

From (7.44) and the fact that $l_n \to \infty$, substituting δ_n and l_n in δ and l, respectively, we obtain $\varepsilon(\rho^{\otimes n}, \boldsymbol{M}^{(n),\delta_n,l_n}) \to 0$. ■

Proof of (7.47): If $\|\delta \boldsymbol{j} - \boldsymbol{a}\| \geq \epsilon'$ and $\boldsymbol{x}^n(i) \in U_{\delta \boldsymbol{j}, l\delta}$, then $\|bx^n(i) - \boldsymbol{a}\| \geq \epsilon' - \sqrt{d}l\delta$. It then follows that

$$\boldsymbol{M}^{(n),\delta,l}\{\|\boldsymbol{x}_\delta^n(\boldsymbol{j}) - \boldsymbol{a}\| \geq \epsilon'\} \leq \boldsymbol{M}^{(n)}\{\|\boldsymbol{x}^n(i) - \boldsymbol{a}\| \geq \epsilon' - \sqrt{d}l\delta\}.$$

Therefore, if δ_n and l_n are chosen such that condition (7.45) is satisfied, then (7.47) is satisfied. ■

Note that similarly we can show that $\varepsilon(p^n, \boldsymbol{\kappa}_{\boldsymbol{M}^{(n),\delta_n,l_n}}) \to 0$, which will be used in Chap. 10.

As discussed in Chap. 6, the asymptotic performance of an estimator can be treated with at least two criteria. One is large deviation, wherein we focus on the decreasing exponential rate of the probability that the estimate does not belong to the neighborhood of the true value with a fixed radius. The other is small deviation, wherein we focus on the asymptotic behavior of mean square error. In mathematical statistics, it is known that its discussion is essentially equivalent to that of the probability that the estimate does not belong to the the neighborhood of the true value with a radius proportional to $\frac{1}{\sqrt{n}}$. That is, the difference between two criteria is essentially expressed by the difference of the asymptotic behavior of the radius of the neighborhood of interest.

As mentioned in Exercise 7.21, if the original POVM is optimal in the sense of a large deviation, the deformed one is also optimal in the same sense. However, even if the original estimator is optimal in the sense of a small deviation, the estimator deformed by the presented method is not necessarily optimal in the same sense. That is, $\lim \varepsilon(\rho^{\otimes n}, \boldsymbol{M}^{(n),\delta_n,l_n})$ affects the accuracy of the deformed estimator in the sense of the small deviation, but not in the sense of the large deviation. Therefore, it is expected that there exists a tradeoff relation between $\lim \varepsilon(\rho^{\otimes n}, \boldsymbol{M}^{(n)})$ and the limit of the mean square error of the estimator.

Moreover, since the measurement with negligible state demolition has not been realized in the experiment, its realization is strongly desired.

Exercises

7.21. Consider the sequence $\boldsymbol{M} = \{(\boldsymbol{M}^n, \hat{\theta}^n)\}$ of estimators for the state family with one parameter $\{\rho_\theta | \theta \in \mathbb{R}\}$. Let $\beta(\boldsymbol{M}, \theta, \epsilon)$ be continuous with respect to ϵ. Show that $\beta(\boldsymbol{M}, \theta, \epsilon) = \beta(\{(\boldsymbol{M}^{n,\delta_n}, x^n_{\delta_n})\}, \theta, \epsilon)$ when $l_n\delta_n \to 0$, where $\boldsymbol{M}^{n,\delta,l}$ is defined in a way similar to that in the above discussion.

7.4 Historical Note

The mathematical description of a measurement process was initiated by von Neumann [408]. In his formulation, the measurement is described by a projection-valued measure. From the mathematical viewpoint, Naĭmark [315] showed that any POVM can be characterized as the restriction of the projection-valued measure. This projection-valued measure is called a Naĭmark extension. Holevo applied this argument to quantum measurements [216].

Further, Davis and Lewis [92] formulated the state reduction as a positive-map-valued measure. Following this research, Ozawa [327] proved that any measurement reduction should be described by a CP-map-valued measure, i.e., an instrument. He also proposed that the indirect measurement model as a description of the interaction with the macroscopic system can be considered to occur after the time evolution U [333, 334]. Indeed, a positive-map-valued measure $\{\kappa_\omega\}$ is a CP-map-valued measure if and only if it can be described by an indirect measurement model [327]. For any POVM, its indirect measurement model gives a Naĭmark extension of this POVM (Theorem 7.1). For example, an indirect measurement model of the joint measurement of the position Q and the momentum P is known (see Holevo [216]). Further, Hayashi et al. [187] constructed indirect measurements for a meaningful POVM for squeezed states. For a more precise description of state reduction, see Holevo [221], who discusses state reductions due to continuous-variable measurements using semigroup theory. Busch et al. [64] discuss the connection between this formulation and experiments. In addition, Ozawa characterized the instrument given by (7.1) as a minimal-disturbance measurement [332, 334]. Furthermore, this book treats state reductions where the input and output systems are different systems because such a reduction is not very uncommon in quantum information processing. Hence, this book focuses on Theorem 7.3 as a generalization of Corollary 7.1 obtained by Ozawa [327].

The uncertainty relation between conjugate observables was discussed in the context of gedanken experiments by Heisenberg [200]. It was first treated mathematically by Robertson [351], who was not, however, concerned with the effect of measurement. Recently, Ozawa [334–338] formulated the disturbance by measurement, and treated the uncertainty relation concerning measurement, mathematically. These are undoubtably the first attempts at a mathematically rigorous treatment of Heisenberg uncertainty. In this book, we mathematically formulate the same problem, but in a different way. In particular, the definition of disturbance in this text is different from that by Ozawa. Hence, the inequality given in this text is a sightly stronger requirement than that of Ozawa. However, the methods of Ozawa's and our proofs are almost identical. For further discussion of the historical perspective of this topic, see Ozawa [338]. Indeed, Ozawa considered inequality (7.29) to be the mathematical formulation of Heisenberg's uncertainty relation, and he gave its counterexample. He also proposed another type of uncertainty relation—(7.31) and (7.35)—due to measurement. However, in this book, inequality (7.30) is treated as the mathematical formulation of Heisenberg's uncertainty relation. Therefore, the discussion in this book is different from that of Ozawa.

Concerning the mixed-state case, Nagaoka [301] generalized inequality (7.26) to inequality (7.27). [Indeed, the RHS of Nagaoka's original inequality has a different expression; however, it is equal to the RHS of (7.27).] This is a stronger inequality than the trivial generalization $\Delta_1(X,\rho)\Delta_1(Y,\rho) \geq \frac{|\operatorname{Tr}\rho[X,Y]|}{2}$. All inequalities in Sect. 7.2 are based on the former, but Ozawa derived several inequalities based on the latter. Hence, inequalities in this book are stronger than the corresponding Ozawa inequality in the mixed-state case.

Further, using inequality (7.27), Nagaoka [301] derived inequality (7.32) in the mixed-state case. The same inequality with the RHS $\frac{|\operatorname{Tr} \rho[X,Y]|}{2}$ has been discussed by many researchers [14, 15, 239, 328]. Nagaoka applied this inequality to the Cramér–Rao-type bound and obtained the bound (6.95) in the two-parameter case, first. Hayashi [172] extended this inequality to the case with more than two parameters.

The study of measurements with a negligible state demolition has been motivated by quantum universal variable-length source coding because quantum universal variable-length source coding requires determination of the compression rate with small state demolition (Sect. 10.5). Hayashi and Matsumoto [185] treated this problem and obtained the main idea of Sect. 7.3. This method is useful for estimating the state of the system without serious state demolition. This method is often called gentle tomography. Bennett et al. [46] considered the complexity of this kind of measurement.

8

Entanglement and Locality Restrictions

Summary. Quantum mechanics violates everyday intuition not only because the measured data can only be predicted probabilistically but also because of a quantum-specific correlation called entanglement. It is believed that this type of correlation does not exist in macroscopic objects. Entanglement can be used to cause nonlocal phenomena. States possessing such correlations are called entangled states (or states that possess entanglement). Among these states, the states with the highest degree of entanglement are called maximally entangled states or EPR states. Historically, the idea of a nonlocal effect due to entanglement was pointed out by Einstein, Podolsky, and Rosen; hence, the name EPR state.

In order to transport a quantum state over a long distance, we have to retain its coherence during its transmission. However, it is often very difficult because the transmitted system can be easily correlated with the environment system. If the sender and receiver share an entangled state, the sender can transport his/her quantum state to the receiver without transmitting it, as is mentioned in Chap. 9. This protocol is called quantum teleportation and clearly explains the effect of entanglement in quantum systems. Many other effects of entanglement have also been examined, some of which are given in Chap. 9.

However, it is difficult to take advantage of entanglement if the shared state is insufficiently entangled. Therefore, one topic of investigation will be to discuss how much of a maximally entangled state can be extracted from a state with a small amount of entanglement. Of course, if we allow quantum operations between two systems, we can always produce maximally entangled states. Therefore, we examine cases where only local quantum operations and classical communications are allowed.

Table 8.1. Denotations used in Chap. 8

$\|\Phi_L\rangle\langle\Phi_L\|$	Maximally entangled state of size L
$\mathcal{H}_s$	Symmetric space
$\mathcal{H}_a$	Antisymmetric space
F	Flip operator $P_s - P_a$
σ_α	Maximally correlated state (8.113)
$\rho_{W,p}$	Werner state (8.188)
$\rho_{I,p}$	Isotropic state (8.191)

Table 8.1. Continued

Characterizations of q-q channel κ	
$F_e(\rho, \kappa)$	Entanglement fidelity for TP-CP κ (8.17)
$F_e(\rho, \boldsymbol{\kappa})$	Entanglement fidelity for an instrument (8.27)
$I(\rho, \kappa)$	Transmission information of q-q channel κ (8.33)
$I_c(\rho, \kappa)$	Coherent information (8.35)
$\tilde{I}_c(\rho, \kappa)$	Pseudocoherent information (8.46)
$H_e(\kappa, \rho)$	Entropy exchange $H((\kappa \otimes \iota_R)(\|x\rangle\langle x\|))$

Class of localized operations ($C =$)	
$\emptyset$	Only local operations
$\dashrightarrow$	Local operations and zero-rate classical communications from A to B
$\rightarrow$	Local operations and classical communications from A to B
$\leftarrow$	Local operations and classical communications from B to A
$\leftrightarrow$	Local operations and two-way classical communications between A and B
S	Separable operations
PPT	Positive partial transpose (PPT) operations

Entanglement quantities	
$E_{d,1}^{C}(\rho)$	Entanglement of distillation (8.69)
$E_{d,1}^{C,\dagger}(\rho)$	Strong converse entanglement of distillation (8.70)
$E_{d,e}^{C}(\rho)$	Entanglement of exact distillation (8.80)
$E_{c}^{C}(\rho)$	Entanglement of cost (8.93)
$E_{c,e}^{C}(\rho)$	Entanglement of exact cost (8.97)
$E_{sq}(\rho)$	Squashed entanglement (8.111)
$E_{c}^{\dashrightarrow}(\rho)$	Entanglement of cost with zero-rate communication (8.124)
$E_f(\rho)$	Entanglement of formation (8.83)
$E_{r,S}(\rho)$	Entanglement of relative entropy concerning separable states ($\stackrel{\mathrm{def}}{=} \min_{\sigma:\mathrm{separable}} D(\rho\\|\sigma)$)
$E_{r,\mathrm{PPT}}(\rho)$	Entanglement of relative entropy concerning PPT states ($\stackrel{\mathrm{def}}{=} \min_{\sigma:\mathrm{PPT}} D(\rho\\|\sigma)$)
$E_{\mathrm{SDP}}(\rho)$	SDP bound ($\stackrel{\mathrm{def}}{=} \min_{\sigma} D(\rho\\|\sigma) + \log \\|\tau^A(\sigma)\\|_1$)
$E_p(\rho)$	Entanglement of purification (8.127)
$E_{m}^{C}(\rho)$	Maximum of negative conditional entropy (8.104)
$E_{sr}(\rho)$	Logarithm of Schmidt rank (8.98)
$C_o(\rho)$	Concurrence (8.184)
$E_{d,L}^{C}(\rho)$	Conclusive teleportation fidelity (8.79)

Other types of correlation	
$I_\rho(X:Y)$	Quantum mutual information (8.32)
$C_d^{A\to B}(\rho)$	See (8.133)
$C_c(\rho)$	See (8.149)
$C(\rho, \delta)$	See (8.151)
$\tilde{C}(\rho, \delta)$	See (8.152)
$C(\rho)$	$\stackrel{\mathrm{def}}{=} C(\rho, 0) = \tilde{C}(\rho, 0)$
$C_k^{A\to B-E}(\rho)$	Optimal generation rate of secret key with one-way communication (9.69)
$C_d^{A\to B-E}(\rho)$	See (9.70)

8.1 Entanglement and Local Quantum Operations

Using quantum teleportation, we can effectively transmit a quantum state with only local quantum operations and classical communications if a maximally entangled state is shared between the two parties. We now consider the reverse problem. What operations are possible on an entangled state if only local quantum operations and classical communications are allowed?

As stated in Sect. 1.2, a pure state on $\mathcal{H}_A$ can be represented by an element u on $\mathcal{H}_A$ with the norm $\|u\|$ equal to 1. A pure entangled state in the composite system is represented by an element x on $\mathcal{H}_A \otimes \mathcal{H}_B$. Using the basis $u_1, \ldots, u_d$ for $\mathcal{H}_A$ and the basis $v_1, \ldots, v_{d'}$ for $\mathcal{H}_B$, x can be written as $x = \sum_{j,i} x^{i,j} |u_i\rangle \otimes |v_j\rangle$. Let us define the matrix form of x as the linear map X_x from $\mathcal{H}_B$ to $\mathcal{H}_A$ with respect to x by $|x\rangle = |X_x\rangle$. Then,

$$X_x = \sum_{j,i} x^{i,j} |u_i\rangle\langle v_j|. \tag{8.1}$$

Therefore, the correspondence $x \mapsto X_x$ gives a one-to-one relationship between these matrices and the elements of $\mathcal{H}_A \otimes \mathcal{H}_B$. From (1.20), any tensor product $u \otimes v$ satisfies

$$\langle u \otimes v | x \rangle = \langle u \otimes v | X_x \rangle = \operatorname{Tr} |u\rangle\langle \overline{v}| X_x = \langle \overline{v} | X_x | u \rangle.$$

Now, let X'_x be the same as X_x but defined in a different basis v'_j for B, and define $U \stackrel{\text{def}}{=} \sum_{j,k,l} \langle v'_k|v_j\rangle\langle v'_k|v_l\rangle |v_j\rangle\langle v_l|$. Since $|v_j\rangle = \sum_k \langle v'_k|v_j\rangle |v'_k\rangle$ and $\langle v'_k| = \sum_l \langle v'_k|v_l\rangle\langle v_l|$, we have

$$X'_x = \sum_{i,j,k} x^{i,j} \langle v'_k|v_j\rangle |u_i\rangle\langle v'_k| = \sum_{i,j,k,l} x^{i,j} \langle v'_k|v_j\rangle\langle v'_k|v_l\rangle |u_i\rangle\langle v_l| = X_x U.$$

That is, the definition of X_x depends on the orthonormal basis of $\mathcal{H}_B$.

Further, we have

$$\begin{aligned}
\rho_x \stackrel{\text{def}}{=} \operatorname{Tr}_B |x\rangle\langle x| =& \sum_j \left(\sum_i x^{i,j} |u_i\rangle \right) \left(\sum_{i'} \overline{x^{i',j}} \langle u_{i'}| \right) \\
=& \sum_{i',i} \left(\sum_j \overline{x^{i',j}} x^{i,j} \right) |u_i\rangle\langle u_{i'}| \qquad (8.2) \\
=& X_x X_x^*. \qquad (8.3)
\end{aligned}$$

Now, let us denote the nonzero eigenvalues of ρ_x in (8.3) by $\lambda_1, \ldots, \lambda_l$. Then, we can apply the arguments given in Sect. A.2 to the matrix $x^{i,j}$. Choosing sets of orthogonal vectors of length 1 as $u'_1, \ldots, u'_l$ and $v'_1, \ldots, v'_l$, we obtain $x = \sum_{i=1}^{l} \sqrt{\lambda_i} |u'_i\rangle \otimes |v'_i\rangle$. The right-hand side (RHS) of this equation is often called the *Schmidt decomposition*, and $\sqrt{\lambda_i}$ is called the *Schmidt coefficient*. Of course,

$$\mathrm{Tr}_B |x\rangle\langle x| = \sum_i \lambda_i |u_i'\rangle\langle u_i'|, \quad \mathrm{Tr}_A |x\rangle\langle x| = \sum_i \lambda_i |v_i'\rangle\langle v_i'|.$$

Hence, both have the same eigenvalues and entropies. However, the above is true only if the state on the composite system is a pure state. The number of nonzero Schmidt coefficients $\sqrt{\lambda_i}$ is called the *Schmidt rank* and is equal to the ranks of both $\mathrm{Tr}_B |x\rangle\langle x|$ and X_x.

Conversely, for a general state ρ^A on $\mathcal{H}_A$, a pure state $|u\rangle\langle u|$ on $\mathcal{H}_A \otimes \mathcal{H}_R$ satisfying $\rho^A = \mathrm{Tr}_R |u\rangle\langle u|$ is called the *purification* of ρ^A. The quantum system $\mathcal{H}_R$ used for the purification is called the *reference* and is denoted R.

Lemma 8.1 *For two pure states $|x\rangle\langle x|$ and $|y\rangle\langle y|$ on the composite system $\mathcal{H}_A \otimes \mathcal{H}_R$, the following two conditions are equivalent.*

① *The Schmidt coefficients of $|x\rangle\langle x|$ and $|y\rangle\langle y|$ coincide, i.e.,*

$$x = \sum_{i=1}^{l} \sqrt{\lambda_i} |u_i\rangle \otimes |v_i\rangle, \quad y = \sum_{i=1}^{l} \sqrt{\lambda_i} |u_i'\rangle \otimes |v_i'\rangle.$$

② *There exist unitary matrices U^A, U^R in A and R such that*

$$x = \left(U^A \otimes U^R\right) y. \tag{8.4}$$

Furthermore, if

$$\mathrm{Tr}_R |x\rangle\langle x| = \mathrm{Tr}_R |y\rangle\langle y|, \tag{8.5}$$

then

$$x = \left(I \otimes U^R\right) y \tag{8.6}$$

for a suitable unitary matrix U^R on R. Therefore, the purification of the mixed state ρ^A on $\mathcal{H}_A$ can be transferred by the operation of the unitary matrix on R.

Proof. ②⇒① by inspection. If unitary matrices U^A and U^R on $\mathcal{H}_A$ and $\mathcal{H}_R$, respectively, are chosen such that $U^A(u_i) = u_i'$ and $U^R(v_i) = v_i'$, then (8.4) is satisfied, and hence ①⇒②. From (8.5), $X_x X_x^* = X_y X_y^*$. From (A.8), choosing appropriate unitary matrices U_x and U_y, we have $X_x = \sqrt{X_x X_x^*} U_x$ and $X_y = \sqrt{X_x X_x^*} U_y$. Therefore, $X_x = X_y U_y^* U_x$. Then, (8.6) can be obtained from (1.19). ■

Therefore, pure entangled states can be classified according to their Schmidt coefficients. In particular, when all the Schmidt coefficients are equal

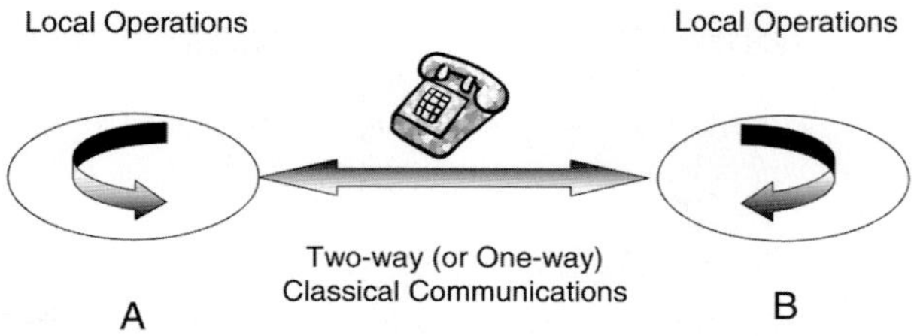

Fig. 8.1. Two-way LOCC (or one-way LOCC)

to $\sqrt{\frac{1}{L}}$, the state is called a *maximally entangled state* of size L. Any maximally entangled state of size L may be transformed from a maximally entangled state $|\Phi_L\rangle\langle\Phi_L|$ by local operations. Hence, we can examine the properties of maximally entangled states of size L by treating a typical maximally entangled state $|\Phi_L\rangle\langle\Phi_L|$. A maximally entangled state is separated from separable states as follows. We can relate this to the fidelity according to

$$\max_{\sigma:\text{separable}} \langle\Phi_L|\sigma|\Phi_L\rangle = \frac{1}{L}. \tag{8.7}$$

For the proof of this relation, it is sufficient to show that $\langle\Phi_L|\sigma|\Phi_L\rangle \le \frac{1}{L}$ for a separable pure state $|u\otimes v\rangle\langle u\otimes v|$. This inequality can be shown from the monotonicity of the partial trace of the fidelity

$$\langle\Phi_L|u\otimes v\rangle\langle u\otimes v|\Phi_L\rangle \le \langle u|\left(\mathrm{Tr}_B\,|\Phi_L\rangle\langle\Phi_L|\right)|u\rangle = \langle u|\rho_{\mathrm{mix}}|u\rangle = \frac{1}{L}.$$

When u is equal to $\sqrt{L}X_{\Phi_L}v$, $\langle\Phi_L|u\otimes v\rangle = \frac{1}{\sqrt{L}}$, which implies (8.7).

Next, we discuss state operations consisting of local operations (LO) and classical communications (CC). This can be classified into three classes: (i) only classical communications from A to B are allowed (*one-way LOCC* simply $\rightarrow$); (ii) classical communications from A to B and B to A are allowed (*two-way LOCC*, simply $\leftrightarrow$); and (iii) no classical communications are allowed (only local quantum operations are allowed) (it is simply denoted by $\emptyset$). In terms of the Choi–Kraus representation of the TP-CP map given in ⑥ of Theorem 5.1, the state evolutions may be written

$$\kappa(\rho) = \sum_i \left(E_{A,i}\otimes E_{B,i}\right)\rho\left(E_{A,i}^*\otimes E_{B,i}^*\right). \tag{8.8}$$

If a TP-CP map can be written in the form (8.8), it is called a *separable TP-CP map* (*S-TP-CP map*) [399]. A TP-CP map κ is an S-TP-CP map if and only if the matrix $K(\kappa)$ defined in (5.2) can be regarded as a separable state in the composite system $(\mathcal{H}_A\otimes\mathcal{H}_{A'})\otimes(\mathcal{H}_B\otimes\mathcal{H}_{B'})$, where we assume that the map $E_{A,i}$ ($E_{B,i}$) is a map from $\mathcal{H}_A$ ($\mathcal{H}_B$) to $\mathcal{H}_{A'}$ ($\mathcal{H}_{B'}$).

Theorem 8.1 *(Lo and Popescu [276]) Let the initial state of a composite system $\mathcal{H}_A \otimes \mathcal{H}_B$ be a known pure state $|x\rangle\langle x|$. LOCC state operations consisting of two-way classical communications can be realized by state operations consisting of one-way classical communications from A to B.*

Proof. For the proof of this theorem, it is sufficient to show that any final state realized by operation (1) can be realized by operation (2), where operations (1) and (2) are given as follows. In operation (1), we (i) perform a measurement in system B, (ii) transmit B's measured data to system A, and (iii) finally apply state evolutions to each system. In operation (2), we (i) perform a measurement in system A, (ii) transmit A's measured data to system B, and (iii) finally apply state evolutions to each system.

From Theorem 7.2, any operation with class (1) can be described by the the state reduction

$$I_A \otimes \sqrt{M_i^B}|x\rangle\langle x|I_A \otimes \sqrt{M_i^B}, \tag{8.9}$$

and local TP-CP maps on A and B depend on the measurement datum i. Hence, it is sufficient to prove that the state reduction (8.9) can be realized by a state reduction by a measurement on A and local TP-CP maps on A and B depending on the measurement datum i.

Using (1.19) and (A.7), we have

$$\begin{aligned} I_A \otimes \sqrt{M_i^B}|X_x\rangle &= |X_x \sqrt{M_i^B}^T\rangle = |U_i \overline{\sqrt{M_i^B}(X_x)^* U}\rangle \\ &= |U_i \overline{\sqrt{M_i^B} V^* X_x V^* U_i}\rangle = U_i V_i^* (V_i \overline{\sqrt{M_i^B}} V_i^*) \otimes (V_i^* U_i)^T |X_x\rangle, \end{aligned}$$

where U_i and V_i are unitary matrices satsfying $X_x \sqrt{M_i^B}^T = U_i|X_x \sqrt{M_i^B}^T|$ and $X_x = V|X_x|$. This equation implies that the state reduction (8.9) is realized by the state reduction on A by the instrument $\{V_i \overline{\sqrt{M_i^B}} V_i^*\}_i$ and the local unitaries $U_i V_i^*$ and $(V_i^* U_i)^T$ depending on the datum i on A and B, respectively. ∎

Exercises

8.1. Let x be the purification of state ρ on $\mathcal{H}_A$.
Show that $H(\rho) = H(\mathrm{Tr}_A |x\rangle\langle x|)$.

8.2 Fidelity and Entanglement

We can characterize the fidelity of two states on $\mathcal{H}_A$ using the purification of mixed states in the following way.

Lemma 8.2 *(Uhlmann [392]) Consider two mixed states ρ_1 and ρ_2 on $\mathcal{H}_A$. Let $|u_1\rangle\langle u_1|$ and $|u_2\rangle\langle u_2|$ be their purifications, respectively. Then,*

$$F(\rho_1, \rho_2)(\stackrel{\text{def}}{=} \operatorname{Tr} |\sqrt{\rho_1}\sqrt{\rho_2}|) = \max_{u_1,u_2} |\langle u_1|u_2\rangle|, \tag{8.10}$$

where the max *on the RHS is with respect to the purifications of ρ_1 and ρ_2.*

Proof. First, we choose the matrix X_{u_i} according to (8.1) in the previous section as a matrix from the reference system $\mathcal{H}_R$ to the system $\mathcal{H}_A$. (Note that the map $u \mapsto X_u$ depends upon the basis of $\mathcal{H}_R$). Since $\rho_i = X_{u_i} X_{u_i}^*$, from (A.8) we obtain $X_{u_i} = \sqrt{\rho_i} U_i$ choosing an appropriate unitary matrix U_i on $\mathcal{H}_R$. From (1.20) and (A.12) we have

$$\begin{aligned} |\langle u_1|u_2\rangle| &= |\operatorname{Tr} X_{u_2} X_{u_1}^*| = |\operatorname{Tr} \sqrt{\rho_2} U_2 U_1^* \sqrt{\rho_1}| \\ &= |\operatorname{Tr} \sqrt{\rho_1}\sqrt{\rho_2} U_2 U_1^*| \le \operatorname{Tr} |\sqrt{\rho_1}\sqrt{\rho_2}|, \end{aligned} \tag{8.11}$$

which proves the $\ge$ part of (8.10). The equality follows from the existence of $U_2 U_1^*$ satisfying the equality of (8.11). ■

From (8.6), for an arbitrary purification x of ρ_1, there exists a purification y of ρ_2 such that

$$F(\rho_1, \rho_2) = |\langle x|y\rangle| = \langle x|y\rangle, \tag{8.12}$$

where the second equation follows from choosing suitable phase factor $e^{i\theta}$ in y. Vectors $v_1, \dots v_n$ satisfying $\sum_{i=1}^n |v_i\rangle\langle v_i| = \rho$ are called a *decomposition* of ρ. Using this fact, we obtain the following corollary regarding decompositions.

Corollary 8.1 *Let ρ_1 and ρ_2 be two mixed states on $\mathcal{H}_A$. For an arbitrary decomposition $u_1, \dots, u_l$ of ρ_1, there exists a decomposition $v_1, \dots v_l$ of ρ_2 such that $F(\rho_1, \rho_2) = \sum_{i=1}^l \langle u_i|v_i\rangle$.*

Proof. Let $w_1, \dots, w_l$ be an orthonormal basis for the space $\mathcal{H}_R$. Let $x = \sum_{i=1}^l u_i \otimes w_i$. Choose a purification $y \in \mathcal{H}_A \otimes \mathcal{H}_R$ of ρ_2 satisfying (8.12). Since $w_1, \dots, w_l$ is an orthonormal basis, there exists an appropriate element $v_1, \dots, v_l$ of $\mathcal{H}_A$ such that $y = \sum_{i=1}^l v_i \otimes w_i$. Therefore, $|\langle x|y\rangle| = \sum_{i=1}^l \langle u_i|v_i\rangle$. ■

Corollary 8.2 *(Uhlmann [392]) Let $\rho = \sum_i p_i \rho_i$ for the states ρ_i and σ, and the probability p_i. The following concavity holds:*

$$F^2(\rho, \sigma) \ge \sum_i p_i F^2(\rho_i, \sigma). \tag{8.13}$$

If σ is a pure state, then

$$F^2(\rho, |u\rangle\langle u|) = \langle u|\rho|u\rangle, \tag{8.14}$$

and the equality in (8.13) holds.

Proof. The validity of (8.14) follows from the fact that $F(\rho, |u\rangle\langle u|) = \mathrm{Tr}\sqrt{|u\rangle\langle u|\rho|u\rangle\langle u|}$. Let y be the purification of σ, and x_i be the purification of ρ_i satisfying $\langle x_i|y\rangle = F(\rho_i, \sigma)$. Then,

$$\sum_i p_i F^2(\rho_i, \sigma) = \sum_i p_i \langle y|x_i\rangle\langle x_i|y\rangle = F^2(\sum_i p_i |x_i\rangle\langle x_i|, |y\rangle\langle y|) \le F^2(\rho, \sigma)$$

completes the proof. The last inequality can be proved by considering the partial trace. ■

A stronger statement (*strong concavity of the fidelity*) holds regarding the concavity of $F(\rho, \sigma)$.

Corollary 8.3 *(Nielsen and Chuang [316]) For states ρ_i and σ_i and probabilities $\{p_i\}$ and $\{q_i\}$, the following concavity property holds:*

$$F\left(\sum_i p_i \rho_i, \sum_i q_i \sigma_i\right) \ge \sum_i \sqrt{p_i q_i} F(\rho_i, \sigma_i). \tag{8.15}$$

Proof. Let x_i and y_i be the purifications of ρ_i and σ_i, respectively, satisfying $F(\rho_i, \sigma_i) = \langle x_i|y_i\rangle$. Consider the space spanned by the orthonormal basis $\{u_i\}$. The purifications of $\sum_i p_i \rho_i$ and $\sum_i q_i \sigma_i$ are then $x \stackrel{\mathrm{def}}{=} \sum_i \sqrt{p_i} x_i \otimes u_i$ and $y \stackrel{\mathrm{def}}{=} \sum_i \sqrt{q_i} y_i \otimes u_i$. Therefore,

$$F\left(\sum_i p_i \rho_i, \sum_i q_i \sigma_i\right) \ge |\langle x|y\rangle| = \sum_i \sqrt{p_i q_i} \langle x_i|y_i\rangle,$$

completing the proof. ■

Monotonicity is the subject of the following corollary.

Corollary 8.4 *For an arbitrary TP-CP map κ from $\mathcal{H}_A$ to $\mathcal{H}_{A'}$,*

$$F(\rho_1, \rho_2) \le F(\kappa(\rho_1), \kappa(\rho_2)). \tag{8.16}$$

This corollary is called the *monotonicity*. Further, the *monotonicity* (5.37), i.e., $b(\rho, \sigma) \ge b(\kappa(\rho), \kappa(\sigma))$, can be derived from this.

Proof. Choose the Stinespring representation $(\mathcal{H}_C, |u\rangle\langle u|, U)$ of κ, i.e., choose $(\mathcal{H}_C, |u\rangle\langle u|, U)$, such that it satisfies $\kappa(\rho) = \mathrm{Tr}_{A,C} U(\rho \otimes |u\rangle\langle u|)U^*$. Let two pure states u_1 and u_2 be purifications of ρ_1 and ρ_2 on $\mathcal{H}_A \otimes \mathcal{H}_R$ maximizing the RHS of (8.10). Since

$$\kappa(\rho_i) = \mathrm{Tr}_{A,C,R}(U \otimes I_R)(|u_i\rangle\langle u_i| \otimes |u\rangle\langle u|)(U \otimes I_R)^*,$$

$(U \otimes I_R)(u_i \otimes u)$ is the purification of $\kappa(\rho_i)$; therefore, it satisfies $|\langle u_1 \otimes u|u_2 \otimes u\rangle| = |\langle u_1|u_2\rangle| = F(\rho_1, \rho_2)$. Then, (8.16) can be obtained from (8.10). ■

Let us next examine a quantity called the *entanglement fidelity*, which expresses how much entanglement is preserved in a TP-CP map κ from $\mathcal{H}_A$ to $\mathcal{H}_A$ [361]. Let R be the reference system with respect to the CP map κ and the mixed state ρ on $\mathcal{H}_A$. The entanglement fidelity is then defined as

$$F_e(\rho, \kappa) \stackrel{\text{def}}{=} \sqrt{\langle x|\kappa \otimes \iota_R(|x\rangle\langle x|)|x\rangle}, \tag{8.17}$$

where x is the purification of ρ. At first glance, this definition seems to depend on the choice of the purification x. Using the Choi–Kraus representation $\{E_j\}_j$ of κ, we can show that Ex. 8.4 [31]

$$F_e^2(\rho, \kappa) = \langle x|\kappa \otimes \iota_R(|x\rangle\langle x|)|x\rangle = \sum_j |\operatorname{Tr} E_j \rho|^2. \tag{8.18}$$

Hence, $F_e(\rho, \kappa)$ is independent of the purification x and of the Choi–Kraus representation $\{E_i\}_i$. From the monotonicity of the fidelity, we have

$$F_e(\rho, \kappa) \leq F(\rho, \kappa(\rho)). \tag{8.19}$$

The equality holds if ρ is a pure state. The entanglement fidelity satisfies the following properties,[1] which will be applied in later sections.

① Let κ' be a TP-CP map from $\mathcal{H}_A$ to $\mathcal{H}_B$, and κ be a TP-CP map from $\mathcal{H}_B$ to $\mathcal{H}_A$ with $\dim \mathcal{H}_A \geq \dim \mathcal{H}_B$. Given a state ρ on $\mathcal{H}_A$, there exists an isometric matrix U on $\mathcal{H}_A$ such that [31]

$$F_e^2(\rho, \kappa \circ \kappa') \leq F_e(\rho, \kappa \circ \kappa_U). \tag{8.20}$$

② If $\rho = \sum_i p_i \rho_i$, we have [31] Ex. 8.6

$$F_e^2(\rho, \kappa) \leq \sum_i p_i F_e^2(\rho_i, \kappa). \tag{8.21}$$

In particular, when all the ρ_i are pure states, the following holds [361]:

$$F_e^2(\rho, \kappa) \leq \sum_i p_i F^2(\rho_i, \kappa(\rho_i)). \tag{8.22}$$

③ Let $\mathcal{H}_B$ be a subspace of $\mathcal{H}_A$. Given a real number a such that $1 > a > 0$, there exists a subspace $\mathcal{H}_C$ of $\mathcal{H}_B$ with a dimension $\lfloor (1-a) \dim \mathcal{H}_B \rfloor$ such that [31, 154]

$$\max_{x \in \mathcal{H}_C} \{1 - F^2(x, \kappa(x))\} \leq \frac{1 - F_e^2(\rho_{\text{mix}}^B, \kappa)}{a}. \tag{8.23}$$

---

[1] A large part of the discussion relating to entanglement fidelity and information quantities relating to entanglement (to be discussed in later sections) was first done by Schumacher [361].

④ Let the support of ρ be included in the subspace $\mathcal{H}_B$ of $\mathcal{H}_A$. The following then holds [31]:

$$\frac{2}{3}\left(1 - F_e^2(\rho,\kappa)\right) \le \max_{x\in\mathcal{H}_B^1}\left\{1 - F^2(x,\kappa(x))\right\}, \tag{8.24}$$

where $\mathcal{H}_B^1$ is the set of normalized vectors of $\mathcal{H}_B$, i.e., $\{x \in \mathcal{H}_B \| \|x\| = 1\}$. The completely mixed-state ρ_{mix} on $\mathcal{H}$ satisfies

$$\frac{d}{d+1}\left(1 - F_e^2(\rho_{\mathrm{mix}},\kappa)\right) = \mathrm{E}_x\left[1 - F^2(x,\kappa(x))\right], \tag{8.25}$$

where E_x denotes the expectation with respect to a uniform distribution on $\mathcal{H}^1$ and d is the dimension of $\mathcal{H}$.

From the definition, for a general CP map κ_i and a positive real number f_i, we have

$$\sum_i f_i F_e^2(\rho,\kappa_i) = F_e^2(\rho,\sum_i f_i\kappa_i). \tag{8.26}$$

Therefore, we can define the entanglement fidelity $F_e(\rho,\boldsymbol{\kappa})$ as

$$F_e^2(\rho,\boldsymbol{\kappa}) \stackrel{\mathrm{def}}{=} \sum_\omega F_e^2(\rho,\kappa_\omega)\left(= F_e^2(\rho,\sum_\omega \kappa_\omega)\right) \tag{8.27}$$

for an instrument $\boldsymbol{\kappa} = \{\kappa_\omega\}$ with an input and output $\mathcal{H}_A$ and a state ρ on $\mathcal{H}_A$. Since $\varepsilon(\rho,\boldsymbol{\kappa}) \le 1 - F_e^2(\rho,\boldsymbol{\kappa})$ from (8.22), combining these properties gives (7.41) and (7.42).

In fact, the purification is useful only for treating a single state. In order to analyze a mixed-state ρ on $\mathcal{H}_A$, we often focus on the *probabilistic decomposition* of ρ; this is defined as the set $\{(p_i,\rho_i)\}$ satisfying

$$\rho = \sum_i p_i\rho_i,$$

where p_i is a probability distribution and ρ_i is a state on $\mathcal{H}_A \otimes \mathcal{H}_B$. In a quantum system, the probabilistic decomposition is not uniqe for a given mixed state ρ. Now, we let $|X\rangle$ be a purification of ρ with the reference system $\mathcal{H}_R$. (Here, we choose the reference $\mathcal{H}_R$ whose dimension is equal to the rank of ρ.) When we perform a POVM $\boldsymbol{M} = \{M_i\}$ on the reference $\mathcal{H}_R$, the outcome i is obtained with the probability:

$$p_i \stackrel{\mathrm{def}}{=} \langle X|(M_i \otimes I_A)|X\rangle = \mathrm{Tr}\, M_i\rho = \mathrm{Tr}\, XM_iX^*. \tag{8.28}$$

The final state on $\mathcal{H}_A$ is given as

$$\rho_i \stackrel{\mathrm{def}}{=} \frac{1}{p_i}\mathrm{Tr}_R(\sqrt{M_i}\otimes I_A)|x\rangle\langle x|(\sqrt{M_i}\otimes I_A) = \frac{1}{p_i}XM_iX^*. \tag{8.29}$$

Since

$$\sum_i p_i\rho_i = \sum_i XM_iX^* = X(\sum_i M_i)X^* = XX^* = \rho,$$

any POVM $\boldsymbol{M}$ on $\mathcal{H}_R$ gives a probabilistic decomposition. Conversely, for any probabilistic decomposition $\{(p_i, \rho_i)\}$ of ρ, the matrix $M_i = X^{-1}p_i\rho_i(X^*)^{-1}$ on $\mathcal{H}_R$ forms a POVM as

$$\sum_i M_i = \sum_i X^{-1}p_i\rho_i(X^*)^{-1} = X^{-1}\rho(X^*)^{-1} = I.$$

Moreover, this POVM $\{M_i\}$ satisfies (8.28) and (8.29). Hence, we obtain the following lemma.

Lemma 8.3 *Any probabilistic decomposition $\{(p_i, \rho_i)\}$ of ρ is given by a POVM $\boldsymbol{M}$ on the reference system as (8.28) and (8.29).*

Indeed, using this discussion, we can characterize the TP-CP map to the environment based on the output state $(\kappa \otimes \iota_R)(|\Phi_d\rangle\langle\Phi_d|)$ of the given channel κ as follows. In this case, since the initial state of the total system of the reference system, the output system, and environment system is pure, its final state is also pure. That is, the final state of the total system is given as the purification $|u\rangle\langle u|$ of $(\kappa \otimes \iota_R)(|\Phi_d\rangle\langle\Phi_d|)$. Since any state ρ can be described as $d_A \operatorname{Tr}_R I_A \otimes \rho^T |\Phi_d\rangle\langle\Phi_d|$, the output state with the input state ρ on $\mathcal{H}_A$ is given as

$$d_A \operatorname{Tr}_{A,R}(I_{A,E} \otimes \rho^T)|u\rangle\langle u|. \tag{8.30}$$

Exercises

8.2. Show that $1 - \left(\operatorname{Tr}|\sqrt{\rho}\sqrt{\sigma}|\right)^2 \ge d_1^2(\rho,\sigma)$ using (8.10) and Exercise 2.23 for two mixed states ρ and σ.

8.3. Show that

$$F^2(\rho,\sigma) \le \operatorname{Tr}\sqrt{\rho}\sqrt{\sigma} \le F(\rho,\sigma) \tag{8.31}$$

using the purifications and the monotonicity of $\phi(1/2, \rho, \sigma)$.

8.4. Prove (8.18) noting that $\operatorname{Tr}(E_i \otimes I)|x\rangle\langle x| = \operatorname{Tr}_E E_i\rho$.

8.5. Prove property ① of the entanglement fidelity by following the steps below.

a Show that there exist Choi–Kraus representations $\{E_i\}_i$ and $\{A_i\}_j$ of κ and κ', respectively, such that the matrix $\{\operatorname{Tr} E_iA_j\rho\}_{i,j}$ can be written in diagonal form with positive and real diagonal elements.

b Using **a** and (8.18), show that there exists a Choi–Kraus representation $\{E_i\}_i$ of κ and a matrix A such that $\operatorname{Tr} A\rho A^* = 1$ and $F_e^2(\rho, \kappa \circ \kappa') \le |\operatorname{Tr} E_1 A \rho|^2$.

c Let E be a matrix from $\mathcal{H}_B$ to $\mathcal{H}_A$. Let $E^*E \le I$ and $\operatorname{Tr} A\rho A^* = \operatorname{Tr} \rho = 1$. Take U^* to be isometric under the polar decomposition $E = U|E|$. Show that $|\operatorname{Tr} EA\rho|^2 \le \operatorname{Tr} U|E|U^*\rho = \operatorname{Tr} EU^*\rho$.

d Take the polar decomposition $E_1 = U|E_1|$ such that U^* is isometric on $\mathcal{H}_B$. Show that $F_e(\rho, \kappa \circ \kappa_{U^*}) \ge F_e^2(\rho, \kappa \circ \kappa')$.

8.6. Show ②, using (8.18) and the fact that the function $\rho \mapsto |\operatorname{Tr} A\rho|^2$ is a convex function.

8.7. Prove ③ by following the steps below. As the first step, determine the orthogonal basis $x_1, \ldots, x_d$ of $\mathcal{H}_B$ inductively. Let x_1 be the vector $\operatorname{argmax}_{x \in \mathcal{H}_B^1} \{1 - F^2(x, \kappa(x))\}$. Given $x_1, \ldots, x_j$, let $\mathcal{H}_j$ be the orthogonal complement space to the space spanned by $x_1, \ldots, x_j$. Let x_{j+1} be $\operatorname{argmax}_{x \in \mathcal{H}_j^1} \{1 - F^2(x, \kappa(x))\}$. Then, let $\mathcal{H}_C$ be the space spanned by $x_{d_B}, \ldots, x_{d_B - d_C + 1}$, where $d_C = \lfloor (1-a) \dim \mathcal{H}_B \rfloor$. Show that the space $\mathcal{H}_C$ satisfies (8.23) using Markov's inequality and ②.

8.8. Show (8.24) in ④ by following the steps below.

a Show that $F_e^2(\rho, \kappa) = \sum_{i,j} p_i p_j \langle u_i | \kappa(|u_i\rangle\langle u_j|) | u_j \rangle$ for $\rho = \sum_i p_i |u_i\rangle\langle u_i|$, where $p_1 \ge p_2 \ge \ldots \ge p_d$.

b Let $\phi = (\phi_1, \ldots, \phi_d)$. Define $u(\phi) \stackrel{\text{def}}{=} \sum_j \sqrt{p_j} e^{i\phi_j} u_j$. Show that $F_e^2(\rho, \kappa) + \sum_{j \ne k} p_j p_k \langle u_k | \kappa(|u_j\rangle\langle u_j|) | u_k \rangle$ is equal to the expectation of $F^2(u(\phi), \kappa(u(\phi)))$ under the uniform distribution with respect to $\phi = (\phi_1, \ldots, \phi_d)$.

c Let δ be the RHS of (8.24). Show that

$$\sum_{k=2}^{d} p_k \langle u_k | \kappa(|u_1\rangle\langle u_1|) | u_k \rangle \le p_2 \delta,$$

$$\sum_{j=2}^{d} \sum_{k \ne j} p_j p_k \langle u_k | \kappa(|u_j\rangle\langle u_j|) | u_k \rangle \le \sum_{j=2}^{d} p_k p_1 \delta.$$

d Show (8.24) using **a** to **c**.

8.9. Show that the equality of (8.24) in ④ holds when κ is a depolarizing channel for a quantum two-level system and ρ is the completely mixed-state ρ_{mix}.

8.10. Prove (8.25) following the steps below.

a Prove (8.25) when κ is a depolarizing channel.

b Define a depolarizing channel $\kappa_{d,\lambda}$ for any channel κ as

$$\kappa_{d,\lambda}(\rho) = \int_{\mathrm{SU}(d_A)} U^* \kappa(U\rho U^*) U \nu(dU),$$

where $\nu(dU)$ is the uniform distribution. Show that $\mathrm{E}_x F^2(x, \kappa(x)) = \mathrm{E}_x F^2(x, \kappa_{d,\lambda}(x))$, where E_x is the expectation concerning the uniform distribution.

c Show that $F_e(\rho_{\mathrm{mix}}, \kappa) = F_e(\rho_{\mathrm{mix}}, \kappa_{d,\lambda})$.

d Prove (8.25) for any channel κ.

8.11. Verify (8.26).

8.12. Prove inequalities (7.41) and (7.42) by combining the formulas given in this section.

8.13. Show that the states ρ_i in (8.29) are pure and orthogonal to each other if and only if the POVM $\boldsymbol{M} = \{M_i\}$ is a PVM and commutative with ρ and $\operatorname{rank} M_i = 1$.

8.14. Let κ be a TP-CP map from $\mathbb{C}^d$ to $\mathbb{C}^{d'}$ and κ' be a TP-CP map from $\mathbb{C}^{d'}$ to $\mathbb{C}^d$. Show that $F_e(\rho_{\mathrm{mix}}, \kappa' \circ \kappa) \le \sqrt{\frac{d'}{d}}$.

8.3 Entanglement and Information Quantities

So far, we have examined the transmission information for a classical-quantum channel, but not the quantum version of the mutual information $I(X:Y)$, defined by (2.21) in Sect. 2.1.1. In Sect. 5.5, we defined the *quantum mutual information* $I_\rho(A:B)$ as

$$I_\rho(A:B) = H_\rho(A) + H_\rho(B) - H_\rho(AB) = D(\rho \| \rho^A \otimes \rho^B) \tag{8.32}$$

with respect to a state ρ on $\mathcal{H}_{A,B}$ for quantum systems $\mathcal{H}_A$ and $\mathcal{H}_B$. We used the notation introduced in Sect. 5.5 for the second expression above. Confusingly, the transmission information $I(p, W)$ for classical-quantum channels is also occasionally called the quantum mutual information. However, since $I_\rho(A:B)$ is a more natural generalization of the mutual information defined in (2.21), we shall call this the quantum mutual information in this text.

As discussed in Sect. 2.1.1, there is a precise relationship between the mutual information and the transmission information for classical systems. Similarly, there is a relationship between the classical-quantum transmission information and the quantum mutual information. To see this relation, let us consider a classical-quantum channel W with an input system $\mathcal{X}$ and an output system $\mathcal{H}_A$. Let $\{u_x\}$ be the orthonormal basis states of the Hilbert space $\mathcal{H}_X$. Let us consider a state on the composite system $\mathcal{H}_X \otimes \mathcal{H}_A$ given by $\rho = \sum_x p_x |u_x\rangle\langle u_x| \otimes W_x$, where p is a probability distribution in $\mathcal{X}$. The quantum mutual information is then given by $I_\rho(X : A) = I(p, W)$. Therefore, this is equal to the transmission information of a classical-quantum channel.

It is possible to find a connection between the transmission information and the quantum mutual information of a classical-quantum channel by appropriately defining the composite system. Let us now define the transmission information of the quantum-quantum channel κ (which is a TP-CP map) from the quantum mutual information using a similar method. Here it is necessary to find the quantum-mechanical correlation between the input and output systems. Therefore, we consider the purification x of the state ρ on the input system $\mathcal{H}_A$. The *transmission information* $I(\rho, \kappa)$ of the quantum-quantum channel κ can then be defined using the quantum mutual information as [3]

$$I(\rho, \kappa) \stackrel{\text{def}}{=} I_{(\kappa \otimes \iota_R)(|x\rangle\langle x|)}(R : B), \tag{8.33}$$

where R is the reference system and B is the output system. Since $H(\rho)$ is equal to the entropy of the reference system, this can also be written as

$$I(\rho, \kappa) = H(\kappa(\rho)) + H(\rho) - H(\kappa \otimes \iota_R(|x\rangle\langle x|)). \tag{8.34}$$

This quantity will play an important role in Sect. 9.3.

Let us now consider the following quantity called the *coherent information*, which expresses how much coherence is preserved through a quantum-quantum channel κ [362].

$$I_c(\rho, \kappa) \stackrel{\text{def}}{=} H(\kappa(\rho)) - H(\kappa \otimes \iota_R(|x\rangle\langle x|)) = -H_{\kappa \otimes \iota_R(|x\rangle\langle x|)}(R|B) \tag{8.35}$$

for a TP-CP map κ from $\mathcal{H}_A$ to $\mathcal{H}_B$, a state ρ on $\mathcal{H}_A$, and a purification x of ρ. Therefore, the coherent information is equal to the negative conditional entropy. Of course, in the classical case, the conditional entropy can only take either positive values or 0. Therefore, a negative conditional entropy indicates the existence of some quantum features in the system. For example, in an entanglement-breaking channel, the conditional entropy is nonnegative, as can be seen in (8.59).

The coherent information can be related to the entanglement fidelity if $\sqrt{2(1 - F_e(\rho, \kappa))} \le 1/e$ as follows Ex. 8.15,8.16 [31]:

$$0 \le H(\rho) - I_c(\rho, \kappa) \le 2\sqrt{2(1 - F_e(\rho, \kappa))}\left(\log d - \log\sqrt{2(1 - F_e(\rho, \kappa))}\right). \tag{8.36}$$

The first inequality holds without any assumption. Therefore, we can expect that the difference between $H(\rho)$ and the coherent information $I_c(\rho, \kappa)$ will express how the TP-CP map κ preserves the coherence. This will be justified in Sect. 9.6.

The above information quantities also satisfy the *monotonicity* [3,362]

$$I_c(\rho, \kappa' \circ \kappa) \leq I_c(\rho, \kappa), \tag{8.37}$$
$$I(\rho, \kappa' \circ \kappa) \leq I(\rho, \kappa), \tag{8.38}$$
$$I(\rho, \kappa \circ \kappa') \leq I(\kappa'(\rho), \kappa). \tag{8.39}$$

If U is an isometric matrix, then the coherent information satisfies [32,362]

$$I_c(\rho, \kappa \circ \kappa_U) = I_c(U\rho U^*, \kappa). \tag{8.40}$$

If $\kappa = \sum_i p_i \kappa_i$, these quantities satisfy the *convexity* for channels [3,32]

$$I_c(\rho, \kappa) \leq \sum_i p_i I_c(\rho, \kappa_i), \tag{8.41}$$
$$I(\rho, \kappa) \leq \sum_i p_i I(\rho, \kappa_i). \tag{8.42}$$

The transmission information satisfies the *concavity* for states [3]

$$I\Big(\sum_{i=1}^k p_i\rho_i, \kappa\Big) \geq \sum_{i=1}^k p_i I(\rho_i, \kappa). \tag{8.43}$$

Conversely, the following reverse inequality also holds:

$$I\Big(\sum_{i=1}^k p_i\rho_i, \kappa\Big) \leq \sum_{i=1}^k p_i I(\rho_i, \kappa) + 2\log k. \tag{8.44}$$

Let κ^A (κ^B) be a TP-CP map from $\mathcal{H}_A$ ($\mathcal{H}_B$) to $\mathcal{H}_{A'}$ ($\mathcal{H}_{B'}$). Let $\rho^{A,B}$ be a state on $\mathcal{H}_A \otimes \mathcal{H}_B$. Let ρ^A and ρ^B be the partially traced state of $\rho^{A,B}$. The transmission information of a quantum-quantum channel then satisfies

$$I(\rho^{A,B}, \kappa^A \otimes \kappa^B) \leq I(\rho^A, \kappa^A) + I(\rho^B, \kappa^B) \tag{8.45}$$

in a similar way to (4.4) for the transmission information of a classical-quantum channel [3].

In addition to the types of information defined up until now, we may also define the *pseudocoherent information*

$$\tilde{I}_c(\rho, \kappa) \stackrel{\text{def}}{=} H(\rho) - H(\kappa \otimes \iota_E(|x\rangle\langle x|)). \tag{8.46}$$

Although it is difficult to interpret the above quantity as information, it does possess the following useful properties [220], which will be used in Sect. 9.3.

$$\tilde{I}_c(\rho, \kappa \circ \kappa') \leq \tilde{I}_c(\kappa'(\rho), \kappa), \tag{8.47}$$
$$\tilde{I}_c\Big(\sum_j p_j\rho_j, \kappa\Big) \geq \sum_j p_j \tilde{I}_c(\rho_j, \kappa). \tag{8.48}$$

The first property (8.47) is the *monotonicity*, and can be derived immediately from property (8.39) and definitions. The second inequality (8.48) is the *concavity* with respect to a state. The following reverse inequality also holds, i.e.,

$$\tilde{I}_c\Big(\sum_{j=1}^{k} p_j\rho_j, \kappa\Big) \le \sum_{j=1}^{k} p_j\tilde{I}_c(\rho_j, \kappa) + \log k. \tag{8.49}$$

The derivations for (8.48) and (8.49) are rather difficult (Exercises 8.24 and 8.25). We can also obtain the following relationship by combining (8.47) and (8.48):

$$\tilde{I}_c\Big(\sum_{j} p_j\kappa_j(\rho), \kappa\Big) \ge \sum_{j} p_j\tilde{I}_c(\rho, \kappa\circ\kappa_j). \tag{8.50}$$

Finally, we focus on the entropy $H((\kappa\otimes\iota_R)(|x\rangle\langle x|))$, which is called the *entropy exchange* [361] and is denoted by $H_e(\kappa,\rho)$. This is equal to the entropy of the environment system $\mathcal{H}_E$ after the state ρ is transmitted. Its relationship to the entanglement fidelity $F_e(\rho,\kappa)$ is given by the *quantum Fano inequality* as [361][2]

$$H_e(\rho,\kappa) \le h(F_e^2(\rho,\kappa)) + (1 - F_e^2(\rho,\kappa))\log(d^2 - 1), \tag{8.51}$$

where d is the dimension of $\mathcal{H}$.

Exercises

8.15. Show the first inequality in (8.36) by considering the Stinespring representation of κ and (5.58) with respect to the composite system of the environment system E and the reference system R.

8.16. Show the second inequality of (8.36) by considering the purification of ρ and Fannes inequality (Theorem 5.9).

8.17. Prove (8.37) based on the Stinespring representation of κ', the strong subadditivity (5.55) of the von Neumann entropy.

8.18. Prove (8.38) using the fact that the RHS of (8.34) is equal to

$$D(\kappa\otimes\iota_E(|x\rangle\langle x|)\|\kappa(\rho)\otimes \mathrm{Tr}_A |x\rangle\langle x|). \tag{8.52}$$

8.19. Prove (8.41) and (8.42) using the concavity (5.60) of the conditional entropy.

[2] Since the form of this inequality is similar to the Fano inequality, it is called quantum Fano inequality. However, it cannot be regarded as a quantum extension of the Fano inequality (2.26). The relationship between the two formulas is still unclear.

8.20. Prove (8.39) based on (8.52) and the monotonicity of the quantum relative entropy by considering the Stinespring representation of κ'.

8.21. Prove (8.40) based on the relationship between the purification of ρ and the purification of $U\rho U^*$.

8.22. Let $\mathcal{H}_{E'}$ be the environment system after performing a state evolution given by the TP-CP map κ from $\mathcal{H}_A$ to $\mathcal{H}_{A'}$. Let x be the purification of the state ρ on $\mathcal{H}_A$. Let the reference system be $\mathcal{H}_R$. Show that

$$\begin{aligned} I_c(\rho,\kappa) &= H_{x'}(A') - H_{x'}(E'), \\ I(\rho,\kappa) &= H_{x'}(A') + H_{x'}(A'E') - H_{x'}(E'), \end{aligned} \tag{8.53}$$

where x' is the final state of x.

8.23. Let x be the purification of ρ with respect to the reference system $\mathcal{H}_R$. Let $\mathcal{H}_{E'_A}$ and $\mathcal{H}_{E'_B}$ be the environment systems after the state evolutions κ^A and κ^B. Let x' be the final state of x. Show (8.45) by following the steps below.

a Show the following, using Exercise 8.22.

$$\begin{aligned} I(\rho^A,\kappa^A) &= H_{x'}(A') + H_{x'}(A'E'_A) - H_{x'}(E'_A) \\ I(\rho,\kappa^A\otimes\kappa^B) &= H_{x'}(A'B') + H_{x'}(A'B'E'_AE'_B) - H_{x'}(E'_AE'_B). \end{aligned}$$

b Show that

$$\begin{aligned} &I(\rho^A,\kappa^A) + I(\rho^B,\kappa^B) - I(\rho,\kappa^A\otimes\kappa^B) \\ &= H_{x'}(A')+H_{x'}(B')-H_{x'}(A'B')-(H_{x'}(E'_A)+H_{x'}(E'_B)-H_{x'}(E'_AE'_B)) \\ &\quad + (H_{x'}(A'E'_A) + H_{x'}(B'E'_B) - H_{x'}(A'E'_AB'E'_B)). \end{aligned}$$

c Prove (8.45) by combining (8.32) and (5.75) with **b**.

8.24. Let κ be the state evolution from $\mathcal{H}_A$ and from $\mathcal{H}_{A'}$. Show (8.48) following the steps below.

a Let x_j be the purification of ρ_j with respect to the reference system $\mathcal{H}_R$. Let $\{u_j\}$ be an orthonormal basis of another system $\mathcal{H}_{R'}$. Show that the pure state $x \stackrel{\text{def}}{=} \sum_j \sqrt{p_j}x_j \otimes u_j$ on $\mathcal{H}_A \otimes \mathcal{H}_R \otimes \mathcal{H}_{R'}$ is the purification of $\rho \stackrel{\text{def}}{=} \sum_j p_j\rho_j$.

b Show that the pinching $\kappa_{\boldsymbol{E}}$ of the measurement $\boldsymbol{E} = \{|u_j\rangle\langle u_j|\}$ on $\mathcal{H}_{R'}$ satisfies

$$\begin{aligned} &D((\kappa_{\boldsymbol{E}}\otimes\iota_{A,R})(\kappa\otimes\iota_{R,R'})(|x\rangle\langle x|)\|(\kappa_{\boldsymbol{E}}\otimes\iota_{A,R})(\kappa(\rho)\otimes\mathrm{Tr}_A(|x\rangle\langle x|))) \\ &= H(\kappa(\rho)) + \sum_j p_jH(\rho_j) - \sum_j p_jH((\kappa\otimes\iota_{R,R'})(|x_j\rangle\langle x_j|)). \end{aligned}$$

c Prove (8.48) by considering the monotonicity of the quantum relative entropy for the pinching $\kappa_{\boldsymbol{E}}$.

8.25. Prove (8.49) using the same symbols as Exercise 8.24 by following the steps below.

a Show that

$$\sum_{j=1}^{k} p_j \tilde{I}_c(\rho_j, \kappa) \\ = H(\kappa_{\boldsymbol{E}}(\mathrm{Tr}_A(|x\rangle\langle x|))) - H(\kappa_{\boldsymbol{E}} \otimes \iota_{A,R})(\kappa \otimes \iota_{R,R'})(|x\rangle\langle x|)).$$

b Verify that

$$\begin{aligned} &\tilde{I}_c(\rho, \kappa) - \sum_{j=1}^{k} p_j \tilde{I}_c(\rho_j, \kappa) \\ &= H(\mathrm{Tr}_A |x\rangle\langle x|) - H(\kappa_{\boldsymbol{E}}(\mathrm{Tr}_A |x\rangle\langle x|)) - H(\kappa \otimes \iota_{R,R'})(|x\rangle\langle x|)) \\ &\quad + H(\kappa_{\boldsymbol{E}} \otimes \iota_{A,R})(\kappa \otimes \iota_{R,R'})(|x\rangle\langle x|)) \\ &\le H(\kappa_{\boldsymbol{E}} \otimes \iota_{A,R})(\kappa \otimes \iota_{R,R'})(|x\rangle\langle x|)) - H(\kappa \otimes \iota_{R,R'})(|x\rangle\langle x|)). \end{aligned} \tag{8.54}$$

c Prove (8.49) using (5.53) and the above results.

8.26. Prove (8.43) using (8.24) and (5.49).

8.27. Prove (8.44) using (8.49) and (5.51).

8.28. Show that

$$\max\{H(\rho) | \langle u|\rho|u\rangle = f\} = h(f) + (1-f)\log(d-1)$$

for a pure state $|u\rangle\langle u|$ on $\mathcal{H}$ ($\dim \mathcal{H} = d$). Prove (8.51) using this result.

8.29. Show that

$$H_e(\kappa, \rho_{\mathrm{mix}}) = H(p), \quad H_e(\kappa, |e_0\rangle\langle e_0|) = H_e(\kappa, |e_1\rangle\langle e_1|) = h(p_0 + p_3)$$

for the entropy exchange of a Pauli channel κ_p.

8.4 Entanglement and Majorization

In this section we consider what kind of state evolutions are possible using only local quantum operations and classical communications given an entangled state between two systems. Before tackling this problem, let us first consider a partial ordering called *majorization* defined between two d-dimensional vectors $a = (a_i), b = (b_i)$ with positive real-number components. This will be useful in the discussion that follows. If a and b satisfy

$$\sum_{j=1}^{k} a_j^{\downarrow} \le \sum_{j=1}^{k} b_j^{\downarrow}, \quad (1 \le k \le n), \quad \sum_{j=1}^{n} a_j^{\downarrow} = \sum_{j=1}^{n} b_j^{\downarrow},$$

we say that b majorizes a, which we denote as $a \preceq b$. In the above, $(a_j^{\downarrow})$ and $(b_j^{\downarrow})$ are the reordered versions of the elements of a and b, respectively, largest first. If $x \preceq y$ and $y \preceq x$, we represent it as $x \cong y$. If $\frac{1}{\sum_i x_i} x \cong \frac{1}{\sum_i y_i} y$, we write $x \approx y$. If $\frac{1}{\sum_i x_i} x = \frac{1}{\sum_i y_i} y$, we represent it as $x \propto y$. The following theorem discusses the properties of this partial ordering. The relation with entanglement will be discussed after this theorem.

Theorem 8.2 *The following conditions for two d-dimensional vectors $x = (x_i)$ and $y = (y_i)$ with positive real components are equivalent [49].*

① $x \preceq y$.
② *There exists a finite number of T-transforms $T_1, \ldots, T_n$ such that $x = T_n \cdots T_1 y$. A T-transform is defined according to a matrix $A = (a^{i,j})$ satisfying $a^{i_1,i_1} = a^{i_2,i_2} = 1 - t$ and $a^{i_1,i_2} = a^{i_2,i_1} = t$ for some pair i_1 and i_2, and $a^{i,j} = \delta_{i,j}$ otherwise, where t is a real number between $0 \le t \le 1$.*
③ *There exists a double stochastic matrix A such that $x = Ay$.*
④ *There exists a stochastic matrix $B = (b^{i,j})$ such that $(B^j)^T \circ x \approx y$ for all integers j. $(B^j)^T$ is the column vector obtained by transposing B^j. The product of the two vectors x and y is defined as $(y \circ x)_i \stackrel{\text{def}}{=} y_i x_i$.*

The product $\circ$ satisfies the associative law. A vector e with each of its components equal to 1 satisfies $e \circ x = x$ and $\sum_j (B^j)^T = e$.

From the concavity of the entropy, we can show that a T-transform T and a probability distribution satisfy $H(T(p)) \ge H(p)$. Therefore, if $q \preceq p$, then

$$H(q) \ge H(p). \tag{8.55}$$

Since a double stochastic matrix Q and a probability distribution p satisfy $Q(p) \preceq p$, we have

$$H(Q(p)) \ge H(p), \tag{8.56}$$

from which we obtain (2.20).

Further, any double stochastic matrix A can be written by a distribution p on the permutations S_k as $(Ax)_i = \sum_{s \in S_k} p_s x_{s^{-1}(i)}$. Thus, when two positive-valued vectors x and y has decreasing ordered elements, we can show that

$$\langle x, y \rangle \ge \langle x, Ay \rangle. \tag{8.57}$$

Let us now consider how majorization can be defined for two density matrices ρ and σ. The eigenvalues of ρ and σ form the respective vectors with real-number components. Therefore, majorization can be defined with respect to these vectors. Letting $\rho = \sum_i a_i|u_i\rangle\langle u_i|$ and $\sigma = \sum_i b_i|v_i\rangle\langle v_i|$, we can write $\rho \preceq \sigma$ if $a \preceq b$. If ρ and σ come from different Hilbert spaces, let us define $\rho \preceq \sigma$ by adding zero eigenvalues to the smaller Hilbert space until the size of the spaces are identical. The relations $\rho \cong \sigma$ and $\rho \approx \sigma$ can be defined in a similar way.

As this is a partial ordering, if $\rho \preceq \rho'$ and $\rho' \preceq \sigma$, then $\rho \preceq \sigma$. Since the entropy $H(\rho)$ of a density matrix ρ depends only on its eigenvalues, if $\rho \preceq \sigma$, then $H(\rho) \geq H(\sigma)$ due to (8.55). Further, we can also show that for a unital channel κ (e.g., pinching),

$$\kappa(\rho) \preceq \rho. \tag{8.58}$$

From this we find that $H(\kappa(\rho)) \geq H(\rho)$, and therefore the first inequality in (5.54) is satisfied even if $\boldsymbol{M}$ is a general POVM. Thus, the following theorem can be shown from Theorem 8.2.

Theorem 8.3 *(Nielsen and Kempe [317]) Let $\rho^{A,B}$ be a separable state on $\mathcal{H}_A \otimes \mathcal{H}_B$. Then, $\rho^{A,B} \preceq \rho^A \stackrel{\mathrm{def}}{=} \mathrm{Tr}_B\, \rho^{A,B}$.*

Combining (8.55) with this theorem, we find that $H(\rho^{A,B}) \geq H(\rho^A)$ [i.e., (5.50)] if $\rho^{A,B}$ is separable. This shows that any separable state ρ satisfies [68]

$$H_\rho(B|A) \geq 0. \tag{8.59}$$

The following theorem shows how two entangled states can be transformed between each other.

Theorem 8.4 *(Nielsen [313], Vidal [402]) Let $|u\rangle\langle u|$ and $|v_j\rangle\langle v_j|$ be pure states on $\mathcal{H}_A \otimes \mathcal{H}_B$. It is possible to transform the state $|u\rangle\langle u|$ into $|v_j\rangle\langle v_j|$ using a two-way LOCC with probability p_j if and only if the condition*

$$\sum_{i=1}^{k} \lambda_i^{\downarrow} \leq \sum_{i=1}^{k} \sum_j p_j \lambda_i^{j,\downarrow}, \quad \forall k \tag{8.60}$$

holds, where $\sqrt{\lambda_i^j}$ is the Schmidt coefficient of $|v_j\rangle$ and $\sqrt{\lambda_i}$ is the Schmidt coefficient of $|u\rangle$. This operation can be realized by performing a measurement at A and then performing a unitary state evolution at B corresponding to the measurement value i at A. Of course,

$$H(\mathrm{Tr}_B\, |u\rangle\langle u|) \geq \sum_j p_j H(\mathrm{Tr}_B\, |v_j\rangle\langle v_j|). \tag{8.61}$$

In particular, it is possible to transform $|u\rangle\langle u|$ into $|v\rangle\langle v|$ using a two-way LOCC with probability 1 if and only if the condition

$$\mathrm{Tr}_B\, |u\rangle\langle u| \preceq \mathrm{Tr}_B\, |v\rangle\langle v| \tag{8.62}$$

holds. These conditions still hold even if the two-way LOCC is restricted to a one-way LOCC.

Proof. First, we show that (8.60) holds if it is possible to transform the pure state $|u\rangle\langle u|$ into $|v_j\rangle\langle v_j|$ with probability p_j. According to the discussion concerning instruments in Sect. 7.1, an arbitrary state evolution κ can be regarded as an instrument given by the Choi–Kraus representation $\{A_j\}_j$. Therefore, we see that if the initial state is a pure state, the final state for each measurement value i must also be a pure state.

Now, consider local operations and two-way communications from A to B and from B to A. This operation consists of repetitions of the following procedure. First, A performs a measurement $\{A'_j\}_j$ and then sends this measurement value j to B. Then, B performs a measurement $\{B_i^j\}_i$ at B corresponding to A's measurement value j. Finally, B sends his or her measurement value i to A. Since the final state after the measurement is also a pure state, the measurement at B may be written as A's measurement and a unitary operation at B corresponding to A's measurement value, according to Theorem 8.1.

Therefore, we see that the whole operation is equivalent to performing a measurement $\{A_j\}_j$ at A and then performing a unitary state operation at B, corresponding to the measurement value at A. By defining $\rho_u \stackrel{\text{def}}{=} \mathrm{Tr}_B |u\rangle\langle u|$, the probability of obtaining the measurement value i is then $p_j \stackrel{\text{def}}{=} \mathrm{Tr}\, A_j \rho_u A_j^*$. The final state is a pure state, and the partially traced state is equal to $\frac{1}{\mathrm{Tr}\, A_j \rho_u A_j^*} A_j \rho_u A_j^*$. Taking the unitary matrix U_j giving the polar decomposition $\sqrt{\rho_u} A_j^* = U_j \sqrt{A_j \rho_u A_j^*}$, we obtain

$$U_j A_j \rho_u A_j^* U_j^* = U_j \sqrt{A_j \rho_u A_j^*} \sqrt{A_j \rho_u A_j^*} U_j^* = \sqrt{\rho_u} A_j^* A_j \sqrt{\rho_u}.$$

If P is a projection with rank k and satisfies the equation $\mathrm{Tr}\, \rho_u P = \sum_{i=1}^k \lambda_i^{\downarrow}$, then

$$\sum_{i=1}^k \sum_j p_j \lambda_i^{j,\downarrow} = \sum_j \max\{\mathrm{Tr}\, A_j \rho_u A_j^* P_j | P_j \text{ is a projection of rank } k\}$$
$$\geq \sum_j \mathrm{Tr}\, U_j A_j \rho_u A_j^* U_j^* P = \sum_j \sqrt{\rho_u} A_j^* A_j \sqrt{\rho_u} P = \mathrm{Tr}\, \rho_u P = \sum_{i=1}^k \lambda_i^{\downarrow}.$$

Therefore, we obtain (8.60).

Next, let us construct the operation that evolves $|u\rangle\langle u|$ into $|v\rangle\langle v|$ with probability 1 when (8.62) is satisfied. Let the Schmidt coefficients of $|u\rangle$ and $|v\rangle$ be $\sqrt{\lambda_i}$ and $\sqrt{\lambda'_i}$, respectively. Let a stochastic matrix (b^{ij}) satisfy Condition ④ of Theorem 8.2 when $x = \lambda = (\lambda_i)$ and $y = \lambda' = (\lambda'_i)$. Now, let us define an orthonormal basis $\{u_i\}$ and E_j by

$$\rho_u = \sum_i \lambda_i |u_i\rangle\langle u_i|, \quad E_j \stackrel{\text{def}}{=} \sum_i b^{i,j} |u_i\rangle\langle u_i|. \tag{8.63}$$

Then, we have $\sum_j E_j = I$ because $B = (b^{i,j})$ is a stochastic matrix. The probability of obtaining the measurement value j for the measurement $\{E_j\}$ is $\operatorname{Tr} \rho_u E_i$. The final state for this measurement value is a pure state, and the partially traced state is $\frac{1}{\operatorname{Tr} \rho_u E_i} \sqrt{E_i} \rho_u \sqrt{E_i}$. From $(B^j)^T \circ \lambda \approx \lambda'$ we have $\frac{1}{\operatorname{Tr} \rho_u E_i} \sqrt{E_i} \rho_u \sqrt{E_i} \cong \operatorname{Tr}_B |v\rangle\langle v|$. Therefore, if an appropriate unitary state evolution is applied corresponding to the measurement value j, the final state will be $|v\rangle\langle v|$ for every measurement value i with probability 1.

Finally, we construct the operation that evolves the pure state $|u\rangle\langle u|$ on $\mathcal{H}_A \otimes \mathcal{H}_B$ into $|v_j\rangle\langle v_j|$ with probability p_j when inequality (8.60) holds. Let $\lambda' = (\lambda'_i)$ be a probability distribution such that

$$\sum_{i=1}^k \lambda'_i = \sum_{i=1}^k \sum_j p_j \lambda_i^{j,\downarrow}, \quad \forall k.$$

The above discussion guarantees that there exists an LOCC operation transforming $|u\rangle\langle u|$ into $|v\rangle\langle v|$ with the Schmidt coefficient $\sqrt{x_i}$. Therefore, it is sufficient to construct the required operation when the equality of (8.62) holds. Let us define $b^{i,j} \stackrel{\text{def}}{=} p_j \lambda_i^{j,\downarrow} / \lambda_i^{\downarrow}$. Then, $B = (b^{i,j})$ is a stochastic matrix. Defining E_j using (8.63), we have

$$p_j = \operatorname{Tr} \rho_u E_j, \quad \operatorname{Tr}_B |v_j\rangle\langle v_j| \cong \sum_i \lambda_i^{j,\downarrow} |u_i\rangle\langle u_i| \cong \frac{1}{\operatorname{Tr} \rho_u E_j} \sqrt{E_j} \rho_u \sqrt{E_j}.$$

This completes the proof. ∎

Using this theorem, we obtain the following characterization.

Lemma 8.4 *(Vidal et al. [404]) Let v and u be entangled pure states with Schmidt coefficient $\sqrt{p_j}$ and $\sqrt{q_j}$ in decreasing order. Then, we have*

$$|\langle u|v\rangle|^2 \le \Big(\sum_j \sqrt{p_j}\sqrt{q_j}\Big)^2. \tag{8.64}$$

The equality holds when vectors v and u have the Schmidt decompositions $v = \sum_i \sqrt{p_j} |e_i^A\rangle \otimes |e_i^B\rangle$ and $u = \sum_i \sqrt{q_j} |e_i^A\rangle \otimes |e_i^B\rangle$ by the same Schmidt basis. Further, we have

$$\max_{\kappa \in \leftrightarrow} \langle u|\kappa(|v\rangle\langle v|)|u\rangle = \max_{q \preceq q'} \Big(\sum_j \sqrt{p_j}\sqrt{q'_j}\Big)^2. \tag{8.65}$$

Proof. Let ρ and σ be the reduced density matrix on $\mathcal{H}_A$ of v and u. Then, there exisis a unitary matrix U such that

$$\langle u|v\rangle = \mathrm{Tr}\,\sqrt{\rho}\sqrt{\sigma}U.$$

Assume that σ is diagonalized as $\sigma = \sum_j q_j |e_j\rangle\langle e_j|$. Thus,

$$\begin{aligned}
|\mathrm{Tr}\,\sqrt{\rho}\sqrt{\sigma}U|^2 &= \Big|\sum_j \sqrt{q_j}\langle e_j|\sqrt{\rho}U|e_j\rangle\Big|^2 \le \Big(\sum_j \sqrt{q_j}\Big|\langle e_j|\sqrt{\rho}U|e_j\rangle\Big|\Big)^2 \\
&\le \Big(\sum_j \sqrt{\sqrt{q_j}\langle e_j|\sqrt{\rho}|e_j\rangle}\sqrt{\sqrt{q_j}\langle e_j|U^*\sqrt{\rho}U|e_j\rangle}\Big)^2 \\
&\le \Big(\sum_j \sqrt{q_j}\langle e_j|\sqrt{\rho}|e_j\rangle\Big)\Big(\sum_j \sqrt{q_j}\langle e_j|U^*\sqrt{\rho}U|e_j\rangle\Big).
\end{aligned}$$

Now, we diagonalize ρ as $\rho = \sum_i p_i |f_i\rangle\langle f_i|$. Hence,

$$\sum_j \sqrt{q_j}\langle e_j|\sqrt{\rho}|e_j\rangle = \sum_{i,j} \sqrt{q_j}\sqrt{p_i}|\langle e_j|f_i\rangle|^2.$$

Since $|\langle e_j|f_i\rangle|^2$ is a double stochastic matrix, (8.57) implies $\sum_{i,j}\sqrt{q_j}\sqrt{p_i}|\langle e_j|f_i\rangle|^2 \le \sum_i \sqrt{q_i}\sqrt{p_i}$. Thus,

$$\sum_j \sqrt{q_j}\langle e_j|\sqrt{\rho}|e_j\rangle \le \sum_i \sqrt{q_i}\sqrt{p_i}.$$

Similarly, we have

$$\sum_j \sqrt{q_j}\langle e_j|U^*\sqrt{\rho}U|e_j\rangle \le \sum_i \sqrt{q_i}\sqrt{p_i}.$$

Therefore, we obtain (8.64).

Next, we prove (8.65). From the equality condition of (8.64) and Theorem 8.4, we can easily verify the $\ge$ part of (8.65). Assume that the LOCC operation κ generates the state v_j with probability r_j from the initial pure state v. When the Schmidt coefficient of v_j is $(\sqrt{p_i^j})_i$, Corollary 8.2 and (8.64) imply

$$\begin{aligned}
&\langle u|\kappa(|v\rangle\langle v|)|u\rangle = \sum_j r_j|\langle u|v_j\rangle|^2 \le \sum_j r_j\Big(\sum_i \sqrt{p_i^j}\sqrt{q_i}\Big)^2 \\
\le&\Big(\sum_i \sqrt{\sum_j r_j p_i^j}\sqrt{q_i}\Big)^2.
\end{aligned}$$

Since Theorem 8.4 guarantees that $(p_i)_i \preceq (\sum_j r_j p_i^j)_i$, we obtain the $\le$ part of (8.65). ■

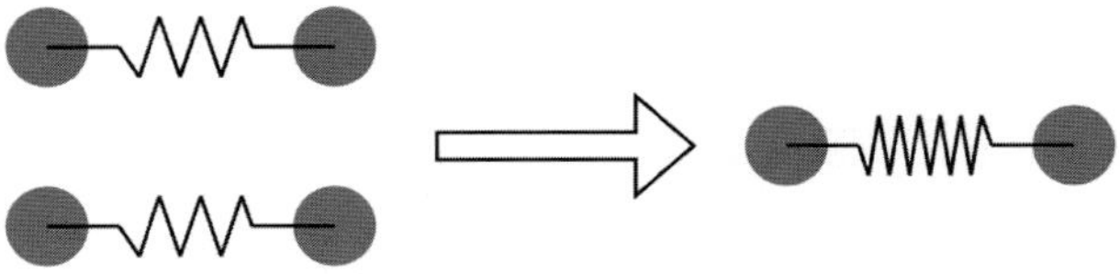

Fig. 8.2. Two partially entangled states (*left*) and one completely entangled state (*right*)

Exercises

8.30. Prove (8.58) by diagonalizing $\rho=\sum_i p_i|u_i\rangle\langle u_i|$ and $\kappa(\rho)=\sum_j q_j|v_j\rangle\langle v_j|$ and examining the map $q=(q_i)\mapsto (\langle v_i|\kappa(\sum_j q_j|u_j\rangle\langle u_j|)|v_i\rangle)$.

8.31. Show that a maximally entangled state of size less than $1/\lambda_1^{\downarrow}$ can be produced with error probability 0 if the initial state is a pure entangled state with Schmidt coefficients $\sqrt{\lambda_i}$.

8.5 Distillation of Maximally Entangled States

In order to use the merit of entanglement, we often require maximally entangled states, not partially entangled states. Then, one encounters the distillation problem of maximally entangled states from partially entangled states. Such an operation is called *entanglement distillation* and is one of the established fields in quantum information theory. If the initial state is pure, it is called entanglement concentration. It has also been verified experimentally [342, 428]. Other experimental models have also been proposed by combining other protocols [410].

Consider the problem of creating a maximally entangled state $|\Phi_L\rangle\langle\Phi_L|$ on $\mathbb{C}^L\otimes\mathbb{C}^L$ from a pure state $|u\rangle\langle u|$ on the composite system $\mathcal{H}_A\otimes\mathcal{H}_B$. If the relation

$$\mathrm{Tr}_B\,|u\rangle\langle u| \preceq \mathrm{Tr}_B\,|\Phi_L\rangle\langle\Phi_L|$$

does not hold, it is impossible to create $|\Phi_L\rangle\langle\Phi_L|$ with a probability 1. Therefore, we must allow some failure probability in our scheme for creating $|\Phi_L\rangle\langle\Phi_L|$.

Theorem 8.5 *(Hayashi [176]) Consider the two-way LOCC operation κ converting the initial state $|u\rangle\langle u|$ to a maximally entangled state $|\Phi_L\rangle\langle\Phi_L|$. The optimal failure probability $\varepsilon_1(\kappa,|u\rangle\langle u|)$ is less than $f(x)\stackrel{\mathrm{def}}{=}\mathrm{Tr}(\rho_u-xI)\{\rho_u-xI\geq 0\}$ if and only if*

$$L\leq\frac{1-f(x)}{x}. \tag{8.66}$$

Proof. Since our operation has two outcomes "success" and "failure," the distribution of the outcome is described by the two-valued POVM $\{T, I-T\}$. Hence, from Theorem 7.2 our operation is given by the combination of the state evolution (7.1) due to a measurement $\{T, I-T\}$ and the TP-CP map depending on its outcome of "success" or "failure," The final state $|v\rangle\langle v|$ corresponding to "success" should satisfy $\rho_v (\stackrel{\text{def}}{=} \operatorname{Tr}_B |v\rangle\langle v|) \preceq \operatorname{Tr}_B |\Phi_L\rangle\langle\Phi_L|$ because of Theorem 8.4. Thus, Theorem 8.1 characterizes the minimum probability that the creation of $|\Phi_L\rangle\langle\Phi_L|$ fails as

$$
\begin{aligned}
&\min_{T \geq 0 \text{ on } \mathcal{H}_A} \left\{ \operatorname{Tr} \rho_u (I-T) \middle| \frac{1}{\operatorname{Tr} \rho_u P} \sqrt{T} \rho_u \sqrt{T} \preceq \operatorname{Tr}_B |\Phi_L\rangle\langle\Phi_L| \right\} \\
&= \min_{0 \leq T \leq I: \text{ on } \mathcal{H}_A} \left\{ \operatorname{Tr} \rho_u (I-T) \,|\sqrt{\rho_u} T \sqrt{\rho_u} \leq x \right\},
\end{aligned}
$$

where we used (A.6) to rewrite the above equation: henceforth, we abbreviate xI to x. Now, let L be the size of the maximally entangled state to be created and the ratio $\frac{\operatorname{Tr} \rho_u P}{L}$ be fixed to x. Since $\operatorname{Tr} \rho_u P$ is the success probability, the minimum failure probability can be calculated from the following equation:

$$
\begin{aligned}
&\min_{0 \leq T \leq I: \text{ on } \mathcal{H}_A} \left\{ \operatorname{Tr} \rho_u (I-T) \,|\sqrt{\rho_u} T \sqrt{\rho_u} \leq x \right\} \\
&= \min_{0 \leq S \leq \rho: \text{ on } \mathcal{H}_A} \left\{ 1 - \operatorname{Tr} S \,|S \leq x \right\} \\
&= \min_{S \text{ on } \mathcal{H}} \left\{ 1 - \sum_i \langle u_i | S | u_i \rangle \middle| \langle u_i | S | u_i \rangle \leq \lambda_i, x \right\} \\
&= 1 - \sum_{i: \lambda_i \leq x} \lambda_i - \sum_{i: \lambda_i > x} x = \operatorname{Tr} (\rho_u - x) \{ \rho_u - x \geq 0 \} = f(x),
\end{aligned}
\tag{8.67}
$$

where $S = \sqrt{\rho_u} T \sqrt{\rho_u}$. Therefore, if the failure probability is less than $f(x)$, the size L of the maximally entangled state satisfies

$$
L \leq \max_{x'} \left\{ \frac{1}{x'} (1 - f(x')) \middle| f(x') \leq f(x) \right\} = \frac{1}{x} (1 - f(x)).
$$

In the last equality, we used the fact that $f(x)$ is strictly monotonically increasing and continuous.

Conversely, if (8.66) is true, then by choosing a projection P that attains the minimum value in (8.67) and performing a two-valued projective measurement $\{P, I-P\}$ on the system $\mathcal{H}_A$, the outcome corresponding to P will be obtained with a probability $1 - f(x)$. Since the final state u satisfies

$$
\frac{1}{1 - f(x)} P \rho_u P \leq \frac{x}{1 - f(x)} \leq \frac{I}{L},
$$

we may construct a maximally entangled state of size L according to Theorem 8.4. ■

On the other hand, Lo and Popescu [276] characterized the optimal success probability $P^{opt}(u \to |\Phi_L\rangle)$ for obtaining a maximally entangled state $|\Phi_L\rangle$ as follows:

$$P^{opt}(u \to |\Phi_L\rangle) = \max_{r:1\le r\le L} \frac{L}{L-r-1} \sum_{i=r}^{L} \lambda_i^{\downarrow}. \tag{8.68}$$

Next, we consider the problem of determining how large a maximally entangled state we can distill from a tensor product state $\rho^{\otimes n}$ of a partially entangled state ρ on $\mathcal{H}_A \otimes \mathcal{H}_B$, in the asymptotic case. Here, we formulate this problem in the mixed-state case as well as in the pure-state case. In such problems, we require that our operation κ_n be optimized for a given partially entangled state $\rho^{\otimes n}$ and hence treat the first type of *entanglement of distillation*:

$$E_{d,1}^{C}(\rho) \stackrel{\text{def}}{=} \sup_{\{\kappa_n\}\subset C} \left\{ \underline{\lim} \frac{1}{n} \log L(\kappa_n) \middle| \lim \varepsilon_1(\kappa_n,\rho) = 0 \right\} \tag{8.69}$$

$$E_{d,1}^{C,\dagger}(\rho) \stackrel{\text{def}}{=} \sup_{\{\kappa_n\}\subset C} \left\{ \underline{\lim} \frac{1}{n} \log L(\kappa_n) \middle| \lim \varepsilon_1(\kappa_n,\rho) < 1 \right\}, \tag{8.70}$$

where C denotes the set of local operations, i.e., $C = \to$, $C = \emptyset$, $C = \leftarrow$, $C = \leftrightarrow$, and $C = S$ imply the set of one-way ($\mathcal{H}_A \to \mathcal{H}_B$) LOCC operations, only local operations, one-way ($\mathcal{H}_A \leftarrow \mathcal{H}_B$) LOCC operations, two-way LOCC operations, S-TP-CP maps, respectively. Here, we denote the size of the maximally entangled state produced by the operation κ by $L(\kappa)$. If ρ is a mixed state, it is extremely difficult to produce a maximally entangled state perfectly, even allowing some failure probability. Therefore, let us relax our conditions and aim to produce a state close to the desired maximally entangled state. Hence, for our operation κ', we will evaluate the error $\varepsilon_2(\kappa',\rho) \stackrel{\text{def}}{=} 1-\langle \Phi_L|\kappa'(\rho)|\Phi_L\rangle$ between the final state $\kappa'(\rho)$ and the maximally entangled state $|\Phi_L\rangle\langle\Phi_L|$ of size L. When the initial state is a pure state v with Schmidt coefficient $\sqrt{\lambda_i^{\downarrow}}$, Lemma 8.4 gives the optimum fidelity:

$$\max_{\kappa\in\leftrightarrow} \langle \Phi_L|\kappa(|v\rangle\langle v|)|\Phi_L\rangle = \left(\max_{p\preceq p'} \sum_{i=1}^{L} \sqrt{\frac{p_i'}{L}} \right)^2. \tag{8.71}$$

In the asymptotic case, we optimize the operation κ_n' for a given $\rho^{\otimes n}$; thus, we focus on the second type of *entanglement of distillation*:

$$E_{d,2}^{C}(\rho) \stackrel{\text{def}}{=} \sup_{\{\kappa_n\}\subset C} \left\{ \underline{\lim} \frac{1}{n} \log L(\kappa_n) \middle| \lim \varepsilon_2(\kappa_n,\rho) = 0 \right\},$$

$$E_{d,2}^{C,\dagger}(\rho) \stackrel{\text{def}}{=} \sup_{\{\kappa_n\}\subset C} \left\{ \underline{\lim} \frac{1}{n} \log L(\kappa_n) \middle| \lim \varepsilon_2(\kappa_n,\rho) < 1 \right\}.$$

The following trivial relations follow from their definitions:

$$E_{d,2}^{C}(\rho) \geq E_{d,1}^{C}(\rho), \ E_{d,2}^{C,\dagger}(\rho) \geq E_{d,1}^{C,\dagger}(\rho), \ E_{d,i}^{C,\dagger}(\rho) \geq E_{d,i}^{C}(\rho),$$

for $i = 1, 2$. The following theorem holds under these definitions.

Theorem 8.6 *(Bennett et al. [38]) The two kinds of entanglement of distillation of any pure state $|u\rangle\langle u|$ in the composite system $\mathcal{H}_A \otimes \mathcal{H}_B$ can be expressed by the reduced density $\rho_u = \operatorname{Tr}_B |u\rangle\langle u|$ as*

$$E_{d,i}^{C}(|u\rangle\langle u|) = E_{d,i}^{C,\dagger}(|u\rangle\langle u|) = H(\rho_u),$$

for $i = 1, 2$ and $C = \emptyset, \rightarrow, \leftarrow, \leftrightarrow, S$.

The proof of this theorem will be given later, except for the case of $C = \emptyset$. This case is proved in Exercise 8.33. This theorem states that the entropy of the reduced density matrix $\rho_u = \operatorname{Tr}_B |u\rangle\langle u|$ gives the degree of entanglement when the state of the total system is a pure state. Further, as shown by Hayashi and Matsumoto [183], there exists an LO protocol that attains this bound without any knowledge about the pure state u, as long as the given state is its tensor product state. That is, there exists a local operation protocol (without any communication) that produces a maximally entangled state of size $e^{nH(\rho_n)}$ and is independent of u. This protocol is often called a *universal concentration protocol.*

For a general mixed state ρ on the composite system $\mathcal{H}_A \otimes \mathcal{H}_B$, the entropy of the reduced density does not have the same implication. Consider

$$E_{r,S}(\rho) \stackrel{\text{def}}{=} \min\{D(\rho\|\sigma)|\sigma : \text{the separable state on } \mathcal{H}_A \otimes \mathcal{H}_B\}$$

as its generalization for a mixed state ρ. This is called the *entanglement of relative entropy.* Any pure state $|u\rangle\langle u|$ satisfies

$$E_{r,S}(|u\rangle\langle u|) = H(\operatorname{Tr}_B |u\rangle\langle u|). \tag{8.72}$$

Lemma 8.5 *(Vedral and Plenio [399]) The entanglement of the relative entropy satisfies the monotonicity property*

$$E_{r,S}(\kappa(\rho)) \leq E_{r,S}(\rho) \tag{8.73}$$

for any S-TP-CP map κ. Hence, any LOCC operation satisfies the above monotonicity because it is an S-TP-CP map.

Proof. Let σ be a separable state such that $D(\rho\|\sigma) = E_r(\rho)$, then $\kappa(\sigma)$ is separable. From the monotonicity of the relative entropy (5.30),

$$E_{r,S}(\kappa(\rho)) \leq D(\kappa(\rho)\|\kappa(\sigma)) \leq D(\rho\|\sigma) = E_{r,S}(\rho),$$

which gives (8.73). ∎

The following theorem may be proved by using the method in the proof of Lemma 3.5.

Theorem 8.7 *(Vedral and Plenio [399]) Any mixed state ρ on the composite system $\mathcal{H}_A \otimes \mathcal{H}_B$ and any separable state σ satisfy*

$$E_{d,2}^{S,\dagger}(\rho) \le D(\rho\|\sigma). \tag{8.74}$$

Hence, we obtain

$$E_{d,2}^{S,\dagger}(\rho) \le E_{r,S}(\rho), \tag{8.75}$$

$$E_{d,2}^{S,\dagger}(\rho) \le E_{r,S}^{\infty}(\rho) \stackrel{\text{def}}{=} \lim \frac{E_{r,S}(\rho^{\otimes n})}{n}. \tag{8.76}$$

Proof. Consider an S-TP-CP map κ'_n on $\mathcal{H}_A \otimes \mathcal{H}_B$ and a real number $r > D(\rho\|\sigma)$. Since $\kappa'_n(\sigma^{\otimes n})$ is also separable, equation (8.7) implies that $\langle \Phi_{e^{nr}}|\kappa'_n(\sigma^{\otimes n})|\Phi_{e^{nr}}\rangle \le e^{-nr}$. From $I - |\Phi_{e^{nr}}\rangle\langle\Phi_{e^{nr}}| \ge 0$ we have

$$\begin{aligned} I - (\kappa'_n)^*(|\Phi_{e^{nr}}\rangle\langle\Phi_{e^{nr}}|) &= (\kappa'_n)^*(I) - (\kappa'_n)^*(|\Phi_{e^{nr}}\rangle\langle\Phi_{e^{nr}}|) \\ &= (\kappa'_n)^*(I - |\Phi_{e^{nr}}\rangle\langle\Phi_{e^{nr}}|) \ge 0, \end{aligned}$$

where $(\kappa'_n)^*$ is the dual map of κ'_n (see ④ of Theorem 5.1). Moreover,

$$\begin{aligned} &(\kappa'_n)^*(|\Phi_{e^{nr}}\rangle\langle\Phi_{e^{nr}}|) \ge 0, \\ &\operatorname{Tr}\sigma^{\otimes n}(\kappa'_n)^*(|\Phi_{e^{nr}}\rangle\langle\Phi_{e^{nr}}|) = \langle\Phi_{e^{nr}}|\kappa'_n(\sigma^{\otimes n})|\Phi_{e^{nr}}\rangle \le e^{-nr}. \end{aligned}$$

Since the matrix $(\kappa'_n)^*(|\Phi_{e^{nr}}\rangle\langle\Phi_{e^{nr}}|)$ satisfies the condition for the test $0 \le (\kappa'_n)^*(|\Phi_{e^{nr}}\rangle\langle\Phi_{e^{nr}}|) \le I$, the inequality (3.38) in Sect. 3.7 yields

$$\langle\Phi_{e^{nr}}|\kappa'_n(\rho^{\otimes n})|\Phi_{e^{nr}}\rangle = \operatorname{Tr}\rho^{\otimes n}(\kappa'_n)^*(|\Phi_{e^{nr}}\rangle\langle\Phi_{e^{nr}}|) \le e^{n\frac{-\phi(s)-sr}{1-s}},$$

for $s \le 0$, where $\phi(s) \stackrel{\text{def}}{=} \log\operatorname{Tr}\rho^{1-s}\sigma^s$. Using arguments similar to those used for the proof of Lemma 3.5, we have $\langle\Phi_{e^{nr}}|\kappa'_n(\rho^{\otimes n})|\Phi_{e^{nr}}\rangle \to 0$. We thus obtain (8.74). Applying the same arguments to $\rho^{\otimes k}$, we have

$$E_{d,2}^{S,\dagger}(\rho) \le \frac{E_{r,S}(\rho^{\otimes k})}{k}.$$

Combining this relation with Lemma A.1 in Appendix A, we obtain (8.76). ∎

Conversely, the following lemma holds for a pure state.

Lemma 8.6 *Any pure state $|u\rangle\langle u|$ on the composite system $\mathcal{H}_A \otimes \mathcal{H}_B$ satisfies*

$$E_{d,1}^{\to}(|u\rangle\langle u|) \ge H(\rho_u). \tag{8.77}$$

Proof. When $R < H(\rho_u)$, according to Theorem 8.5, there exists an operation κ_n satisfying

$$L(\kappa_n) = \frac{1 - \mathrm{Tr}(\rho_u^{\otimes n} - e^{-nR}) \left\{\rho_u^{\otimes n} - e^{-nR} \geq 0\right\}}{e^{-nR}} \tag{8.78}$$

$$\varepsilon_1(\kappa_n, |u\rangle\langle u|^{\otimes n}) = \mathrm{Tr}(\rho_u^{\otimes n} - e^{-nR}) \left\{\rho_u^{\otimes n} - e^{-nR} \geq 0\right\}.$$

Define $\psi(s) \stackrel{\mathrm{def}}{=} \log \mathrm{Tr}\, \rho_u^{1-s}$. The failure probability $\varepsilon_1(\kappa_n, |u\rangle\langle u|^{\otimes n})$ can then be calculated as

$$\begin{aligned}\varepsilon_1(\kappa_n, |u\rangle\langle u|^{\otimes n}) &\leq \mathrm{Tr}\, \rho_u^{\otimes n} \left\{\rho_u^{\otimes n} - e^{-nR} \geq 0\right\} \leq \mathrm{Tr}\left(\rho_u^{\otimes n}\right)^{1+s} e^{snR} \\ &= e^{n(\psi(-s)+sR)},\end{aligned}$$

for $s \geq 0$. Since $R < H(\rho_u)$, we can show that $\varepsilon(\kappa_n) \to 0$ using arguments similar to those given in Sect. 2.1. Based on this relation, we can show that

$$1 - \mathrm{Tr}(\rho_u^{\otimes n} - e^{-nR}) \left\{\rho_u^{\otimes n} - e^{-nR} \geq 0\right\} \to 1,$$

which proves that $\lim \frac{\log L(\kappa_n)}{n} = R$ for $L(\kappa_n)$. Hence, we obtain (8.77). ■

Proofs of Theorem 8.6 and equation (8.72). Let $|u\rangle = \sum_i \sqrt{p_i} |u_i \otimes u_i'\rangle$ and $\sigma = \sum_i p_i |u_i \otimes u_i'\rangle\langle u_i \otimes u_i'|$. Since u_i and u_i' are orthogonal,

$$E_{r,S}(|u\rangle\langle u|) \leq D(|u\rangle\langle u| \| \sigma) = H(\rho_u).$$

Combining (8.75) and (8.77), we obtain

$$E_{r,S}(|u\rangle\langle u|) \leq H(\mathrm{Tr}_B |u\rangle\langle u|) \leq E_{d,1}^{\to}(|u\rangle\langle u|) \leq E_{d,2}^{S,\dagger}(|u\rangle\langle u|) \leq E_{r,S}(|u\rangle\langle u|).$$

This proves Theorem 8.6 and (8.72). ■

When we treat the optimum outcome case, the following value is important:

$$E_{d,L}^{C}(\rho) \stackrel{\mathrm{def}}{=} \max_{\boldsymbol{\kappa}=\{\kappa_\omega\}\in C} \max_{\omega} \frac{\langle \Phi_L | \kappa_\omega(\rho) | \Phi_L \rangle}{\mathrm{Tr}\, \kappa_\omega(\rho)}.$$

It can easily be checked that

$$E_{d,L}^{C}(\rho) = \max_{A,B} \frac{\langle \Phi_L | (A \otimes B)\rho(A \otimes B)^* | \Phi_L \rangle}{\mathrm{Tr}\, \rho(A^* A \otimes B^* B)} \tag{8.79}$$

for $C = \to \leftrightarrow, S$. This value is called conclusive teleportation fidelity and was introduced by Horodecki et al. [230]; it describes the relation between this value and the conclusive teleportation.

Exercises

8.32. Define the *entanglement of exact distillation*

$$E^C_{d,e}(\rho) \stackrel{\rm def}{=} \lim \frac{\tilde{E}^C_{d,e}(\rho^{\otimes n})}{n}, \quad \tilde{E}^C_{d,e}(\rho) \stackrel{\rm def}{=} \max_{\kappa \in C:} \{ \log L(\kappa) | \varepsilon_2(\kappa, \rho) = 0 \}, \quad (8.80)$$

and show [176, 181]

$$E^{\leftrightarrow}_{d,e}(|u\rangle\langle u|) = -\log \lambda^{\downarrow}_1.$$

(This bound can be attained by a tournamentlike method for $d = 2$, but such a method is known to be impossible for $d > 2$ [294, 295].)

8.33. Let $u = \sum_i \sqrt{\lambda_i} u^A_i \otimes u^B_i$ be a Schmidt decomposition, and define the POVM $\boldsymbol{M}^{X,n} = \{M^{X,n}_q\}_{q \in T_n}$ as $M^{X,n}_q \stackrel{\rm def}{=} \sum_{\boldsymbol{i} \in T^n_q} |u^X_{\boldsymbol{i}}\rangle\langle u^X_{\boldsymbol{i}}|$ for $X = A, B$. Show that the final state with the measurement data q is a maximally entangled state with the size $|T^n_q|$. Using this protocol, show $E^{\emptyset}_{d,1}(|u\rangle\langle u|) \geq H(\rho_u)$. This protocol is called a Procrustean method [38].

8.34. Define the generalized Bell states $u^{A,B}_{i,j} \stackrel{\rm def}{=} (I_A \otimes \mathbf{X}^i_B \mathbf{Z}^j_B) u^{A,B}_{0,0}$, and the generalized Bell diagonal states $\rho_p \stackrel{\rm def}{=} \sum_{i,j} p_{i,j} |u^{A,B}_{i,j}\rangle\langle u^{A,B}_{i,j}|$, where $u^{A,B}_{0,0} \stackrel{\rm def}{=} \sum_i u^A_i \otimes u^B_i$. Show that $E_{r,S}(\rho_p) \leq \log d - H(p)$.

8.35. Define the quantity

$$E^C_{d,i}(R|\rho) \stackrel{\rm def}{=} \sup_{\{\kappa_n\} \subset C} \left\{ \underline{\lim} \frac{-1}{n} \log \varepsilon_i(\kappa_n, \rho) \middle| \underline{\lim} \frac{-1}{n} \log L(\kappa_n) \geq R \right\},$$

and show [176, 181]

$$E^C_{d,1}(R||u\rangle\langle u|) = \max_{s \leq 0} -\psi(s) + sR \text{ for } C = \rightarrow, \leftarrow, \leftrightarrow .$$

8.36. Define the quantity

$$E^{C,*}_{d,i}(R|\rho) \stackrel{\rm def}{=} \inf_{\{\kappa_n\} \subset C} \left\{ \overline{\lim} \frac{-1}{n} \log(1 - \varepsilon_i(\kappa_n, \rho)) \middle| \underline{\lim} \frac{-1}{n} \log L(\kappa_n) \geq R \right\},$$

and show [176, 181]

$$E^{S,*}_{d,2}(R||u\rangle\langle u|) \geq \max_{0 \leq t \leq 1} \frac{-\psi(t) + (1-t)R}{t}.$$

8.37. Show the following equation:

$$E^S_{d,e}(|u\rangle\langle u|) = -\log \lambda^{\downarrow}_1. \quad (8.81)$$

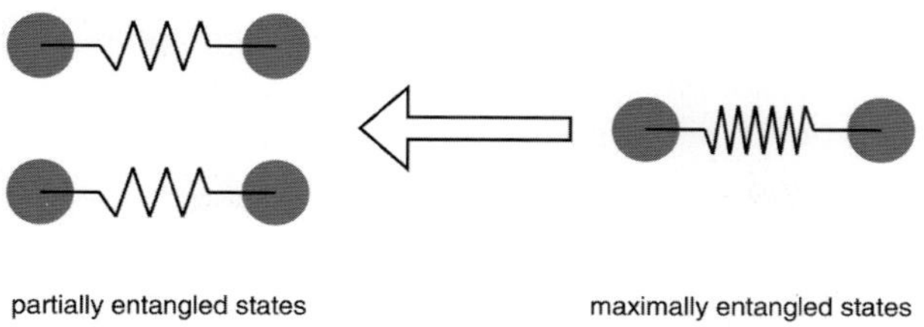

Fig. 8.3. Entanglement dilution

a Check the following relation:

$$\begin{aligned}
&\max_{\kappa\in S}\{d|\operatorname{Tr}\kappa(|u\rangle\langle u|)|\Phi_d\rangle\langle\Phi_d|=1\} \\
&=\max_{\kappa\in S}\{\min_{\sigma\in S}(\operatorname{Tr}|\Phi_d\rangle\langle\Phi_d|\kappa(\sigma))^{-1}|\operatorname{Tr}\kappa(|u\rangle\langle u|)|\Phi_d\rangle\langle\Phi_d|=1\} \\
&=\max_{\kappa\in S}\{\min_{\sigma\in S}(\operatorname{Tr}\kappa^*(|\Phi_d\rangle\langle\Phi_d|)\sigma)^{-1}|\operatorname{Tr}|u\rangle\langle u|\kappa^*(|\Phi_d\rangle\langle\Phi_d|)=1\} \\
&\le\max_{0\le T\le I}\{\min_{\sigma\in S}(\operatorname{Tr}T\sigma)^{-1}|\operatorname{Tr}|u\rangle\langle u|T=1\} \\
&=\min_{\sigma\in S}(\operatorname{Tr}|u\rangle\langle u|\sigma)^{-1}=(\lambda_1^{\downarrow})^{-1}.
\end{aligned} \tag{8.82}$$

b Prove equation (8.81).

8.6 Dilution of Maximally Entangled States

In the previous section, we considered the problem of producing a maximally entangled state from the tensor product of a particular entangled state. In this section, we examine the converse problem, i.e., to produce a tensor product state of a particular entangled state from a maximally entangled state in the composite system $\mathcal{H}_A \otimes \mathcal{H}_B$. In this book, we call this problem *entanglement dilution* even if the required state is mixed while historically it has been called this only for the pure-state case.

For an analysis of the mixed state ρ on the composite system $\mathcal{H}_A \otimes \mathcal{H}_B$, we define the *entanglement of formation* $E_f(\rho)$ for a state ρ in the composite system $\mathcal{H}_A \otimes \mathcal{H}_B$ based on the probabilistic decomposition $\{(p_i, \rho_i)\}$ of ρ [40]:

$$E_f(\rho) \stackrel{\text{def}}{=} \min_{\{(p_i,\rho_i)\}} \sum_i p_i H(\operatorname{Tr}_B \rho_i). \tag{8.83}$$

Since this minimum value is attained when all ρ_i are pure, this minimization can be replaced by the minimization for probabilistic decompositions by pure states. From the above definition, a state ρ_1 on $\mathcal{H}_{A_1} \otimes \mathcal{H}_{B_1}$ and a state ρ_2 on $\mathcal{H}_{A_2} \otimes \mathcal{H}_{B_2}$ satisfy

$$E_f(\rho_1) + E_f(\rho_2) \ge E_f(\rho_1 \otimes \rho_2). \tag{8.84}$$

Theorem 8.8 *Perform an operation corresponding to the S-TP-CP map* κ *with respect to a maximally entangled state* $|\Phi_L\rangle\langle\Phi_L|$ *of size L (the initial state). The fidelity between the final state and the target pure state* $|x\rangle\langle x|$ *on* $\mathcal{H}_A \otimes \mathcal{H}_B$ *then satisfies*

$$\max_{\kappa \in S} F(\kappa(|\Phi_L\rangle\langle\Phi_L|), |x\rangle\langle x|) \\ = \max_{\kappa \in \to} F(\kappa(|\Phi_L\rangle\langle\Phi_L|), |x\rangle\langle x|) = \sqrt{P(x, L)}, \tag{8.85}$$

where $P(u, L)$ *is defined using the Schmidt coefficients* $\sqrt{\lambda_i}$ *of* $|u\rangle$ *as follows:*

$$P(u, L) \stackrel{\text{def}}{=} \sum_{i=1}^{L} \lambda_i^{\downarrow}. \tag{8.86}$$

Note the similarity between $P(u, L)$ *and* $P(p, L)$ *given in Sect. 2.1.4. Furthermore, the fidelity between the final state and a general mixed state* ρ *on* $\mathcal{H}_A \otimes \mathcal{H}_B$ *satisfies*

$$\max_{\kappa \in S} F(\kappa(|\Phi_L\rangle\langle\Phi_L|), \rho) = \max_{\{(p_i, x_i)\}} \sqrt{\sum_i p_i P(x_i, L)}, \tag{8.87}$$

where $\{(p_i, x_i)\}$ *is the probabilistic decomposition of* ρ.

Using (2.39), we obtain

$$P^c(x, [e^R]) \leq e^{\frac{\log \operatorname{Tr} \rho_x^{1-s} - sR}{1-s}}, \tag{8.88}$$

where $0 \leq s \leq 1$ and $\rho_x \stackrel{\text{def}}{=} \operatorname{Tr}_B |x\rangle\langle x|$.

Proof. The proof only considers the case of a pure state $|x\rangle\langle x|$. Let $\{E_{A,i} \otimes E_{B,i}\}_i$ be the Choi–Kraus representation of the S-TP-CP map κ. Then

$$\kappa(|\Phi_L\rangle\langle\Phi_L|) = \sum_i (E_{A,i} \otimes E_{B,i}) |\Phi_L\rangle\langle\Phi_L| (E_{A,i} \otimes E_{B,i})^* .$$

Taking the partial trace inside the summation $\sum_i$ on the RHS, we have

$$\operatorname{Tr}_B (E_{A,i} \otimes E_{B,i}) |\Phi_L\rangle\langle\Phi_L| (E_{A,i} \otimes E_{B,i})^* \\ = (E_{A,i} X_{\Phi_L} E_{B,i}^T)(E_{A,i} X_{\Phi_L} E_{B,i}^T)^*$$

from (1.19). Its rank is less than L. Let y be a pure state on the composite system such that the rank of the reduced density of y is equal to L. Thus, by proving that

$$|\langle y|x\rangle| \leq \sqrt{P(x, L)}, \tag{8.89}$$

the proof can be completed. To this end, we define the pure state $|y_i\rangle\langle y_i|$ as

$$q_i|y_i\rangle\langle y_i| = (E_{A,i} \otimes E_{B,i})\,|\Phi_L\rangle\langle\Phi_L|\,(E_{A,i} \otimes E_{B,i})^*,$$

where q_i is a normalized constant. Then,

$$F^2(\kappa(|\Phi_L\rangle\langle\Phi_L|), |x\rangle\langle x|) = \sum_i q_i F^2(|y_i\rangle\langle y_i|, |x\rangle\langle x|) \le \sum_i q_i P(x, L).$$

We can use this relation to show that the fidelity does not exceed $\sqrt{P(x,L)}$. In a proof of (8.89), it is sufficient to show that $F(\rho_x, \sigma) \le \sqrt{P(x,L)}$ for $\rho_x \stackrel{\text{def}}{=} \operatorname{Tr}_B |x\rangle\langle x|$ and a density matrix σ of rank L. First, let $\sigma \stackrel{\text{def}}{=} \sum_{i=1}^L p_i|v_i\rangle\langle v_i|$ and let P be the projection to the range of σ. Since the rank of P is L, choosing an appropriate unitary matrix U, we obtain the following relation:

$$\begin{aligned}
\operatorname{Tr}|\sqrt{\rho_x}\sqrt{\sigma}| &= \operatorname{Tr}\sqrt{\rho_x}\sqrt{\sigma}U = \operatorname{Tr}\sqrt{\rho_x}\left(\sum_{i=1}^L \sqrt{p_i}|v_i\rangle\langle v_i|U\right) \\
&= \sum_{i=1}^L \sqrt{p_i}\langle v_i|U\sqrt{\rho_x}|v_i\rangle \le \sqrt{\sum_{i=1}^L p_i}\sqrt{\sum_{i=1}^L |\langle v_i|U\sqrt{\rho_x}|v_i\rangle|^2} &(8.90)\\
&= \sqrt{\sum_{i=1}^L \langle v_i|\sqrt{\rho_x}U^*|v_i\rangle\langle v_i|U\sqrt{\rho_x}|v_i\rangle} \le \sqrt{\sum_{i=1}^L \langle v_i|\sqrt{\rho_x}\sqrt{\rho_x}|v_i\rangle} &(8.91)\\
&= \sqrt{\operatorname{Tr} P\rho_x} \le \sqrt{P(x,L)}. &(8.92)
\end{aligned}$$

This evaluation can be checked as follows. Inequality (8.90) follows from the Schwarz inequality. Inequality (8.91) follows from $U^*|v_i\rangle\langle v_i|U \le I$. The final inequality (8.92) can be derived from the fact that P is a projection of rank L. Thus, we obtain

$$\max_{\kappa \in S} F(\kappa(|\Phi_L\rangle\langle\Phi_L|), |x\rangle\langle x|) \le \sqrt{P(x,L)}.$$

Conversely, we can verify the existence of an S-TP-CP map with a fidelity of $\sqrt{P(x,L)}$ by the following argument. There exists a pure state y satisfying the equality in (8.89); this can be confirmed by considering the conditions for equality in the above inequalities. Since the pure state $|y\rangle\langle y|$ satisfies $\operatorname{Tr}_B |\Phi_L\rangle\langle\Phi_L| \preceq \operatorname{Tr}_B |y\rangle\langle y|$, according to Theorem 8.4, there exists a one-way LOCC that produces the pure state $|y\rangle\langle y|$ from the maximally entangled state $|\Phi_L\rangle\langle\Phi_L|$, i.e., it attains the RHS of (8.85). This proves the existence of an S-TP-CP map with a fidelity of $\sqrt{P(x,L)}$. ∎

Next, let us consider how large a maximally entangled state is required for producing n tensor products of the entangled state ρ. In order to examine its asymptotic case, we focus on the S-TP-CP map κ_n to produce the state

$\rho^{\otimes n}$. The following *entanglement of cost* $E_c^C(\rho)$ expresses the asymptotic conversion rate

$$E_c^C(\rho) \stackrel{\text{def}}{=} \inf_{\{\kappa_n\}\subset C} \left\{ \overline{\lim} \frac{1}{n} \log L_n \middle| \lim F(\rho^{\otimes n}, \kappa_n(|\Phi_{L_n}\rangle\langle\Phi_{L_n}|)) = 1 \right\}, \quad (8.93)$$

which is the subject of the following theorem.

Theorem 8.9 *(Bennett et al. [38], Hayden et al. [196]) For any state ρ on the composite system $\mathcal{H}_A \otimes \mathcal{H}_B$,*

$$E_c^C(\rho) = \lim \frac{E_f(\rho^{\otimes n})}{n} = \inf_n \frac{E_f(\rho^{\otimes n})}{n}, \quad (8.94)$$

for $C = \rightarrow, \leftarrow, \leftrightarrow, S$.

For a proof of the mixed-state case, see Appendix B.5. The above theorem implies that the entanglement cost $E_c^C(\rho)$ has the same value for $C = \rightarrow$, $\leftarrow$, $\leftrightarrow$, S. Hence, we denote it by $E_c(\rho)$.

Proof of Pure-State Case. We prove Theorem 8.9 by analyzing the pure state $|x\rangle\langle x|$ in greater detail and by noting that $\psi(s) = \psi(s| \operatorname{Tr}_B |x\rangle\langle x|)$. For any $R > H(x)$, we can calculate how fast the quantity (1 − Optimal fidelity) approaches 0 according to

$$\lim \frac{-1}{n} \log\left(1 - \sqrt{P(p^n, e^{nR})}\right) = \lim \frac{-1}{n} \log(1 - P(p^n, e^{nR}))$$
$$= - \min_{0\le s\le 1} \frac{\psi(s) - sR}{1-s},$$

where we used (2.145) for $P(x^{\otimes n}, e^{nR})$ and $1-\sqrt{1-\epsilon} \cong \frac{1}{2}\epsilon$. If $R < H(x)$, the fidelity approaches zero for any LOCC (or separable) operation. The speed of this approach is

$$\lim \frac{-1}{n} \log \sqrt{P(p^n, e^{nR})} = -\frac{1}{2} \min_{s\le 0} \frac{\psi(s) - sR}{1-s},$$

where we used (2.147). From these inequalities, we have $E_c^C(\rho) = H(\rho_u)$ for $C = \rightarrow, \leftarrow, \leftrightarrow, S$ using the relationship between $H(p)$ and $\psi(s)$ given in Sect. 2.1.4. That is, we may also derive Theorem 8.9 in the pure-state case. ■

Since the additivity relation

$$\frac{E_f(\rho^{\otimes n})}{n} = E_f(\rho) \quad (8.95)$$

holds in the pure-state case, the entanglement of cost has a simple expression:

$$E_c^C(\rho) = E_f(\rho), \quad (8.96)$$

for $C = \to, \leftarrow, \leftrightarrow, S$. However, it is not known whether this formula holds for mixed states, except in a few cases, which will be treated later. Certainly, this problem is closely connected with other open problems, as will be discussed in Sect. 9.2.

Similar to (8.80), we define the *entanglement of exact cost*

$$E^C_{c,e}(\rho) \stackrel{\text{def}}{=} \lim \frac{\tilde{E}^C_{c,e}(\rho^{\otimes n})}{n}, \quad \tilde{E}^C_{c,e}(\rho) \stackrel{\text{def}}{=} \min_{\kappa \in C} \{\log L | \, F(\rho, \kappa(|\Phi_L\rangle\langle\Phi_L|)) = 1\} \tag{8.97}$$

and the logarithm of the *Schmidt rank* for a mixed state ρ:

$$E_{sr}(\rho) \stackrel{\text{def}}{=} \min_{\{(p_i,\rho_i)\}} \max_{i:p_i>0} \log \operatorname{rank} \rho_i. \tag{8.98}$$

From Theorem 8.4 we have $\tilde{E}^C_{c,e}(\rho) = E_{sr}(\rho)$. Hence,

$$E^C_{c,e}(\rho) = \lim \frac{E_{sr}(\rho^{\otimes n})}{n} \quad \text{for } C = \to, \leftarrow, \leftrightarrow, S. \tag{8.99}$$

Any pure state $|u\rangle\langle u|$ satisfies the additivity $E_{sr}(|u\rangle\langle u|^{\otimes n}) = nE_{sr}(|u\rangle\langle u|)$. However, the quantity $E_{sr}(\rho)$ with a mixed state ρ does not necessarily satisfy the additivity. Moreover, as is mentioned in Sect. 8.11, there exists an example ρ such that $E_{sr}(\rho) = E_{sr}(\rho^{\otimes 2})$ [385].

Exercises

8.38. Let $\mathcal{H}_A = \mathcal{H}_B = \mathbb{C}^3$. Let the support of a state ρ on $\mathcal{H}_A \otimes \mathcal{H}_B$ be $\{v \otimes u - u \otimes v | u, v \in \mathbb{C}^3\}$. Show that $E_f(\rho) = \log 2$ [403].

8.39. Show that E_f satisfies the *monotonicity* for a two-way LOCC κ.

$$E_f(\rho) \geq E_f(\kappa(\rho)). \tag{8.100}$$

8.40. Verify the strong converse theorem concerning Theorem 8.9, i.e., show that $E^{C,\dagger}_c(|u\rangle\langle u|) \geq E(\mathrm{Tr}_B\,|u\rangle\langle u|))$ for any pure state $|u\rangle\langle u|$ by defining $E^{C,\dagger}_c(\rho)$ in a similar way to Theorem 8.7.

8.7 Unified Approach to Distillation and Dilution

In this section, we derive the converse parts of distillation and dilution based on the following unified method. In this method, for a class of local operations C, we focus on the function $E^C(\rho)$ of a state $\rho \in \mathcal{S}(\mathcal{H}_A \otimes \mathcal{H}_B)$ that satisfies the following conditions.

E1 (*Normalization*) $E^C(\rho) = \log d$ when ρ is a maximally entangled state of size d.

E2C (*Monotonicity*) $E^C(\kappa(\rho)) \le E^C(\rho)$ holds for any local operation κ in class C.

E3 (*Continuity*) When any two states ρ_n and σ_n of system $\mathcal{H}_n$ satisfy $\|\rho_n - \sigma_n\|_1 \to 0$, the convergence $\frac{|E^C(\rho_n) - E^C(\sigma_n)|}{\log \dim \mathcal{H}_n} \to 0$ holds.

E4 (*Convergence*) The quantity $\frac{E^C(\rho^{\otimes n})}{n}$ converges as $n \to \infty$.

Based on the above conditions only, we can prove the following theorem.

Theorem 8.10 *(Donald et al. [103]) When the quantity $E^C(\rho)$ satisfies the above conditions,*

$$E^C_{d,2}(\rho) \le E^{C,\infty}(\rho) \left(\stackrel{\text{def}}{=} \lim \frac{E^C(\rho^{\otimes n})}{n} \right) \le E^C_c(\rho). \tag{8.101}$$

Proof. Let κ_n be a local operation κ_n in class C from $(\mathcal{H}_A)^{\otimes n} \otimes (\mathcal{H}_B)^{\otimes n}$ to $\mathbb{C}^{d_n} \otimes \mathbb{C}^{d_n}$ such that[3]

$$\| \, |\Phi_{d_n}\rangle\langle\Phi_{d_n}| - \kappa_n(\rho^{\otimes n})\|_1 \to 0, \tag{8.102}$$

where $\frac{\log d_n}{n} \to E^C_{d,2}(\rho)$. From Conditions **E1** and **E3** and (8.102), we have

$$\left| \frac{E^C(\kappa_n(\rho^{\otimes n}))}{n} - E^C_{d,2}(\rho) \right| \\ \le \frac{|E^C(\kappa_n(\rho^{\otimes n})) - E^C(|\Phi_{d_n}\rangle\langle\Phi_{d_n}|)|}{n} + \left| \frac{\log d_n}{n} - E^C_{d,2}(\rho) \right| \to 0.$$

Hence, Condition **E2**C guarantees that

$$\lim \frac{E^C(\rho^{\otimes n})}{n} \ge \lim \frac{E^C(\kappa_n(\rho^{\otimes n}))}{n} = E^C_{d,2}(\rho). \tag{8.103}$$

We obtain the first inequality.

Next, we choose a local operation κ_n in class C from $\mathbb{C}^{d_n} \otimes \mathbb{C}^{d_n}$ to $(\mathcal{H}_A)^{\otimes n} \otimes (\mathcal{H}_B)^{\otimes n}$ such that

$$\|\kappa_n(|\Phi_{d_n}\rangle\langle\Phi_{d_n}|) - \rho^{\otimes n}\|_1 \to 0,$$

where $\frac{\log d_n}{n} \to E^C_c(\rho)$. Similarly, we can show $\left|\frac{E^C(\kappa_n(|\Phi_{d_n}\rangle\langle\Phi_{d_n}|))}{n} - \frac{E^C(\rho^{\otimes n})}{n}\right| \to 0$. Since $\frac{E^C(\kappa_n(|\Phi_{d_n}\rangle\langle\Phi_{d_n}|))}{n} \le \frac{\log d_n}{n}$, we obtain

$$\lim \frac{E^C(\rho^{\otimes n})}{n} \le E^C_c(\rho).$$

∎

[3] If the operation κ_n in C has a larger output system than $\mathbb{C}^{d_n} \otimes \mathbb{C}^{d_n}$, there exists an operation κ'_n in C with the output system $\mathbb{C}^{d_n} \otimes \mathbb{C}^{d_n}$ such that $\| \, |\Phi_{d_n}\rangle\langle\Phi_{d_n}| - \kappa_n(\rho^{\otimes n})\|_1 \ge \| \, |\Phi_{d_n}\rangle\langle\Phi_{d_n}| - \kappa'_n(\rho^{\otimes n})\|_1$.

For example, the entanglement of formation $E_f(\rho)$ satisfies Conditions **E1**, **E2**$\leftrightarrow$(Exercise 8.39), **E3** (Exercise 8.42), and **E4** (Lemma A.1). The entanglement of relative entropy $E_{r,S}(\rho)$ also satisfies Conditions **E1**, **E2**s (Lemma 8.5), and **E4** (Lemma A.1). Further, Donald and Horodecki [102] showed Condition **E3** for entanglement of relative entropy $E_{r,S}(\rho)$. In addition, the *maximum of the negative conditional entropy*

$$E_m^C(\rho) \stackrel{\text{def}}{=} \max_{\kappa \in C} -H_{\kappa(\rho)}(A|B) \tag{8.104}$$

satisfies Conditions **E1**, **E2**C, **E3** (Exercise 8.44), and **E4** (LemmaA.1) for $C =\rightarrow, \leftrightarrow, S$. Thus,

$$E_{d,2}^C(\rho) \leq \lim \frac{E_m^C(\rho^{\otimes n})}{n} \leq E_c^C(\rho) \tag{8.105}$$

for $C =\rightarrow, \leftrightarrow, S$. Conversely, as will be proved in Sect. 9.6, the opposite inequality (Hashing inequality)

$$E_{d,2}^{\rightarrow}(\rho) \geq -H_\rho(A|B) \tag{8.106}$$

holds, i.e., there exists a sequence of one-way LOCC operations producing an approximate maximally entangled state of an approximate size of $e^{-nH_\rho(A|B)}$. Performing the local operation κ_n in class C, we can prepare the state $\kappa_n(\rho^{\otimes n})$. Applying this sequence of one-way LOCC operations to the state $\kappa_n(\rho^{\otimes n})$, we can show that $E_{d,2}^C(\rho) \geq E_{d,2}^C(\rho^{\otimes n}) \geq -H_{\kappa_n(\rho^{\otimes n})}(A|B)$, which implies $E_{d,2}^C(\rho) \geq \frac{E_m^C(\rho^{\otimes n})}{n}$. Thus, we obtain

$$E_{d,2}^C(\rho) = \lim \frac{E_m^C(\rho^{\otimes n})}{n}. \tag{8.107}$$

Therefore, since the relation $E_{r,S}(\rho) \leq E_f(\rho)$ holds $^{\text{Ex. 8.41}}$, we obtain

$$\begin{aligned} E_{d,2}^C(\rho) &= \lim \frac{E_m^C(\rho^{\otimes n})}{n} \leq E_{d,2}^{C,\dagger}(\rho) \leq \lim \frac{E_{r,S}(\rho^{\otimes n})}{n} \\ &\leq \lim \frac{E_f(\rho^{\otimes n})}{n} = E_c^C(\rho). \end{aligned}$$

We also have the following relations without the limiting forms:

$$-H_\rho(A|B) \leq E_m^C(\rho) \leq E_{d,2}^C(\rho) \leq E_{d,2}^{C,\dagger}(\rho) \leq E_{r,S}(\rho) \leq E_f(\rho). \tag{8.108}$$

The above quantities are the same in the pure-state case. However, the equalities do not necessarily hold in the mixed-state case.

Indeed, the expression of $E_m^C(\rho)$ can be slightly simplified as follows. Consider a TP-CP κ with the Choi–Kraus representation $\{F_i\}$. This operation is realized by the following process. First, we perform the measurement $\boldsymbol{M} = \{M_i\}_{i=1}^k$ and obtain the datum i with probability $p_i = \mathrm{P}_\rho^{\boldsymbol{M}}(i)$, where

$F_i = U_i\sqrt{M_i}$. Next, we perform the isometry matrix U_i depending on i. All data are sent to system B. Finally, we take the partial trace concerning the data space on $\mathcal{H}_B$. Hence, inequality (5.78) yields

$$\begin{aligned}-H_{\kappa(\rho)}(A|B) &\leq -H_{\bigoplus_{i=1}^k U_i\sqrt{M_i}\rho\sqrt{M_i}U_i^*}(A|B)\\ &= -\sum_i p_i H_{\frac{U_i\sqrt{M_i}\rho\sqrt{M_i}U_i^*}{p_i}}(A|B).\end{aligned}$$

Since operation κ is separable at least, the unitary U_i has the form $U_i^A \otimes U_i^B$. Hence,

$$H_{\frac{U_i\sqrt{M_i}\rho\sqrt{M_i}U_i^*}{p_i}}(A|B) = H_{\frac{\sqrt{M_i}\rho\sqrt{M_i}}{p_i}}(A|B).$$

Therefore,

$$E_m^C(\rho) = \max_{M\in C} -\sum_i p_i H_{\frac{\sqrt{M_i}\rho\sqrt{M_i}}{p_i}}(A|B) = \max_{M\in C} -H_{\hat{\kappa}_M(\rho)}(A|BE), \quad (8.109)$$

where $p_i \stackrel{\text{def}}{=} \mathrm{P}_\rho^{\boldsymbol{M}}(i)$, and $\mathcal{H}_E$ is the space spanned by $\{e_i^E\}$ because

$$\hat{\kappa}_{\boldsymbol{M}}(\rho) = \sum_i p_i \frac{\sqrt{M_i}\rho\sqrt{M_i}}{p_i} \otimes |e_i^E\rangle\langle e_i^E|. \quad (8.110)$$

As another measure of entanglement, Christandl and Winter [78] introduced *squashed entanglement*:

$$E_{sq}(\rho) \stackrel{\text{def}}{=} \inf\left\{ \frac{1}{2} I_{\rho_{A,B,E}}(A:B|E) \middle| \rho_{A,B,E} : \mathrm{Tr}_E\, \rho_{A,B,E} = \rho \right\}. \quad (8.111)$$

It satisfies Conditions **E1**, **E2**$\leftrightarrow$ (see [78]), **E3** (Exercise 8.43), and **E4** and the additivity (Exercise 8.46)

$$E_{sq}(\rho) + E_{sq}(\sigma) = E_{sq}(\rho \otimes \sigma). \quad (8.112)$$

Hence, we obtain

$$E_{d,2}^{\leftrightarrow}(\rho) = \lim \frac{E_m^{\leftrightarrow}(\rho^{\otimes n})}{n} \leq E_{sq}(\rho) \leq \lim \frac{E_f(\rho^{\otimes n})}{n} = E_c^C(\rho).$$

Next, a state on the composite system $\mathcal{H}_A \otimes \mathcal{H}_B$ is assumed to have the form $\rho_\alpha \stackrel{\text{def}}{=} \sum_{i,j} \alpha_{i,j} |u_i^A \otimes u_i^B\rangle\langle u_j^A \otimes u_j^B|$, where $(\alpha_{i,j})$ is a matrix and $\{u_i^A\}(\{u_i^B\})$ is an orthonormal basis of $\mathcal{H}_A(\mathcal{H}_B)$. In this case, such a state is called a *maximally correlated state*. A state ρ is maximally correlated if and only if there exist CONSs $\{u_i^A\}$ and $\{u_i^B\}$ of $\mathcal{H}_A$ and $\mathcal{H}_B$ such that the data of the measurement $\{|u_i^A\rangle\langle u_i^A|\}$ coincide with those of $\{|u_i^B\rangle\langle u_i^B|\}$. The separable state

$$\sigma_\alpha \stackrel{\text{def}}{=} \sum_i \alpha_{i,i} |u_i^A \otimes u_i^B\rangle\langle u_i^A \otimes u_i^B| \tag{8.113}$$

satisfies

$$E_{r,S}(\rho_\alpha) \leq D(\rho_\alpha \| \sigma_\alpha) = H(\sigma_\alpha) - H(\rho_\alpha) = -H_{\rho_\alpha}(A|B).$$

Hence, we obtain

$$-H_{\rho_\alpha}(A|B) = E_m^C(\rho_\alpha) = E_{d,2}^C(\rho_\alpha) = E_{d,2}^{C,\dagger}(\rho_\alpha) = E_{r,S}(\rho_\alpha), \tag{8.114}$$

for $C = \rightarrow, \leftarrow, \leftrightarrow, S$. Regarding the entanglement formation, the equation

$$E_f(\rho_\alpha) + E_f(\sigma) = E_f(\rho_\alpha \otimes \sigma) \tag{8.115}$$

holds for any maximally correlated state ρ_α on $\mathcal{H}_{A,1} \otimes \mathcal{H}_{B,1}$ and any state σ on $\mathcal{H}_{A,2} \otimes \mathcal{H}_{B,2}$. Hence, any maximally correlated state ρ_α satisfies $E_f(\rho_\alpha) = E_c(\rho_\alpha)$. A state ρ is maximally correlated if and only if it has a probabilistic decomposition of pure states $(p_i, |x_i\rangle)$ such that all $|x_i\rangle$ have the common Schmidt bases on $\mathcal{H}_A$ and $\mathcal{H}_B$. Its necessary and sufficient condition was obtained by Hiroshima and Hayashi [209]. For example, any mixture of two maximally entangled states is maximally correlated. We also have another characterization of maximally correlated states.

Lemma 8.7 *Let $|x\rangle$ be a pure state on the composite system $\mathcal{H}_A \otimes \mathcal{H}_B \otimes \mathcal{H}_R$. Then, the following conditions are equivalent.*

① $\rho^{AB} \stackrel{\text{def}}{=} \mathrm{Tr}_R |x\rangle\langle x|$ *is maximally correlated.*
② $\rho^{BR} \stackrel{\text{def}}{=} \mathrm{Tr}_A |x\rangle\langle x|$ *has the following form*

$$\rho^{BR} = \sum_i p_i |u_i^B \otimes x_i^R\rangle\langle u_i^B \otimes x_i^R|, \tag{8.116}$$

where $\{u_i^B\}$ is a CONS of $\mathcal{H}_B$, but $\{x_i^R\}$ is not necessarily a CONS of $\mathcal{H}_R$.

Proof of (8.115): Let x_1 and x_2 be the purifications of ρ_α and σ with the reference systems $\mathcal{H}_{R,1}$ and $\mathcal{H}_{R,2}$, respectively. Then, any probabilistic decomposition of $\rho_\alpha \otimes \sigma$ is given by a POVM $\boldsymbol{M} = \{M_i\}$ on $\mathcal{H}_{R,1} \otimes \mathcal{H}_{R,2}$ (Lemma 8.3). From the definition of maximally correlated states, we have the expression of the conditional state on system $\mathcal{H}_B$ as follows:

$$\begin{aligned}
&\mathrm{Tr}_{R,A}(\sqrt{M_i} \otimes I_{A,B})|x_1 \otimes x_2\rangle\langle x_1 \otimes x_2|(\sqrt{M_i} \otimes I_{A,B}) \\
&= \mathrm{Tr}_R \sum_i I_{R,B} \otimes |u_i^A\rangle\langle u_i^A|(\sqrt{M_i} \otimes I_{A,B})|x_1 \otimes x_2\rangle\langle x_1 \otimes x_2|(\sqrt{M_i} \otimes I_{A,B}) \\
&= p_i \sum_j Q_j^i |u_j^B\rangle\langle u_j^B| \otimes \sigma_j^i,
\end{aligned} \tag{8.117}$$

where $p_i = \langle x_1 \otimes x_2 | M_i \otimes I_{A,B} | x_1 \otimes x_2 \rangle$, and σ_j^i is a density on $\mathcal{H}_{B,2}$. From the strong concavity of von Neumann entropy (5.77),

$$H(\sum_j Q_j^i |u_j^B\rangle\langle u_j^B| \otimes \sigma_j^i) \geq H(\sum_j Q_j^i |u_j^B\rangle\langle u_j^B|) + \sum_j Q_j^i H(\sigma_j^i). \qquad (8.118)$$

Hence, the probabilistic decomposition by the POVM $\boldsymbol{M}$ yields the following average entropy on $\mathcal{H}_{B,1} \otimes \mathcal{H}_{B,2}$:

$$\begin{aligned} &\sum_i p_i H(\sum_j Q_j^i |u_j^B\rangle\langle u_j^B| \otimes \sigma_j^i) \\ &\geq \sum_i p_i H(\sum_j Q_j^i |u_j^B\rangle\langle u_j^B|) + \sum_{i,j} p_i Q_j^i H(\sigma_j^i). \end{aligned}$$

Next, we define the POVMs $\boldsymbol{M}^1 = \{M_i^1\}$ on $\mathcal{H}_{R,1}$ and $\boldsymbol{M}^2 = \{M_{i,j}^2\}$ on $\mathcal{H}_{R,2}$ as

$$\begin{aligned} M_i^1 &\stackrel{\text{def}}{=} \mathrm{Tr}_{\mathcal{H}_{R,2}} \sqrt{M_i}(I_{\mathcal{H}_{R,1}} \otimes \sigma)\sqrt{M_i}, \\ M_{i,j}^2 &\stackrel{\text{def}}{=} \mathrm{Tr}_{\mathcal{H}_{R,1}} \sqrt{M_i}(|u_j^B\rangle\langle u_j^B| \otimes I_{\mathcal{H}_{R,2}})\sqrt{M_i}. \end{aligned}$$

Then, the probabilistic decompositions by the POVMs $\boldsymbol{M}^1$ and $\boldsymbol{M}^2$ give the average entropies $\sum_i p_i H(\sum_j Q_j^i |u_j^B\rangle\langle u_j^B|)$ and $\sum_{i,j} p_i Q_j^i H(\sigma_j^i)$, respectively. ∎

Proof of Lemma 8.7. Assume Condition ①. Perform the POVM $\{|u_i^A\rangle\langle u_i^A|\}$. The final state on $\mathcal{H}_B \otimes \mathcal{H}_R$ is a pure state $|u_i^B \otimes x_i^R\rangle\langle u_i^B \otimes x_i^R|$. In this case, $\{u_i^B\}$ is a CONS of $\mathcal{H}_B$. Since any measurement on $\mathcal{H}_A$ gives a probabilistic decomposition on $\mathcal{H}_B \otimes \mathcal{H}_R$ (Lemma 8.3), we have (8.116).

Next, assume Condition ②. There exists a CONS $\{u_i^A\}$ of $\mathcal{H}_A$ such that

$$|x\rangle = \sum_i \sqrt{p_i} u_i^A \otimes u_i^B \otimes x_i^R. \qquad (8.119)$$

Thus, when we perform the measurements $\{|u_i^A\rangle\langle u_i^A|\}$ and $\{|u_i^B\rangle\langle u_i^B|\}$, we obtain the same data. That is, $\rho^{A,B}$ is maximally correlated. ∎

Evidently, equation (8.117) is essential to the proof of (8.115) [403]. When the vectors $u_1^A, \ldots, u_{d_A}^A$ in equation (8.113) are orthogonal, the state ρ_α satisfies equation (8.117). As is shown in Sect. 9.2, such a state is essentially related to entanglement-breaking channels.

Further, we focus on the following condition for a state ρ on the system $\mathcal{H}_A \otimes \mathcal{H}_B$. Let $|x\rangle$ be a purification of ρ with the reference system $\mathcal{H}_R$. There exists a TP-CP map from system $\mathcal{H}_B$ to system $\mathcal{H}_R$ such that the equation

$$\begin{aligned} &\kappa(\mathrm{Tr}_{A,R}(\sqrt{M_i} \otimes I_{B,R})|x\rangle\langle x|(\sqrt{M_i} \otimes I_{B,R})) \\ &= \mathrm{Tr}_{A,B}(\sqrt{M_i} \otimes I_{B,R})|x\rangle\langle x|(\sqrt{M_i} \otimes I_{B,R}) \end{aligned} \qquad (8.120)$$

holds for any POVM $\boldsymbol{M} = \{M_i\}$ on $\mathcal{H}_A$. In particular, any maximally correlated state satisfies this condition Ex. 8.45. If condition (8.120) holds, the quantity $E_m^{\rightarrow}(\rho)$ has the form

$$E_m^{\rightarrow}(\rho) = -H_\rho(A|B). \tag{8.121}$$

When the state ρ satisfies condition (8.120), the tensor product state $\rho^{\otimes n}$ also satisfies this condition. Hence, we obtain

$$-H_\rho(A|B) = \lim \frac{E_m^{\rightarrow}(\rho^{\otimes n})}{n} = E_{d,2}^{\rightarrow}(\rho). \tag{8.122}$$

Thus, condition (8.120) is satisfied in the following case. The system $\mathcal{H}_B$ can be decomposed as a composite system $\mathcal{H}_{B,1} \otimes \mathcal{H}_{B,2}$ such that the system $\mathcal{H}_{B,1}$ is unitary equivalent to $\mathcal{H}_R$ by a unitary U. Moreover, the state $\operatorname{Tr}_{\mathcal{H}_A,\mathcal{H}_{B,2}} |x\rangle\langle x|$ commutes the projection to the symmetric subspace of $\mathcal{H}_R \otimes \mathcal{H}_{B,1}$, which is spanned by the set $\{U(x) \otimes y + U(y) \otimes x | x, y \in \mathcal{H}_{B,1}\}$. In this case, any state ρ_s on the symmetric subspace and any state ρ_a on the antisymmetric subspace satisfy $U \operatorname{Tr}_R \rho_i U^* = \operatorname{Tr}_{B,2} \rho_i$ for $i = s, a$. Note that the antisymmetric subspace is spanned by the set $\{U(x) \otimes y - U(y) \otimes x | x, y \in \mathcal{H}_{B,1}\}$. Hence, the map κ satisfying (8.120) is given as the partial trace concerning $\mathcal{H}_{B,2}$.

Proof of (8.122): For any one-way ($\mathcal{H}_A \to \mathcal{H}_B$) LOCC operation κ', the local operation on $\mathcal{H}_A$ can be described by the Choi–Kraus representation $\{F_i\}$, and the operation on $\mathcal{H}_B$ can be described by a set of TP-CP maps $\{\kappa_i\}$ on $\mathcal{H}_B$. In this case, the measured datum i is obtained with the probability $p_i = \operatorname{Tr} F_i(\operatorname{Tr}_B \rho) F_i^*$, and the final state with the measured datum i is the state $\rho_i = \frac{1}{p_i} \operatorname{Tr}_A F_i \rho F_i^*$. The entropy of ρ_i is equal to $H\left(\operatorname{Tr}_{A,B}(\sqrt{M_i} \otimes I_{B,R})|x\rangle\langle x|(\sqrt{M_i} \otimes I_{B,R})\right)$, where $M_i \stackrel{\text{def}}{=} F_i^* F_i$. The monotonicity of transmission information (Exercise 5.26) and the relation (8.120) imply

$$\begin{aligned}
&- H_\rho(A|B) \\
&\quad = H\Big(\sum_i p_i \operatorname{Tr}_{A,R}(\sqrt{M_i} \otimes I_{B,R})|x\rangle\langle x|(\sqrt{M_i} \otimes I_{B,R}\Big) \\
&- H\Big(\sum_i p_i \operatorname{Tr}_{A,B}(\sqrt{M_i} \otimes I_{B,R})|x\rangle\langle x|(\sqrt{M_i} \otimes I_{B,R}\Big) \\
&\quad \geq \sum_i p_i H(\operatorname{Tr}_{A,R}(\sqrt{M_i} \otimes I_{B,R})|x\rangle\langle x|(\sqrt{M_i} \otimes I_{B,R}) \\
&- \sum_i p_i H(\operatorname{Tr}_{A,B}(\sqrt{M_i} \otimes I_{B,R})|x\rangle\langle x|(\sqrt{M_i} \otimes I_{B,R}) \\
&\quad = -\sum_i p_i H_{\rho_i}(A|B).
\end{aligned}$$

Further, from inequality (5.78)

$$-H_{\kappa'(\rho)}(A|B) = -\sum_i p_i H_{(\iota_A\otimes\kappa_i)(\rho_i)}(A|B) \le -\sum_i p_i H_{\rho_i}(A|B).$$

Hence, we obtain $-H_{\kappa'(\rho)}(A|B) \le -H_\rho(A|B)$. ∎

Exercises

8.41. Show that $E_{r,S}(\rho) \le E_f(\rho)$ using the joint convexity of the quantum relative entropy.

8.42. Show that the entanglement of formation $E_f(\rho)$ satisfies Condition **E3** (continuity) (Nielsen [314]) following the steps below.

a Let states ρ_n and σ_n satisfy $\|\rho_n - \sigma_n\|_1 \to 0$. Show that there exists a probabilistic decomposition $\{p_x^i, \rho_x^i\}$ of ρ_n such that $\frac{1}{n}|\sum_x p_x^i H(\mathrm{Tr}_B\, \rho_x^i) - E_f(\sigma_n)| \to 0$. Here, choose their purifications x_n and y_n based on Lemma 8.2.

b Prove Condition **E3** (continuity).

8.43. Show that the squashed entanglement $E_{sq}(\rho)$ satisfies Condition **E3** following the steps below.

a Let states ρ_n and σ_n satisfy $\|\rho_n - \sigma_n\|_1 \to 0$. Show that there exists an extension ρ_n^{ABE} of ρ_n such that $\frac{1}{n}|\frac{1}{2}I_{\rho_n^{ABE}}(A:B|E) - E_{sq}(\sigma_n)| \to 0$ using (5.73).

b Show Condition **E3** (continuity).

8.44. Show that $E_m^C(\rho)$ satisfies Condition **E3** (continuity) for $C = \to, \leftrightarrow, S$ using (8.109), (5.71), and the monotonicity of trace norm.

8.45. Let $\rho_\alpha = \sum_j p_j |x_j\rangle\langle x_j|$ be the spectral decomposition of the maximally correlated state ρ_α, and $\sum_j \sqrt{p_j}|x_j \otimes x_j'\rangle$ be a purification of ρ_α. Show that the map $\rho \mapsto \kappa(\rho) \stackrel{\mathrm{def}}{=} \sum_{j,i} p_j |x_j'\rangle\langle x_j'| |\langle x_j|u_i \otimes u_i'\rangle|^2 \frac{\langle u_i'|\rho|u_i'\rangle}{\alpha_{i,i}}$ satisfies condition (8.120).

8.46. Show the additivity of squashed entanglement (8.112) using chain rule (5.76) for quantum conditional mutual information.

8.47. Let $|x\rangle\langle x|$ be a purification of ρ with the reference system $\mathcal{H}_R$. Assume that the state $\mathrm{Tr}_B\, |x\rangle\langle x|$ is separable between $\mathcal{H}_A$ and $\mathcal{H}_R$. Prove the equation

$$E_f(\rho) + E_f(\sigma) = E_f(\rho \otimes \sigma) \tag{8.123}$$

under the following assumption using a similar discussion to the proof of (8.115) [403].

8.48. Show that the inequality $E_{d,2}^C(\rho) \le \lim \frac{E^C(\rho^{\otimes n})}{n}$ holds even though we replace Condition **E3** in Theorem 8.10 by the following condition:

E3′ (Weak lower continuity) When the sequence of states ρ_n on $\mathbb{C}^{d_n} \otimes \mathbb{C}^{d_n}$ satisfies $\|\rho_n - |\Phi_{d_n}\rangle\langle\Phi_{d_n}|\|_1 \to 0$, then $\lim \frac{E^C(\rho^{\otimes n})}{n} \ge \lim \frac{\log d_n}{n}$.

8.8 Dilution with Zero-Rate Communication

In this section, we treat entanglement dilution with small communication costs. For this analysis, we focus on the *entanglement of cost with zero-rate communication*:

$$E_c^{--\to}(\rho) \stackrel{\rm def}{=} \inf_{\{\kappa_n\}\subset C} \left\{ \overline{\lim} \frac{\log L_n}{n} \,\middle|\, \begin{array}{l} \lim F(\rho^{\otimes n}, \kappa_n(|\Phi_{L_n}\rangle\langle\Phi_{L_n}|)) = 1 \\ \frac{\log \mathrm{CC}(\kappa_n)}{n} \to 0 \end{array} \right\}, \tag{8.124}$$

where $\mathrm{CC}(\kappa)$ is the size of classical communication. This value is calculated in the pure-state case as follows.

Lemma 8.8 *(Lo and Popescu [275])*

$$E_c^{--\to}(|u\rangle\langle u|) = H(\mathrm{Tr}_B |u\rangle\langle u|)$$

Proof. To prove this, we first assume that there exist sets Ω_n and Ω'_n and a distribution p'_n on Ω'_n for a given distribution p on Ω and $\epsilon > 0$ such that

$$d_1((p^n)^\downarrow, (p_{\mathrm{mix},\Omega_n} \times p'_n)^\downarrow) \to 0, \ \underline{\lim} \frac{\log |\Omega'_n|}{n} < \epsilon, \ \lim \frac{\log |\Omega_n|}{n} \le H(p). \tag{8.125}$$

Indeed, the pure state with the Schmidt coefficients $\sqrt{(p_{\mathrm{mix},\Omega_n} \times p'_n)_i^\downarrow}$ can be realized from the maximally entangled state with the size $|\Omega_n| \times |\Omega'_n|$ by classical communication with a size of at most $|\Omega'_n|$. Therefore, if the state $|u\rangle\langle u|$ has the Schmidt coefficients $\sqrt{p_i}$, its n-fold tensor product $|u\rangle\langle u|^{\otimes n}$ can be asymptotically realized from the maximally entangled state with asymptotically zero-rate classical communication.

It is sufficient to prove (8.125) by replacing a distribution p'_n on Ω'_n by a positive measure p'_n on Ω'_n. Letting $l_n \le l'_n$ be integers, we construct a measure $\tilde{p}_n$ on Ω^n as follows. For a type $q \in T_n$ satisfying $l_n \le |T_q^n| \le l'_n$, we choose a subset $T_q^{n\prime} \subset T_q^n$ such that $|T_q^n \setminus T_q^{n\prime}| < l_n$. We define a measure $\tilde{p}_n \stackrel{\rm def}{=} p^n 1_{\Omega'_n}$, where $\Omega'_n \stackrel{\rm def}{=} \cup_{q\in T_n: l_n \le |T_q^n| \le l'_n} T_q^{n\prime}$. Then,

$$\begin{aligned} d(\tilde{p}_n, p^n) \le & \sum_{q\in T_n: l_n \le |T_q^n| \le l'_n} l_n e^{n \sum_\omega q_\omega \log p_\omega} \\ & + \sum_{q\in T_n: |T_q^n| < l_n} p^n(T_q^n) + \sum_{q\in T_n: |T_q^n| > l'_n} p^n(T_q^n). \end{aligned} \tag{8.126}$$

In this case, the measure $\tilde{p}_n$ has the form $p_{\mathrm{mix},\Omega_n} \times p'_n$ with $|\Omega_n| = l_n$ and $|\Omega'_n| = \frac{l'_n}{l_n}|T^n|$. When we choose $l_n = e^{n(H(p)-\epsilon)}$ and $l'_n = e^{n(H(p)+\epsilon)}$, $\lim \frac{\log |\Omega'_n|}{n} = 2\epsilon$ and $\lim \frac{\log |\Omega_n|}{n} = H(p) - \epsilon$. From the discussion in Sect. 2.5.1, the right-hand side (RHS) of (8.126) goes to 0. ∎

Next, we focus on the mixed state $\mathrm{Tr}_{A_2,B_2}|u\rangle\langle u|$. Using this theorem, we can check that

$$E_c^{-\to}(\mathrm{Tr}_{A_2,B_2}|u\rangle\langle u|) \le H(\mathrm{Tr}_B|u\rangle\langle u|).$$

Hence, defining the *entanglement of purification* for a state ρ on $\mathcal{H}_{A_1}\otimes\mathcal{H}_{B_1}$:

$$E_p(\rho) \stackrel{\mathrm{def}}{=} \min_{u:\mathrm{Tr}_{A_2,B_2}|u\rangle\langle u|=\rho} H(\mathrm{Tr}_B|u\rangle\langle u|), \tag{8.127}$$

we obtain

$$E_c^{-\to}(\rho) \le \lim \frac{E_p(\rho^{\otimes n})}{n}.$$

Conversely, the opposite inequality follows from generalizing Theorem 8.10. Hence, we have the following theorem.

Theorem 8.11 *(Terhal et al. [387])*

$$E_c^{-\to}(\rho) = \lim \frac{E_p(\rho^{\otimes n})}{n} \tag{8.128}$$

To generalize Theorem 8.10, we prepare the following condition.

E2′ (*Weak monotonicity*) Let κ be an operation containing quantum communication with size d. Then,

$$E(\kappa(\rho)) \le E(\rho) + \log d.$$

Lemma 8.9 *When the quantity $E(\rho)$ satisfies Conditions* **E1**, **E2′**, **E3**, *and* **E4**,

$$E^\infty(\rho) \left(\stackrel{\mathrm{def}}{=} \lim \frac{E(\rho^{\otimes n})}{n} \right) \le E_c^{-\to}(\rho). \tag{8.129}$$

This inequality holds even if we replace the one-way classical communication in the definition of $E_c^{-\to}(\rho)$ with quantum communication.

In fact, the entanglement of purification $E_p(\rho)$ satisfies Conditions **E1**, **E2∅** (Exercise 8.49), **E2′** (Exercise 8.49), **E3** (Exercise 8.50), and **E4**. Hence, equation (8.128) holds.

Further, using relation (8.128), we can slightly modify Lemma 8.9. For this purpose, we introduce the following condition.

E1′ (*Strong normalization*)

$$E(|u\rangle\langle u|) = H(\mathrm{Tr}_B|u\rangle\langle u|)$$

Lemma 8.10 *When the quantity $E(\rho)$ satisfies Conditions* **E1′** *and* **E2∅**,

$$E(\rho) \le E_p(\rho). \tag{8.130}$$

Hence,

$$E^{\infty}(\rho) \left(\stackrel{\text{def}}{=} \lim \frac{E(\rho^{\otimes n})}{n} \right) \le \lim \frac{E_p(\rho^{\otimes n})}{n} = E_c^{-\to}(\rho). \tag{8.131}$$

Proof of Lemma 8.9. We choose a local operation κ_n with one-way classical communication with a size l_n such that

$$\|\kappa_n(|\Phi_{d_n}\rangle\langle\Phi_{d_n}|) - \rho^{\otimes n}\|_1 \to 0, \qquad \frac{\log l_n}{n} \to 0,$$

where $\frac{\log d_n}{n} \to E_c^{-\to}(\rho)$. Condition **E3** guarantees that $|\frac{E_c^{-\to}(\kappa_n(|\Phi_{d_n}\rangle\langle\Phi_{d_n}|))}{n} - \frac{E_c^{-\to}(\rho^{\otimes n})}{n}| \to 0$. Combining Conditions **E2∅** and **E2′**, we have

$$E(\kappa_n(|\Phi_{d_n}\rangle\langle\Phi_{d_n}|)) \le \log d_n + \log l_n.$$

Therefore, we obtain (8.129). ∎

Proof of Lemma 8.10. Let $|u\rangle$ be a purification of ρ attaining the minimum of $H(\operatorname{Tr}_B |u\rangle\langle u|)$. Hence, from Conditions **E1′** and **E2∅**,

$$E(\rho) \le E(|u\rangle\langle u|) = H(\operatorname{Tr}_B |u\rangle\langle u|) = E_p(\rho).$$

∎

The entanglement of purification $E_p(\rho)$ is expressed as follows.

Lemma 8.11 *(Terhal et al. [387])*

$$\begin{aligned} E_p(\rho) & \left(= \min_u \{ H(\operatorname{Tr}_B |u\rangle\langle u|) | \operatorname{Tr}_{A_2,B_2} |u\rangle\langle u| = \rho \} \right) \\ & \le \min\{H(\rho^{A_1}), H(\rho^{B_1})\}, \end{aligned} \tag{8.132}$$

where $\dim \mathcal{H}_{A_2} \le d_{A_1} d_{B_1}$, $\dim \mathcal{H}_{B_2} \le (d_{A_1} d_{B_1})^2$, *and* $\rho^{A_1} = \operatorname{Tr}_{B_1} \rho, \rho^{B_1} = \operatorname{Tr}_{A_1} \rho$.

Proof. Let $|u\rangle$ be a purification of ρ and $\mathcal{H}'_{B_2}$ be its reference space. Then, any purification is given by an isometry U from $\mathcal{H}'_{B_2} \otimes \mathcal{H}_{A_2}$ to $\mathcal{H}_{B_2} \otimes \mathcal{H}_{A_2}$ as

$$U \otimes I_{A_1,B_1}(|u\rangle\langle u| \otimes \rho_0)(U \otimes I_{A_1,B_1})^*,$$

where ρ_0 is a pure state on $\mathcal{H}_{A_2}$. Hence,

$$E_p(\rho) = \min_\kappa H(\kappa(\operatorname{Tr}_{A_1} |u\rangle\langle u|)),$$

where κ is a TP-CP map from $\mathcal{H}'_{B_2}$ to $\mathcal{H}_{B_2}$. Since the minimum value is attained with an extremal point, from Corollary 5.2 we can restrict $\mathcal{H}_{B_2}$ to a $(d_{A_1}d_{B_1})^2$-dimensional space. Further, we can restrict $\mathcal{H}_{A_2}$ to a $d_{A_1}d_{B_1}$-dimensional space.

In addition, substituting $\kappa = \iota$, we obtain $E_p(\rho) \le H(\rho^{A_1})$. Similarly, the inequality $E_p(\rho) \le H(\rho^{B_1})$ holds. ■

Indeed, the quantum mutual information $\frac{I_\rho(A:B)}{2}$ satisfies Conditions **E1′** and **E2′** (Exercise 5.43) and the additivity $\frac{I_{\rho_1}(A:B)}{2} + \frac{I_{\rho_2}(A:B)}{2} = \frac{I_{\rho_1 \otimes \rho_2}(A:B)}{2}$. Hence,

$$\frac{I_\rho(A:B)}{2} \le E_c^{\dashrightarrow}(\rho).$$

As another quantity satisfying these two conditions, we focus on

$$\begin{aligned} C_d^{A\to B}(\rho) &\stackrel{\text{def}}{=} \max_{\boldsymbol{M}} H(\rho^B) - \sum_i \mathrm{P}_{\rho^A}^{\boldsymbol{M}}(i) H(\rho_i^B) \\ &= \max_{\boldsymbol{M}} H(\rho^B) - H_{\mathrm{Tr}_A \hat{\kappa}_{\boldsymbol{M}} \otimes \iota_B(\rho)}(B|E), \end{aligned} \tag{8.133}$$

$$\rho_i^B \stackrel{\text{def}}{=} \frac{1}{\mathrm{P}_{\rho^A}^{\boldsymbol{M}}(i)} \mathrm{Tr}_A (M_i \otimes I_B)\rho,$$

where $\boldsymbol{M} = \{M_i\}$ is a POVM on the system $\mathcal{H}_A$ [205]. For the derivation of this equation, see (8.110). It is easily checked that it satisfies Conditions **E1′** (Exercise 8.52) and **E2∅** (Exercise 8.52). Thus,

$$\lim \frac{C_d^{A\to B}(\rho^{\otimes n})}{n} \le E_c^{\dashrightarrow}(\rho). \tag{8.134}$$

In fact, $C_d^{A\to B}(\rho)$ satisfies Condition **E3** (continuity) (Exercise 8.54).

For example, when state ρ is a maximally correlated state, $C_d^{A\to B}(\rho)$ is calculated as

$$C_d^{A\to B}(\rho) = H(\rho^B)$$

because there exists a POVM $\boldsymbol{M}$ such that $H(\rho_i^B) = 0$. Moreover,

$$C_d^{A\to B}(\rho^{\otimes n}) = nH(\rho^B).$$

Hence, from (8.132),

$$E_c^{\dashrightarrow}(\rho) = H(\rho^B) = C_d^{A\to B}(\rho).$$

Indeed, when $\mathcal{H}_A = \mathcal{H}_B$, we can define the *flip operator* F as $F(u \otimes v) \stackrel{\text{def}}{=} v \otimes u$. Operator F has the form $F = P_s - P_a$, where $P_s(P_a)$ is the projection to the *symmetric space* $\mathcal{H}_s$ (the *antisymmetric space* $\mathcal{H}_a$), which is spanned by $\{u \otimes v + v \otimes u\}$ ($\{u \otimes v - v \otimes u\}$). If the support $\mathrm{supp}(\rho)$ is contained by $\mathcal{H}_s$ or $\mathcal{H}_a$, the equation

$$E_p(\rho) = H(\rho^B) = H(\rho^A) \tag{8.135}$$

holds as follows [79]. Since $\rho^{\otimes n}$ also satisfies this condition, we have

$$E_c^{-\to}(\rho) = E_p(\rho) = H(\rho^B) = H(\rho^A). \tag{8.136}$$

Proof of (8.135). Let $|u\rangle$ be a purification of ρ with the reference systems A_2 and B_2. Then, $F|u\rangle\langle u|F^* = |u\rangle\langle u|$. $H_{|u\rangle\langle u|}(B_1B_2) = H_{F|u\rangle\langle u|F}(B_1B_2) = H_{|u\rangle\langle u|}(A_1B_2) = H_{|u\rangle\langle u|}(B_1A_2)$. Hence, from inequality (5.67) we have

$$\begin{aligned} &H_{|u\rangle\langle u|}(A_1A_2) + H_{|u\rangle\langle u|}(B_1B_2) = H_{|u\rangle\langle u|}(A_1A_2) + H_{|u\rangle\langle u|}(B_1A_2) \\ \geq &H_{|u\rangle\langle u|}(A_1) + H_{|u\rangle\langle u|}(B_1). \end{aligned}$$

Since $H_{|u\rangle\langle u|}(A_1A_2) = H_{|u\rangle\langle u|}(B_1B_2)$ and $H_{|u\rangle\langle u|}(A_1) = H_{|u\rangle\langle u|}(B_1)$, we obtain

$$H_{|u\rangle\langle u|}(A_1A_2) \geq H_{|u\rangle\langle u|}(A_1),$$

which implies (8.135). ∎

In what follows, we calculate $E_c^{-\to}(\rho)$ and $E_p(\rho)$ in another case by treating $C_d^{A\to B}(\rho)$. Using the monotonicity of quantum relative entropy for the measurement, we can show that

$$C_d^{A\to B}(\rho) \leq I_\rho(A:B). \tag{8.137}$$

In particular, any separable state ρ satisfies (Exercise 8.53)

$$I_\rho(A:B) \leq E_p(\rho). \tag{8.138}$$

Using this inequality, we can calculate $C_d^{A\to B}(\rho)$ in the following case. Consider the case where ρ has the following specific separable form:

$$\rho = \sum_i p_i |u_i^A\rangle\langle u_i^A| \otimes \rho_i^B, \tag{8.139}$$

where $\{u_i^A\}$ is a CONS on $\mathcal{H}_A$. In this case, the optimal POVM $\boldsymbol{M}$ on $\mathcal{H}_A$ is $|u_i^A\rangle\langle u_i^A|$ because $H(\rho^B) - \sum_i p_i H(\rho_i^B) = I_\rho(A:B)$. Thus,

$$C_d^{A\to B}(\rho) = H(\rho^B) - \sum_i p_i H(\rho_i^B). \tag{8.140}$$

In particular, when ρ_i^B is pure, we have $C_d^{A\to B}(\rho) = H(\rho^B)$. In this case, we also have $C_d^{A\to B}(\rho^{\otimes n}) = nH(\rho^B)$. Hence,

$$E_c^{-\to}(\rho) = C_d^{A\to B}(\rho) = H(\rho^B) \le H(\rho^A). \tag{8.141}$$

Since $E_c^{-\to}(\rho) \le E_p(\rho) \le H(\rho^B)$, we have

$$E_p(\rho) = H(\rho^B).$$

Further, an interesting characterization of $C_d^{A\to B}(\rho)$ holds when $|x\rangle$ is a purification of ρ with the reference system $\mathcal{H}_R$. Since any probabilistic decomposition of the state on $\mathcal{H}_B \otimes \mathcal{H}_R$ can be given by a POVM on $\mathcal{H}_A$ [Lemmas (8.3) and (8.29)], the relation

$$C_d^{A\to B}(\rho) = H(\rho^B) - E_f(\rho^{B,R}) \tag{8.142}$$

holds, where $\rho^{B,R} \stackrel{\text{def}}{=} \operatorname{Tr}_A |x\rangle\langle x|$ [259]. As is mentioned in Sect. 9.5, this quantity is different from the usual entanglement measure in representing not only the amount of entanglement but also the amount of classical correlation.

Finally, we consider the equality condition of (8.137).

Lemma 8.12 *The equality of (8.137) holds if and only if ρ has the specific separable form of (8.139).*

Proof. It is trivial that a state with the form of (8.139) satisfies the equality in (8.137). Hence, we will prove (8.139) from the equality in (8.137). From the equality in (8.137), there exists a POVM $\boldsymbol{M} = \{M_i\}$ such that

$$H(\rho^B) - \sum_i \mathrm{P}_{\rho^A}^{\boldsymbol{M}}(i) H(\rho_i^B) = I_\rho(A:B). \tag{8.143}$$

In what follows, we can assume $\rho^A > 0$ without loss of generality. Now, we focus on the entanglement-breaking channel $\tilde{\kappa}_{\boldsymbol{M}}$ from system $\mathcal{H}_A$ to system $\mathbb{C}^k$ for a POVM $\boldsymbol{M} = \{M_i\}_{i=1}^k$ on $\mathcal{H}_A$:

$$\tilde{\kappa}_{\boldsymbol{M}}(\rho) \stackrel{\text{def}}{=} \sum_i (\operatorname{Tr} \rho M_i)|u_i\rangle\langle u_i|,$$

where $\{u_i\}$ is a CONS of $\mathbb{C}^k$. Then, the left-hand side (LHS) of (8.143) is equal to $I_{(\tilde{\kappa}_{\boldsymbol{M}} \otimes \iota_B)(\rho)}(A:B)$, i.e.,

$$D((\tilde{\kappa}_{\boldsymbol{M}} \otimes \iota_B)(\rho) \| (\tilde{\kappa}_{\boldsymbol{M}} \otimes \iota_B)(\rho^A \otimes \rho^B)) = D(\rho \| \rho^A \otimes \rho^B),$$

where $\rho^A = \operatorname{Tr}_B \rho$, $\rho^B = \operatorname{Tr}_A \rho$. Applying Theorem 5.6, we have

$$\rho = \sum_i (\operatorname{Tr}_A \rho (M_i \otimes I_B)) \otimes \frac{\sqrt{M_i}\rho^A \sqrt{M_i}}{\operatorname{Tr} M_i \rho^A}. \tag{8.144}$$

Now, we define map W from a pure state on $\mathcal{H}_A$ to a state on $\mathcal{H}_B$ as follows.

$$W_x \stackrel{\text{def}}{=} \frac{1}{p_x} \operatorname{Tr}_A \rho(|x\rangle\langle x| \otimes I_B), \quad p_x \stackrel{\text{def}}{=} \langle x|\rho^A|x\rangle.$$

Let $W_{x'}$ be an extremal point of the convex hull of $\{W_x\}$. Let $\mathcal{H}_{A,x'}$ be the space spanned by the set $\{x|W_x = W_{x'}\}$. From (8.144),

$$W_{x'} = \sum_i (\operatorname{Tr}_A \rho(M_i \otimes I_B)) \frac{\langle x'|\sqrt{M_i}\rho^A\sqrt{M_i}|x'\rangle}{\operatorname{Tr} M_i\rho^A}.$$

From the above discussion, $\sqrt{M_i}|x\rangle$ belongs to $\mathcal{H}_{A,x'}$. Thus, ρ is commutative with $|x'\rangle\langle x'| \otimes I_B$. Hence, the support of $\rho - p_{x'}|x'\rangle\langle x'| \otimes W_{x'}$ is included by $\mathcal{H}_{A,x'}^{\perp} \otimes \mathcal{H}_B$, where $\mathcal{H}_{A,x'}^{\perp}$ is the orthogonal space of $|x'\rangle$. Repeating this discussion, we obtain (8.139). Note that we must redefine $\mathcal{H}_B$ such that $\rho^B > 0$ before repeating this. ∎

Exercises

8.49. Show that the entanglement of purification $E_p(\rho)$ satisfies Conditions **E2∅** and **E2′**.

8.50. Show that the entanglement of purification $E_p(\rho)$ satisfies Condition **E3** based on the discussion in the proof of Lemma 8.11.

8.51. Show that

$$C_d^{A\to B}(\rho) + C_d^{A\to B}(\sigma) = C_d^{A\to B}(\rho \otimes \sigma) \tag{8.145}$$

for a separable state ρ using Exercise 8.47.

8.52. Show that the quantity $C_d^{A\to B}(\rho)$ satisfies Conditions **E1′** and **E2∅**.

8.53. Prove inequality (8.138) following the steps below.

a Let $|X\rangle\langle X|$ be a pure entangled state on $\mathcal{H}_A \otimes \mathcal{H}_B$ and $\boldsymbol{M} = \{M_i\}$ be a rank-one PVM on $\mathcal{H}_A$. Show that $H(\operatorname{Tr}_A |u\rangle\langle u|) = I_\rho(A:B)$, where $\rho = \sum M_i \otimes X M_i^T X^*$.

b Show inequality (8.138) using Theorem 8.1.

8.54. Show that the quantity $C_d^{A\to B}(\rho)$ satisfies Condition **E3** using (8.133) and (5.71).

8.9 State Generation from Shared Randomness

In this section, we treat in an asymptotic formulation the state generation from minimum shared random numbers. If desired state ρ is nonseparable between $\mathcal{H}_A$ and $\mathcal{H}_B$, it is impossible to generate state ρ only from shared random numbers. Hence, we treat a separable state:

$$\rho = \sum_i p_i \rho_i^A \otimes \rho_i^B. \tag{8.146}$$

In particular, when the conditions

$$[\rho_i^A, \rho_j^A] = 0 \quad \forall i, j, \tag{8.147}$$

$$[\rho_i^B, \rho_j^B] = 0 \quad \forall i, j \tag{8.148}$$

hold, the problem is essentially classical. In this problem, our operation is described by the size of shared random numbers M and the local states σ_i^A and σ_i^B depending on the shared random number $i = 1, \ldots, M$, i.e., we focus on our operation $\Phi \stackrel{\text{def}}{=} \{\sigma_i^A \otimes \sigma_i^B\}_{i=1}^M$ for approximating state ρ. Its performance is characterized by the size $|\Phi| \stackrel{\text{def}}{=} M$ and the quality of the approximation $\left\| \frac{1}{|\Phi|} \sum_{i=1}^{|\Phi|} \sigma_i^A \otimes \sigma_i^B - \rho \right\|_1$. Hence, the minimum size of shared random numbers is asymptotically characterized by $C_c(\rho)$[4]

$$C_c(\rho) \stackrel{\text{def}}{=} \inf_{\{\Phi_n\}} \left\{ \overline{\lim} \frac{1}{n} \log |\Phi_n| \,\middle|\, \lim \left\| \frac{1}{|\Phi_n|} \sum_{i=1}^{|\Phi_n|} \rho_{n,i}^A \otimes \rho_{n,i}^B - \rho^{\otimes n} \right\|_1 = 0 \right\}, \tag{8.149}$$

where $\Phi_n = \{\sigma_{n,i}^A \otimes \sigma_{n,i}^B\}$. Since a shared random number with size M can be simulated by a maximally entangled state with size M, we have

$$C_c(\rho) \geq E_c^{-\to}(\rho). \tag{8.150}$$

For this analysis, we define the quantities

$$C(\rho, \delta) \stackrel{\text{def}}{=} \inf \left\{ I_{\rho_{ABE}}(AB : E) \,\middle|\, \begin{array}{l} \mathrm{Tr}_E \, \rho_{ABE} = \rho, I_{\rho_{ABE}}(A : B|E) \leq \delta \\ \rho_{ABE} = \sum_x p_x \rho_x^{A,B} \otimes |u_x^E\rangle\langle u_x^E| \end{array} \right\}, \tag{8.151}$$

$$\tilde{C}(\rho, \delta) \stackrel{\text{def}}{=} \inf \left\{ I_{\rho_{ABE}}(AB : E) \,|\mathrm{Tr}_E \, \rho_{ABE} = \rho, I_{\rho_{ABE}}(A : B|E) \leq \delta \right\}, \tag{8.152}$$

where $\{u_x^E\}$ is a CONS on $\mathcal{H}_E$. From the definitions, the inequality

$$C(\rho, \delta) \geq \tilde{C}(\rho, \delta) \tag{8.153}$$

holds. In particular, we can prove Ex. 8.57

$$C(\rho, 0) = \tilde{C}(\rho, 0) \tag{8.154}$$

and denote it by $C(\rho)$. Further, this quantity satisfies the following properties.

[4] The subscript c denotes "common randomness."

① (*Monotonicity*) The operations κ_A and κ_B on $\mathcal{H}_A$ and $\mathcal{H}_B$ satisfy the monotonicity Ex. 8.55

$$C(\rho,\delta) \geq C((\kappa_A \otimes \kappa_B)(\rho),\delta), \ \tilde{C}(\rho,\delta) \geq \tilde{C}((\kappa_A \otimes \kappa_B)(\rho),\delta). \quad (8.155)$$

② (*Additivity*) The quantity $C(\rho)$ satisfies the *additivity* Ex. 8.56:

$$C(\rho \otimes \sigma) = C(\rho) + C(\sigma) \quad (8.156)$$

③ (*Continuity*) The former quantity $C(\rho,\delta)$ satisfies two kinds of continuity, i.e., if ρ_n is separable and $\rho_n \to \rho$, then

$$\lim_{n\to\infty} C(\rho_n) = C(\rho), \quad (8.157)$$
$$\lim_{\delta\to 0} C(\rho,\delta) = C(\rho,0). \quad (8.158)$$

In particular, the convergence in (8.158) is locally uniform concerning ρ.

④ (*Asymptotic weak-lower-continuity*) When $\|\rho_n - \rho^{\otimes n}\|_1 \to 0$, the inequality

$$\underline{\lim} \frac{C(\rho_n)}{n} \geq C(\rho) \quad (8.159)$$

holds.

⑤ $C(\rho)$ satisfies

$$C(\rho) \geq I_\rho(A:B) \quad (8.160)$$

because

$$I_{\rho^{ABE}}(AB:E) \geq I_{\rho^{ABE}}(A:E) = I_{\rho^{ABE}}(A:E) + I_{\rho^{ABE}}(A:B|E)$$
$$= I_{\rho^{ABE}}(A:BE) \geq I_{\rho^{ABE}}(A:B)$$

for any extension ρ^{ABE} of ρ satisfying $I_{\rho^{ABE}}(A:B|E) = 0$.

⑥ When condition (8.147) holds, $C(\rho)$ is upper bounded as

$$C(\rho) \leq H_\rho(A). \quad (8.161)$$

This can be checked by substituting $\mathcal{H}_A$ into $\mathcal{H}_E$ in the definition of $\tilde{C}(\rho,0)$.

Using the quantity $C(\rho)$, we can characterize $C_c(\rho)$ as follows.

Theorem 8.12 *(Wyner [426]) When ρ is separable, then*

$$C_c(\rho) = C(\rho). \quad (8.162)$$

Hence, from (8.150),

$$E_c^{-\rightarrow}(\rho) \leq C(\rho). \tag{8.163}$$

Further, there exists an example of separable states ρ such that conditions (8.147) and (8.148) hold and $C_c(\rho) > E_c^{-\rightarrow}(\rho)$ [421].

Proof. Since the direct part follows from the discussion in Sect. 9.4, its proof will be given in Sect. 9.4. Hence, we only prove the converse part here. Now, we choose the state $\rho_n \stackrel{\text{def}}{=} \frac{1}{|\Phi_n|}\sum_{i=1}^{|\Phi_n|} \sigma_{n,i}^A \otimes \sigma_{n,i}^B \otimes |u_i^E\rangle\langle u_i^E|$ such that $\|\operatorname{Tr}_E \rho_n - \rho^{\otimes n}\|_1 \rightarrow 0$. Then, we have

$$\log|\Phi_n| \geq I_{\rho_n}(AB:E) \geq C(\operatorname{Tr}_E \rho_n). \tag{8.164}$$

Hence, combining (8.159), we obtain

$$\underline{\lim} \frac{1}{n} \log|\Phi_n| \geq C(\rho).$$

∎

Proof of (8.158). We first characterize the quantity $C(\rho)$ as follows. Since the state $\rho^{(AB)E}$ is restricted to a separable state between AB and E, the state $\rho^{(AB)E}$ is given by a probabilistic decomposition $(p_i\rho_i)$ of ρ. Now, recall that any probabilistic decomposition of ρ on $\mathcal{H}_A \otimes \mathcal{H}_B$ is given by POVM $\boldsymbol{M}$ on the reference system as (8.28) and (8.29). In order to satisfy the condition $I_{\rho^{ABE}}(A:B|E) = 0$, any component ρ_i has a tensor product form. Hence,

$$C(\rho) = \inf_{\boldsymbol{M}} \left\{ I_{\rho_{\boldsymbol{M}}^{ABE}}(AB:E) \middle| I_{\rho_{\boldsymbol{M}}^{ABE}}(A:B|E) = 0 \right\},$$

where

$$\rho_{\boldsymbol{M}}^{ABE} \stackrel{\text{def}}{=} \sum_i \operatorname{Tr}_R(\sqrt{M_i} \otimes I)|x\rangle\langle x|(\sqrt{M_i} \otimes I) \otimes |u_i^E\rangle\langle u_i^E|.$$

Therefore, from Lemma A.9 we can restrict the range of the above infimum to the POVM $\boldsymbol{M}$ with at most $2(\dim \mathcal{H}_{A,B})^2$ elememnts. Since the set of POVMs with at most $2(\dim \mathcal{H}_{A,B})^2$ elements is compact, the above infimum can be replaced by the maximum. Further, we define

$$C(\rho, \delta) = \inf_{\boldsymbol{M}} \left\{ I_{\rho_{\boldsymbol{M}}^{ABE}}(AB:E) \middle| I_{\rho_{\boldsymbol{M}}^{ABE}}(A:B|E) = \delta \right\}. \tag{8.165}$$

Since $I_{\rho_{\boldsymbol{M}}^{ABE}}(A:B|E)$ is written as

$$I_{\rho_{\boldsymbol{M}}^{ABE}}(A:B|E) = \sum_i p_i I_\rho(M_i), \quad p_i \stackrel{\text{def}}{=} \operatorname{Tr}(M_i \otimes I)|x\rangle\langle x|$$

$$I_\rho(M_i) \stackrel{\text{def}}{=} I_{\frac{\operatorname{Tr}_R(\sqrt{M_i}\otimes I)|x\rangle\langle x|(\sqrt{M_i}\otimes I)}{p_i}}(A:B),$$

from Lemma A.9 we can restrict the range of the infimum in (8.165) to the POVMs $\boldsymbol{M}$ such that $|\boldsymbol{M}| \leq 2(\dim \mathcal{H}_{A,B})^2$. Since the set of POVMs with at most $2(\dim \mathcal{H}_{A,B})^2$ elements is compact, from Lemma A.4 we have

$$\lim_{\delta \to 0} C(\rho, \delta) = C(\rho). \tag{8.166}$$

Indeed, the above convergence is locally uniform for ρ. From (5.73), the functions $I_{\rho_{\boldsymbol{M}}^{ABE}}(AB : E)$ and $I_{\rho_{\boldsymbol{M}}^{ABE}}(A : B|E)$ satisfy

$$|I_{\rho_{\boldsymbol{M}}^{ABE}}(AB : E) - I_{\sigma_{\boldsymbol{M}}^{ABE}}(AB : E)| \leq 5\epsilon \log \dim \mathcal{H}_{A,B} + \eta_0(\epsilon) + 2h(\epsilon)$$
$$|I_{\rho_{\boldsymbol{M}}^{ABE}}(A : B|E) - I_{\sigma_{\boldsymbol{M}}^{ABE}}(A : B|E)| \leq 8\epsilon \log \dim \mathcal{H}_{A,B} + 6h(\epsilon),$$

where $\epsilon = \|\sigma - \rho\|_1$. Hence, the local uniformality follows by checking the discussion in the proof of Lemma A.4. ■

Proof of (8.157). Now, we prove equation (8.157). Let $|x_n\rangle$ ($|x\rangle$) be a purification of ρ_n (ρ) such that $|\langle x|x_n\rangle| = F(\rho, \rho_n)$. From (5.42),

$$\| \, |x\rangle\langle x| - |x_n\rangle\langle x_n| \, \|_1 \to 0. \tag{8.167}$$

We choose a POVM $\boldsymbol{M}_n$ with at most $2(\dim \mathcal{H}_{A,B})^2$ elememnts such that

$$I_{\rho_{n,\boldsymbol{M}_n}^{ABE}}(AB : E) = C(\rho_n), \quad I_{\rho_{n,\boldsymbol{M}_n}^{ABE}}(A : B|E) = 0.$$

From (8.167), (5.72), and (5.73),

$$\delta_n \stackrel{\text{def}}{=} I_{\rho_{\boldsymbol{M}_n}^{ABE}}(A : B|E) \to 0,$$
$$\delta_n' \stackrel{\text{def}}{=} |I_{\rho_{n,\boldsymbol{M}_n}^{ABE}}(AB : E) - I_{\rho_{\boldsymbol{M}_n}^{ABE}}(AB : E)| \to 0.$$

Hence,

$$C(\rho_n) + \delta_n' \geq C(\rho, \delta_n).$$

From (8.166) we obtain the inequality $\lim C(\rho_n) \geq C(\rho)$.

Conversely, we choose a POVM $\boldsymbol{M}$ with at most $2(\dim \mathcal{H}_{A,B})^2$ elememnts such that

$$I_{\rho_{\boldsymbol{M}_n}^{ABE}}(AB : E) = C(\rho), \quad I_{\rho_{\boldsymbol{M}_n}^{ABE}}(A : B|E) = 0.$$

From (8.167), (5.72), and (5.73),

$$\epsilon_n \stackrel{\text{def}}{=} I_{\rho_{n,\boldsymbol{M}}^{ABE}}(A : B|E) \to 0,$$
$$\epsilon_n' \stackrel{\text{def}}{=} |I_{\rho_{n,\boldsymbol{M}}^{ABE}}(AB : E) - I_{\rho_{\boldsymbol{M}}^{ABE}}(AB : E)| \to 0.$$

Hence,

$$C(\rho) + \epsilon_n' \geq C(\rho_n, \epsilon_n).$$

Since the convergence of (8.166) is locally uniform, we obtain the opposite inequality $\lim C(\rho_n) \leq C(\rho)$. ■

Proof of (8.159). Let ρ_n^{ABE} be a state satisfying $\mathrm{Tr}_E\,\rho_n^{ABE} = \rho_n$, $I_{\rho_n^{ABE}}(A : B|E) = 0$, and $I_{\rho_n^{ABE}}(AB : E) = C(\rho_n)$. From (5.66), the state ρ_n^{ABE} satisfies $H_{\rho_n^{ABE}}(AB|E) \le \sum_i H_{\rho_n^{ABE}}(A_iB_i|E)$. Hence,

$$\begin{aligned} C(\rho_n) &= H_{\rho_n^{ABE}}(AB) - H_{\rho_n^{ABE}}(AB|E) \\ &\ge H_{\rho_n^{ABE}}(AB) - \sum_i H_{\rho_n^{ABE}}(A_iB_i|E) \\ &= H_{\rho_n}(AB) - \sum_i H_{\rho_n^{ABE}}(A_iB_i) + \sum_i (H_{\rho_n}(A_iB_i) - H_{\rho_n^{ABE}}(A_iB_i|E)) \\ &\ge H_{\rho_n}(AB) - \sum_i H_{\rho_n}(A_iB_i) + \sum_i C(\rho_{n,i}), \end{aligned}$$

where $\rho_{n,i}$ is the reduced density matrix on A_iB_i. The final inequality follows from the definition of $C(\rho_{n,i})$. Since $\rho_{n,i}$ approaches ρ, properties (8.157) and (5.64) yield

$$\underline{\lim}\, \frac{C(\rho_n)}{n} \ge C(\rho).$$

■

Exercises

8.55. Prove inequality (8.155).

8.56. Prove equation (8.156) following the steps below.

a Assume that an extension ρ^{ABE} of $\rho^{A_1B_1} \otimes \rho^{A_2B_2}$ satisfies $I_{\rho^{ABE}}(A_1A_2 : B_1B_2|E) = 0$. Show that $I_{\rho^{ABE}}(A_1 : B_1|E) = I_{\rho^{ABE}}(A_2 : B_2|A_1B_1E) = 0$ using (5.76).
b Prove equation (8.156) using **a**.

8.57. Prove equation (8.154) following the steps below.

a Assume that an extension ρ^{ABE} of ρ satisfies $I_{\rho^{ABE}}(A : B|E) = 0$. Show that $I_{(\kappa_M \otimes \iota_{AB})(\rho^{ABE})}(A : B|E) = 0$ for any PVM $\boldsymbol{M}$ on $\mathcal{H}_E$.
b Prove equation (8.154) using **a**.

8.10 Positive Partial Transpose (PPT) Operations

In this section, we treat the class of positive partial transpose (PPT) maps (operations) as a wider class of local operations than the class of S-TP-CP maps. Remember that τ^A is defined as a transpose concerning the system $\mathcal{H}_A$ with the basis $\{u_1, \ldots, u_d\}$, as defined in Example 5.7 in Sect. 5.2. As was mentioned in Sect. 5.2, any separable state ρ satisfies $\tau^A(\rho) = (\tau^A \otimes$

$\iota_B)(\rho) \geq 0$. These states are called *positive partial transpose (PPT) states.* Note that the PPT condition $\tau^A(\rho) \geq 0$ does not depend on the choice of the basis of $\mathcal{H}_A$ [Ex. 8.58]. A TP-CP map κ from a system $\mathcal{H}_A \otimes \mathcal{H}_B$ to another system $\mathcal{H}_{A'} \otimes \mathcal{H}_{B'}$ is called a *positive partial transpose (PPT) map (operation)* if the map $\tau^{A'} \circ \kappa \circ \tau^A$ is a TP-CP map. As is easily checked, any PPT map κ can transform a PPT state into another PPT state. This condition is equivalent to the condition that the matrix $K(\kappa)$ defined in (5.2) has a PPT state form similar to a state on the composite system $(\mathcal{H}_A \otimes \mathcal{H}_{A'}) \otimes (\mathcal{H}_B \otimes \mathcal{H}_{B'})$. Hence, any PPT state can be produced by a PPT operation without any entangled state. Note that S-TP-CP maps also have a similar characterization. Since any separable state on the composite system $(\mathcal{H}_A \otimes \mathcal{H}_{A'}) \otimes (\mathcal{H}_B \otimes \mathcal{H}_{B'})$ is a PPT state on the composite system $(\mathcal{H}_A \otimes \mathcal{H}_{A'}) \otimes (\mathcal{H}_B \otimes \mathcal{H}_{B'})$, all S-TP-CP maps are PPT maps [353]. Hence, the class of PPT maps $C = \mathrm{PPT}$ is the largest class of local operations among $C = \emptyset, \rightarrow, \leftarrow, \leftrightarrow, S, \mathrm{PPT}$. Further, the definition of PPT maps does not depend on the choice of the basis [Ex. 8.59]. In addition, Cirac et al. [80] showed that any PPT operation could be realized by a bound entangled state and an LOCC operation.

As an entanglement measure related to PPT maps, we often focus on the *log negativity* $\log \|\tau^A(\rho)\|_1$, which does not depend on the choice of the basis [Ex. 8.60]. For a pure state $u = \sum_i \sqrt{\lambda_i} u_i^A \otimes u_i^B$,

$$\tau^A(|u\rangle\langle u|) = \sum_{i,j} \sqrt{\lambda_i}\sqrt{\lambda_j} |u_j^A \otimes u_i^B\rangle\langle u_i^A \otimes u_j^B|.$$

Then,

$$\begin{aligned} |\tau^A(|u\rangle\langle u|)| &= \sqrt{\tau^A(|u\rangle\langle u|)^* \tau^A(|u\rangle\langle u|)} \\ &= \sum_{i,j} \sqrt{\lambda_i}\sqrt{\lambda_j} |u_i^A \otimes u_j^B\rangle\langle u_i^A \otimes u_j^B| = (\sum_i \sqrt{\lambda_i} |u_i\rangle\langle u_i|)^{\otimes 2}. \end{aligned} \tag{8.168}$$

Therefore, $\|\tau^A(|u\rangle\langle u|)\|_1 = (\sum_i \sqrt{\lambda_i})^2$. In particular,

$$\tau^A(|\Phi_L\rangle\langle\Phi_L|) = \frac{1}{L} F, \quad |\tau^A(|\Phi_L\rangle\langle\Phi_L|)| = \frac{1}{L}, \tag{8.169}$$

where F is the flip operator $P_s - P_a$. Moreover, the log negativity satisfies the additivity $\log \|\tau^A(\rho \otimes \sigma)\|_1 = \log \|\tau^A(\rho)\|_1 + \log \|\tau^A(\sigma)\|_1$ and the *monotonicity* regarding the PPT operations κ

$$\|\tau^A(\kappa(\rho))\|_1 \leq \|\tau^A(\rho)\|_1, \tag{8.170}$$

i.e.,

$$\log \|\tau^A(\kappa(\rho))\|_1 \leq \log \|\tau^A(\rho)\|_1. \tag{8.171}$$

Using (8.169), we can generalize relation (8.7) as

$$\begin{aligned}\langle \Phi_L|\rho|\Phi_L\rangle &= \operatorname{Tr}\rho|\Phi_L\rangle\langle\Phi_L| = \operatorname{Tr}\tau^A(\rho)\tau^A(|\Phi_L\rangle\langle\Phi_L|) \\ &\le \|\tau^A(\rho)\|_1\|\tau^A(|\Phi_L\rangle\langle\Phi_L|)\| = \frac{\|\tau^A(\rho)\|_1}{L}. \end{aligned} \tag{8.172}$$

This relation implies that

$$E_{d,2}^{\mathrm{PPT},\dagger}(\rho) \le D(\rho\|\sigma) + \log\|\tau^A(\sigma)\|_1. \tag{8.173}$$

The RHS is called an SDP (semidefinite programming) bound [353] and satisfies the *monotonicity*, i.e.,

$$D(\rho\|\sigma) + \log\|\tau^A(\sigma)\|_1 \ge D(\kappa(\rho)\|\kappa(\sigma)) + \log\|\tau^A(\kappa(\sigma))\|_1$$

for a PPT operation κ. It implies inequality (8.170). As a consequence, we have

$$E_{d,2}^{\mathrm{PPT},\dagger}(\rho) \le \log\|\tau^A(\rho)\|_1, \tag{8.174}$$

$$E_{d,2}^{\mathrm{PPT},\dagger}(\rho) \le D(\rho\|\sigma), \text{ for a PPT state } \sigma. \tag{8.175}$$

Hence, the information quantities $E_{\mathrm{SDP}}(\rho) \stackrel{\mathrm{def}}{=} \min_\sigma D(\rho\|\sigma) + \log\|\tau^A(\sigma)\|_1$ and $E_{r,\mathrm{PPT}}(\rho) \stackrel{\mathrm{def}}{=} \min_{\sigma:\mathrm{PPT}} D(\rho\|\sigma)$ do not increase for a PPT operation, i.e., satisfy the *monotonicity*. Further, from (8.173) we obtain

$$E_{d,2}^{\mathrm{PPT},\dagger}(\rho) \le \lim \frac{E_{\mathrm{SDP}}(\rho^{\otimes n})}{n} \le \lim \frac{E_{r,\mathrm{PPT}}(\rho^{\otimes n})}{n}.$$

Regarding the direct part, since the quantity $E_{d,2}^{\mathrm{PPT}}(\rho)$ satisfies Condition **E3'** (weak lower continuity) because of Fannes' inequality (5.64), from Exercise 8.48 and the Hashing inequality (8.106) we can show

$$E_{d,2}^{\mathrm{PPT}}(\rho) = \lim \frac{E_m^{\mathrm{PPT}}(\rho^{\otimes n})}{n}. \tag{8.176}$$

Proof of (8.173). Consider a PPT operation κ_n' on $\mathcal{H}_A \otimes \mathcal{H}_B$ and a real number $r > D(\rho\|\sigma) + \log\|\tau^A(\sigma)\|_1$. Inequalities (8.172) and (8.170) imply that $\langle \Phi_{e^{nr}}|\kappa_n'(\sigma^{\otimes n})|\Phi_{e^{nr}}\rangle \le e^{-nr}\|\tau^A(\kappa_n'(\sigma^{\otimes n}))\|_1 \le e^{-nr}\|\tau^A(\sigma)\|_1^n$. From $I - |\Phi_{e^{nr}}\rangle\langle\Phi_{e^{nr}}| \ge 0$ we have

$$\begin{aligned} I - (\kappa_n')^*(|\Phi_{e^{nr}}\rangle\langle\Phi_{e^{nr}}|) &= (\kappa_n')^*(I) - (\kappa_n')^*(|\Phi_{e^{nr}}\rangle\langle\Phi_{e^{nr}}|) \\ &= (\kappa_n')^*(I - |\Phi_{e^{nr}}\rangle\langle\Phi_{e^{nr}}|) \ge 0, \end{aligned}$$

where $(\kappa_n')^*$ is the dual map of κ_n' (see ④ of Theorem 5.1). Moreover,

$$\begin{aligned} &(\kappa_n')^*(|\Phi_{e^{nr}}\rangle\langle\Phi_{e^{nr}}|) \ge 0, \\ &\operatorname{Tr}\sigma^{\otimes n}(\kappa_n')^*(|\Phi_{e^{nr}}\rangle\langle\Phi_{e^{nr}}|) = \langle\Phi_{e^{nr}}|\kappa_n'(\sigma^{\otimes n})|\Phi_{e^{nr}}\rangle \le e^{-n(r-\log\|\tau^A(\sigma)\|_1)}. \end{aligned}$$

Since the matrix $(\kappa_n')^*(|\Phi_{e^{nr}}\rangle\langle\Phi_{e^{nr}}|)$ satisfies the condition of test $0 \le (\kappa_n')^*(|\Phi_{e^{nr}}\rangle\langle\Phi_{e^{nr}}|) \le I$, inequality (3.38) in Sect. 3.7 yields

$$\langle\Phi_{e^{nr}}|\kappa_n'(\rho^{\otimes n})|\Phi_{e^{nr}}\rangle = \operatorname{Tr}\rho^{\otimes n}(\kappa_n')^*(|\Phi_{e^{nr}}\rangle\langle\Phi_{e^{nr}}|) \le e^{n\frac{-\phi(s)-s(r-\log\|\tau^A(\sigma)\|_1)}{1-s}}$$

for $s \le 0$, where $\phi(s) \stackrel{\text{def}}{=} \log \operatorname{Tr}\rho^{1-s}\sigma^s$. Using arguments similar to those used for the proof of Lemma 3.5, the condition $r - \log\|\tau^A(\sigma)\|_1 > D(\rho\|\sigma)$ implies $\langle\Phi_{e^{nr}}|\kappa_n'(\rho^{\otimes n})|\Phi_{e^{nr}}\rangle \to 0$. We thus obtain (8.173). ∎

Further, using the log negativity, Rains [353] showed that

$$E_{d,2}^{\mathrm{PPT}}(\rho_1) + E_{d,2}^{\mathrm{PPT}}(\rho_2) \le E_{d,2}^{\mathrm{PPT}}(\rho_1\otimes\rho_2) \le E_{d,2}^{\mathrm{PPT}}(\rho_1) + \log\|\tau^A(\rho_2)\|_1. \tag{8.177}$$

Indeed, Donald and Horodecki [102] proved Condition **E3** for $E_{r,\mathrm{PPT}}(\rho)$. Therefore, since $E_{r,\mathrm{PPT}}(\rho)$ satisfies Conditions **E1**, **E2**PPT, and **E4** in a manner similar to $E_{r,S}(\rho)$, Theorem 8.10 guarantees the inequality

$$\lim \frac{E_{r,\mathrm{PPT}}(\rho^{\otimes n})}{n} \le E_c^{\mathrm{PPT}}(\rho). \tag{8.178}$$

In inequality (8.174), the log negativity gives the upper bound of the entanglement of distillation; however, it does not give the lower bound of the entanglement of cost because $\log\|\tau^A(|u\rangle\langle u|)\|_1 = 2\log(\sum_i\sqrt{\lambda_i}) > -\sum_i\lambda_i\log\lambda_i = E_c^S(|u\rangle\langle u|)$. Thus, it does not satisfy Condition **E3** (continuity) because Theorem 8.10 leads to a contradiction if it holds (Exercise 8.63). In this case, we can show the following lemma as its alternative.

Lemma 8.13 *When the quantity $\tilde{E}_C(\rho)$ satisfies Conditions* **E1** *and* **E2**C*, the entanglement of exact distillation and the entanglement of exact cost are evaluated as*

$$\tilde{E}_{d,e}^C(\rho) \le \tilde{E}^C(\rho) \le \tilde{E}_{c,e}^C(\rho).$$

Further, if it satisfies Condition **E4** *also, their limits are evaluated as*

$$E_{d,e}^C(\rho) \le \lim \frac{\tilde{E}^C(\rho^{\otimes n})}{n} \le E_{c,e}^C(\rho).$$

Hence, from (8.171) we have the following formula for the exact cost with PPT operations [19]:

$$\log\|\tau^A(\rho)\|_1 \le E_{c,e}^{\mathrm{PPT}}(\rho). \tag{8.179}$$

Further, Audenaert et al. [19] showed the opposite inequality

$$E_{c,e}^{\mathrm{PPT}}(\rho) \le \log(\|\tau^A(\rho)\|_1 + d_A d_B \max(0, -\lambda_{\min}(\tau^A(|\tau^A(\rho)|))),$$

where $\lambda_{\min}(X)$ denotes the minimum eigenvalue of X. Hence, when

$$\tau^A(|\tau^A(\rho)|) \geq 0, \tag{8.180}$$

we obtain

$$\log \|\tau^A(\rho)\|_1 = E_{c,e}^{\mathrm{PPT}}(\rho).$$

For example, from (8.168) any pure state satisfies condition (8.180). Further, Ishizaka [241] proved that all states on the system $\mathbb{C}^2 \otimes \mathbb{C}^2$ satisfy this condition. Therefore, the entanglement measures for a pure state $\rho = |u\rangle\langle u|$ are summarized as follows. Let λ be a probability distribution of the eigenvalues of the reduced density $\mathrm{Tr}_B\, \rho$. Then, each entanglement measure is described by the Rényi entropy $\psi(s|\lambda) = \log \sum_i \lambda^{1-s}$ as Ex. 8.32,8.62

$$\begin{array}{ccccccccc}
E_{d,e}^{C_1}(\rho) & \leq & E_{d,i}^{C_2}(\rho) & = E_c^{C_3}(\rho) \leq & E_{c,e}^{\mathrm{PPT}}(\rho) & \leq & E_{c,e}^{C_4}(\rho) \\
\| & & \| & & \| & & \| \\
\lim_{s\to-\infty} \frac{\psi(s|\lambda)}{s} & \leq & \lim_{s\to 0} \frac{\psi(s|\lambda)}{s} & \leq & \frac{\psi(1/2|\lambda)}{1/2} & \leq & \lim_{s\to 1-0} \frac{\psi(s|\lambda)}{s},
\end{array} \tag{8.181}$$

where $i = 1, 2$, $C_1 = \rightarrow, \leftrightarrow, S$, PPT, $C_2 = \emptyset, \rightarrow, \leftrightarrow, S$, PPT, $C_3 = \dashrightarrow, \rightarrow, \leftrightarrow$, S, PPT, and $C_4 = \rightarrow, \leftrightarrow, S$. Remember that the quantity $\frac{\psi(s|\lambda)}{s}$ is monotone increasing for s (Sect. 2.1.4).

To conclude this section, we briefly discuss the relationship between $E_{d,2}^C(\rho^{A,B})$ and Theorem 8.3 [233]. In Theorem 8.3, we derived $\rho^{A,B} \preceq \rho^A$ from the fact that $\rho^{A,B}$ is separable. In fact, there exist several other conditions regarding separability:

① $\rho^{A,B}$ is separable.
② $\tau^A \otimes \iota_B(\rho^{A,B}) \geq 0$. (PPT state)
③ $E_{d,2}^{\leftrightarrow}(\rho^{A,B}) = 0$. (nondistillable state)
④ $\rho^A \otimes I_B \geq \rho^{A,B}$ and $I_A \otimes \rho^B \geq \rho^{A,B}$.
⑤ $\rho^{A,B} \preceq \rho^A$ and $\rho^{A,B} \preceq \rho^B$.

The relations between these conditions can be summarized as follows:

① ⇒ ② $\overset{\text{Horodecki [227]}}{\Rightarrow}$ ③ $\overset{\text{Horodecki [229]}}{\Rightarrow}$ ④ $\overset{\text{Hiroshima [208]}}{\Rightarrow}$ ⑤

In particular, a nondistillable state is called a bound entangled state when it is not separable. The relation ②⇒① (Theorem 5.3) has been shown only for $\mathbb{C}^2 \otimes \mathbb{C}^2$ and $\mathbb{C}^2 \otimes \mathbb{C}^3$. Hence, there is no bound entangled state on the $\mathbb{C}^2 \otimes \mathbb{C}^2$ system. However, a counterexample, i.e., a bound entangled state, exists for ①⇐② on $\mathbb{C}^2 \otimes \mathbb{C}^4$ and $\mathbb{C}^3 \otimes \mathbb{C}^3$ [236]. Since any PPT state can be produced by a PPT operation without any entangled state, this counterexample provides an interesting insight. That is, there exists a separable state ρ' and PPT operation κ' such that $\kappa'(\rho')$ is not separable. Further, it is known that the relation ②⇐④ holds for $\mathbb{C}^2 \otimes \mathbb{C}^n$ Ex. 8.64 [69, 101, 229].

As is easily checked, Condition ① is equivalent to the conditions $\tilde{E}^s_{c,e}(\rho) = 0$ and $E_f(\rho) = 0$. Since $E_f(\kappa'(\rho'))$ is not 0, E_f is not monotone for PPT operations. Further, examining the quantity $C_d^{A\to B}(\rho)$, Yang et al. [430] showed that if the entanglement of cost $E_c(\rho)$ is zero, ρ is separable. That is, a nonseparable state has nonzero entanglement of cost. Hence, E_c is not monotone for PPT operations.

Further, for any nonseparable state σ, there exists a state ρ and an integer L such that [280]

$$E^C_{d,L}(\rho) < E^C_{d,L}(\rho \otimes \sigma),$$

which implies that $E_c(\sigma) > 0$.

In addition, a counterexample also exists for ④⇐⑤ when $\mathbb{C}^2 \otimes \mathbb{C}^2$ [240]. However, it is an open problem whether the opposite relation ②⇐③ holds. To discuss it in greater detail, we focus on the following relation:

$$\begin{aligned} E^{\leftrightarrow}_{d,2}(\rho) \le E^S_{d,2}(\rho) \le E^{\mathrm{PPT}}_{d,2}(\rho) \le \lim \frac{E_{\mathrm{SDP}}(\rho^{\otimes n})}{n} \le \lim \frac{E_{r,\mathrm{PPT}}(\rho^{\otimes n})}{n} \\ \le E^{\mathrm{PPT}}_c(\rho) \le E^{\mathrm{PPT}}_{c,e}(\rho) \le \tilde{E}^{\mathrm{PPT}}_{c,e}(\rho). \end{aligned}$$

Since any PPT state can be produced by a PPT operation without any entangled state, Condition ② is equivalent to the condition $\tilde{E}^{\mathrm{PPT}}_{c,e}(\rho) = 0$. From (8.179) it is also equivalent to the condition $E^{\mathrm{PPT}}_{c,e}(\rho) = 0$. Therefore, if Condition ② is equivalent to Condition ③, these conditions hold if and only if one of the above values is equal to zero.

Exercises

8.58. Let $\tilde{\tau}^A$ be a partial transpose concerning another (the second) basis, and U be a unitary matrix mapping every base of the first basis to every base of the second basis. Show that $\tilde{\tau}^A = U\tau^A U^*$.

8.59. Show that $\tau^{A'} \circ \kappa \circ \tau^A$ is TP-CP if and only if $\tilde{\tau}_{A'} \circ \kappa \circ \tilde{\tau}_A$ is TP-CP in the above assumption.

8.60. Show that $\|\tilde{\tau}^A(\rho)\|_1 = \|\tau^A(\rho)\|_1$ in the above assumption.

8.61. Show

$$E^{\mathrm{PPT},*}_{d,2}(R\||u\rangle\langle u|) \ge \max_{0\le t\le 1} \frac{-\psi(t) + (1-t)R}{t}.$$

8.62. Prove the following equation:

$$E^{\mathrm{PPT}}_{d,e}(|u\rangle\langle u|) = -\log \lambda_1^{\downarrow}. \tag{8.182}$$

a Prove the following inequality as a generalization of (8.172):

$$\operatorname{Tr}\sigma\rho \leq \|\tau^A(\sigma)\|_1 \|\tau^A(\rho)\|. \tag{8.183}$$

b Prove equation (8.182) by combining (8.183), (8.168), and an inequality similar to (8.82).

8.63. Check the following counterexample of the continuity of $2\log\|\tau^A(|u\rangle\langle u|)\|_1$ as follows [273].

a Show that $\|\rho^{\otimes n} - \rho_n\|_1 \to 0$, i.e., $F(\rho^{\otimes n}, \rho_n) \to 1$, where $\rho_n \stackrel{\text{def}}{=} \frac{\{e^{-n(H(\rho)+\epsilon)} \leq \rho^{\otimes n} \leq e^{-n(H(\rho)-\epsilon)}\}\rho^{\otimes n}}{\operatorname{Tr}\{e^{-n(H(\rho)+\epsilon)} \leq \rho^{\otimes n} \leq e^{-n(H(\rho)-\epsilon)}\}\rho^{\otimes n}}$.

b Show that $H(\rho) - \epsilon \leq \frac{1}{sn}\psi(s|\rho_n) \leq H(\rho) + \epsilon$.

b Check that the purifications x_n, y_n of $\rho^{\otimes n}, \rho_n$ give a counterexample of the continuity of $2\log\|\tau^A(|u\rangle\langle u|)\|_1$.

8.64. Let A, B, and C be $n \times n$ matrices. Show that $\begin{pmatrix} A & B^* \\ B & C \end{pmatrix} \geq 0$ when $\begin{pmatrix} A+C & 0 \\ 0 & A+C \end{pmatrix} \geq \begin{pmatrix} A & B \\ B^* & C \end{pmatrix} \geq 0$ [69].

8.11 Examples

In this section, we summarize the preceding calculation of entanglement measures using several examples in the mixed-state case.

8.11.1 2 × 2 System

In the case of $\mathbb{C}^2 \otimes \mathbb{C}^2$, Wootters [424] calculated the entanglement of formation as

$$E_f(\rho) = h\left(\frac{1+\sqrt{1-C_o(\rho)^2}}{2}\right), \quad C_o(\rho) \stackrel{\text{def}}{=} \max\{0, \lambda_1 - \lambda_2 - \lambda_3 - \lambda_4\}, \tag{8.184}$$

where λ_i is the square root of the eigenvalue of $\rho(S_2 \otimes S_2)\bar{\rho}(S_2 \otimes S_2)$ in decreasing order. The function $C_o(\rho)$ is called *concurrence*. When we perform an instrument $\{\kappa_\omega\}_\omega$ with the separable form $\kappa_\omega(\rho) = (A_\omega \otimes B_\omega)\rho(A_\omega \otimes B_\omega)^*$, the final state $\frac{(A_\omega \otimes B_\omega)\rho(A_\omega \otimes B_\omega)^*}{\operatorname{Tr}(A_\omega \otimes B_\omega)\rho(A_\omega \otimes B_\omega)^*}$ has the following concurrence [272, 401] Ex. 8.65:

$$C_o\left(\frac{(A_\omega \otimes B_\omega)\rho(A_\omega \otimes B_\omega)^*}{\operatorname{Tr}(A_\omega \otimes B_\omega)\rho(A_\omega \otimes B_\omega)^*}\right) = C_o(\rho)\frac{\det A_\omega \det B_\omega}{\operatorname{Tr}(A_\omega \otimes B_\omega)\rho(A_\omega \otimes B_\omega)^*}. \tag{8.185}$$

For example, the concurrence of the Bell diagonal state $\rho_{\mathrm{Bell},p} \stackrel{\mathrm{def}}{=} \sum_{i=0}^{3} p_i |e_i^{A,B}\rangle\langle e_i^{A,B}|$ is calculated as [Ex. 8.66]

$$C_o(\rho_{\mathrm{Bell},p}) = 2 \max_i p_i - 1, \tag{8.186}$$

and it does not increase by any stochastic operation [250][Ex. 8.67]:

$$C_o(\rho_{\mathrm{Bell},p}) \geq C_o \left(\frac{(A_\omega \otimes B_\omega)\rho_{\mathrm{Bell},p}(A_\omega \otimes B_\omega)^*}{\mathrm{Tr}(A_\omega \otimes B_\omega)\rho_{\mathrm{Bell},p}(A_\omega \otimes B_\omega)^*} \right). \tag{8.187}$$

This state satisfies

$$-H_{\rho_{\mathrm{Bell},p}}(A|B) = \log 2 - H(p), \quad I_{\rho_{\mathrm{Bell},p}}(A:B) = 2\log 2 - H(p).$$

Further, the maximally correlated state $\rho_{a,b} \stackrel{\mathrm{def}}{=} a|00\rangle\langle 00| + b|00\rangle\langle 11| + b|11\rangle\langle 00| + (1-a)|11\rangle\langle 11|$[Ex. 8.68] has the concurrence $2b$ [Ex. 8.69]. Hence,

$$E_c(\rho_{a,b}) = E_f(\rho_{a,b}) = h\left(\frac{1+\sqrt{1-4b^2}}{2}\right).$$

Regarding distillation, from (8.114) and (8.173) we have [Ex. 8.70]

$$\begin{aligned} &E_{d,2}^{C,\dagger}(\rho_{a,b}) = E_{d,2}^{C}(\rho_{a,b}) = E_{r,S}(\rho_{a,b}) = E_{r,\mathrm{PPT}}(\rho_{a,b}) = -H_{\rho_{a,b}}(A|B) \\ =& h(a) - h\left(\frac{1+\sqrt{(2a-1)^2+4b^2}}{2}\right), \end{aligned}$$

for $C = \rightarrow, \leftarrow, \leftrightarrow, S$, and PPT. Further,

$$\begin{aligned} I_{\rho_{a,b}}(A:B) =& 2h(a) - h\left(\frac{1+\sqrt{(2a-1)^2+4b^2}}{2}\right) \\ C_d^{A\rightarrow B}(\rho_{a,b}) =& E_c^{--\rightarrow}(\rho_{a,b}) = h(a). \end{aligned}$$

Since Ishizaka [241] proved $\tau^A(|\tau^A(\rho)|) \geq 0$ for the 2×2 case, the relation

$$E_{c,e}^{\mathrm{PPT}}(\rho) = \log \|\tau^A(\rho_{a,b})\|_1 = \log(1+2b)$$

holds. Hence, comparing these values, we obtain the inequality

$$\log(1+2b) \geq h\left(\frac{1+\sqrt{1-4b^2}}{2}\right) \geq h(a) - h\left(\frac{1+\sqrt{(2a-1)^2+4b^2}}{2}\right)$$

for $\sqrt{a(1-a)} \geq b$. In particular, the second equality holds only when $\sqrt{a(1-a)} = b$, i.e., the state $\rho_{a,b}$ is pure.

8.11.2 Werner State

Next, we consider the *Werner state*:

$$\rho_{W,p} \stackrel{\text{def}}{=} (1-p)\rho_{\text{mix}}^{s} + p\rho_{\text{mix}}^{a} = \frac{p}{d(d-1)}(I-F) + \frac{1-p}{d(d+1)}(I+F), \quad (8.188)$$

where ρ_{mix}^{s} (ρ_{mix}^{a}) is the completely mixed state on the symmetric space (antisymmetric space). We can easily check that

$$\begin{aligned} -H_{\rho_{W,p}}(A|B) &= \log d + p\log\frac{2p}{d(d-1)} + (1-p)\log\frac{2(1-p)}{d(d+1)} \\ I_{\rho_{W,p}}(A:B) &= 2\log d + p\log\frac{2p}{d(d-1)} + (1-p)\log\frac{2(1-p)}{d(d+1)}. \end{aligned} \quad (8.189)$$

Further, any pure state $|u\rangle\langle u|$ on $\mathcal{H}_A$ satisfies

$$\operatorname{Tr}_A(|u\rangle\langle u| \otimes I_B)\rho_{\text{mix}}^{a} = \frac{I_B - |u\rangle\langle u|}{d-1}, \quad \operatorname{Tr}_A(|u\rangle\langle u| \otimes I_B)\rho_{\text{mix}}^{s} = \frac{I_B + |u\rangle\langle u|}{d+1}.$$

Thus,

$$\operatorname{Tr}_A(|u\rangle\langle u| \otimes I_B)\rho_{W,p} = p\frac{I_B - |u\rangle\langle u|}{d-1} + (1-p)\frac{I_B + |u\rangle\langle u|}{d+1},$$

which has entropy $-\frac{(d+1)p+(d-1)(1-p)}{d+1}\log\frac{(d+1)p+(d-1)(1-p)}{d^2-1} - \frac{2(1-p)}{d+1}\log\frac{2(1-p)}{d+1}$. Since this entropy is independent of $|u\rangle$,

$$\begin{aligned} C_d^{A\to B}(\rho_{W,p}) = &\log d + \frac{2(1-p)}{d+1}\log\frac{2(1-p)}{d+1} \\ &+ \frac{(d+1)p+(d-1)(1-p)}{d+1}\log\frac{(d+1)p+(d-1)(1-p)}{d^2-1}. \end{aligned}$$

Using the symmetry of this state, Vollbrecht and Werner [407] showed that

$$E_f(\rho_{W,p}) = \begin{cases} h\left(\frac{1+2\sqrt{p(1-p)}}{2}\right) & \text{if } p \geq \frac{1}{2} \\ 0 & \text{if } p < \frac{1}{2}. \end{cases}$$

Rains [352] showed

$$E_{r,S}(\rho_{W,p}) = E_{r,\text{PPT}}(\rho_{W,p}) = \log 2 - h(p).$$

Rains [353] and Audenaert et al. [18] proved

$$\lim \frac{1}{n}E_{r,S}((\rho_{W,p})^{\otimes n}) = E_{\text{SDP}}(\rho_{W,p}) = \begin{cases} 0 & \text{if } p \leq \frac{1}{2} \\ \log 2 - h(p) & \text{if } \frac{1}{2} < p \leq \frac{1}{2} + \frac{1}{d} \\ \log\frac{d-2}{d} + p\log\frac{d+2}{d-2} & \text{if } \frac{1}{2} + \frac{1}{d} < p \leq 1, \end{cases}$$

where d is the dimension of the local system. Note that $\frac{1}{2} + \frac{1}{d} = 1$ when $d = 2$. Hence, $E_{r,S}(\rho)$ does not satisfy the additivity. This also implies $\lim \frac{1}{n} E_{r,\mathrm{PPT}}((\rho_{W,p})^{\otimes n}) = E_{\mathrm{SDP}}(\rho_{W,p})$.

Further, Rains [353] also showed that

$$E_{d,2}^{\mathrm{PPT}}(\rho_{W,1}) = \log \frac{d+2}{d}.$$

Since $\tau^A(|\tau^A(\rho_{W,p})|) \geq 0^{\mathrm{Ex.\ 8.72}}$ [19], we obtain

$$E_{c,e}^{\mathrm{PPT}}(\rho_{W,p}) = \log \|\tau^A(\rho_{W,p})\|_1 = \log \left(\frac{2(2p-1)}{d} + 1 \right).$$

In particular,

$$E_{d,2}^{\mathrm{PPT}}(\rho_{W,1}) = E_{d,2}^{\mathrm{PPT},\dagger}(\rho_{W,1}) = E_{\mathrm{SDP}}(\rho_{W,1}) = E_{c}^{\mathrm{PPT}}(\rho_{W,1})$$
$$= E_{c,e}^{\mathrm{PPT}}(\rho_{W,1}) = \log \frac{d+2}{d} \leq \log 2 = E_f(\rho_{W,1}).$$

The equality of the inequality $\log \frac{d+2}{d} \leq \log 2$ holds only when $d = 2$. From (8.177) the entanglement of distillation of the state $\rho_{W,1}$ satisfies the additivity

$$E_{d,2}^{\mathrm{PPT}}(\rho_{W,1}) + E_{d,2}^{\mathrm{PPT}}(\rho) = E_{d,2}^{\mathrm{PPT}}(\rho_{W,1} \otimes \rho)$$

for any state ρ.

On the other hand, Yura [433] calculated the entanglement of cost of any state ρ in the antisymmetric space of the system $\mathbb{C}^3 \otimes \mathbb{C}^3$ as

$$E_c(\rho) = \log 2,$$

which is equal to its entanglement of formation $E_f(\rho)^{\mathrm{Ex.\ 8.38}}$. Hence, in this case,

$$E_{c}^{\mathrm{PPT}}(\rho_{W,1}) = E_{c,e}^{\mathrm{PPT}}(\rho_{W,1}) = \log \frac{5}{3} < \log 2 = E_c(\rho_{W,1}).$$

Further, Matsumoto and Yura [291] focused on $\rho_{W,1}^{B,R} \stackrel{\mathrm{def}}{=} \mathrm{Tr}_A |x\rangle\langle x|$, where $|x\rangle$ is a purification of $\rho_{W,1}$ with the reference system $\mathcal{H}_R$, and showed that

$$E_c(\rho_{W,1}^{B,R}) = \frac{E_f((\rho_{W,1}^{B,R})^{\otimes n})}{n} = E_f(\rho_{W,1}^{B,R}) = \log(d-1).$$

Hence, using (8.142), we have

$$C_d^{A \to B}(\rho_{W,1}^{\otimes n}) = \log \frac{d}{d-1}.$$

Since $\rho_{W,1}$ and $\rho_{W,0}$ satisfy the condition for (8.135), the relation (8.136)holds, i.e.,

$$E_c^{-\to}(\rho_{W,1}) = E_c^{-\to}(\rho_{W,0}) = \log d. \tag{8.190}$$

The entanglement purification $E_p(\rho_{W,p})$ of the other cases has been numerically calculated by Terhal et al. [387].

8.11.3 Isotropic State

Next, we consider the *isotropic state*

$$\begin{aligned}\rho_{I,p} &\stackrel{\text{def}}{=} (1-p)\frac{I-|\Phi_d\rangle\langle\Phi_d|}{d^2-1} + p|\Phi_d\rangle\langle\Phi_d| \\ &= \frac{(1-p)d^2}{d^2-1}\rho_{\text{mix}} + \frac{d^2p-1}{d^2-1}|\Phi_d\rangle\langle\Phi_d|,\end{aligned} \tag{8.191}$$

where $|\Phi_d\rangle = \frac{1}{\sqrt{d}}\sum_i u_i \otimes u_i$. We can easily check that

$$\begin{aligned}-H_{\rho_{I,p}}(A|B) &= \log d + p\log p + (1-p)\log\frac{1-p}{d^2-1} \\ I_{\rho_{I,p}}(A:B) &= 2\log d + p\log p + (1-p)\log\frac{1-p}{d^2-1}.\end{aligned} \tag{8.192}$$

Further, any pure state $|u\rangle\langle u|$ on $\mathcal{H}_A$ satisfies

$$\begin{aligned}\mathrm{Tr}_A\, |u\rangle\langle u| \otimes I_B\rho_{I,p} &= \frac{(1-p)d^2}{d^2-1}\rho^B_{\text{mix}} + \frac{d^2p-1}{d^2-1}|u\rangle\langle u| \\ &= \frac{(1-p)d}{d^2-1}(I-|u\rangle\langle u|) + \frac{dp+1}{d+1}|u\rangle\langle u|,\end{aligned}$$

which has entropy $-\frac{(1-p)d}{d^2-1}\log\frac{(1-p)d}{d+1} - \frac{dp+1}{d+1}\log\frac{dp+1}{d+1}$. Since this entropy is independent of $|u\rangle$,

$$C_d^{A\to B}(\rho_{W,p}) = \log d + \frac{(1-p)d}{d^2-1}\log\frac{(1-p)d}{d+1} + \frac{dp+1}{d+1}\log\frac{dp+1}{d+1}.$$

Further, King [254] showed that

$$C_d^{A\to B}((\rho_{W,p})^{\otimes n}) = nC_d^{A\to B}(\rho_{W,p}).$$

Define $\rho_{I,p}^{B,R} \stackrel{\text{def}}{=} \mathrm{Tr}_A\, |x\rangle\langle x|$, where $|x\rangle$ is a purification of $\rho_{I,p}$ with the reference system $\mathcal{H}_R$. Then, using (8.142), we have

$$\begin{aligned}E_c(\rho_{I,p}^{B,R}) &= \frac{E_f((\rho_{I,p}^{B,R})^{\otimes n})}{n} = E_f(\rho_{I,p}^{B,R}) \\ &= -\frac{(1-p)d}{d^2-1}\log\frac{(1-p)d}{d+1} - \frac{dp+1}{d+1}\log\frac{dp+1}{d+1}.\end{aligned}$$

Using the symmetry of this state, Terhal and Vollbrecht [386] showed that

$$\begin{aligned}E_f(\rho_{I,p}) = \min_{x,y,p\ge 0} & p(h(\gamma(x)) + (1-\gamma(x))\log(d-1)) \\ & + (1-p)(h(\gamma(y)) + (1-\gamma(y))\log(d-1)),\end{aligned}$$

where we take the minimum with the condition $p = px + (1-p)y$, and

$$\gamma(p) = \frac{1}{d}(\sqrt{p} + \sqrt{(d-1)(1-p)})^2.$$

They also showed the following relation for the $d = 3$ case and conjectured it in the $d > 3$ case as follows:

$$E_f(\rho_{I,p}) = \begin{cases} 0 & \text{if } p \le \frac{1}{d} \\ h(\gamma(p)) + (1-\gamma(p))\log(d-1)) & \text{if } \frac{1}{d} < p \le \frac{4(d-1)}{d^2} \\ \frac{(p-1)d}{d-2}\log(d-1) + \log d & \text{if } \frac{4(d-1)}{d^2} < p \le 1. \end{cases}$$

Note that the isotropic state is locally unitary equivalent to the Werner state when $d = 2$.

Further, Rains [352] showed that

$$E_{r,S}(\rho_{I,p}) = E_{r,\mathrm{PPT}}(\rho_{I,p}) = \begin{cases} \log d - (1-p)\log(d-1) - h(p) & \text{if } p \ge \frac{1}{d} \\ 0 & \text{if } p < \frac{1}{d}. \end{cases}$$

Rains [353] also proved

$$E_{d,2}^{\mathrm{PPT}}(\rho_{I,p}) \ge \log d - (1-p)\log(d+1) - h(p).$$

Since $\tau^A(|\tau^A(\rho_{I,p})|) \ge 0^{\text{Ex. 8.73}}$, we obtain

$$E_{c,e}^{\mathrm{PPT}}(\rho_{I,p}) = \log \|\tau^A(\rho_{I,p})\|_1 = \begin{cases} \log dp & \text{if } p \ge \frac{1}{d} \\ 0 & \text{if } p < \frac{1}{d}. \end{cases} \tag{8.193}$$

In the system $\mathbb{C}^2 \otimes \mathbb{C}^2$, Terhal and Horodecki [385] proved

$$E_{sr}(\rho_{I,\frac{1}{\sqrt{2}}}) = \log 2, \quad E_{sr}(\rho_{I,\frac{1}{\sqrt{2}}}^{\otimes 2}) = \log 2.$$

Since $E_{c,e}^{\mathrm{PPT}}(\rho_{I,p}) \le E_{c,e}^{S}(\rho_{I,p})$, from (8.193) and (8.99) we obtain

$$E_{c,e}^{C}(\rho_{I,\frac{1}{\sqrt{2}}}) = \log\sqrt{2} \text{ for } C = \rightarrow, \leftarrow, \leftrightarrow, S, \mathrm{PPT}.$$

Exercises

8.65. Prove equation (8.185) following the steps below.

a Show that $A^T S_2 A = S_2 \det A$.

b Show that $(A \otimes B)\rho(A \otimes B)^*(S_2 \otimes S_2)(A \otimes B)\bar{\rho}(A \otimes B)^*(S_2 \otimes S_2) = |\det A|^2 |\det B|^2 \rho(S_2 \otimes S_2)\bar{\rho}(S_2 \otimes S_2)$.

c Show that $C_o(\frac{(A\otimes B)\rho(A\otimes B)^*}{\mathrm{Tr}(A\otimes B)\rho(A\otimes B)^*)}) = \frac{(\det A)(\det B)C_o(\rho)}{\mathrm{Tr}(A\otimes B)\rho(A\otimes B)^*)}$.

d Prove (8.185).

8.66. Prove (8.186) following the steps below.

a Show that $\overline{\rho_{\mathrm{Bell},p}} = \rho_{\mathrm{Bell},p}$.

b Show that $(S_2 \otimes S_2)\rho_{\mathrm{Bell},p}(S_2 \otimes S_2) = \rho_{\mathrm{Bell},p}$.
c Prove (8.186).

8.67. Prove equation (8.187) following the steps below.

a Show that $\operatorname{Tr}(A_\omega \otimes B_\omega)\rho_{\mathrm{Bell},p}(A_\omega \otimes B_\omega)^*) = \sum_{i=0}^{3} \frac{p_i}{2} \operatorname{Tr} A^* A S_i^T B^T \bar{B} S_i^T$.
b Show that $\operatorname{Tr} A^* A S_i^T B^T \bar{B} S_i^T \geq |\det A|^2 |\det B|^2$.
c Prove (8.187).

8.68. Show that any maximally entangled state can be written as $\rho_{a,b} \stackrel{\mathrm{def}}{=} a|00\rangle\langle 00| + b|00\rangle\langle 11| + b|11\rangle\langle 00| + (1-a)|11\rangle\langle 11|$ in a 2×2 system by choosing suitable bases.

8.69. Show that $C_o(\rho_{a,b}) = 2b$ following the steps below.

a Show that $(S_2 \otimes S_2)\overline{\rho_{a,b}}(S_2 \otimes S_2) = \rho_{1-a,b}$.
b Show that $\rho_{a,b}(S_2 \otimes S_2)\overline{\rho_{a,b}}(S_2 \otimes S_2) = (a(1-a)+b^2)|00\rangle\langle 00| + 2ab|00\rangle\langle 11| + 2(1-a)b|11\rangle\langle 00| + (a(1-a)+b^2)|11\rangle\langle 11|$.
c Show that $C_o(\rho_{a,b}) = 2b$.

8.70. Show that $H(\rho_{a,b}) = h\left(\frac{1+\sqrt{(2a-1)^2+4b^2}}{2}\right)$.

8.71. Assume that the input state is the maximally entangled state $|\Phi_d\rangle\langle\Phi_d|$ with the reference system. Show that the output state of depolarizing channel $\kappa_{d,\lambda}$ (transpose depolarizing channel $\kappa_{d,\lambda}^T$) is equal to the isotropic state (Werner state) as

$$(\kappa_{d,\lambda} \otimes \iota_R)(|\Phi_d\rangle\langle\Phi_d|) = \rho_{I,\frac{1-\lambda(d^2-1)}{d^2}} \tag{8.194}$$

$$(\kappa_{d,\lambda}^T \otimes \iota_R)(|\Phi_d\rangle\langle\Phi_d|) = \rho_{W,\frac{(1-(d+1)\lambda)(d-1)}{2d}}. \tag{8.195}$$

8.72. Show that $\tau^A(|\tau^A(\rho)|) \geq 0$ following the steps below.

a Show that $\rho_{W,p} = qI + r\tau^A(\Phi_d\rangle\langle\Phi_d|)$, where $q = \frac{1-p}{d(d+1)} + \frac{p}{d(d-1)}$, $r = \frac{1-p}{d+1} - \frac{p}{d-1}$.
b Show that $\tau^A(\rho_{W,p}) = q(I - |\Phi_d\rangle\langle\Phi_d|) + (q+r)|\Phi_d\rangle\langle\Phi_d|$.
c Show that $\tau^A(|\tau^A(\rho_{W,p})|) = q(I - \frac{1}{d}F) + \frac{q+r}{d}F$.

8.73. Show that $\tau^A(|\tau^A(\rho_{I,p})|) \geq 0$ for $p > \frac{1}{d}$ following the steps below. (This inequality is trivial when $p \leq \frac{1}{d}$ because $\tau^A(\rho_{I,p}) \geq 0$.)

a Show that $\tau^A(\rho_{I,p}) = \frac{1-p}{d^2-1}I + \frac{d^2p-1}{d(d^2-1)}F = \frac{1-p}{d^2-1}(I+F) + \frac{dp-1}{d(d-1)}F$.
b Show that $|\tau^A(\rho_{I,p})| = \frac{1-p}{d^2-1}(I+F) + \frac{dp-1}{d(d-1)}I$.
c Show that $\tau^A(|\tau^A(\rho_{I,p})|) = \frac{1-p}{d^2-1}(I + d|\Phi_d\rangle\langle\Phi_d|) + \frac{dp-1}{d(d-1)}I$.

8.74. Show that $(1-\lambda)\rho_{\mathrm{mix}} + \lambda\tau^A(|\Phi_d\rangle\langle\Phi_d|) \geq 0$ if and only if $\frac{-1}{d-1} \leq \lambda \leq \frac{1}{d+1}$, where $\mathcal{H}_A = \mathcal{H}_B = \mathbb{C}^d$.

8.12 Historical Note

The study of conversion among entangled states in an asymptotic setting was initiated by Bennett et al. [38]. These researchers derived the direct and converse parts of Theorem 8.6 in the pure-state case (Exercise 8.33). After this research, Lo and Popescu [276] considered convertibility among two pure states with LOCC. They found that the two-way LOCC could be simulated by the one-way LOCC (Theorem 8.1) in the finite regime when the initial state was pure. They also obtained the optimal value of the probability that we will succeed in converting a given pure partially entangled state into a desired maximally entangled state by LOCC (8.68). Following this research, Nielsen [313] completely characterized the LOCC convertibility between two pure states by use of majorization (pure-state case of Theorem 8.4). Vidal [402] extended this result to the mixed-state case, i.e., he showed the mixed-state case of Theorem 8.4. Using Nielsen's condition, Morikoshi and Koashi [295] proved that the optimal deterministic distillation with an initial pure state can be realized only by two-pair collective manipulations in each step. Applying the method of type to the optimal failure probability (the optimal successful probability) for distillation with an initial pure state, Hayashi et al. [181] derived the optimal generation rate with an exponential constraint for the failure probability (for the successful probability). They also treated this problem with the fidelity criteria. Further, Hayashi [176] extended this result to the non-i.i.d. case. Regarding the mixed-state case, Bennett et al. [40] discussed the relation between distillation and quantum error correction, which will be mentioned in Sect. 9.6. They derived several characterizations of the two-way LOCC distillation as well as of the one-way LOCC distillation. They also conjectured the Hashing inequality (8.106). Rains [352] showed this inequality in the maximally correlated case and the relation (8.114). Horodecki et al. [232] showed that equation (8.107) holds if this inequality holds. They also initiated a unified approach, which has been established by Donald et al. [103] as Theorem 8.10. Modifying the discussion by Devetak [95], Devetak and Winter [98] proved the inequality for any mixed state.

Concerning the converse part, Bennett et al. [38] proved the converse part of the pure-state case by constructing the dilution protocol attaining the entropy rate. Then, proposing the entanglement of relative entropy $E_r(\rho)$, Vedral and Plenio [399] proved the inequality $E^S_{d,2}(\rho) \leq E_r(\rho)$. In this book, its improved version (Theorem 8.7) is derived by combining their idea and the strong converse of quantum Stein's lemma. Horodecki et al. [232] obtained the first inequality in (8.105). Further, establishing a unified approach, Donald et al. [103] simplified its proof.

Christandl and Winter [78] introduced squashed entanglement and proved the inequality $E^{\leftrightarrow}_{d,2}(\rho) \leq E_{sq}(\rho)$. Concerning PPT operations, Rains [353] proved inequality (8.173).

Regarding the dilution, as mentioned above, Bennett et al. [38] proved Theorem 8.9 in the pure-state case. Bennett et al. [40] introduced the entanglement formation. Following these results, Hayden et al. [196] proved Theorem 8.9. In this book, we prove Theorem 8.9 in a little different way to that given in [196]. In Appendix B.5.2, we rigorously optimize the fidelity with the finite regime and prove Theorem 8.9 by taking its limit.

Further, Lo and Popescu [275] showed that the bound can be attained by classical communication with the square root of n bits in the pure-state case. Further, Hayden and Winter [198] and Harrow and Lo [163] proved the optimality of Lo and

Popescu's protocol. Using their results, Terhal et al. [387] showed that the optimal rate of dilution with zero-rate communication can be characterized by the entanglement of purification. They also showed that it is lower bounded by the quantity $C_d^{A\to B}(\rho)$, which was introduced by Henderson and Vedral [205]. As a problem related to dilution with zero-rate communication, we may consider the problem generating a given separable state from common randomness. This problem with the classical setting has been solved by Wyner [426]. Theorem 8.12 is its quantum extension.

Concerning entanglement of exact cost for PPT operations, Audenaert et al. [19] derived its lower bound. Concerning entanglement of exact cost for LOCC operations, Terhal and Horodecki [385] focused on the Schmidt rank and calculated it for the two-tensor product of the two-dimensional isotropic state. Joining these, we derived the entanglement of exact cost for these settings in this example.

As a related problem, we often consider how to characterize a pure entangled state producing a given state with nonzero probability by LOCC. This problem is called stochastic convertibility. Owari et al. [340] treated this problem in infinite-dimensional systems using the partial order. Miyake [293] treated this problem in tripartite systems using a hyperdeterminant. Ishizaka [242] focused on PPT operations and showed that any pure entangled state can be stochastically converted from another pure entangled state by PPT operations.

9

Analysis of Quantum Communication Protocols

Summary. The problems of transmitting a classical message via a quantum channel (Chap. 4) and estimating a quantum state (Chaps. 3 and 6) have a classical analog. These are not intrinsically quantum-specific problems but quantum extenstions of classical problems. These problems are difficult due to the noncommutativity of quantum mechanics.

However, quantum information processing is not merely a noncommuting version of classical information processing. There exist many quantum protocols without any classical analog. In this context, quantum information theory covers a greater field than a noncommutative analog of classical information theory. The key to these additional effects is the advantage of using entanglement treated in Chap. 8, where we examined only the quantification of entanglement. In this chapter, we will introduce several quantum communication protocols that are possible only by using entanglement and are therefore classically impossible. (Some of protocols introduced in this section have classical analogs.) We also examine problems such as the transmission of quantum states (quantum error correction), communication in the presence of eavesdroppers, and several other types of communication that we could not handle in Chap. 4. As seen in this chapter, the transmission of a quantum state is closely retated to communication with no information leakage to eavesdroppers. The noise in the transmission of a quantum state clearly corresponds to the eavesdropper in a quantum communication.

Table 9.1. Denotations used in Chap. 9

$\chi_\kappa(\rho)$	Holevo information (9.4)
$H_\kappa(\rho)$	Minimum average output entropy (9.5)
Chanel capacities	
$C_c(\kappa)$	Channel capacity for when no entanglement is used in input
$C_c^e(\kappa)$	Channel capacity for when entanglement is used in input (9.1)
$C_a(\rho^{A,B})$	Amount of assistance for sending information by state $\rho^{A,B}$ (9.32)
$C_{c,e}^e(\kappa)$	Entanglement-assisted channel capacity (9.37)
$C_r(W,\sigma)$	Quantum-channel resolvability capacity (9.50)

Table 9.1. Continued

Chanel capacities	
$C_c^{B,E}(W)$	Wiretap channel capacity (9.60)
$C_{q,1}$	Capacity for quantum-state transmission in worst case (9.81)
$C_{q,2}$	Capacity for quantum-state transmission with entanglement fidelity (9.81)

9.1 Quantum Teleportation

The strange properties of entangled states were first examined by Einstein et al. [104] in an attempt to show that quantum mechanics was incomplete. Recently, the entangled states have been treated in a manner rather different than when it was first introduced. For example, by regarding these states as the source of a quantum advantage, Bennett et al. [36] proposed quantum teleportation. Since this topic can be understood without any complicated mathematics, we introduce it in this section.

In quantum teleportation, an entangled state is first shared between two parties. Then, by sending a classical message from one party to the other, it is possible to transmit a quantum state without directly sending it. Let us look at this protocol in more detail. First, we prepare an entangled state $e_0^{A,B} = \frac{1}{\sqrt{2}}(u_0^A \otimes u_0^B + u_1^A \otimes u_1^B)$ on the composite system $\mathcal{H}_A \otimes \mathcal{H}_B$ comprising two qubits $\mathcal{H}_A$ and $\mathcal{H}_B$ spanned by u_0^A, u_1^A and u_0^B, u_1^B, respectively. The sender possesses a qubit $\mathcal{H}_C$ spanned by u_0^C, u_1^C as well as the quantum system $\mathcal{H}_A$. The sender sends qubit $\mathcal{H}_C$ to the receiver. The receiver possesses the other qubit $\mathcal{H}_B$. Then, we have the following theorem.

Theorem 9.1 *(BBCJPW [36]) Let the sender perform a measurement corresponding to the CONS* $e_i^{A,C} \stackrel{\text{def}}{=} (I_A \otimes S_i^C)e_0^{A,C} (i = 0, 1, 2, 3)$ *on the composite system* $\mathcal{H}_A \otimes \mathcal{H}_C$ *and its result be sent to the receiver. [From (1.18), it satisfies the conditions for a PVM.] Let the receiver perform a unitary time evolution corresponding to* $\overline{S_i^B}$ *on the quantum system* $\mathcal{H}_B$*. Then, the final state on* $\mathcal{H}_B$ *is the same state as the initial state on* $\mathcal{H}_C$*.*

This argument holds irrespective of the initial state on $\mathcal{H}_C$ and measurement value i, as proved below.

Proof. Let us first consider the case where the measurement data are 0. Let the initial state on $\mathcal{H}_C$ be the pure state $x = \sum_i x^i u_i^C$. Then, the state on the composite system $\mathcal{H}_A \otimes \mathcal{H}_B \otimes \mathcal{H}_C$ is $\frac{1}{\sqrt{2}} \sum_{i,j,k} x^i \delta^{j,k} u_j^A \otimes u_k^B \otimes u_i^C$. Therefore, the final state on $\mathcal{H}_B$ is $\sum_k \sum_{i,j} \frac{1}{\sqrt{2}} x^i \delta^{j,k} \frac{1}{\sqrt{2}} \delta^{i,j} u_k^B = \sum_k \frac{1}{2} x^k u_k^B$, following Exercise 7.4. Normalizing this vector, we can prove that the final state on $\mathcal{H}_B$ equals $\sum_k x^k u_k^B$, which is the same state as the initial state on $\mathcal{H}_C$.

Now, consider the case in which the measurement datum i is obtained. Since $(S_i^C)^* = S_i^C$,

$$\begin{aligned}
&\mathrm{Tr}_{A,C}\left(|e_i^{A,C}\rangle\langle e_i^{A,C}| \otimes I_B\right)|x \otimes e_0^{A,B}\rangle\langle x \otimes e_0^{A,B}| \\
&= \mathrm{Tr}_{A,C}(\overline{S_i^C} \otimes I_{A,B})\left(|e_0^{A,C}\rangle\langle e_0^{A,C}| \otimes I_B\right)(\overline{S_i^C} \otimes I_{A,B})|x \otimes e_0^{A,B}\rangle\langle x \otimes e_0^{A,B}| \\
&= \mathrm{Tr}_{A,C}\left(|e_0^{A,C}\rangle\langle e_0^{A,C}| \otimes I_B\right)|(\overline{S_i}x) \otimes e_0^{A,B}\rangle\langle(\overline{S_i}x) \otimes e_0^{A,B}| = \frac{1}{4}|\overline{S_i}x\rangle\langle\overline{S_i}x|.
\end{aligned}$$

Operating $\overline{S_i}$ on $\mathcal{H}_B$ ($\overline{S_i}$ is its own inverse), the final state on $\mathcal{H}_B$ is x. ∎

It is noteworthy that this protocol has been experimentally demonstrated [57, 139, 341]. Other protocols that combine quantum teleportation with cloning have also been proposed [297] and demonstrated.

Exercises

9.1. Show that quantum teleportation in any dimension is also possible by following the steps below. Let $\mathcal{H}_A, \mathcal{H}_B, \mathcal{H}_C$ be the spaces spanned by $u_1^A, \ldots, u_d^A, u_1^B, \ldots, u_d^B, u_1^C, \ldots, u_d^C$, respectively. Prepare an entangled state $u_{0,0}^{A,B} \stackrel{\mathrm{def}}{=} \frac{1}{\sqrt{d}}\sum_{i=1}^d u_i^A \otimes u_i^B$ in $\mathcal{H}_A \otimes \mathcal{H}_B$. Now perform a measurement corresponding to $\{u_{i,j}^{A,C} \stackrel{\mathrm{def}}{=} (I_A \otimes \mathbf{X}_C^i \mathbf{Z}_C^j) u_{0,0}^{A,C}\}_{i,j}$, and then an operation $\overline{\mathbf{X}_B^i \mathbf{Z}_B^j}$ depending on the measurement data (i,j). Show that the final state on $\mathcal{H}_B$ is the same as the initial state on $\mathcal{H}_C$.

9.2 C-Q Channel Coding with Entangled Inputs

In this section, we treat classical message transmission via a quantum channel κ from $\mathcal{H}_A$ to $\mathcal{H}_B$. When we use only tensor product states in $\mathcal{S}(\mathcal{H}_A^{\otimes n})$, the problem becomes that of classical-quantum (c-q) channel coding discussed in Chap. 4 by setting $\mathcal{X}$ to $\mathcal{S}(\mathcal{H}_A)$ and $W_\rho \stackrel{\mathrm{def}}{=} \kappa(\rho)$. However, when we are allowed to use any state in $\mathcal{S}(\mathcal{H}_A^{\otimes n})$ as input state, our problem cannot be regarded as a special case of Chap. 4. In this case, the channel capacity is given by

$$C_c(\kappa) = \max_{p \in \mathcal{P}(\mathcal{S}(\mathcal{H}_A))} I(p, \kappa).$$

When we are allowed to send any state entangled between n systems of the input, the channel capacity is $C_c(\kappa^{\otimes n})/n$. When any entangled state is available as an input state, the code $\hat{\Phi}^{(n)} = (N_n, \hat{\varphi}^{(n)}, Y^{(n)})$ is expressed by the triplet of the size N_n, the encoder $\hat{\varphi}^{(n)}$ mapping from $\{1, \ldots, N_n\}$ to $\mathcal{S}(\mathcal{H}_A^{\otimes n})$, and the POVM $Y^{(n)} = \{Y_i^{(n)}\}_{i=1}^{N_n}$ taking values in $\{1, \ldots, N_n\}$ on the output space $\mathcal{H}_B^{\otimes n}$. The error probability is given by

$$\varepsilon[\hat{\Phi}^{(n)}] \stackrel{\mathrm{def}}{=} \frac{1}{N_n}\sum_{i=1}^{N_n}\left(1 - \mathrm{Tr}\, Y_i^{(n)} \kappa^{\otimes n}(\hat{\varphi}^{(n)}(i))\right).$$

Theorem 9.2 *Define the capacity* $C_c^e(\kappa)$*:*[1]

$$C_c^e(\kappa) \stackrel{\text{def}}{=} \sup_{\{\hat{\Phi}^{(n)}\}} \left\{ \underline{\lim} \frac{1}{n} \log |\hat{\Phi}^{(n)}| \,\middle|\, \lim \varepsilon[\hat{\Phi}^{(n)}] = 0 \right\} \tag{9.1}$$

for a quantum channel κ *from* $\mathcal{H}_A$ *to* $\mathcal{H}_B$*. Then, we have*

$$C_c^e(\kappa) = \sup_n \frac{C_c(\kappa^{\otimes n})}{n} = \lim \frac{C_c(\kappa^{\otimes n})}{n}.$$

The definition implies that $C_c^e(\kappa) \geq C_c(\kappa)$. To date, however, no example of $C_c^e(\kappa) > C_c(\kappa)$ has been found. Therefore it is currently conjectured that $C_c^e(\kappa) = C_c(\kappa)$. To prove this we must show that

$$nC_c(\kappa) = C_c(\kappa^{\otimes n}). \tag{9.2}$$

Equation (9.2) may be derived from the *additivity* of the channel capacity for two arbitrary TP-CP maps κ^1 and κ^2:

$$C_c(\kappa^1) + C_c(\kappa^2) = C_c(\kappa^1 \otimes \kappa^2). \tag{9.3}$$

Here, remember the relation (8.142). This relation indicates the relation between the classical capacity and entanglement formation. Indeed, the classical capacity $C_c(\kappa)$ is described by

$$C_c(\kappa) = \max_\rho \chi_\kappa(\rho),$$

where *Holevo information* $\chi_\kappa(\rho)$ and *minimum average output entropy* $H_\kappa(\rho)$ are defined by

$$\chi_\kappa(\rho) \stackrel{\text{def}}{=} H(\kappa(\rho)) - H_\kappa(\rho), \tag{9.4}$$

$$H_\kappa(\rho) \stackrel{\text{def}}{=} \min_{(p_x,\rho_x):\sum_x p_x \rho_x = \rho} \sum_x p_x H(\kappa(\rho_x)). \tag{9.5}$$

When κ is the partial trace from the system $\mathcal{H}_A \otimes \mathcal{H}_B$ to $\mathcal{H}_B$, the relation (*MSW correspondence* [290])

$$H_\kappa(\rho) = E_f(\rho) \tag{9.6}$$

holds, i.e.,

$$\chi_{\operatorname{Tr}_A}(\rho) = H(\operatorname{Tr}_A \rho) - E_f(\rho). \tag{9.7}$$

Indeed, the additivity of $C_c(\kappa)$ is equivalent to many other conditions, e.g., the additivity of entanglement of formation.

---

[1] The superscript e of C_c^e indicates that "entangled" input is allowed.

Theorem 9.3 *(Matsumoto et al. [290], Shor [374], Pomeransky [350]) The following 14 conditions are equivalent.*

HM *Additivity of classical capacity of q-q channel (additivity of maximum Holevo information):*

$$\max_{\rho^1} \chi_{\kappa^1}(\rho^1) + \max_{\rho^2} \chi_{\kappa^2}(\rho^2) = \max_{\rho^{1,2}} \chi_{\kappa^1 \otimes \kappa^2}(\rho^{1,2}) \tag{9.8}$$

holds for arbitrary channels κ^1 and κ^2.

HA *Additivity of Holevo information:*

$$\chi_{\kappa^1}(\rho^1) + \chi_{\kappa^2}(\rho^2) = \chi_{\kappa^1 \otimes \kappa^2}(\rho^1 \otimes \rho^2) \tag{9.9}$$

holds for arbitrary channels κ^1 and κ^2 and arbitrary states ρ^1 and ρ^2.

HL *Additivity of classical capacity of q-q channel with linear cost constraint (additivity of maximum Holevo information with linear cost constraint):*

$$\max_{\lambda} C_{X^1 \le \lambda K}(\kappa^1) + C_{X^2 \le (1-\lambda) K}(\kappa^2) = C_{X^1 + X^1 \le K}(\kappa^1 \otimes \kappa^2), \tag{9.10}$$

i.e.,

$$\begin{aligned} &\max_{\lambda} \max_{\rho^1 : \operatorname{Tr} \rho^1 X^1 \le \lambda K} \chi_{\kappa^1}(\rho^1) + \max_{\rho^2 : \operatorname{Tr} \rho^2 X^2 \le (1-\lambda) K} \chi_{\kappa^2}(\rho^2) \\ &= \max_{\rho^{1,2} : \operatorname{Tr} \rho^{1,2} (X^1 + X^2) \le K} \chi_{\kappa^1 \otimes \kappa^2}(\rho^{1,2}) \end{aligned} \tag{9.11}$$

holds for arbitrary channels κ^1 and κ^2, arbitrary Hermitian matrices X^1 and X^2 on the respective input system, and an arbitrary constant K. Here we identify X^1 (X^2) with $X^1 \otimes I^2$ ($I^1 \otimes X^2$). Note that the classical capacity of a q-q channel with linear cost constraint has the form $C_{X \le \lambda K}(\kappa) = \max_{\rho : \operatorname{Tr} \rho X \le K} \chi_{\kappa}(\rho)$.

HC *Additivity of conjugate Holevo information:*

$$\chi^*_{\kappa^1}(X^1) + \chi^*_{\kappa^2}(X^2) = \chi^*_{\kappa^1 \otimes \kappa^2}(X^1 + X^2) \tag{9.12}$$

holds for Hermitian matrices X^1 and X^2 on systems $\mathcal{H}_1$ and $\mathcal{H}_2$, where conjugate Holevo information $\chi^*_{\kappa}(X)$ is defined as the Legendre transform of $\chi_{\kappa}(\rho)$ as

$$\chi^*_{\kappa}(X) \stackrel{\text{def}}{=} \max_{\rho} \operatorname{Tr} X\rho + \chi_{\kappa}(\rho).$$

HS *Subadditivity of Holevo information:*

$$\chi_{\kappa^1}(\rho^1) + \chi_{\kappa^2}(\rho^2) \le \chi_{\kappa^1 \otimes \kappa^2}(\rho^{1,2}) \tag{9.13}$$

holds for arbitrary channels κ^1 and κ^2 and arbitrary states $\rho^{1,2}$, where $\rho^1 = \operatorname{Tr}_2 \rho^{1,2}$ and $\rho^2 = \operatorname{Tr}_1 \rho^{1,2}$.

EM *Additivity of minimum output entropy:*

$$\min_{\rho^1} H(\kappa^1(\rho^1)) + \min_{\rho^2} H(\kappa^2(\rho^2)) = \min_{\rho^{1,2}} H(\kappa^1 \otimes \kappa^2(\rho^{1,2})), \tag{9.14}$$

i.e.,

$$\min_{\rho^1} H_{\kappa^1}(\rho^1) + \min_{\rho^2} H_{\kappa^2}(\rho^2) = \min_{\rho^{1,2}} H_{\kappa^1 \otimes \kappa^2}(\rho^{1,2}) \tag{9.15}$$

holds for arbitrary channels κ^1 and κ^2. Note that the minimum output entropy has the form $\min_\rho H(\kappa(\rho)) = \min_\rho H_\kappa(\rho)$.

EA *Additivity of minimum average output entropy:*

$$H_{\kappa^1}(\rho^1) + H_{\kappa^2}(\rho^2) = H_{\kappa^1 \otimes \kappa^2}(\rho^1 \otimes \rho^2) \tag{9.16}$$

holds for arbitrary channels κ^1 and κ^2 and arbitrary states ρ^1 and ρ^2.

EL *Additivity of minimum average output entropy with linear cost constraint):*

$$\begin{aligned} &\min_\lambda \min_{\rho^1 : \operatorname{Tr} \rho^1 X^1 \le \lambda K} H_{\kappa^1}(\rho^1) + \min_{\rho^2 : \operatorname{Tr} \rho^2 X^2 \le (1-\lambda) K} H_{\kappa^2}(\rho^2) \\ &= \min_{\rho^{1,2} : \operatorname{Tr} \rho^{1,2}(X^1 + X^2) \le K} H_{\kappa^1 \otimes \kappa^2}(\rho^{1,2}) \end{aligned} \tag{9.17}$$

holds for arbitrary channels κ^1 and κ^2, arbitrary Hermitian matrices X^1 and X^2 on the respective input system, and an arbitrary constant K.

EC *Additivity of conjugate minimum average output entropy:*

$$H^*_{\kappa^1}(X^1) + H^*_{\kappa^2}(X^2) = H^*_{\kappa^1 \otimes \kappa^2}(X^1 + X^2) \tag{9.18}$$

holds for Hermitian matrices X^1 and X^2 on systems $\mathcal{H}_1$ and $\mathcal{H}_2$, where the conjugate minimum average output entropy $H^*_\kappa(X)$ is defined as the Legendre transform of $H_\kappa(\rho)$ as

$$H^*_\kappa(X) \stackrel{\text{def}}{=} \max_\rho \operatorname{Tr} X\rho - H_\kappa(\rho).$$

ES *Superadditivity of minimum average output entropy:*

$$H_{\kappa^1}(\rho^1) + H_{\kappa^2}(\rho^2) \ge H_{\kappa^1 \otimes \kappa^2}(\rho^{1,2}) \tag{9.19}$$

holds for arbitrary channels κ^1, κ^2 and arbitrary states $\rho^{1,2}$.

FA *Additivity of entanglement of formation:*

$$E_f(\rho^1) + E_f(\rho^2) = E_f(\rho^1 \otimes \rho^2) \tag{9.20}$$

holds for a state ρ^1 on $\mathcal{H}_1 \stackrel{\text{def}}{=} \mathcal{H}_{A_1} \otimes \mathcal{H}_{B_1}$ and a state ρ^2 on $\mathcal{H}_2 \stackrel{\text{def}}{=} \mathcal{H}_{A_2} \otimes \mathcal{H}_{B_2}$.

FL *Additivity of minimum entanglement of formation with linear cost constraint:*

$$\begin{aligned}&\min_{\lambda} \min_{\rho^1:\mathrm{Tr}\,\rho^1 X^1 \le \lambda K} E_f(\rho^1) + \min_{\rho^2:\mathrm{Tr}\,\rho^2 X^2 \le (1-\lambda)K} E_f(\rho^2)\\ &= \min_{\rho^{1,2}:\mathrm{Tr}\,\rho^{1,2}(X^1+X^2)\le K} E_f(\rho^{1,2}) \end{aligned} \tag{9.21}$$

holds for arbitrary Hermitian matrices X^1 and X^2 and an arbitrary constant K.

FC *Additivity of conjugate entanglement of formation:*

$$E_f^*(X^1) + E_f^*(X^2) = E_f^*(X^1 + X^2) \tag{9.22}$$

holds for Hermitian matrices X^1 and X^2 on systems $\mathcal{H}_1$ and $\mathcal{H}_2$, where the conjugate entanglement of formation $E_f^*(X)$ is defined as the Legendre transform of $E_f(\rho)$ as

$$E_f^*(X) \stackrel{\mathrm{def}}{=} \max_{\rho} \mathrm{Tr}\, X\rho - E_f(\rho).$$

FS *Superadditivity of entanglement of formation:*

$$E_f(\rho^1) + E_f(\rho^2) \le E_f(\rho^{1,2}) \tag{9.23}$$

holds for a state $\rho^{1,2}$ on $(\mathcal{H}_{A_1} \otimes \mathcal{H}_{A_2}) \otimes (\mathcal{H}_{B_1} \otimes \mathcal{H}_{B_2})$.

However, whether all these conditions hold remains an open problem. Some numerical verifications of **HM** [193, 325] and **FS** [369] have been performed. No counterexample has been found yet. Concerning the additivity of the channel capacity, only the following cases have been proved.

a When $C_c(\kappa)$ is equal to the dimension of the output system, additivity (9.2) holds (trivial case).

b Any entanglement-breaking channel κ_1 (Example 5.4) satisfies additivity (9.3) with an arbitrary channel κ^2 [372].

c Any depolarizing channel $\kappa_{d,\lambda}$ (Example 5.3) satisfies additivity (9.3) with $\kappa^1 = \kappa_{d,\lambda}$ and an arbitrary channel κ^2 [254].

d Any unital qubit channel κ^1 satisfies additivity (9.3) with an arbitrary channel κ^2 [253].

e Any antisymmetric channel $\kappa^T_{d,\frac{-1}{d-1}}$ (Werner–Holevo channels, Example 5.9) satisfies additivity (9.2) with $\kappa = \kappa^T_{d,\frac{-1}{d-1}}$ [291].

f All transpose-depolarizing channels $\kappa^T_{d,\lambda}$ and $\kappa^T_{d',\lambda}$ satisfy additivity (9.3) with $\kappa^1 = \kappa^T_{d,\lambda}$ and $\kappa^2 = \kappa^T_{d',\lambda}$ [90, 112].

g1 Channels $\kappa_{d,\lambda} \circ \kappa_D^{\mathrm{PD}}$ satisfy additivity (9.3) with $\kappa^1 = \kappa_{d,\lambda} \circ \kappa_D^{\mathrm{PD}}$ and an arbitrary channel κ^2 [91, 138].

g2 Channels $\kappa^T_{d,\frac{-1}{d-1}} \circ \kappa_D^{\mathrm{PD}}$ and $\kappa_D^{\mathrm{PD}} \circ \kappa^T_{d,\frac{-1}{d-1}}$ satisfy additivity (9.2) with $\kappa = \kappa^T_{d,\frac{-1}{d-1}} \circ \kappa_D^{\mathrm{PD}}$ or $\kappa_D^{\mathrm{PD}} \circ \kappa^T_{d,\frac{-1}{d-1}}$ [138, 423].

g3 Channels $\kappa_{d,\lambda}^T \circ \kappa_D^{\mathrm{PD}}$ and $\kappa_D^{\mathrm{PD}} \circ \kappa_{d,\lambda}^T$ satisfy additivity (9.3) with $\kappa_1 = \kappa_{d,\lambda}^T \circ \kappa_D^{\mathrm{PD}}$ or $\kappa_D^{\mathrm{PD}} \circ \kappa_{d,\lambda}^T$ and $\kappa_2 = \kappa_{d',\lambda}^T \circ \kappa_{D'}^{\mathrm{PD}}$ or $\kappa_{D'}^{\mathrm{PD}} \circ \kappa_{d',\lambda}^T$ [138, 423].

Therefore, we obtain $C_c^e(\kappa) = C_c(\kappa)$ in the cases **a**, **b**, **c**, **d**, **e**, **g1**, and **g2**. Indeed, since $C_c(\kappa_{d,\lambda} \circ \kappa_D^{\mathrm{PD}}) = C_c(\kappa_{d,\lambda})$, **c** yields that $C_c(\kappa_{d,\lambda} \circ \kappa_D^{\mathrm{PD}} \otimes \kappa_2) \le C_c(\kappa_{d,\lambda} \otimes \kappa_2) = C_c(\kappa_{d,\lambda}) + C_c(\kappa_2) = C_c(\kappa_{d,\lambda} \circ \kappa_D^{\mathrm{PD}}) + C_c(\kappa_2)$, which implies **g1**. Similarly, we can show **g2** and **g3** from **e** and **f**.

Moreover, the additivity of minimum output entropy holds not only in the above cases but also in the more extended cases of **c**, **e**, and **f** as opposed to **g1**, **g2**, and **g3** [138, 423]. As a stronger condition than **EM**, we often focus on the following condition.

RM The multiplicativity of the maximal output Rényi entropy (p-norm):

$$\psi_{\max}(s|\kappa^1 \otimes \kappa^2) = \psi_{\max}(s|\kappa^1) + \psi_{\max}(s|\kappa^2) \text{ for } -1 \le s \le 0 \quad (9.24)$$

holds, where $\psi_{\max}(s|\kappa) \stackrel{\mathrm{def}}{=} \max_\rho \psi(s|\kappa(\rho))$.

This is because the above multiplicativity is sometimes easier to check than the additivity of minimum output entropy. Indeed, condition (9.24) implies **EM**; however, the converse is not true. In particular, Werner and Holevo [413] showed that (9.24) does not hold for $s \le -3.79$ when κ^1 and κ^2 are antisymmetric channels (Werner–Holevo channels).

Among the above conditions, the relations **HC** ⇒ **HM** and **EC** ⇒ **EM** are trivial. From MSW correspondence (9.6) we obtain **HA** ⇒ **FA** and **EA** ⇒ **FA**. Next, we focus on the Stinespring representation $(\mathcal{H}_C, \rho_0, U_\kappa)$ of κ mapping from a system $\mathcal{H}_A$ to another system $\mathcal{H}_B$. In this case, the MSW correspondence (9.6) can be generalized as

$$H_\kappa(\rho) = \min_{(p_i,\rho_i):\sum_i p_i\rho_i = \rho} \sum_i p_i H(\mathrm{Tr}_{A,C}\, U_\kappa(\rho_i \otimes \rho_0)U_\kappa^*) = E_f(\bar{\kappa}(\rho)),$$
$$\bar{\kappa}(\rho) \stackrel{\mathrm{def}}{=} U_\kappa(\rho \otimes \rho_0)U_\kappa^*,$$

i.e.,

$$\chi_\kappa(\rho) = H(\kappa(\rho)) - E_f(\bar{\kappa}(\rho)), \quad (9.25)$$

where we use the notation E_f as the entanglement of formation between the output system $\mathcal{H}_B$ and the environment $\mathcal{H}_A \otimes \mathcal{H}_C$. Hence, if Condition **FS** holds, for $\rho^{1,2}$, we have

$$\begin{aligned}
\chi_{\kappa^1\otimes\kappa^2}(\rho^{1,2}) &= H(\kappa^1 \otimes \kappa^2(\rho^{1,2})) - E_f(\overline{\kappa^1 \otimes \kappa^2}(\rho^{1,2})) \\
&\le H(\kappa^1(\rho^1)) + H(\kappa^2(\rho^2)) - E_f(\overline{\kappa^1 \otimes \kappa^2}(\rho^{1,2})) \\
&\le H(\kappa^1(\rho^1)) + H(\kappa^2(\rho^2)) - (E_f(\overline{\kappa^1}(\rho^1)) + E_f(\overline{\kappa^2}(\rho^2))) \\
&= \chi_{\kappa^1}(\rho^1) + \chi_{\kappa^2}(\rho^2).
\end{aligned}$$

Hence, we have **FS** ⇒ **HS**. Similarly, the relation **FS** ⇒ **MS** holds.

The following lemma is useful for proofs of the remaining relations.

Lemma 9.1 *Let f^i be a convex function defined on $\mathcal{S}(\mathcal{H}_i)$ $(i = 1, 2)$ and $f^{1,2}$ be a convex function defined on $\mathcal{S}(\mathcal{H}_1 \otimes \mathcal{H}_2)$ satisfying*

$$f^1(\rho^1) + f^2(\rho^2) \geq f^{1,2}(\rho^1 \otimes \rho^2). \tag{9.26}$$

The relations **L**⇔**C**⇔**S** ⇒ **A** *hold among the following conditions.*

S *Superadditivity:*

$$f^1(\rho^1) + f^2(\rho^2) \leq f^{1,2}(\rho^{1,2}) \tag{9.27}$$

holds for a state $\rho^{1,2}$ on $(\mathcal{H}_{A_1} \otimes \mathcal{H}_{A_2}) \otimes (\mathcal{H}_{B_1} \otimes \mathcal{H}_{B_2})$.

C *Additivity of conjugate function:*

$$f^{1*}(X^1) + f^{2*}(X^2) = f^{1,2*}(X^1 + X^2) \tag{9.28}$$

holds for Hermitian matrices X^1 and X^2 on the systems $\mathcal{H}_1$ and $\mathcal{H}_2$, where conjugate entanglement of formation $f^*(X)$ is defined as the Legendre transform of $f(\rho)$ as

$$f^*(X) \stackrel{\text{def}}{=} \max_{\rho} \operatorname{Tr} X\rho - f(\rho).$$

L *Additivity of minimum value with linear cost constraint:*

$$\begin{aligned} &\min_{\lambda} \min_{\rho^1:\operatorname{Tr}\rho^1 X^1 \leq \lambda K} f^1(\rho^1) + \min_{\rho^2:\operatorname{Tr}\rho^2 X^2 \leq (1-\lambda)K} f^2(\rho^2) \\ &= \min_{\rho^{1,2}:\operatorname{Tr}\rho^{1,2}(X^1+X^2)\leq K} f^{1,2}(\rho^{1,2}) \end{aligned} \tag{9.29}$$

holds for arbitrary Hermitian matrices X^1 and X^2 and an arbitrary constant K.

A *Additivity:*

$$f^1(\rho^1) + f^2(\rho^2) = f^{1,2}(\rho^1 \otimes \rho^2) \tag{9.30}$$

holds for a state ρ^1 on $\mathcal{H}_1 \stackrel{\text{def}}{=} \mathcal{H}_{A_1} \otimes \mathcal{H}_{B_1}$ and a state ρ^2 on $\mathcal{H}_2 \stackrel{\text{def}}{=} \mathcal{H}_{A_2} \otimes \mathcal{H}_{B_2}$.

Lemma 9.1 yields the relations **HL**⇔**HC**⇔**HS** ⇒ **HA**, **EL**⇔**EC**⇔**ES** ⇒ **EA**, and **FL**⇔**FC**⇔**FS** ⇒ **FA**.

Hence, if we prove the relations **HM** ⇒ **HC**, **EM** ⇒ **EC**, and **FA** ⇒ **FS**, we obtain the equivalence among the above 14 conditions. These proofs will be given in Sect. B.6. The relations are summarized as follows.

Finally, we prove the additivity for the channel capacity for entanglement-breaking channels using inequality (5.77). From the definition, any entanglement-breaking channel κ^1 has the form of the output state for any input state ρ_x as

$$(\kappa_1 \otimes \iota)(\rho_x) = \sum_y Q^x_y \rho^1_{x,y} \otimes \rho^2_{x,y}.$$

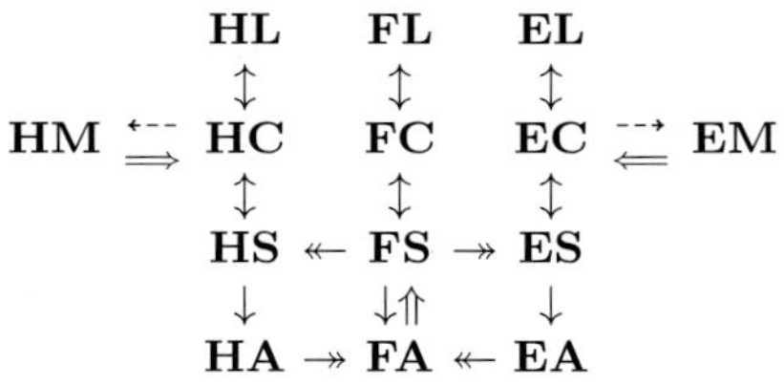

Fig. 9.1. $\rightarrow$: Lemma 9.1, $\dashrightarrow$: easy, $\twoheadrightarrow$: MSW correspondence, $\Longrightarrow$: hard

Hence, using (5.77) and (5.58), we have

$$\begin{aligned}
& C_c(\kappa^1 \otimes \kappa^2) \\
&= \max_{\rho} H((\kappa^1 \otimes \kappa^2)(\rho)) - \min_{(p_x,\rho_x):\sum_x p_x\rho_x=\rho} \sum_x p_x H((\kappa^1 \otimes \kappa^2)(\rho_x)) \\
&= \max_{\rho} H((\kappa^1 \otimes \kappa^2)(\rho)) \\
&\quad - \min_{(p_x,\rho_x):\sum_x p_x\rho_x=\rho} \sum_x p_x H\Big(\sum_y Q_y^x \rho_{x,y}^1 \otimes \kappa^2(\rho_{x,y}^1)\Big) \\
&\le H(\mathrm{Tr}_2(\kappa^1 \otimes \kappa^2)(\rho)) + H(\mathrm{Tr}_1(\kappa^1 \otimes \kappa^2)(\rho)) \\
&\quad - \min_{(p_x,\rho_x):\sum_x p_x\rho_x=\rho} \sum_x p_x \left(\sum_y Q_y^x H(\kappa^2(\rho_{x,y}^2)) + H\Big(\sum_y Q_y^x \rho_{x,y}^1\Big)\right) \\
&= H(\kappa^1(\mathrm{Tr}_2\,\rho)) - \min_{(p_x,\rho_x):\sum_x p_x\rho_x=\rho} \sum_x p_x H\Big(\sum_y Q_y^x \rho_{x,y}^1\Big) \\
&\quad + H(\kappa^2(\mathrm{Tr}_1\,\rho)) - \min_{(p_x,\rho_x):\sum_x p_x\rho_x=\rho} \sum_x p_x \sum_y Q_y^x H(\kappa^2(\rho_{x,y}^2)) \\
&\le C_c(\kappa^1) + C_c(\kappa^2),
\end{aligned} \tag{9.31}$$

which implies (9.3) for entanglement channel κ_1 and arbitrary channel κ_2.

Exercises

9.2. Using a discussion similar to (9.31), show that the additivity of minimum output entropy when κ^1 is entanglement breaking.

9.3. Prove Theorem 9.2 by referring to the proof of Theorem 4.1.

9.3 C-Q Channel Coding with Shared Entanglement

In the preceding section, we considered the effectiveness of using the input state entangled between systems that are to be sent. In this section, we will consider the usefulness of entangled states $\rho^{A,B}$ on a composite system

$\mathcal{H}_A \otimes \mathcal{H}_B$ that is a priori shared between the sender and the receiver. If the sender wishes to send some information corresponding to an element i of $\{1,\ldots,N\}$, he or she must perform an operation $\varphi_e(i)$ on the system $\mathcal{H}_A$ according to the element i, then send the system $\mathcal{H}_A$ to the receiver using the quantum channel κ. Then, the receiver performs a measurement (POVM) $Y=\{Y_i\}_{i=1}^N$ on the composite system $\mathcal{H}_{A'}\otimes\mathcal{H}_B$. Note that this measurement is performed not only on the output system $\mathcal{H}_{A'}$ of the quantum channel κ but also on the composite system $\mathcal{H}_{A'}\otimes\mathcal{H}_B$.

Consider the simple case in which the systems $\mathcal{H}_A$, $\mathcal{H}_{A'}$, and $\mathcal{H}_B$ are all quantum two-level systems. Let the initial state $\rho^{A,B}$ be a pure state $\frac{1}{\sqrt{2}}\left(|u_0^A\otimes u_0^B\rangle+|u_1^A\otimes u_1^B\rangle\right)$. Assume that there is no noise in the quantum channel, which enables the perfect transmission of the quantum state. In this case, we send the message $i\in\{0,\ldots,3\}$ by applying the unitary transformation S_i^A on system $\mathcal{H}_A$. Then, the receiver possesses the transmitted system as well as the initially shared system. The state of the composite system $(\mathbb{C}^2)^{\otimes 2}$ of the receiver is given by $(S_i^A\otimes I_B)\frac{1}{\sqrt{2}}\left(|u_0^A\otimes u_0^B\rangle+|u_1^A\otimes u_1^B\rangle\right)$. Since the vectors form an orthogonal basis with $i=0,1,2,3$, we can perform a measurement Y comprising this basis. Hence, this measurement provides error-free decoding. According to this protocol, two bits of information may be sent through only one qubit channel. We observe that by sharing an entangled state between two parties a priori, more information can be sent than simply by sending a quantum state [43]. This protocol is often called superdense coding.

However, the initially shared entangled state is not necessarily a maximally entangled state such as $\frac{1}{\sqrt{2}}(|u_0^A\otimes u_0^B\rangle+|u_1^A\otimes u_1^B\rangle)$ in general. Hence, it is an important question to determine how much sharing a partially entangled state improves the channel capacity. This will give a quantitative measure of the utilizable entanglement of a partially entangled state.

Assume that the sender and the receiver share the partially entangled state $(\rho^{A,B})^{\otimes n}$ on $\mathcal{H}_A^{\otimes n}\otimes\mathcal{H}_B^{\otimes n}$. The *code* is then given by the set $\Phi_e^{(n)}=(N_n,\mathcal{H}_{A'_n},\varphi_e^{(n)},Y^{(n)})$ consisting of its size N_n, the quantum system $\mathcal{H}_{A'_n}$ transmitted by the sender to the receiver, the operation $\varphi_e^{(n)}(i)$ from the quantum system $\mathcal{H}_A^{\otimes n}$ to $\mathcal{H}_{A'_n}$ depending on each message i, and the measurement $Y^{(n)}$ on the composite system $\mathcal{H}_{A'_n}\otimes\mathcal{H}_B^{\otimes n}$.

Further, the effectiveness of an entangled state ρ, i.e., the increase of the transmitted message, is given by

$$|\Phi_e^{(n)}| \stackrel{\text{def}}{=} \frac{N_n}{\dim\mathcal{H}_{A'_n}},$$

and the error probability is given by

$$\varepsilon[\Phi_e^{(n)}] \stackrel{\text{def}}{=} \frac{1}{N_n}\sum_{i=1}^{N_n}\left(1-\mathrm{Tr}\left[\varphi_e^{(n)}(i)\otimes\iota_B^{\otimes n}((\rho^{A,B})^{\otimes n})\right]\right).$$

Hence, the amount of assistance for sending information by the state $\rho^{A,B}$ can be quantified as[2]

$$C_a(\rho^{A,B}) \stackrel{\text{def}}{=} \sup \left\{ \overline{\lim} \frac{1}{n} \log |\Phi_e^{(n)}| \,\middle|\, \lim \varepsilon[\Phi_e^{(n)}] = 0 \right\}. \tag{9.32}$$

Then, we obtain the following theorem.

Theorem 9.4 *The quantity $\frac{1}{n} \min_\kappa H_{\kappa \otimes \iota_B((\rho^{A,B})^{\otimes n})}(A|B)$ converges as $n \to \infty$ and*

$$C_a(\rho^{A,B}) = -\lim \frac{1}{n} \min_\kappa H_{\kappa \otimes \iota_B((\rho^{A,B})^{\otimes n})}(A|B), \tag{9.33}$$

where κ is a TP-CP map from $\mathcal{H}_A^{\otimes n}$ to $\mathcal{H}_{A'_n}$ [56, 58, 234, 237, 419]. We assume that the output system $\mathcal{H}_{A'_n}$ can be chosen depending on κ.

When the initial state $\rho^{A,B}$ is a maximally correlated state, $\min_\kappa H_{\kappa \otimes \iota_B((\rho^{A,B})^{\otimes n})}(A|B) = nH_{\rho^{A,B}}(A|B)$, i.e., $C_a(\rho^{A,B}) = -H_{\rho^{A,B}}(A|B)$. Certainly, this equation holds when condition (8.120) is satisfied. In particular, if $\rho^{A,B}$ is a pure state, we have $C_a(\rho^{A,B}) = H(\rho^B)$.

Proof. We first show that

$$C_a(\rho^{A,B}) \geq H(\rho^B) - \min_\kappa H((\kappa \otimes \iota_B)(\rho^{A,B})) \tag{9.34}$$

in order to obtain the $\geq$ part of (9.33). Let κ_m be the channel $\operatorname{argmin}_\kappa H((\kappa \otimes \iota_B)(\rho^{A,B}))$. We denote the output system of κ_m and its dimension by $\mathcal{H}_{A'}$ and d, respectively. Now, we focus on the c-q channel $(i,j) \mapsto W_{(i,j)} \stackrel{\text{def}}{=} (\mathbf{X}_{A'}^i \mathbf{Z}_{A'}^j \otimes I_B)^* (\kappa \otimes \iota_B)(\rho^{A,B}) (\mathbf{X}_{A'}^i \mathbf{Z}_{A'}^j \otimes I_B)$ with the set of input signals $\mathcal{X} \stackrel{\text{def}}{=} \{(i,j)\}_{1 \leq i,j \leq d}$. Using Theorem 4.1 and Exercise 5.8, we see that the capacity of this channel is larger than

$$\begin{aligned}
& H\left(\sum_{(i,j)} \frac{1}{d^2} (\mathbf{X}_{A'}^i \mathbf{Z}_{A'}^j \otimes I_B)^* (\kappa \otimes \iota_B)(\rho^{A,B}) (\mathbf{X}_{A'}^i \mathbf{Z}_{A'}^j \otimes I_B) \right) \\
& \qquad - \sum_{(i,j)} \frac{1}{d^2} H\left((\mathbf{X}_{A'}^i \mathbf{Z}_{A'}^j \otimes I_B)^* (\kappa \otimes \iota_B)(\rho^{A,B}) (\mathbf{X}_{A'}^i \mathbf{Z}_{A'}^j \otimes I_B) \right) \\
& = H\left(\rho_{\text{mix}}^{A'} \otimes (\operatorname{Tr}_{A'} (\kappa \otimes \iota_B)(\rho^{A,B})) \right) - \sum_{(i,j)} \frac{1}{d^2} H\left((\kappa \otimes \iota_B)(\rho^{A,B}) \right) \\
& = H\left(\rho_{\text{mix}}^{A'} \otimes \operatorname{Tr}_A \rho^{A,B} \right) - H\left((\kappa \otimes \iota_B)(\rho^{A,B}) \right) \\
& = \log d + H\left(\operatorname{Tr}_A \rho^{A,B} \right) - H\left((\kappa \otimes \iota_B)(\rho^{A,B}) \right).
\end{aligned}$$

[2] The subscript a expresses "assistance."

From the definition of $|\Phi_e^{(n)}|$, we immediately obtain (9.34). Fixing n and applying the same argument to $\kappa_n \stackrel{\text{def}}{=} \operatorname{argmin}_\kappa H((\kappa \otimes \iota_B^{\otimes n})((\rho^{A,B})^{\otimes n}))$, we obtain $C_a(\rho^{A,B}) \geq H(\operatorname{Tr}_A \rho^{A,B}) - \frac{1}{n} \min_\kappa H((\kappa \otimes \iota_B^{\otimes n})((\rho^{A,B})^{\otimes n}))$. Therefore, we have $C_a(\rho^{A,B}) \geq H(\operatorname{Tr}_A \rho^{A,B}) - \inf_n \frac{1}{n} \min_\kappa H((\kappa \otimes \iota_B^{\otimes n})((\rho^{A,B})^{\otimes n}))$. Since $nH(\operatorname{Tr}_A \rho^{A,B}) - \min_\kappa H((\kappa \otimes \iota_B^{\otimes n})((\rho^{A,B})^{\otimes n}))$ satisfies the assumptions of Lemma A.1, this converges with $n \to \infty$. We therefore obtain (9.33) with the $\geq$ sign.

Next, we prove the $\leq$ part of (9.33). Let X be a random variable taking values in $\{1, \ldots, N_n\}$ and following a uniform distribution. Let Y be the decoded message at the receiver as the random variable taking values in $\{1, \ldots, N_n\}$. Since $H(X) = \log N_n$,

$$\begin{aligned} I(X:Y) &\geq H(X) - \log 2 - \varepsilon[\Phi^{(n)}] \log N_n \\ &= -\log 2 + \log N_n (1 - \varepsilon[\Phi_e^{(n)}]), \end{aligned} \tag{9.35}$$

from the Fano inequality. Using the monotonicity of the quantum relative entropy and (5.58), it can be shown that [Ex. 9.4]

$$I(X:Y) \leq nH\left(\operatorname{Tr}_A \rho^{A,B}\right) + \log \dim \mathcal{H}_{A'_n} - \min_\kappa H((\kappa \otimes \iota_B^{\otimes n})((\rho^{A,B})^{\otimes n})). \tag{9.36}$$

Combining this inequality with (9.35), we obtain

$$\begin{aligned} & H\left(\operatorname{Tr}_A \rho^{A,B}\right) - \frac{1}{n} \min_\kappa H((\kappa \otimes \iota_B^{\otimes n})((\rho^{A,B})^{\otimes n})) + \frac{\log 2}{n} \\ & \geq \frac{\log N_n}{n} (1 - \varepsilon[\Phi_e^{(n)}]) - \frac{\log \dim \mathcal{H}'_{A,n}}{n}. \end{aligned}$$

Taking the limit $n \to \infty$, we have

$$\begin{aligned} & H\left(\operatorname{Tr}_A \rho^{A,B}\right) - \overline{\lim} \frac{1}{n} \min_\kappa H((\kappa \otimes \iota_B^{\otimes n})((\rho^{A,B})^{\otimes n})) \\ & \geq \overline{\lim} \frac{\log N_n - \log \dim \mathcal{H}_{A'_n}}{n}, \end{aligned}$$

which gives the $\leq$ part of (9.33). ∎

We assumed above that there was no noise in the quantum channel. Since real quantum channels always contain some noise, we must often restrict our channel to a given TP-CP map κ. Now, consider the case in which the quantum channel κ has some noise, but the sender and the receiver are allowed access to any entangled state. Let us also say that the quantum channel κ can be used n times (i.e., $\kappa^{\otimes n}$), as considered previously.

First, we prepare an entangled pure state $x^{(n)}$ on the composite system $\mathcal{H}_{A'_n} \otimes \mathcal{H}_{R_n}$, comprising quantum system $\mathcal{H}_{A'_n}$ at the sender and quantum system $\mathcal{H}_{R_n}$ at the receiver. Let the size of the code be N_n, and let an element

$i \in \{1, \dots, N_n\}$ be transmitted. Next, the sender performs the operation $\varphi_e^{(n)}(i)$ from the system $\mathcal{H}_{A'_n}$ to the other system $\mathcal{H}_A^{\otimes n}$ depending on $i = 1, \dots, N_n$. Then, the state on $\mathcal{H}_A^{\otimes n}$ is transmitted to the receiver via the given quantum channel $\kappa^{\otimes n}$. The receiver performs a measurement $Y^{(n)}$ on the composite system $\mathcal{H}_B^{\otimes n} \otimes \mathcal{H}_{Rn}$, thereby recovering the original signal i. In this case, our code can be described by the set $(\mathcal{H}_{A'_n}, \mathcal{H}_{R_n}, x^{(n)}, N_n, \varphi_e^{(n)}, Y^{(n)})$, and is denoted by $\Phi_e^{(n),2}$. Hence, the size of the code and its error probability are given by

$$|\Phi_e^{(n),2}| \stackrel{\text{def}}{=} N_n, \quad \varepsilon[\Phi_e^{(n),2}] \stackrel{\text{def}}{=} \frac{1}{N_n} \sum_{i=1}^{N_n} \left(1 - \operatorname{Tr}\left[\varphi_e^{(n)}(i) \otimes \iota_{R,n}((\rho^{A,B})^{\otimes n})\right]\right).$$

The entanglement-assisted channel capacity $C_{c,e}^e(\kappa)$[3] is given by

$$C_{c,e}^e(\kappa) \stackrel{\text{def}}{=} \sup\left\{ \underline{\lim} \frac{1}{n} \log |\Phi_e^{(n),2}| \,\middle|\, \lim \varepsilon[\Phi_e^{(n),2}] = 0 \right\}. \tag{9.37}$$

Theorem 9.5 *(Bennett et al. [41, 42], Holevo [220]) The entanglement-assisted channel capacity $C_{c,e}^e(\kappa)$ of a quantum-quantum channel κ from $\mathcal{H}_A$ to $\mathcal{H}_B$ is*

$$C_{c,e}^e(\kappa) = \max_\rho I(\rho, \kappa), \tag{9.38}$$

where $I(\rho, \kappa)$ is the transmission information of a quantum-quantum channel defined in (8.34).

In a manner similar to $J(p, \sigma, W)$, we define $J(\rho, \sigma, \kappa)$ as

$$\begin{aligned} J(\rho, \sigma, \kappa) &\stackrel{\text{def}}{=} \operatorname{Tr}(\kappa \otimes \iota_R)(|x\rangle\langle x|)(\log(\kappa \otimes \iota_R)(|x\rangle\langle x|)) - \log \rho \otimes \sigma) \\ &= H(\rho) - \operatorname{Tr} \kappa(\rho) \log \sigma - H_e(\rho, \kappa) = \tilde{I}_c(\rho, \kappa) - \operatorname{Tr} \kappa(\rho) \log \sigma, \end{aligned}$$

where x is a purification of ρ. Then, $J(\rho, \sigma, \kappa)$ is concave for ρ because of (8.48) and convex for σ. Since

$$J(\rho, \sigma, \kappa) = I(\rho, \kappa) + D(\kappa(\rho) \| \sigma), \tag{9.39}$$

in a manner similar to (4.43), Lemma A.7 guarantess that

$$C_{c,e}^e(\kappa) = \max_\rho I(\rho, \kappa) = \max_\rho \min_\sigma J(\rho, \sigma, \kappa) = \min_\sigma \max_\rho J(\rho, \sigma, \kappa). \tag{9.40}$$

Proof. We first construct a code attaining the right-hand side (RHS) of (9.38), i.e., we prove the $\geq$ part in (9.38) for $\operatorname{argmax}_\rho I(\rho, \kappa) = \rho_{\text{mix}}^A$. Let

[3] The second subscript, e, of $C_{c,e}^e$ indicates the shared "entanglement." The superscript e indicates "entangled" operations between sending systems.

$\rho^{A,R}$ be the purification of ρ^A_{mix}. Perform the encoding operation using the operation $\rho^{A,R} \mapsto \rho^{A,R}_{(i,j)} \stackrel{\mathrm{def}}{=} (\mathbf{X}^i_A \mathbf{Z}^j_A \otimes I)(\rho^{A,R})(\mathbf{X}^i_A \mathbf{Z}^j_A \otimes I)^*$ at A, as in the case of a noise-free channel. Since

$$\sum_{i,j} \frac{1}{d^2}(\kappa \otimes \iota_R)(\rho^{A,R}_{(i,j)}) = (\kappa \otimes \iota_R)\Big(\sum_{i,j} \frac{1}{d^2}(\rho^{A,R}_{(i,j)})\Big)$$
$$= (\kappa \otimes \iota_R)(\rho^A_{\mathrm{mix}} \otimes \rho^R_{\mathrm{mix}}) = \kappa(\rho^A_{\mathrm{mix}}) \otimes \rho^R_{\mathrm{mix}},$$

we obtain

$$\sum_{i,j} \frac{1}{d^2} D\Big((\kappa \otimes \iota_R)(\rho^{A,R}_{(i,j)}) \Big\| \sum_{i,j} \frac{1}{d^2}(\kappa \otimes \iota_R)(\rho^{A,R}_{(i,j)})\Big)$$
$$= \sum_{i,j} \frac{1}{d^2} D((\kappa \otimes \iota_R)(\rho^{A,R}_{(i,j)}) \| \kappa(\rho^A_{\mathrm{mix}}) \otimes \rho^R_{\mathrm{mix}}) = I(\rho^A_{\mathrm{mix}}, \kappa).$$

Combining this equation with the argument given in Theorem 4.1, we find a code attaining $I(\rho^A_{\mathrm{mix}}, \kappa)$.

Now, consider the case in which $I(\rho^A_{\mathrm{mix}}, \kappa) = \max_\rho I(\rho, \kappa)$ is not true. Let $\rho^{\mathcal{K}_n}_{\mathrm{mix}}$ be a completely mixed state on a subspace $\mathcal{K}_n$ of $\mathcal{H}^{\otimes n}$. If we can take the state $\rho^{\mathcal{K}_n}_{\mathrm{mix}}$ such that

$$\lim \frac{1}{n} I(\rho^{\mathcal{K}_n}_{\mathrm{mix}}, \kappa^{\otimes n}) = \max_\rho I(\rho, \kappa), \tag{9.41}$$

we can construct a code satisfying $\max_\rho I(\rho, \kappa)$. To choose such a subspace $\mathcal{K}_n$, let $\rho_M \stackrel{\mathrm{def}}{=} \mathrm{argmax}_\rho I(\rho, \kappa)$, and we take the spectral decomposition $\rho_M^{\otimes n} = \sum_{j=1}^{v_n} \lambda_{j,n} E_{j,n}$, where v_n represents the number of eigenvalues of $\rho_M^{\otimes n}$. Let $\rho^{j,n}_{\mathrm{mix}}$ be a completely mixed state in the range of $E_{j,n}$, and let $p_{j,n} \stackrel{\mathrm{def}}{=} \lambda_{j,n} \operatorname{rank} E_{j,n}$. Then, we have $\rho_M^{\otimes n} = \sum_{j=1}^{v_n} p_{j,n} \rho^{j,n}_{\mathrm{mix}}$; therefore, $p_{j,n}$ is a probability distribution. Applying (8.44), we have

$$\sum_{j=1}^{v_n} p_{j,n} I(\rho^{j,n}_{\mathrm{mix}}, \kappa^{\otimes n}) + 2 \log v_n \geq I(\rho_M^{\otimes n}, \kappa^{\otimes n}).$$

Thus, there exists an integer $j_n \in [1, v_n]$ such that

$$I(\rho^{j_n,n}_{\mathrm{mix}}, \kappa^{\otimes n}) + 2 \log v_n \geq I(\rho_M^{\otimes n}, \kappa^{\otimes n}).$$

From Lemma 3.7, since $\frac{2}{n} \log v_n \to 0$, we obtain (9.41). This shows the existence of a code attaining the bound.

Next, we show that there is no code that exceeds the RHS of (9.38). Any pure state $\rho^{A',R}$ on $\mathcal{H}_{A'} \otimes \mathcal{H}_R$, a set of TP-CP maps $\{\varphi_e(j)\}$ from $\mathcal{H}_{A'}$ to $\mathcal{H}_A$, and a probability distribution p_j satisfy

$$\sum_j p_j D((\kappa\circ\varphi_e(j)\otimes\iota_R)(\rho^{A',R})\|\sum_j p_j(\kappa\circ\varphi_e(j)\otimes\iota_R)(\rho^{A',R}))$$
$$\leq \max_\rho I(\rho,\kappa), \tag{9.42}$$

where we used (4.6) and (8.50). From (8.45) we obtain

$$\max_\rho I(\rho,\kappa^{\otimes n}) = n \max_\rho I(\rho,\kappa). \tag{9.43}$$

Using the Fano inequality appropriately as in (9.35), we show the $\leq$ part in (9.38). ∎

Next, we examine the relation between the entanglement-assisted capacity $C^e_{c,e}(\kappa)$ and the capacity $C_c(\kappa)$. Let $\mathcal{H}_B$ be the output system of κ, $\mathcal{H}_R$ be the input system of κ, and $\mathcal{H}_R$ be a reference system of $\mathcal{H}_A$. From relation (8.142) the channel capacity $C_c(\kappa)$ can be expressed by the quantity $C_d^{R\to B}(\rho)$ as

$$C_c(\kappa) = \sup_{|x\rangle\langle x|} C_d^{R\to B}((\kappa\otimes\iota_R)(|x\rangle\langle x|)).$$

Hence, from (8.137)

$$C_d^{R\to B}((\kappa\otimes\iota_R)(|x\rangle\langle x|)) \leq I_{(\kappa\otimes\iota_R)(|x\rangle\langle x|)}(R:B), \tag{9.44}$$

i.e.,

$$C_c(\kappa) \leq C^e_{c,e}(\kappa). \tag{9.45}$$

Concerning the equality condition, the following lemma holds.

Theorem 9.6 *When channel κ is entanglement breaking and is written by a CONS $\{u_i^A\}$ on $\mathcal{H}_A$ as*

$$\kappa(\rho) = \sum_i \langle u_i^A|\rho|u_i^A\rangle \rho_i^B, \tag{9.46}$$

the equality of (9.45) holds. Conversely, when the equality of (9.45) holds, the channel essentially has the form of (9.46), i.e., there exists a state $\rho_{\max}$ such that $I(\rho_{\max},\kappa) = C^e_{c,e}(\kappa)$ and $\kappa|_{\mathrm{supp}(\rho_{\max})}$ has the form of (9.46), where $\mathrm{supp}(\rho_{\max})$ *is the support of $\rho_{\max}$. Further, in the case of (9.46), the capacity is calculated as*

$$C^e_{c,e}(\kappa) = \max_p H(\sum_i p_i\rho_i) - \sum_i p_i H(\rho_i). \tag{9.47}$$

There exists a counterexample that does not satisfy (9.46) while satisfying the equality of (9.45) (Exercise 9.8). From the above lemma we see that even if channel κ is entanglement breaking, the equality does not necessarily hold, i.e., it is advantageous to use shared entanglement. For example, channel κ is assumed to have the form

$$\kappa(\rho) = \sum_i (\operatorname{Tr} M_i \rho)|u_i^B\rangle\langle u_i^B|, \tag{9.48}$$

where $\{u_i^B\}$ is a CONS on $\mathcal{H}_B$ and $\boldsymbol{M} = \{M_i\}$ is a POVM one rank on $\mathcal{H}_A$. Then, the capacity $C_{c,e}^e(\kappa)$ is calculated as

$$\begin{aligned} C_{c,e}^e(\kappa) =& \sup_\rho H(\rho) + H(\sum_i (\operatorname{Tr} M_i \rho)|u_i^B\rangle\langle u_i^B|) \\ &- H(\sum_i \operatorname{Tr}_A (M_i \otimes I_R |x\rangle\langle x|) \otimes |u_i^B\rangle\langle u_i^B|) \\ =& \sup_\rho H(\rho) = \log d_A, \end{aligned}$$

where $|x\rangle\langle x|$ is a purification of ρ. However, when the POVM $\boldsymbol{M}$ is given as

$$M_0 = \frac{1}{2}|0\rangle\langle 0|,\ M_1 = \frac{1}{2}|1\rangle\langle 1|,\ M_2 = \frac{1}{2}|+\rangle\langle +|,\ M_3 = \frac{1}{2}|-\rangle\langle -|,$$

the capacity without shared entanglement is calculated as Ex. 9.7

$$C_c(\kappa) = C_c^e(\kappa) = \frac{1}{2}\log 2. \tag{9.49}$$

Proof of Theorem 9.6. Assume that condition (9.46) holds. Let $U_{\theta_1,\dots,\theta_{d_A}}$ be defined by $U_\theta \stackrel{\text{def}}{=} \sum_j e^{i\theta_j}|u_i^A\rangle\langle u_i^A|, \quad \theta = (\theta_1,\dots,\theta_{d_A})$. Then, channel κ has the invariance $\kappa(\rho) = \kappa(U_\theta \rho U_\theta^*)$. Hence, $I(\rho,\kappa) = I(U_\theta\rho U_\theta^*,\kappa)$. From (8.43)

$$I(\rho,\kappa) \le I(\int U_\theta \rho U_\theta^* \, d\theta, \kappa).$$

Since $\int U_\theta\rho U_\theta^* \, d\theta$ has eigenvectors $\{u_j^A\}$, we have

$$\begin{aligned} C_{c,e}^e(\kappa) =& \sup_p H\Big(\sum_j p_j |u_j^A\rangle\langle u_j^A|\Big) + H\Big(\sum_i \langle u_i^A|(\sum_j p_j |u_j^A\rangle\langle u_j^A|)|u_i^A\rangle \rho_i\Big) \\ &- H\Big(\sum_i (\operatorname{Tr}_A |u_i^A\rangle\langle u_i^A| \otimes I_R |x\rangle\langle x|) \otimes \rho_i\Big) \\ =& \sup_p H(p) + H\Big(\sum_i p_i\rho_i\Big) - H\Big(\sum_i p_i |u_i^R\rangle\langle u_i^R| \otimes \rho_i\Big) \\ =& \sup_p H(p) + H\Big(\sum_i p_i\rho_i\Big) - H(p) - \sum_i p_i H(\rho_i) \\ =& \sup_p H\Big(\sum_i p_i \rho_i\Big) - \sum_i p_i H(\rho_i), \end{aligned}$$

where $|x\rangle$ is a purification of $\sum_j p_j |u_j^A\rangle\langle u_j^A|$. Hence, we obtain (9.47). In particular, the capacity is equal to $C_c(\kappa)$. That is, the equality of inequality (9.45) holds.

Next, we assume that the equality of (9.45) holds. Then, there exists a state $\rho_{\max}$ such that $I(\rho_{\max}, \kappa) = C^e_{c,e}(\kappa)$ and its purification $|x\rangle$ satisfies the equality in (9.44). Lemma 8.12 guarantees that there exist a CONS $\{u_i^R\}$ on $\mathcal{H}_R$, states ρ_i on $\mathcal{H}_B$, and a probability distribution p such that

$$\sum_i p_i |u_i^R\rangle\langle u_i^R| \otimes \rho_i^B = (\kappa \otimes \iota_R)(|x\rangle\langle x|).$$

Now, we let ρ^R be the reduced density of $|x\rangle\langle x|$. Using relation (5.6),

$$(\kappa|_{\mathrm{supp}(\rho_{\max})} \otimes \iota_R)(|\Phi_d\rangle\langle\Phi_d|) = \sum_i dp_i \big(\sqrt{\rho^R}^{-1} |u_i^R\rangle\langle u_i^R| \sqrt{\rho^R}^{-1}\big) \otimes \rho_i^B,$$

where d is the dimension of $\mathrm{supp}(\rho_{\max})$. Since $\sum_i dp_i \sqrt{\rho^R}^{-1} |u_i^R\rangle\langle u_i^R| \sqrt{\rho^R}^{-1}$ is the completely mixed state on $\mathrm{supp}(\rho_{\max})$, each u_i^R is an eigenvector of ρ^R with the eigenvalue q_i. Hence,

$$(\kappa|_{\mathrm{supp}(\rho_{\max})} \otimes \iota_R)(|\Phi_d\rangle\langle\Phi_d|) = \sum_i \frac{dp_i}{q_i} |u_i^R\rangle\langle u_i^R| \otimes \rho_i^B.$$

From the discussion in Theorem 5.1 we can conclude that channel κ has the form (9.46). ∎

Exercises

9.4. Show (9.36) using (5.58) and the monotonicity of the quantum relative entropy.

9.5. Show (9.42) using (8.50) and the inequality $D(\sum_j p_j (\kappa \circ \varphi_e(j) \otimes \iota_R)(\rho^{A',R}) \| \sum_j p_j (\kappa \circ \varphi_e(j)(\rho^{A'}) \otimes \rho^R) \geq 0$.

9.6. Show that the $\leq$ part of (9.38) by combining (9.42) and (9.43) with the Fano inequality.

9.7. Show equation (9.49) following the steps below.

a Show that $C_c(\kappa) = \log 4 - \min_\theta f(\theta)$, where $f(\theta) \stackrel{\mathrm{def}}{=} -\frac{1+\cos\theta}{4} \log \frac{1+\cos\theta}{4} - \frac{1-\cos\theta}{4} \log \frac{1-\cos\theta}{4} - \frac{1+\sin\theta}{4} \log \frac{1+\sin\theta}{4} - \frac{1-\sin\theta}{4} \log \frac{1-\sin\theta}{4}$.

b Show that $\frac{df}{d\theta}(\theta) = \frac{\sin\theta}{4} \log \frac{1+\cos\theta}{1-\cos\theta} + \frac{\cos\theta}{4} \log \frac{1-\sin\theta}{1+\sin\theta}$ and $\frac{d^2 f}{d\theta^2}(\theta) = \frac{\cos\theta}{4} \log \frac{1+\cos\theta}{1-\cos\theta} + \frac{\sin\theta}{4} \log \frac{1+\sin\theta}{1-\sin\theta} - 1$.

c Show the following table.

θ	0		$\frac{\pi}{4}$		$\frac{\pi}{2}$
$f(\theta)$	$\frac{3}{2}\log 2$	$\nearrow$	$*$	$\searrow$	$\frac{3}{2}\log 2$
$\frac{df}{d\theta}(\theta)$	0	$\curvearrowright$	0	$\curvearrowleft$	0
$\frac{d^2 f}{d\theta^2}(\theta)$	$+\infty$	$\searrow$	$-$	$\nearrow$	$+\infty$

9.8. Let κ_1 and κ_2 be channels from systems $\mathcal{H}_{A,1}$ and $\mathcal{H}_{A,2}$ to system $\mathcal{H}_B$, respectively. Assume that the states $\rho_{\max,1} \stackrel{\rm def}{=} \operatorname{argmax}_\rho I(\rho,\kappa_1)$ and $\rho_{\max,2} \stackrel{\rm def}{=} \operatorname{argmax}_\rho I(\rho,\kappa_2)$ satisfy that $\kappa_1(\rho_{\max,1}) = \kappa_2(\rho_{\max,2})$. Define channel κ from system $\mathcal{H}_{A,1} \oplus \mathcal{H}_{A,2}$ ($\mathcal{H}_{A,i} \cong \mathcal{H}_A$) to system $\mathcal{H}_B$ as $\kappa(\rho) = \kappa_1(P_1 \rho P_1) + \kappa_2(P_2 \rho P_2)$, where P_i is the projection to $\mathcal{H}_{A,i}$. Show that $C^e_{c,e}(\kappa_1) = C^e_{c,e}(\kappa)$ using (9.40).

Further, show the equality of (9.45), i.e., $C_c(\kappa) = C^e_{c,e}(\kappa)$, even though κ_2 does not satisfy (9.46) if κ_1 satisfies (9.46).

9.4 Quantum Channel Resolvability

In this section, we examine the problem of approximating a given quantum state on the output of a c-q channel. In this problem, we choose a finite number of input signals and approximate a desired quantum state by the average output state with the uniform distribution on the chosen input signals. Then, the task of this problem is to choose the support of the uniform distribution at the input system as small as possible while approximating the desired state by the average output state as accurately as possible.

The classical version of this problem is called *channel resolvability*. It was proposed by Han and Verdú [161] in order to examine another problem called the *identification code* given by Ahlswede and Dueck [5]. The problem of approximating a quantum state at the output system of a c-q channel is analogously called *quantum-channel resolvability*. Hence, quantum-channel resolvability is expected to be useful for examining identification codes [6] for (classical-) quantum channels. Indeed, this problem essentially has been treated by Wyner [425] in order to evaluate the information of the eavesdropper. Hence, it is also a fundamental tool for the discussion of communications in the presence of an eavesdropper for the following reason. Regarding the channel connecting the sender to the eavesdropper, approximating two states on the output system is almost equivalent to making these two states indistinguishable for the eavesdropper. Its detail will be discussed in the next section.

Quantum-channel resolvability may be formulated as follows. Consider a c-q channel $W : \mathcal{X} \to \mathcal{S}(\mathcal{H})$ and a quantum state $\sigma \in W(\mathcal{P}(\mathcal{X}))$, and prepare a map φ from $\{1,\ldots,M\}$ to the alphabet set $\mathcal{X}$. Now, the sender chooses an element i of $\{1,\ldots,M\}$ according to the uniform distribution and sends the state $W_{\varphi(i)}$. The problem is then to detrmine how many (M) elements are required for sufficienltly approximating the quantum state σ by the output average state $W_\varphi \stackrel{\mathrm{def}}{=} \frac{1}{M}\sum_{j=1}^M W_{\varphi(j)}$ of the c-q channel W. (Here, we are allowed to use input elements duplicately.) The quality of the approximation is evaluated by the trace norm $\|W_\varphi - \sigma\|_1$. Here, we choose the trace norm as the criterion of the approximation because it represents how well two states can be discriminated, as seen in Lemma 3.1. If the number M is sufficiently large, we can easily approximate the state $W_p = \sigma$ by the output average state W_φ. However, our aim is to approximate the state $W_p = \sigma$ with a small number M. One of the features of this problem is that even when the distribution $p_\varphi(x) = \#\{\varphi^{-1}\{x\}\}/M$ at the input system is not close to p, the state $\sigma = W_p$ can be approximated by $W_{p_\varphi} = W_\varphi$ using the noise of channel W. In this case, our protocol is represented by $\Phi \stackrel{\mathrm{def}}{=} (M, \varphi)$, and its performance by $M = |\Phi|$ and $\varepsilon[\Phi] \stackrel{\mathrm{def}}{=} \left\| \left(\frac{1}{M}\sum_{j=1}^M W_{\varphi(j)} \right) - \sigma \right\|_1$. Let $\Phi^{(n)}$ be a protocol producing approximately the tensor product state $\sigma^{\otimes n}$ for a given state $\sigma \in W(\mathcal{P}(\mathcal{X}))$ by using the n-fold stationary memoryless channel $W^{(n)}$ of W. The asymptotic bound of its performance is given as the quantum-channel resolvability capacity;[4]

$$C_r(W, \sigma) \stackrel{\mathrm{def}}{=} \inf_{\{\Phi^{(n)}\}} \left\{ \overline{\lim} \frac{1}{n} \log |\Phi^{(n)}| \,\middle|\, \lim \varepsilon[\Phi^{(n)}] = 0 \right\}. \tag{9.50}$$

Theorem 9.7 *The quantum-channel resolvability capacity $C_r(W, \sigma)$ satisfies*

$$C_r(W, W_p) \le I(p, W). \tag{9.51}$$

We first prove the following lemma in order to prove this theorem.[5]

Lemma 9.2 *There exists a map φ from $\{1, \ldots, M\}$ to $\mathcal{X}$ satisfying*

$$\begin{aligned} &\left\| \left(\frac{1}{M} \sum_{i=1}^M W_{\varphi(i)} \right) - W_p \right\|_1 \\ &\le 2\sqrt{\sum_x p(x) \operatorname{Tr} W_x \{\kappa_{W_p}(W_x) - C W_p \ge 0\}} + \sqrt{\frac{Cv}{M}}, \end{aligned} \tag{9.52}$$

where v is the number of eigenvalues of W_p. Furthermore, for $s \le 0$,

$$\sum_x p(x) \operatorname{Tr} W_x \{\kappa_{W_p}(W_x) - C W_p \ge 0\} \le C^s \exp(\phi(s|W, p)) \tag{9.53}$$

$$\phi(s|W, p) \stackrel{\mathrm{def}}{=} \log \left(\sum_x p(x) \operatorname{Tr} W_x^{1-s} W_p^s \right),$$

---

[4] The subscript r indicates “resolvability”.

[5] This theorem was first derived in this text. Lemma 9.2 is also a new result for the classical case.

where κ_σ is the pinching map concerning the matrix σ, which is defined in (1.12).

Proof. We prove (9.52) by emplying the random coding method. Let $X \stackrel{\rm def}{=} (x_1, \dots, x_M)$ be M independent random variables subject to a probability distribution p in $\mathcal{X}$. Consider a protocol (M, φ) such that $\varphi(i) = x_i$. Denoting the expectation value by E_X, we obtain

$$\mathrm{E}_X \left\| \left(\frac{1}{M} \sum_{i=1}^{M} W_{x_i} \right) - W_p \right\|_1 \\ \le 2 \sqrt{\sum_x p(x) \operatorname{Tr} W_x \{ \kappa_{W_p}(W_x) - C W_p \ge 0 \}} + \sqrt{\frac{Cv}{M}}, \tag{9.54}$$

which will be shown in what follows. Now, define $P_x \stackrel{\rm def}{=} \{\kappa_{W_p}(W_x) - C W_p \ge 0\}$, $P_x^c \stackrel{\rm def}{=} I - P_x$, and $W_p' \stackrel{\rm def}{=} \sum_x p(x) P_x^c W_x$. Since $W_p' - W_p = \sum_x p(x) P_x W_x$, we have

$$\begin{aligned} &\mathrm{E}_X \left\| \left(\frac{1}{M} \sum_{i=1}^{M} W_{x_i} \right) - W_p \right\|_1 \\ &= \mathrm{E}_X \left\| \left(\frac{1}{M} \sum_{i=1}^{M} P_{x_i} W_{x_i} \right) + (W_p' - W_p) + \left(\frac{1}{M} \sum_{i=1}^{M} P_{x_i}^c W_{x_i} \right) - W_p' \right\|_1 \\ &\le \mathrm{E}_X \left\| \frac{1}{M} \sum_{i=1}^{M} (P_{x_i}^c W_{x_i} - W_p') \right\|_1 + \mathrm{E}_x \| P_x W_x \|_1 \\ &\quad + \mathrm{E}_X \left[\frac{1}{M} \sum_{i=1}^{M} \| P_{x_i} W_{x_i} \|_1 \right], \end{aligned} \tag{9.55}$$

where E_x denotes the expectation value for a random variable x. Hence, Exercise 6.5 implies

$$\| P_x W_x \|_1 \le \sqrt{\operatorname{Tr} W_x P_x}. \tag{9.56}$$

Thus, Exercise 6.6 yields

$$\begin{aligned} &\left\| \frac{1}{M} \sum_{i=1}^{M} (P_{x_i}^c W_{x_i} P_{x_i}^c - W_p') \right\|_1 \\ &\le \frac{\sqrt{\sum_{i,j} \operatorname{Tr} W_p^{-1} \left(P_{x_i}^c W_{x_i} - W_p' \right) \left(W_{x_j} P_{x_j}^c - W_p'^{*} \right)}}{M}. \end{aligned}$$

Using these relations as well as Jensen's inequality for $x \mapsto -\sqrt{x}$, we obtain

$$
\begin{aligned}
& \mathrm{E}_X \left\| \frac{1}{M} \sum_{i=1}^{M} (P^c_{x_i} W_{x_i} - W'_p) \right\|_1 + \mathrm{E}_x \|P_x W_x\|_1 \\
& \quad + \mathrm{E}_X \left(\frac{1}{M} \sum_{i=1}^{M} \|P_{x_i} W_{x_i}\|_1 \right) \\
& \le \frac{\sqrt{\mathrm{E}_X \sum_{i,j} \mathrm{Tr}\, W_p{}^{-1} (P^c_{x_i} W_{x_i} - W'_p) \left(W_{x_j} P^c_{x_j} - W'^*_p\right)}}{M} \\
& \quad + \sqrt{\mathrm{E}_X \left(\frac{1}{M} \sum_{i=1}^{M} \mathrm{Tr}\, W_{x_i} P_{x_i} \right)} + \sqrt{\mathrm{E}_x \,\mathrm{Tr}\, W_x P_x} \\
& = \sqrt{\frac{\mathrm{E}_x \,\mathrm{Tr}\, W_p{}^{-1} (P^c_x W_x - W'_p)(W_x P^c_x - W'^*_p)}{M}} \\
& \quad + 2\sqrt{\mathrm{E}_x \,\mathrm{Tr}\, W_x P_x}.
\end{aligned}
\tag{9.57}
$$

The last equality follows from the fact that x_i and x_j for different i and j are independent and the expectation value of $\left(P^c_x W_x - W'_p\right)$ is a zero matrix. From the definition of W'_p,

$$
\begin{aligned}
& \sum_x p(x)\, \mathrm{Tr}\, W_p{}^{-1} (P^c_x W_x - W'_p)(W_x P^c_x - W'^*_p) \\
& = \sum_x p(x)\, \mathrm{Tr}\, \sqrt{W_p}^{-1} P^c_x W_x^2 P^c_x - \mathrm{Tr}\, W_p{}^{-1} W'_p W'^*_p \\
& \le \sum_x p(x)\, \mathrm{Tr}\, W_p{}^{-1} P^c_x W_x^2 P^c_x = \sum_x p(x)\, \mathrm{Tr}\, W_x W_x^{\frac{1}{2}} P^c_x W_p{}^{-1} P^c_x W_x^{\frac{1}{2}} \\
& \le \sum_x p(x)\, \mathrm{Tr}\, W_x \|W_x^{\frac{1}{2}} P^c_x W_p{}^{-1} P^c_x W_x^{\frac{1}{2}}\| = \sum_x p(x) \|W_x^{\frac{1}{2}} P^c_x W_p{}^{-1} P^c_x W_x^{\frac{1}{2}}\|.
\end{aligned}
\tag{9.58}
$$

Since W_p and $\kappa_{W_p}(W_x)$ commute, W_p and $P^c_x = \{\kappa_{W_p}(W_x) - CW_p < 0\}$ also commute. Therefore, we obtain

$$
P^c_x W_x P^c_x \le v P^c_x \kappa_{W_p}(W_x) P^c_x \le v P_x C W_p P_x \le C v W_p.
$$

Hence,

$$
\begin{aligned}
& \|W_x^{\frac{1}{2}} P^c_x W_p{}^{-1} P^c_x W_x^{\frac{1}{2}}\| = \|W_x^{\frac{1}{2}} P^c_x W_p{}^{-\frac{1}{2}} W_p{}^{-\frac{1}{2}} P^c_x W_x^{\frac{1}{2}}\| \\
& = \|W_p{}^{-\frac{1}{2}} P^c_x W_x^{\frac{1}{2}} W_x^{\frac{1}{2}} P^c_x W_p{}^{-\frac{1}{2}}\| = \|W_p{}^{-\frac{1}{2}} P^c_x W_x P^c_x W_p{}^{-\frac{1}{2}}\| \\
& \le \|W_p{}^{-\frac{1}{2}} C v W_p W_p{}^{-\frac{1}{2}}\| = Cv.
\end{aligned}
$$

Thus, (9.54) may be shown using (9.58), (9.57), and (9.55). Then, inequality (9.52) follows from (9.54). In addition, (9.53) may be shown using Exercises 3.16 and 9.9. ∎

Proof of Theorem 9.7. Let $C = e^{n(R-r)}$ and $M = e^{nR}$, and apply Lemma 3.7, (9.52), and (9.53) to $W^{(n)}$ and p^n. Therefore, there exists a protocol $\Phi^{(n)}$ such that

$$|\Phi^{(n)}| = e^{nR} \varepsilon[\Phi^{(n)}] \le 2e^{n(\phi(s|W,p)+s(R-r))/2} + (n+1)^{d/2} e^{-nr/2},$$

where $\phi(s|W^{(n)}, p^n) = n\phi(s|W,p)$, and s is a real negative number. Now, let $R < I(p,W)$. Then, there exist real numbers $s < 0$ and $r > 0$ such that $\phi(s|W,p) + s(R-r) < 0$ because $\lim_{s\to 0} \frac{\phi(s|W,p)}{s} = I(p,W)$. Since $\varepsilon[\Phi^{(n)}]$ approaches 0, we obtain (9.51). ∎

It has been shown that if channel W is a classical channel, then $\max_p C_r(W, W_p)$ is equal to $\max_p I(p,W)$. This is done by examining the connection with the problem of identification codes [6]. The same fact seems to hold for a c-q channel W using a similar method.

Here, we prove the direct part of Theorem 8.12 by using quantum-channel resolvability.

Proof of Direct Part of Theorem 8.12: Choose a probabilistic decomposition $(p_x, \rho_x^A \otimes \rho_x^B)_{x\in\mathcal{X}}$ of the separable state ρ as

$$C(\rho) = I_{\rho^{ABE}}(AB : E), \quad \rho^{ABE} \stackrel{\text{def}}{=} \sum_i p_i \rho_i^A \otimes \rho_i^B \otimes |u_i^E\rangle\langle u_i^E|. \tag{9.59}$$

For any $\epsilon > 0$, we let $M_n = e^{n(C(\rho)+\epsilon)}$. Hence, we can choose M_n indexes $\varphi(1), \dots, \varphi(M_n)$ in $\mathcal{X}^n$ such that

$$\left\| \frac{1}{M_n} \sum_{i=1}^{M_n} \rho_{\varphi(i)}^{A,(n)} \otimes \rho_{\varphi(i)}^{B,(n)} - \rho^{\otimes n} \right\|_0 \to 0,$$

which implies the direct part of Theorem 8.12. ∎

Exercises

9.9. Prove (9.53) by applying the monotonicity of relative Rényi entropy (2.63) to the two-valued POVM $\{\{\kappa_{W_p}(W_x) - CW_p \ge 0\}, \{\kappa_{W_p}(W_x) - CW_p < 0\}\}$.

9.10. Show that

$$\sup_{\{\Phi^{(n)}\}} \left\{ \lim \frac{-\log \varepsilon[\Phi^{(n)}]}{n} \middle| \lim \frac{\log |\Phi^{(n)}|}{n} \le R \right\} \ge \max_{s\le 0} \frac{-\phi(s|W,p) - sR}{2(1-s)},$$

referring to the proof of Theorem 9.7.

9.11. Assume that all the W_x output states commute. Show that

$$\left\| \left(\frac{1}{M} \sum_{i=1}^{M} W_{\varphi(i)} \right) - W_p \right\|_1 \\ \leq 2 \sum_x p(x) \operatorname{Tr} W_x \{ \kappa_{W_p}(W_x) - C W_p \geq 0 \} + \sqrt{\frac{C}{M}}$$

as a modification of (9.52) of Lemma 9.2.

9.5 Quantum-Channel Communications with an Eavesdropper

9.5.1 C-Q Wiretap Channel

The *BB84 protocol* [35] enables us to securely distribute a secret key using a quantum system. Experiments realizing this protocol have been performed [255, 383], with successful transmissions over 150 km [260, 368] via optical fibers. Therefore, the protocol is almost at a practically usable stage. In the original proposal of the BB84 protocol, it was assumed that there was no noise in the channel. However, a real channel always has some noise. In the presence of noise, the noise can be used by an eavesdropper to mask his/her presence while obtaining information from the channel. Therefore, it is necessary to communicate on the assumption that an eavesdropper may obtain a certain amount of information.

This type of communication is called a *wiretap channel* and was first considered by Wyner [425] for the classical case. Its quantum-mechanical extension, i.e., a *classical-quantum wiretap channel* (*c-q wiretap channel*) was examined by Devetak [95]. In this communication, we require a code such that the normal receiver can accurately recover the original message and the eavesdropper cannot obtain any information concerning the original message. Hence, one of the main problems in this communication is to find the bound of the communication rate of the code. Although this problem is not the same problem as the BB84 protocol itself, it will lead us to a proof of its security even in the presence of noise.

Let $\mathcal{H}_B$ be the system received by the normal receiver, $\mathcal{H}_E$ the system received by the eavesdropper, and W_x an output state on the composite system $\mathcal{H}_B \otimes \mathcal{H}_E$ when the sender sends an alphabet $x \in \mathcal{X}$. Hence, the normal receiver receives the state $W_x^B \stackrel{\text{def}}{=} \operatorname{Tr}_E W_x$, and the eavesdropper receives the state $W_x^E \stackrel{\text{def}}{=} \operatorname{Tr}_B W_x$. In this case, we use a probabilistic code as follows. When the sender wishes to send a message $m \in \{1, \ldots, M\}$, he or she transmits the alphabet $x \in \mathcal{X}$ according to the probability distribution Q^m in $\mathcal{X}$ depending on message m. That is, the encoding process is described by a stochastic transition matrix Q from $\{1,\ldots,M\}$ to $\mathcal{X}$. Then, the normal receiver performs the M-valued POVM $Y=\{Y_{m'}\}_{m'=1}^{M}$ and receives signal m'. Therefore, our protocol is described by $\Phi=(M,Q,Y)$ and evaluated by the following three quantities. The first quantity is the size of the protocol $|\Phi| \stackrel{\rm def}{=} M$, and the second is the upper bound of the eavesdropper's information $I_E(\Phi) \stackrel{\rm def}{=} I(p_{\rm mix}^M, W^E Q)$, where $(W^E Q)_m \stackrel{\rm def}{=} \sum_{x\in\mathcal{X}} W_x^E Q_x^m$. The third quantity is the error probability of the normal receiver $\varepsilon[\Phi] \stackrel{\rm def}{=} \frac{1}{M}\sum_{m=1}^{M}\left(1-\mathrm{Tr}(W^B Q)_m Y_m\right)$, where $(W^B Q)_m \stackrel{\rm def}{=} \sum_{x\in\mathcal{X}} W_x^B Q_x^m$. Let us now examine the capacity, i.e., the bound of the communication rate $\frac{1}{n}\log|\Phi^{(n)}|$, for the asymptotically reliable protocol $\{\Phi^{(n)}\}$ with the stationary memoryless channel $W^{(n)}$, i.e., n times use of W. This is given by

$$C_c^{B,E}(W) \stackrel{\rm def}{=} \sup_{\{\Phi^{(n)}\}} \left\{ \underline{\lim}\, \frac{1}{n}\log|\Phi^{(n)}| \,\middle|\, \varepsilon[\Phi^{(n)}]\to 0, I_E(\Phi^{(n)})\to 0 \right\}. \tag{9.60}$$

Theorem 9.8 *(Devetak [95]) The capacity $C_c^{B,E}(W)$ satisfies*

$$C_c^{B,E}(W) = \overline{\lim}\, \frac{1}{n} \sup_Q \sup_p \left(I(p, W_B^{(n)} Q) - I(p, W_E^{(n)} Q) \right). \tag{9.61}$$

If every W_x^E can be written as $W_x^E = \kappa(W_x^B)$, using a completely positive map κ from $\mathcal{H}_B$ to $\mathcal{H}_E$, it is called a *quantum degraded channel*, and it satisfies

$$C_c^{B,E}(W) = \sup_p \left(I(p, W^B) - I(p, W^E) \right). \tag{9.62}$$

It is also proved in Sect. 9.5.6. Further, a quantum degraded channel (W^B, W^E) satisfies Ex. 9.19

$$I(Qp, W^B) - I(Qp, W^E) \geq \sum_i p_i (I(Q^i, W^B) - I(Q^i, W^E)). \tag{9.63}$$

That is, $I(p, W^B) - I(p, W^E)$ satisfies the concavity in this case.

Let us suppose that W^B is given by a TP-CP map κ from $\mathcal{H}_A$ to $\mathcal{H}_B$, and the channel to the eavesdropper W^E is given by a channel κ_E to the environment of κ. Under this assumption, the eavesdropper's state is always a state reduced from the state on the environment. That is, he/she has less information than the environment system. Hence, the eavesdropper's information can be sufficiently estimated by treating the case where the eavesdropper's state is equal to the state on the environment. Now, we consider the set of input signals $\mathcal{X}$ given as the set of pure states on the input system. Then, for any input pure state $|x\rangle$, the states $W_x^{B,(n)}$ and $W_x^{E,(n)}$ are given by $\kappa^{\otimes n}(|x\rangle\langle x|)$ and $(\kappa^E)^{\otimes n}(|x\rangle\langle x|)$, respectively. In this scheme, any entangled state is allowed as the input state. From $H(W_x^B) = H(W_x^E)$ and (8.53), any state $\rho = \sum_i p_i |u_i\rangle\langle u_i|$ satisfies

$$
\begin{aligned}
& I(p, W^B) - I(p, W^E) \\
& = H(\kappa(\rho)) - \sum_x p_x H(W_x^B) - \left(H(\mathrm{Tr}_B(U_\kappa \rho U_\kappa^*)) - \sum_x p_x H(W_x^E) \right) \\
& = H(\kappa(\rho)) - H(\mathrm{Tr}_B(U_\kappa \rho U_\kappa^*)) = I_c(\rho, \kappa). \qquad (9.64)
\end{aligned}
$$

Hence, letting $C_c^{e,B,E}(\kappa)$ be the asymptotic bound of the communication rate when entangled states are used, we can show that

$$
C_c^{e,B,E}(\kappa) \geq \lim \frac{1}{n} \max_{\rho \in \mathcal{S}(\mathcal{H}_A^{\otimes n})} I_c(\rho, \kappa^{\otimes n}). \qquad (9.65)
$$

In addition, the following *monotonicity* also holds with respect to the eavesdropper's information:

$$
I(\{p_i\}, \{\kappa_E' \circ \kappa(\rho_i)\}) \leq I(\{p_i\}, \{(\kappa' \circ \kappa)_E(\rho_i)\}), \qquad (9.66)
$$

$$
I(\{p_i\}, \{\kappa_E(\rho_i)\}) \leq I(\{p_i\}, \{(\kappa' \circ \kappa)_E(\rho_i)\}). \qquad (9.67)
$$

9.5.2 Relation to BB84 Protocol

Let us now relate these arguments to the BB84 protocol discussed earlier. In the BB84 protocol, the sender A transmits a state chosen from e_0, e_1, $e_+ \stackrel{\text{def}}{=} \frac{1}{\sqrt{2}}(e_0 + e_1)$, and $e_- \stackrel{\text{def}}{=} \frac{1}{\sqrt{2}}(e_0 - e_1)$ with an equal probability. The receiver B then chooses one of the two measurement bases $\{|e_0\rangle\langle e_0|, |e_1\rangle\langle e_1|\}$ and $\{|e_+\rangle\langle e_+|, |e_-\rangle\langle e_-|\}$ with an equal probability and performs this measurement on the received quantum system. Then, the normal receiver B sends his/her measurement data to the sender A via a public channel. The sender A tells the normal receiver B whether the original state belongs to the set $\{e_0, e_1\}$ or $\{e_+, e_-\}$ via a public channel. This determines whether the basis used by the sender A coincides with the basis by the normal receiver B. The bases should agree for approximately half number of the transmitted states, which is numbered by n. They choose ϵn bits randomly among these obtained n bits and announce the information of these ϵn bits using the public channel in order to verify whether these bits coincide with each other (ϵ is a suitably chosen positive real number). When they find a bit with disagreement, the sender A and the normal receiver B conclude that an eavesdropper was present. Otherwise, both parties can conclude that they succeeded in sharing a secret key X without divulging information to any third party. Finally, the sender encrypts the information Y_A that he or she wishes to send according to $Z = X + Y_A \pmod 2$. This may be decrypted according to $Y_B = Z + X$, thereby obtaining secure communication.

In reality, the bits held by A and B may not agree due to noise even if an eavesdropper is not present. In this case, we must estimate the quantum channel κ connecting the sender to the receiver based on ϵn bits of information. Consider a case in which the sender sends bits based on the basis $\{e_0, e_1\}$, and the receiver detects the bits through the measurement $E = \{|e_i\rangle\langle e_i|\}_{i=0}^1$. Now, let X_A and X_B be the random bits sent by the sender and by the normal receiver through the measurement, respectively. When the sender transmits bit i, the normal receiver obtains his/her data subject to the distribution $\mathrm{P}^E_{\kappa(e_i)}$. By performing the communication steps as described above, the stochastic transition matrix Q joining Y_A and Y_B is given by

$$Q_0^0 = Q_1^1 = \frac{1}{2}\mathrm{P}^E_{\kappa(e_0)}(0) + \frac{1}{2}\mathrm{P}^E_{\kappa(e_1)}(1),\ Q_1^0 = Q_0^1 = \frac{1}{2}\mathrm{P}^E_{\kappa(e_0)}(1) + \frac{1}{2}\mathrm{P}^E_{\kappa(e_1)}(0),$$

which is the same as that for a noisy classical channel. Using a suitable coding protocol, the sender and the normal receiver can communicate with almost no error and almost no information leakage.

Let us now estimate the amount of information leaked to the eavesdropper. In this case, it is impossible to distinguish the eavesdropping from the noise in the channel. For this reason, we assume that any information lost has been caused by the interception by the eavesdropper. Consider the case in which each bit is independently eavesdropped, i.e., the quantum channel from the state inputted by the sender to the state intercepted by the eavesdropper is assumed to be stationary memoryless. Therefore, if the sender transmits the state e_i, the eavesdropper obtains the state $\kappa^E(e_i)$, where κ^E was defined in (5.5). Since the eavesdropper knows Z, he/she possesses the state on the composite system $\mathcal{H}_E \otimes \mathbb{C}^2$ consisting of the quantum system $\mathcal{H}_E$ and the classical system $\mathbb{C}^2$ corresponding to Z. For example, if $Y_A = i$, the state W_i^E obtained by the eavesdropper is

$$W_0^E = \begin{pmatrix} \frac{1}{2}\kappa^E(e_0) & 0 \\ 0 & \frac{1}{2}\kappa^E(e_1) \end{pmatrix},\ W_1^E = \begin{pmatrix} \frac{1}{2}\kappa^E(e_1) & 0 \\ 0 & \frac{1}{2}\kappa^E(e_0) \end{pmatrix}.$$

We may therefore reduce this problem to the c-q wiretap channel problem discussed previously [98]. In particular, if κ is a Pauli channel κ_p^{GP},

$$I(p_{\mathrm{mix}}, Q) - I(p_{\mathrm{mix}}, W^E) = 1 - H(p), \tag{9.68}$$

which is a known quantity in quantum key distribution [274].

In practice, it is not possible to estimate κ completely using communications that use only e_0, e_1, e_+, e_-. However, it is possible to estimate $I(p, W^E)$.[6] Since the encoding constructed in the proof of Theorem 9.8 depends on the form of W^E, it is desirable to construct a protocol that depends only on the value of $I(p, W^E)$.

---

[6] By adding the states $e_+^* \stackrel{\mathrm{def}}{=} \frac{1}{\sqrt{2}}(e_0 + ie_1)$ and $e_-^* \stackrel{\mathrm{def}}{=} \frac{1}{\sqrt{2}}(e_0 - ie_1)$ in the transmission, and by adding the measurement $\{|e_+^*\rangle\langle e_+^*|, |e_-^*\rangle\langle e_-^*|\}$, it is possible to estimate κ. This is called the six-state method [33, 60, 146].

9.5.3 Secret Sharing

Let us consider an application of the above discussion to a protocol called *secret sharing.* In secret sharing, there are m receivers, and the encoded information sent by the sender can be recovered only by combining the information of m receivers. Therefore, a single receiver cannot obtain the encoded information [54, 365].

Denote the channel from the sender to each receiver by W [81, 151]. The transmission information possessed by one receiver is equal to $I(p, W)$. The transmission information possessed by m receivers is therefore $mI(p, W)$. Theorem 9.8 guarantees that performing the communication n times, the sender can transmit almost $n(m-1)I(p, W)$ bits of information with no leakage to each receiver. That is, the problem is to ensure that the information possessed by an individual receiver approaches zero asymptotically. The random coding method used in the proof of Lemma 9.3 may be used to show the existence of such a code. Let $I_i(\Phi_X)$ be the information possessed by the ith receiver for the code Φ_X. Let $\varepsilon[\Phi_X]$ be the average decoding error probability of combining the m receivers. Then, $\mathrm{E}_X[\varepsilon[\Phi_X]]$ satisfies (9.78), and it can be shown that $\sum_{i=1}^m \mathrm{E}_X[I_i(\Phi_X)] \le m(\epsilon_2 \log d + \eta_0(\epsilon_2))$. Therefore, $\mathrm{E}_X[\varepsilon[\Phi_X]]$ satisfies (9.78), and we can show that there exists a code Φ such that $\varepsilon[\Phi] + \sum_{i=1}^m I_i(\Phi_X) \le \epsilon_1 + m(\epsilon_2 \log d + \eta_0(\epsilon_2))$. Therefore, it is possible to securely transmit $n(m-1)I(p, W)$ bits of information asymptotically. Further, we can consider the capacity with the following requirement: There are m receivers, and the information can be recovered from composite quantum states by any n_1 receivers. However, it can be recovered not only by n_2 receivers. In this case, the capacity is $(n_1 - n_2)C(W)$. It can be shown by the combination of the proofs of Corollary 4.1 and Theorem 9.8.

9.5.4 Distillation of Classical Secret Key

In addition, this approach can be applied to the distillation of a classical secret key from shared state ρ on the the composite system $\mathcal{H}_A \otimes \mathcal{H}_B \otimes \mathcal{H}_E$. In the distillation of a classical secret key, it is our task to generate a secret uniform random number shared by the two systems $\mathcal{H}_A$ and $\mathcal{H}_B$. That is, it is required that the eavesdropper's system $\mathcal{H}_E$ cannot hold any information concerning the distilled random number. Then, the optimal key rate with one-way $(A \to B)$ communication is defined by

$$C_k^{A \to B-E}(\rho) \stackrel{\text{def}}{=} \sup_{\kappa_n} \left\{ \overline{\lim} \frac{\log L_n}{n} \middle| \begin{array}{l} \| \operatorname{Tr}_E \kappa_n(\rho_n) - \rho_{\mathrm{mix},L_n} \|_1 \to 0 \\ I_{\kappa_n(\rho_n)}(AB : E) \to 0 \end{array} \right\}, \tag{9.69}$$

where $\rho_{\mathrm{mix},L} = \frac{1}{L} \sum_{i=1}^L |e_i^A\rangle\langle e_i^A|$. For this analysis, we define the quantity $C^{A\to}(\rho)$:

$$C_d^{A\to B-E}(\rho)$$
$$\stackrel{\text{def}}{=} \max_{\boldsymbol{M}}(H(\rho^B) - \sum_i \mathrm{P}_{\rho^A}^{\boldsymbol{M}}(i)H(\rho_i^B) - H(\rho^E) + \sum_i \mathrm{P}_{\rho^A}^{\boldsymbol{M}}(i)H(\rho_i^E)). \quad (9.70)$$

From this definition, the quantity $C_d^{A\to B-E}(\rho)$ satisfies the monotonicity concerning the one-way $A \to B$ operation. Further, we can show Condition **E2** (continuity) similarly to $C_d^{A\to B}(\rho)$.

Using Theorem 9.8 and a discussion similar to that concerning Theorem 8.10, we obtain the following theorem.

Theorem 9.9 *(Devetak and Winter [97])*

$$C_k^{A\to B-E}(\rho) = \lim \frac{C_d^{A\to B-E}(\rho^{\otimes n})}{n}. \quad (9.71)$$

Further, if there exists a TP-CP map κ from $\mathcal{H}_B$ to $\mathcal{H}_E$ such that

$$\mathrm{Tr}_{AB}\,\rho(M \otimes I_{BE}) = \kappa(\mathrm{Tr}_{AE}\,\rho(M \otimes I_{BE})), \quad \forall M \geq 0, \quad (9.72)$$

we have

$$C_k^{A\to B-E}(\rho) = C_d^{A\to B-E}(\rho). \quad (9.73)$$

In particular, when ρ has the form $\rho^{AB} \otimes \rho^E$,

$$C_k^{A\to B-E}(\rho) = C_d^{A\to B}(\rho^{AB}). \quad (9.74)$$

Proof. First, we prove the direct part:

$$C_k^{A\to B-E}(\rho) \geq \lim \frac{C_d^{A\to B-E}(\rho^{\otimes n})}{n}. \quad (9.75)$$

For this purpose, we consider the following operation. Let M be a POVM on $\mathcal{H}_A$ attaining its maximum on (9.70), and $\{1,\dots,l\}$ be its data set. First, we define the channel W^B, W^E as the sender prepares the classical information $j \in \{1,\dots,l\}$ and perform the measurement $\boldsymbol{M}$ on $\mathcal{H}_A$. Hence, the sender obtains the datum i as its outcome. He sends the classical information $k = i + j \mod l$. Then, systems B and E receive this information k. Since the channel W^B, W^E is described as

$$W_j^B = \sum_i \mathrm{P}_{\rho^A}^{\boldsymbol{M}}(i)\rho_i^B \otimes |e_{i+j}\rangle\langle e_{i+j}|, \quad W_j^E = \sum_i \mathrm{P}_{\rho^A}^{\boldsymbol{M}}(i)\rho_i^E \otimes |e_{i+j}\rangle\langle e_{i+j}|,$$

we obtain

$$I(p_{\mathrm{mix}}, W^B) = I(\mathrm{P}_{\rho^A}^{\boldsymbol{M}}, \rho_\cdot^B) + H(p_{\mathrm{mix}}) - H(\mathrm{P}_{\rho^A}^{\boldsymbol{M}}),$$
$$I(p_{\mathrm{mix}}, W^E) = I(\mathrm{P}_{\rho^A}^{\boldsymbol{M}}, \rho_\cdot^E) + H(p_{\mathrm{mix}}) - H(\mathrm{P}_{\rho^A}^{\boldsymbol{M}}).$$

Hence, Theorem 9.8 yields

$$C_k^{A\to B-E}(\rho) \geq C_d^{A\to B-E}(\rho).$$

Thus, we obtain (9.75).

Next, we prove the converse part:

$$C_k^{A\to B-E}(\rho) \leq \lim \frac{C_d^{A\to B-E}(\rho^{\otimes n})}{n}. \tag{9.76}$$

As was mentioned above, the quantity $C_d^{A\to B-E}(\rho)$ satisfies the monotonicity and the continuity. Hence, from a discussion similar to that concerning Theorem 8.10, we can show inequality (9.76). Further, we can prove (9.73) based on a similar derivation as for (9.62). ∎

9.5.5 Proof of Direct Part of C-Q Wiretap Channel Coding Theorem

In order to show that it is possible to attain the RHS of (9.61), let us first examine the following lemma.

Lemma 9.3 *Let v be the number of eigenvalues of W_p. Define*

$$\begin{aligned}\epsilon_1 \stackrel{\mathrm{def}}{=} & 4\sum_x p(x)\Big[\operatorname{Tr} W_x^B\{W_x^B - 2MLW_p^B < 0\} \\ & + 2ML \operatorname{Tr} W_p^B\{W_x^B - 2MLW_p^B \geq 0\}\Big], \\ \epsilon_2 \stackrel{\mathrm{def}}{=} & 2\sqrt{\sum_x p(x) \operatorname{Tr} W_x^E\{\kappa_{W_p^E}(W_x^E) - CW_p^E \geq 0\}} + \sqrt{\frac{Cv}{L}},\end{aligned}$$

for integers M, L, and $C > 0$. There exists a protocol $\Phi = (M, Q, Y)$ such that

$$\epsilon[\Phi] + I_E(\Phi) \leq \epsilon_1 + \epsilon_2 \log d + \eta_0(\epsilon_2), \quad |\Phi| = M. \tag{9.77}$$

Proof. Consider the map $\varphi(m,l)$ from $(m,l) \in \{1,\ldots,M\} \times \{1,\ldots,L\}$ to $\mathcal{X}$. Define Q^m such that $(WQ)^m = \sum_{l=1}^L \frac{1}{L} W_{\varphi(m,l)}$, where $\Phi = (M,Q,Y)$. Now, apply the random coding method. That is, for each pair (m,l), let $\varphi(m,l)$ be given by the independent and identical random variables $x_{m,l}$ subject to the probability distribution p. Using the random variable $X = (x_{m,l})$, we denote $\varphi = (\varphi(m,l))$ by φ_X. Hence, this protocol is determined by X and is denoted by $\Phi_X = (M, Q_X, Y_X)$. Denoting the expectation value by E_X, we assume that

$$\mathrm{E}_X\left[\varepsilon[\Phi_X]\right] \leq \epsilon_1, \tag{9.78}$$

$$\mathrm{E}_X\left[\sum_{m=1}^M \frac{1}{M} D((W^E Q_X)_m \| W_p^E)\right] \leq \epsilon_2 \log d + \eta_0(\epsilon_2) \tag{9.79}$$

and prove the theorem. Using the above two formulas, we find that

$$\begin{aligned}
\mathrm{E}_X\left[I_E(\Phi_X)\right]=&\mathrm{E}_X\left[H\left(\sum_{m'=1}^{M}\frac{1}{M}(W^E Q_X)_{m'}\right)-\sum_{m=1}^{M}\frac{1}{M}H((W^E Q_X)_m)\right]\\
=&\mathrm{E}_X\left[\sum_{m=1}^{M}\frac{1}{M}D((W^E Q_X)_m\|W^E_p)\right]-\mathrm{E}_X\left[D\left(\sum_{m=1}^{M}\frac{1}{M}(W^E Q_X)_m\middle\|W^E_p\right)\right]\\
\le&\ \epsilon_2\log d+\eta_0(\epsilon_2)
\end{aligned}$$

and therefore

$$\mathrm{E}_X\left[\varepsilon[\Phi_X]+I_E(\Phi_X)\right]\le\epsilon_1+\epsilon_2\log d+\eta_0(\epsilon_2).$$

This proves the existence of a protocol Φ satisfying (9.77).

Next, we prove (9.79). Applying Exercise 5.37 to (9.54), we immediately obtain

$$\mathrm{E}_X\left[\frac{1}{M}D((W^E Q_X)_m\|W^E_p)\right]\le\epsilon_2\log d+\eta_0(\epsilon_2),$$

which verifies (9.79). From (4.36) the RHS of (9.78) gives an upper bound of the mean value of the average error probability in the random coding when ML messages $\{W_{\varphi(m,l)}\}_{(m,l)}$ are transmitted. In this case, since only M messages $\{\sum_{l=1}^{L}W_{\varphi(m,l)}\}_m$ are transmitted and decoded, the error probability can be reduced further. This consideration yields (9.78). ■

We now prove the direct part by applying Lemma 9.3 to $W^{B,(n)}$, $W^{E,(n)}$, and p^n. First, we define $R\stackrel{\mathrm{def}}{=}I(p,W^B)-I(p,W^E)$ and $R_1\stackrel{\mathrm{def}}{=}I(p,W^E)$. Let $M=M_n\stackrel{\mathrm{def}}{=}e^{n(R-3\delta)}$, $C=C_n\stackrel{\mathrm{def}}{=}e^{n(R_1+\delta)}$, and $L=L_n\stackrel{\mathrm{def}}{=}e^{n(R_1+2\delta)}$ for arbitrary $\delta>0$. In this case, we denote the ϵ_1,ϵ_2,d by $\epsilon_1^{(n)},\epsilon_2^{(n)},d_n$, respectively.

We show that the RHS of (9.77) converges to zero under the above conditions. Since $M_nL_n=e^{nI(p,W^B-\delta)}$, the discussion in Sect. 3.6 guarantees that $\epsilon_1^{(n)}$ converges to zero. Using an argument similar to that in Sect. 9.4, we observe that $\epsilon_2^{(n)}$ approaches zero exponentially. Since d_n is the dimension, the equations $\log d_n=\log\dim\mathcal{H}^{\otimes n}=n\log\dim\mathcal{H}$ hold. Hence, the quantity $\epsilon_2^{(n)}\log\dim\mathcal{H}^{\otimes n}+\eta_0(\epsilon_2^{(n)})$ converges to zero. We can therefore show that $C_c^{B,E}(W)\ge I(p,W^B)-I(p,W^E)$. Finally, replacing W^B and W^E with $W^{B,(n)}Q$ and $W^{E,(n)}Q$, respectively, we can show that $C_c^{B,E}(W)\ge\frac{1}{n}\left(I(p,W^{B,(n)}Q)-I(p,W^{E,(n)}Q)\right)$. Therefore, the RHS of (9.61) $\ge$ the LHS of (9.61).

9.5.6 Proof of Converse Part of C-Q Wiretap Channel Coding Theorem

We prove the converse part of the theorem following Devetak [95]. Consider the sequence of protocols $\Phi^{(n)}=(M_n,Q_n,Y_n)$. Let X_n be the random variables taking values in $\{1, \ldots, M_n\}$ subject to the uniform distributions p^n_{mix}. Let Z_n be the random variable corresponding to the message decoded by the receiver.

Then, the equations $\log M_n = H(X_n) = I(X_n : Z_n) + H(X_n : Z_n)$ hold. The Fano inequality yields $H(X_n : Z_n) \le \varepsilon[\Phi^{(n)}] \log M_n + \log 2$. We can evaluate $I(X_n : Z_n)$ to be

$$
\begin{aligned}
&I(X_n : Z_n) = I(Y_n, p^n_{\mathrm{mix}}, (W^{B,(n)} Q_n)) \le I(p^n_{\mathrm{mix}}, (W^{B,(n)} Q_n)) \\
=&I(p^n_{\mathrm{mix}}, (W^{B,(n)} Q_n)) - I(p^n_{\mathrm{mix}}, (W^{E,(n)} Q_n)) + I_E(\Phi^{(n)}) \\
\le&\sup_Q \sup_p I(p, (W^{B,(n)} Q)) - I(p, (W^{E,(n)} Q)) + I_E(\Phi^{(n)}).
\end{aligned}
$$

Therefore,

$$
\begin{aligned}
\frac{1}{n} \log M_n \le& \frac{1}{n} \sup_Q \sup_p I(p, (W^{B,(n)} Q)) - I(p, (W^{E,(n)} Q)) + \frac{1}{n} I_E(\Phi^{(n)}) \\
&+ \varepsilon[\Phi^{(n)}] \frac{1}{n} \log M_n + \frac{1}{n} \log 2.
\end{aligned}
$$

Since $I_E(\Phi^{(n)}) \to 0$ and $\varepsilon[\Phi^{(n)}] \to 0$, the $\le$ part of (9.61) can be shown.

Proof of (9.62). In what follows, we prove (9.62). If we can write $W^E_x = \kappa(W^B_x)$ using a completely positive map κ, then $I(Q^m, W^{B,(n)}) \ge I(Q^m, W^{E,(n)})$. Defining $(Qp)_x \stackrel{\mathrm{def}}{=} \sum_m p(m) Q^m_x$, we have

$$
\begin{aligned}
&I(p, (W^{B,(n)} Q)) - I(p, (W^{E,(n)} Q)) \\
&= H(\sum_x (Qp)_x W^{B,(n)}_x) - \sum_m p(m) H((W^{B,(n)} Q)^m) \\
&\quad - H(\sum_x (Qp)_x W^{E,(n)}_x) + \sum_m p(m) H((W^{E,(n)} Q)^m) \\
&= H(\sum_x (Qp)_x W^{B,(n)}_x) - \sum_m p(m) Q^m_x H(W^{B,(n)}_x) \\
&\quad - \sum_m p(m) I(Q^m, W^{B,(n)}) - H(\sum_x (Qp)_x W^{E,(n)}_x) \\
&\quad + \sum_m p(m) Q^m_x H(W^{E,(n)}_x) + \sum_m p(m) I(Q^m, W^{E,(n)}) \\
&\le I(Qp, W^{B,(n)}) - I(Qp, W^{E,(n)}).
\end{aligned}
\tag{9.80}
$$

Using Exercise 9.13, we can obtain $\frac{1}{n} \sup_p I(p, W^{B,(n)}) - I(p, W^{E,(n)}) \le \sup_p I(p, W^B) - I(p, W^E)$, from which we obtain (9.62). ∎

Exercises

9.12. Show (9.68) using Exercise 8.29.

9.13. Consider two c-q channels W^B and $W^{B'}$ defined in $\mathcal{X}$ and $\mathcal{X}'$, two TP-CP maps κ and κ', and two quantum-classical channels $W_x^E = \kappa(W_x^B)$ and $W_x^{E'} = \kappa'(W_{x'}^{B'})$. Consider an arbitrary probability distribution q in $\mathcal{X} \times \mathcal{X}'$, and let p and p' be marginal distributions in $\mathcal{X}$ and $\mathcal{X}'$, respectively. Show that

$$\begin{aligned}&I(q, W^B \otimes W^{B'}) - I(q, W^E \otimes W^{E'})\\ \le& I(p, W^B) - I(p, W^E) + I(p', W^{B'}) - I(p', W^{E'}).\end{aligned}$$

9.14. Prove (9.65) referring to the discussions in Sects. 9.5.5 and 9.5.6.

9.15. Deform the condition $I_E(\Phi^{(n)}) \to 0$ to another condition $\frac{I_E(\Phi^{(n)})}{n} \to 0$ in the definitions of $C_c^{B,E}(W)$ and $C_c^{e,B,E}(\kappa)$. Show that the capacity is the same as the original one.

9.16. Prove (9.66) and (9.67) by expressing the environment of the composite map $\kappa' \circ \kappa$ in terms of the environment systems of the maps κ' and κ.

9.17. Show that the capacity is equal to the original one even though the condition $I_E(\Phi^{(n)}) \to 0$ in the definition is replaced by another condition $\varepsilon_{E,a}[\Phi^{(n)}] \stackrel{\rm def}{=} \sum_i \sum_{j \neq i} \frac{d_1((W^E Q)_i, (W^E Q)_j)}{M(M-1)} \to 0$. Here, use the Fannes inequality (5.64).

9.18. Show that the capacity is equal to the original one even though the above condition is replaced with another condition $\varepsilon_{E,w}[\Phi] \stackrel{\rm def}{=} \sup_i \sup_j d_1((W^E Q)_i, (W^E Q)_j)$, which converges to 0.

(From a comparision of the sizes of $\varepsilon_{E,w}[\Phi]$ and $\varepsilon_{E,a}[\Phi]$, it is sufficient to prove the direct part of Exercise 9.18 and the converse part of Exercise 9.17.)

9.19. Show that $I(Qp, W^B) - I(Qp, W^E) - \sum_i p_i(I(Q^i, W^B) - I(Q^i, W^E)) = I(p, (W^B Q)) - I(p, (W^E Q))$ for a quantum degraded channel κ. Also show (9.63).

9.6 Channel Capacity for Quantum-State Transmission

Let us consider the problem of finding how a large quantum system can be transmitted with negligible error via a given noisy quantum channel κ using encoding and decoding quantum operations. This problem is important for preventing noise from affecting a quantum state for a reliable quantum computation. Hence, this problem is a crucial one for realizing quantum computers and is called *quantum error correction.*

It is a standard approach in this problem to algebraically construct particular codes [66,67,150,155,256,371,378]. However, the achievability of the optimal rate is shown only by employing the random coding method [95,273,373].

Although this method is not directly applicable in a practical sense, it is still nevertheless an important theoretical result. In the discussion below, we will not discuss the former algebraic approach and concentrate only on the theoretical bounds.

Let us now formally state the problem of transmitting quantum systems accurately via a quantum channel κ from an input quantum system $\mathcal{H}_A$ to an output quantum system $\mathcal{H}_B$. When the quantum system $\mathcal{H}$ is to be sent, the encoding and decoding operations are given as TP-CP maps τ and ν from $\mathcal{H}$ to $\mathcal{H}_A$ and from $\mathcal{H}_B$ to $\mathcal{H}$, respetively. By combining these operations, it is possible to protect the quantum state from noise during transmission. We may therefore express our protocol by $\Phi = (\mathcal{H}, \tau, \nu)$. The quality of our protocol may be measured by the size $|\Phi| \stackrel{\text{def}}{=} \dim \mathcal{H}$ of the system to be sent. The accuracy of transmission is measured by $\varepsilon_1[\Phi] \stackrel{\text{def}}{=} \max_{u \in \mathcal{H}^1} \left[1 - F^2(u, \nu \circ \kappa \circ \tau(u))\right]$ ($\mathcal{H}^1 \stackrel{\text{def}}{=} \{u \in \mathcal{H} | \|u\| = 1\}$). We often focus on $\varepsilon_2[\Phi] \stackrel{\text{def}}{=} \left[1 - F_e^2(\rho_{\text{mix}}, \nu \circ \kappa \circ \tau)\right]$ as another criterion of accuracy. Let us now examine how a large communication rate $\frac{1}{n} \log |\Phi^{(n)}|$ of our code $\Phi^{(n)}$ is possible for a given channel $\kappa^{\otimes n}$ under the condition that $\varepsilon_1[\Phi^{(n)}]$ or $\varepsilon_2[\Phi^{(n)}]$ approaches zero asymptotically. Then, two kinds of channel capacities $C_{q,1}$ and $C_{q,2}$[7] are defined as

$$C_{q,i}(\kappa) \stackrel{\text{def}}{=} \sup_{\{\Phi^{(n)}\}} \left\{ \underline{\lim} \frac{1}{n} \log |\Phi^{(n)}| \,\middle|\, \lim \varepsilon_i[\Phi^{(n)}] = 0 \right\}, \quad i = 1, 2. \tag{9.81}$$

Theorem 9.10 *Two channel capacities $C_{q,1}$ and $C_{q,2}$ are calculated as*

$$C_{q,1}(\kappa) = C_{q,2}(\kappa) = \lim \frac{1}{n} \max_{\rho \in \mathcal{S}(\mathcal{H}_A^{\otimes n})} I_c(\rho, \kappa^{\otimes n}). \tag{9.82}$$

We now give a proof of the above theorem. Our strategy will be to relate this theorem to the problem of transmitting classical information in the

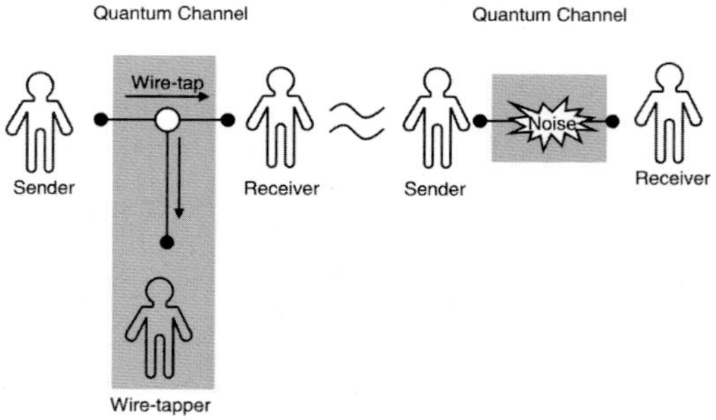

Fig. 9.2. A quantum channel with an eavesdropper and a quantum channel with noise

[7] The subscript q indicates that "quantum" states are to be sent.

presence of an eavesdropper, as examined in Sect. 9.5. In this approach, as shown in the proof of the theorem, protecting a quantum state from noise is equivalent in a sense to sending classical information without wiretapping when regarding the noise as the wiretapper. To this end, consider $(\mathcal{H}_C, \rho_0 = |u_0\rangle\langle u_0|, U_\kappa)$ of κ, a channel $W_u^B \stackrel{\rm def}{=} \kappa(|u\rangle\langle u|)$ to the intended receiver using this κ, and a c-q wiretap channel given by $W_u^E \stackrel{\rm def}{=} \kappa_E(|u\rangle\langle u|)$, where κ^E is the channel to the environment (eavesdropper) defined in (5.5). Then, the following lemma holds; its proof will be given in Appendix B.7.

Lemma 9.4 *Diagonalize* ρ *according to* $\rho = \sum_x p_x |\tilde{u}_x\rangle\langle\tilde{u}_x|$. *Let* M *and* L *be arbitrary integers. Let* v *be the number of eigenvalues of* W_p^E. *There exists an isometric map* V *from* $\mathbb{C}^M$ *to* $\mathcal{H}_A$ *and a TP-CP map* ν *from* $\mathcal{H}_B$ *to* $\mathbb{C}^M$ *such that*

$$\begin{aligned}
&1 - F_e(\rho_{\rm mix}, \nu \circ \kappa \circ \kappa_V) \\
&\leq \sum_x p(x) \Big(24 \operatorname{Tr} W_x^B \{ W_x^B - 2ML W_p^B < 0 \} \\
&\quad + 12ML \operatorname{Tr} W_p^B \{ W_x^B - 2ML W_p^B \geq 0 \} \Big) \\
&\quad + 2 \sqrt{ \sum_x p(x) \operatorname{Tr} W_x^E \{ \kappa_{W_p^E}(W_x^E) - C W_p^E \geq 0 \} + \sqrt{\frac{Cv}{L}} }.
\end{aligned} \tag{9.83}$$

We are now ready to prove the direct part of the theorem, i.e., $C_{q,2}(\kappa) \geq$ (RHS of (9.82)). Using Lemma 9.4 and arguments similar to those in the proof of of the direct part of Theorem 9.8, we obtain $C_{q,2}(\kappa) \geq I(p, W^B) - I(p, W^E)$. Since $I(p, W^B) - I(p, W^E) = I_c(\rho, \kappa)$ from (9.80), we may apply the same argument to $\kappa^{\otimes n}$ to obtain $C_{q,2}(\kappa) \geq$ (RHS of (9.82)), which completes the proof. The proof of $C_{q,2}(\kappa) = C_{q,1}(\kappa)$ is left as Exercises 9.20 and 9.21.

Next, we give a proof of the converse part by a method similar to that in Theorem 8.10, i.e., we show that

$$C_{q,2}(\kappa) \leq \lim \frac{1}{n} \max_\rho I_c(\rho, \kappa^{\otimes n}). \tag{9.84}$$

In this method, for any class of local operations C, we focus on the function $C(\kappa)$ of a channel κ from $\mathcal{H}_A$ to $\mathcal{H}_B$ that satisfies the following conditions.

C1 (*Normalization*) $C(\iota_d) = \log d$ for an identical map ι_d with a size d.

C2 (*Monotonicity*) $C(\kappa' \circ \kappa \circ \kappa_U) \leq C(\kappa)$ holds for any TP-CP map κ' and any isometry U.

C3 (*Continuity*) When any two channels $\kappa_{1,n}$ and $\kappa_{2,n}$ from a system $\mathcal{H}_n$ to another system $\mathcal{H}'_n$ satisfy $\max_x 1 - F^2((\kappa_{1,n} \otimes \iota_R)(|x\rangle\langle x|), (\kappa_{2,n} \otimes \iota_R)(|x\rangle\langle x|)) \to 0$, $\frac{|C(\kappa_{1,n}) - C(\kappa_{2,n})|}{\log(\dim \mathcal{H}_n \dim \mathcal{H}'_n)} \to 0$, where the system $\mathcal{H}_R$ is the reference system of $\mathcal{H}_n$.

C4 (*Convergence*) The quantity $\frac{C(\kappa^{\otimes n})}{n}$ converges as $n \to \infty$.

Based on only the above conditions, we can prove the following theorem.

Lemma 9.5

$$C_{q,2}(\kappa) \le C^{\infty}(\kappa) \left(\stackrel{\text{def}}{=} \lim \frac{C(\kappa^{\otimes n})}{n} \right). \tag{9.85}$$

Since $\max_\rho I_c(\rho, \kappa)$ satisfies Conditions **C1**, **C2** (8.37), **C3** (Exercise 9.22), and **C4** (Lemma A.1), we obtain (9.84).

Proof. According to Condition ④ in Sect. 8.2, we can choose an encoding τ_n and a decoding ν_n with d_n-dimensional space $\mathcal{K}_n$ such that

$$1 - F_e^2(\rho_{\mathrm{mix}}, \nu_n \circ \kappa^{\otimes n} \circ \tau_n) \to 0$$

and $\lim \frac{\log d_n}{n} = C_{q,2}(\kappa)$. Condition ① in Sect. 8.2 guarantees that there exists isometry U_n such that $F_e^2(\rho_{\mathrm{mix}}, \nu_n \circ \kappa^{\otimes n} \circ \tau_n) \le F_e(\rho_{\mathrm{mix}}, \nu_n \circ \kappa^{\otimes n} \circ \kappa_{U_n})$. From Condition ③ in Sect. 8.2 there exists a subspace $\mathcal{K}_n' \subset \mathcal{K}_n$ with the dimension $\frac{d_n}{2}$ such that

$$\max_{x \in \mathcal{K}_n'} 1 - F^2(x, \nu_n \circ \kappa^{\otimes n} \circ \kappa_{U_n}(x)) \le \frac{1 - F_e^2(\rho_{\mathrm{mix}}, \nu_n \circ \kappa^{\otimes n} \circ \kappa_{U_n})}{2}.$$

Therefore, from Condition ④ in Sect. 8.2 we have

$$\frac{2}{3} \max_{\rho \in \mathcal{S}(\mathcal{K}_n')} \left(1 - F^2(\rho, \nu_n \circ \kappa^{\otimes n} \circ \kappa_{U_n})\right) \to 0.$$

Letting $\kappa_{2,n}$ be a noiseless channel, we have $\max_x 1 - F^2((\nu_n \circ \kappa^{\otimes n} \circ \kappa_{U_n} \otimes \iota_R)(|x\rangle\langle x|), (\kappa_{2,n} \otimes \iota_R)(|x\rangle\langle x|)) \to 0$. Thus, Condition **C3** implies

$$\frac{|C(\nu_n \circ \kappa^{\otimes n} \circ \kappa_{U_n}) - C(\iota_{d_n})|}{n} \to 0.$$

From Condition **C1** we have

$$\left| \frac{C(\nu_n \circ \kappa^{\otimes n} \circ \kappa_{U_n})}{n} - C_{q,2}(\kappa) \right|$$
$$\le \frac{|C(\nu_n \circ \kappa^{\otimes n} \circ \kappa_{U_n}) - C(\iota_{\frac{d_n}{2}})|}{n} + \left| \frac{\log d_n - \log 2}{n} - C(\kappa) \right| \to 0.$$

Hence, Condition **C2** guarantees that

$$\lim \frac{C(\kappa^{\otimes n})}{n} \ge \lim \frac{C(\nu_n \circ \kappa^{\otimes n} \circ \kappa_{U_n})}{n} = C_{q,2}(\kappa). \tag{9.86}$$

We obtain the desired inequality. ∎

Moreover, using Exercise 9.13, we can simplify the RHS of (9.82) in the following special case.

Lemma 9.6 *When there exists another channel κ' from the output system $\mathcal{H}_B$ of channel κ to its environment system $\mathcal{H}_E$ such that*

$$\kappa'(\kappa(\rho)) = \kappa^E(\rho) \quad \textit{for } \forall \rho \in \mathcal{S}(\mathcal{H}_A), \tag{9.87}$$

then

$$\max_{\rho \in \mathcal{S}(\mathcal{H}_A)} I_c(\rho, \kappa) = \lim \frac{1}{n} \max_{\rho \in \mathcal{S}(\mathcal{H}_A^{\otimes n})} I_c(\rho, \kappa^{\otimes n}). \tag{9.88}$$

Further, when (9.87) holds, (9.63) implies the concavity

$$\sum_i p_i I_c(\rho_i, \kappa) \le I_c(\sum_i p_i \rho_i, \kappa). \tag{9.89}$$

For example, the following case satisfies the above condition. Suppose that there exist a basis $\{u_1, \dots, u_d\}$ of the input system $\mathcal{H}_A$ of channel κ and the POVM $\boldsymbol{M} = \{M_i\}_{i=1}^d$ on the output system $\mathcal{H}_B$ such that

$$\operatorname{Tr} \kappa(|u_i\rangle\langle u_i|) M_j = \delta_{i,j}. \tag{9.90}$$

As checked as follows, this channel satisfies the condition (9.87). For example, the phase-damping channel κ_D^{PD} satisfies condition (9.90).

Now, consider the Naĭmark–Ozawa extension $(\mathcal{H}_0, \rho_0, U)$ with the ancilla $\mathcal{H}_D$ given in Theorem 7.1. Note that we measure the system $\mathcal{H}_0$ with the measurement basis $\{v_1, \dots, v_d\}$. We also use the Stinespring representation $(\mathcal{H}_C, \rho_0', U_\kappa)$ of κ. Since for any input pure state ρ, the state $(U \otimes I_E)((U_\kappa(\rho \otimes \rho_0')U_\kappa^*) \otimes \rho_0)(U \otimes I_E)^*$ is pure, there exists a basis $\{v_1', \dots, v_d'\}$ on the space $\mathcal{H}_D \otimes \mathcal{H}_E$, where $\mathcal{H}_E = \mathcal{H}_A \otimes \mathcal{H}_C$. This is ensured by (9.90). Hence, the unitary $U' \stackrel{\mathrm{def}}{=} \sum_{i=1}^d |v_i'\rangle\langle v_i|$ satisfies

$$\begin{aligned}
&\operatorname{Tr}_{\mathcal{H}_0} (U \otimes I_E)((U_\kappa(\rho \otimes \rho_0')U_\kappa^*) \otimes \rho_0)(U \otimes I_E)^* \\
&= U' \operatorname{Tr}_{D,E} (U \otimes I_E)((U_\kappa(\rho \otimes \rho_0')U_\kappa^*) \otimes \rho_0)(U \otimes I_E)^* U'^*.
\end{aligned}$$

Therefore,

$$\begin{aligned}
\kappa^E(\rho) &= \operatorname{Tr}_D \operatorname{Tr}_{\mathcal{H}_0} (U \otimes I_E)((U_\kappa(\rho \otimes \rho_0')U_\kappa^*) \otimes \rho_0)(U \otimes I_E)^* \\
&= \operatorname{Tr}_D U' \operatorname{Tr}_{D,E} (U \otimes I_E)((U_\kappa(\rho \otimes \rho_0')U_\kappa^*) \otimes \rho_0)(U \otimes I_E)^* U'^*,
\end{aligned}$$

which implies condition (9.87).

Proof of (8.106). Finally, we prove the Hashing inequality (8.106) based on a remarkable relation between the transmission of the quantum state and the distillation of a partially entangled state. Given a channel κ, we can define the partially entangled state $\kappa \otimes \iota_{A'}(|u_{0,0}^{A,A'}\rangle\langle u_{0,0}^{A,A'}|)$, where $\mathcal{H}_{A'}$ is the reference system of the input system $\mathcal{H}_A$ and $u_{0,0}^{A,A'}$ is a maximally entangled state between $\mathcal{H}_A$ and $\mathcal{H}_{A'}$. Conversely, given a partially entangled state ρ

on the composite system $\mathcal{H}_A \otimes \mathcal{H}_B$, we can define channel κ_ρ with the same dimensional input system $\mathcal{H}_{A'}$ as the system $\mathcal{H}_A$ (with the dimension d) via quantum teleportation as follows [40]. Perform the generalized Bell measurement $\{|u_{i,j}^{A,A'}\rangle\langle u_{i,j}^{A,A'}|\}$ on the composite system $\mathcal{H}_A \otimes \mathcal{H}_{A'}$ and transmit the data (i,j), where $u_{i,j}^{A,A'} \stackrel{\text{def}}{=} (I_{A'} \otimes \mathbf{X}_A^i \mathbf{Z}_A^j) u_{0,0}^{A,A'}$. In this channel, the output system is given by $\mathcal{H}_B \otimes \mathbb{C}^{d^2}$. Hence, considering the quantum state transmission via channel κ_ρ, we obtain $C_{q,2}(\kappa_\rho) \le E_{d,2}^{\rightarrow}(\rho)$. Further, noting the teleportation form A to A'', we have

$$
\begin{aligned}
&\kappa_\rho \otimes \iota_{A''}(|u_{0,0}^{A'',A'}\rangle\langle u_{0,0}^{A'',A'}|) \\
&= \bigoplus_{i,j} (\overline{\mathbf{X}_{A''}^i \mathbf{Z}_{A''}^j} \otimes I_B)\rho(\overline{\mathbf{X}_{A''}^i \mathbf{Z}_{A''}^j} \otimes I_B)^*,
\end{aligned}
$$

which implies $I_c(\rho_{\text{mix}}, \kappa_\rho) = -H_\rho(A|B)$. Thus, we obtain (8.106). ∎

Exercises

9.20. Show that $C_{q,2}(\kappa) \le C_{q,1}(\kappa)$ using (8.23).

9.21. Show that $C_{q,1}(\kappa) \ge C_{q,2}(\kappa)$ using (8.24).

9.22. Show that $\max_\rho I_c(\rho, \kappa)$ satisfies Condition **C3** similarly to Exercise 8.44. Here, use Fannes inequality (Theorem 5.9) for two states $(\kappa_{1,n} \otimes \iota_R)(|x\rangle\langle x|)$ and $(\kappa_{2,n} \otimes \iota_R)(|x\rangle\langle x|)$.

9.23. Give an alternative proof of $C_{q,2}(\kappa) \le$ (RHS of (9.82)) following the steps below [31].

a Let Φ be a code for a channel κ with an input $\mathcal{H}_A$, satisfying $|\Phi| \ge \dim \mathcal{H}_A$. Show that a code Φ' such that $|\Phi'| = \dim \mathcal{H}_A$ and $\varepsilon_1[\Phi'] \le \varepsilon_1[\Phi]$ exists, using property ② of Sect. 8.2.

b Let $\Phi = (\mathcal{H}, \kappa_U, \nu)$ be a code that has an isometric encoding for channel κ (U is isometric). Let ρ_{mix} be a completely mixed state in $\mathcal{H}$. Define $\delta \stackrel{\text{def}}{=} \sqrt{2(1 - F_e(\rho_{\text{mix}}, \nu \circ \kappa \circ \kappa_U))}$. Show that

$$
\begin{aligned}
&\max_{\rho \in \mathcal{S}(\mathcal{H}_A)} I_c(\rho, \kappa) \ge I_c(U \rho_{\text{mix}} U^*, \kappa) \ge I_c(\rho_{\text{mix}}, \nu \circ \kappa \circ \kappa_U) \\
&\ge \log |\Phi| - 2\delta (\log d_A - \log \delta)
\end{aligned}
$$

using (8.37), (8.40), and (8.36).

c Given that a sequence of codes $\Phi^{(n)} = (\mathcal{H}^{\otimes n}, \tau^{(n)}, \nu^{(n)})$ satisfies $\varepsilon_2[\Phi^{(n)}] \to 0$, show that there exists an isometric matrix $U^{(n)}$ such that

$$
\lim \frac{1}{n} \max_{\rho \in \mathcal{S}(\mathcal{H}_A^{\otimes n})} I_c(\rho, \kappa^{\otimes n}) \ge \lim \frac{1}{n} \log \min\{|\Phi^{(n)}|, d_A^n\},
$$

using (8.20). This completes the alternate proof for the case (LHS of above) $< \log d_A$.

d Give an alternate proof for the case $\lim \frac{1}{n} \max_{\rho \in \mathcal{S}(\mathcal{H}_A^{\otimes n})} I_c(\rho, \kappa^{\otimes n}) = \log d_A$ by considering source coding.

9.7 Examples

In this section, we calculate the capacities $\tilde{C}_c(\kappa)$, $C_c(\kappa)$, $C_c^e(\kappa)$, $C_{c,e}^e(\kappa)$, and $C_{q,1}(\kappa)$ in several cases.

9.7.1 Group Covariance Formulas

For this purpose, we derive forlumas for $C_c(\kappa), C_c^e(\kappa), C_{c,e}^e(\kappa)$, and $C_{q,1}(\kappa)$ with the group covariance. Let κ be a TP-CP map from system $\mathcal{H}_A$ to $\mathcal{H}_B$. Assume that there exists an irreducible (projective) representation U_A of a group G on the space $\mathcal{H}_A$ satisfying the following. There exist unitary matrices (not necessarily a representation) $\{U_B(g)\}_{g \in G}$ on $\mathcal{H}_B$ such that

$$\kappa(U_A(g)\rho U_A(g)^*) = U_B(g)\kappa(\rho)U_B(g)^* \tag{9.91}$$

for any density ρ and element $g \in G$. In what follows, we derive useful formulas in the above assumption. Then,

$$I(p, \kappa) = I(p_{U_A(g)}, \kappa),$$

where $p_U(\rho) \stackrel{\text{def}}{=} p(U\rho U^*)$. Hence,

$$\begin{aligned}
I(p, \kappa) = &\int_G I(p_{U_A(g)}, \kappa)\nu(dg) \le I\Big(\int_G p_{U_A(g)}\nu(dg), \kappa\Big) \\
=&H(\kappa(\rho_{\text{mix}})) - \int_G \sum_x p_x H(\kappa(U_A(g)^* \rho_x U_A(g)))\nu(dg) \\
\le&H(\kappa(\rho_{\text{mix}})) - \min_\rho H(\kappa(\rho)).
\end{aligned}$$

This upper bound is attained by the distribution $(\nu(\text{g}), U_A(g)^* \rho_{\min} U_A(g))$. Thus,

$$C_c(\kappa) = H(\kappa(\rho_{\text{mix}})) - \min_\rho H(\kappa(\rho)). \tag{9.92}$$

Next, we define the representation $U_A^{(n)}$ of the group $G^n \stackrel{\text{def}}{=} \underbrace{G \times G \times \cdots \times G}_{n}$ on the n-fold tensor product system $\mathcal{H}_A^{\otimes n}$ as

$$U_A^{(n)}(g_1, \dots, g_n) \stackrel{\text{def}}{=} U_A(g_1) \otimes \cdots \otimes U_A(g_n).$$

Then, the set of unitary matrices

$$U_B^{(n)}(g_1, \cdots, g_n) \stackrel{\text{def}}{=} U_B(g_1) \otimes \cdots \otimes U_B(g_n)$$

satisfies

$$\begin{aligned}
&\kappa^{\otimes n}(U_A^{(n)}(g_1, \cdots, g_n)\rho(U_A^{(n)}(g_1, \cdots, g_n))^*) \\
&= U_B^{(n)}(g_1, \cdots, g_n)\kappa^{\otimes n}(\rho)(U_B^{(n)}(g_1, \cdots, g_n))^*
\end{aligned}$$

for any density ρ on the n-fold tensor product system $\mathcal{H}_A^{\otimes n}$. If U_A is irreducible, then $U_A^{(n)}$ is also irreducible. Hence, we have the formula

$$C_c(\kappa^{\otimes n}) = nH(\kappa(\rho_{\text{mix}})) - \min_\rho H(\kappa^{\otimes n}(\rho)).$$

Thus,

$$C_c^e(\kappa) = H(\kappa(\rho_{\text{mix}})) - \lim \frac{\min_\rho H(\kappa^{\otimes n}(\rho))}{n}. \tag{9.93}$$

Further, relation (9.91) yields that

$$\begin{aligned}
&(\kappa \otimes \iota_R)((U_A(g) \otimes I_R)|u\rangle\langle u|(U_A(g)^* \otimes I_R)) \\
&= (U_B(g) \otimes I_R)(\kappa \otimes \iota_R)(|u\rangle\langle u|)(U_B(g)^* \otimes I_R)).
\end{aligned}$$

Hence, we have

$$I(\rho, \kappa) = I(U_A(g)\rho U_A(g)^*, \kappa), \quad I_c(\rho, \kappa) = I_c(U_A(g)\rho U_A(g)^*, \kappa). \tag{9.94}$$

Concavity (8.43) of the transmission information guarantees that

$$\begin{aligned}
I(\rho, \kappa) &= \int_G I(U_A(g)\rho U_A(g)^*, \kappa)\nu(dg) \\
&\leq I\Big(\int_G U_A(g)\rho U_A(g)^*\nu(dg), \kappa\Big) = I(\rho_{\text{mix}}, \kappa),
\end{aligned} \tag{9.95}$$

which implies

$$C_{c,e}^e(\kappa) = I(\rho_{\text{mix}}, \kappa) = \log d_A + \log d_B - H(\kappa \otimes \iota(|\Phi_d\rangle\langle\Phi_d|)),$$

where $|\Phi_d\rangle\langle\Phi_d|$ is the maximally entangled state.

Next, we consider the quantum capacity $C_{1,q}(\kappa)$ when the wiretap channel (κ, κ^E) is a degraded channel. In this case, concavity (9.89) holds. Hence, using the second relation in (9.94), we have

$$C_{1,q}(\kappa) = \max_\rho I_c(\rho, \kappa) = I_c(\rho_{\text{mix}}, \kappa) = \log d_B - H((\kappa \otimes \iota_R)(|\Phi_d\rangle\langle\Phi_d|)). \tag{9.96}$$

9.7.2 d-Dimensional Depolarizing Channel

When κ is the d-dimensional depolarizing channel $\kappa_{d,\lambda}$, the natural representation of $SU(d)$ satifies the above condition. Hence, using (9.92), we have

$$\begin{aligned} C_c(\kappa_{d,\lambda}) &= H(\rho_{\mathrm{mix}}) - H(\lambda|u\rangle\langle u| + (1-\lambda)\rho_{\mathrm{mix}}) \\ &= \frac{\lambda d + (1-\lambda)}{d}\log(\lambda d + (1-\lambda)) + \frac{(1-\lambda)(d-1)}{d}\log(1-\lambda). \end{aligned}$$

Indeed, we can easily check that this bound is attained by commutative input states. Thus,

$$\tilde{C}_c(\kappa_{d,\lambda}) = \frac{\lambda d + (1-\lambda)}{d}\log(\lambda d + (1-\lambda)) + \frac{(1-\lambda)(d-1)}{d}\log(1-\lambda).$$

Concerning entangled input states, King [254] showed that

$$\min_\rho H(\kappa_{d,\lambda}^{\otimes n}(\rho)) = nH(\lambda|u\rangle\langle u| + (1-\lambda)\rho_{\mathrm{mix}}). \tag{9.97}$$

Thus, formula (9.93) yields [254]

$$C_c^e(\kappa_{d,\lambda}) = C_c(\kappa_{d,\lambda}).$$

Further, from (8.194) and (8.192),

$$\begin{aligned} C_{c,e}^e(\kappa_{d,\lambda}) &= I(\rho_{\mathrm{mix}}, \kappa_{d,\lambda}) = 2\log d - H(\rho_{I,\frac{1-\lambda(d^2-1)}{d^2}}) \\ &= 2\log d + \frac{1-\lambda(d^2-1)}{d^2}\log\frac{1-\lambda(d^2-1)}{d^2} + \frac{(1-\lambda)(d^2-1)}{d^2}\log\frac{1-\lambda}{d^2}. \end{aligned}$$

9.7.3 Transpose-Depolarizing Channel

In the transpose-depolarizing channel $\kappa_{d,\lambda}^T$, the natural representation of $SU(d)$ satisfies the above condition. However, U_B is not a representation. As with a depolarizing channel, using (9.92), we have

$$\begin{aligned} C_c(\kappa_{d,\lambda}^T) &= \tilde{C}_c(\kappa_{d,\lambda}) \\ &= \frac{\lambda d + (1-\lambda)}{d}\log(\lambda d + (1-\lambda)) + \frac{(1-\lambda)(d-1)}{d}\log(1-\lambda). \end{aligned}$$

Further, relation (9.97) yields

$$\begin{aligned} \min_\rho H((\kappa_{d,\lambda}^T)^{\otimes n}(\rho)) &= \min_\rho H((\kappa_{d,\lambda})^{\otimes n}(\rho^T)) = \min_{\rho^T} H((\kappa_{d,\lambda})^{\otimes n}(\rho)) \\ &= nH(\lambda|u\rangle\langle u| + (1-\lambda)\rho_{\mathrm{mix}}) \end{aligned}$$

for $\lambda \geq 0$. Hence,

$$C_c^e(\kappa_{d,\lambda}^T) = C_c(\kappa_{d,\lambda}^T)$$

for $\lambda \geq 0$. Matsumoto and Yura [291] proved this relation for $\lambda = -\frac{1}{1-d}$.
Further, from (8.195) and (8.189), its entanglement-assisted capacity is

$$\begin{aligned}
C_{c,e}^e(\kappa_{d,\lambda}^T) &= I(\rho_{\mathrm{mix}}, \kappa_{d,\lambda}^T) = 2\log d - H(\rho_{W,\frac{(1-(d+1)\lambda)(d-1)}{2d}}) \\
&= 2\log d + \frac{(1-(d+1)\lambda)(d-1)}{2d}\log\frac{1-(d+1)\lambda}{d^2} \\
&\quad + \frac{(1+(d-1)\lambda)(d+1)}{2d}\log\frac{1+(d-1)\lambda}{d^2}.
\end{aligned}$$

9.7.4 Generalized Pauli Channel

In the generalized Pauli channel κ_p^{GP}, the representation of the group $(i,j) \in \mathbb{Z}_d \times \mathbb{Z}_d \mapsto \mathbf{X}_d^i \mathbf{Z}_d^j$ satisfies condition (9.91). Its entanglement-assisted capacity can be calculated as

$$C_{c,e}^e(\kappa_p^{\mathrm{GP}}) = I(\rho_{\mathrm{mix}}, \kappa_p^{\mathrm{GP}}) = 2\log d - H(p).$$

When the dimension d is equal to 2, using (5.28), we can check

$$\tilde{C}_c(\kappa_p^{\mathrm{GP}}) = C_c(\kappa_p^{\mathrm{GP}}) = \log 2 - \min_{i \neq j} h(p_i + p_j).$$

In this case, as is mentioned in **d** in Sect. 9.2, King [253] showed that

$$C_c(\kappa_p^{\mathrm{GP}}) = \tilde{C}_c(\kappa_p^{\mathrm{GP}}) = C_c^e(\kappa_p^{\mathrm{GP}}).$$

When the distribution $p = (p_{i,j})$ satisfies $p_{i,j} = 0$ for $j \neq 0$ in the d-dimensional system, we have

$$C_c(\kappa_p^{\mathrm{GP}}) = \tilde{C}_c(\kappa_p^{\mathrm{GP}}) = C_c^e(\kappa_p^{\mathrm{GP}}) = \log d.$$

In this case, the channel κ_p^{GP} is a phase-damping channel. As is proved in Sect. 9.7.7, it satisfies condition (9.87). Hence, (9.96) yields

$$C_{q,1}(\kappa_p^{\mathrm{GP}}) = I_c(\rho_{\mathrm{mix}}, \kappa_p^{\mathrm{GP}}) = \log d - H(p).$$

9.7.5 PNS Channel

The PNS channel satisfies condition (9.91). Hence, using (9.92), we have

$$C_c(\kappa_{d,n\to m}^{\mathrm{pns}}) = H(\kappa_{d,n\to m}^{\mathrm{pns}}(\rho_{\mathrm{mix}})) - \min_{\rho} H(\kappa_{d,n\to m}^{\mathrm{pns}}(\rho)) = \log\binom{m+d-1}{d-1}.$$

Since $C_c^e(\kappa_{d,n\to m}^{\mathrm{pns}})$ is less than the dimension of the input system, $C_c^e(\kappa_{d,n\to m}^{\mathrm{pns}}) = C_c(\kappa_{d,n\to m}^{\mathrm{pns}})$. Its entanglement-assisted capacity is calculated as

$$C_{c,e}^e(\kappa_{d,n\to m}^{\mathrm{pns}}) = I(\rho_{\mathrm{mix}}, \kappa_{d,n\to m}^{\mathrm{pns}})$$
$$= \log \binom{m+d-1}{d-1} + \log \binom{n+d-1}{d-1} - \log \binom{(n-m)+d-1}{d-1}.$$

From Exercise 5.14, the wiretap channel $(\kappa_{d,n\to m}^{\mathrm{pns}}, (\kappa_{d,n\to m}^{\mathrm{pns}})^E)$ is a degraded channel. Hence, from (9.96), its quantum capacity is calculated as

$$C_{q,1}(\kappa_{d,n\to m}^{\mathrm{pns}}) = I_c(\rho_{\mathrm{mix}}, \kappa_{d,n\to m}^{\mathrm{pns}})$$
$$= -\log \binom{(n-m)+d-1}{d-1} + \log \binom{m+d-1}{d-1}. \qquad (9.98)$$

9.7.6 Erasure Channel

The erasure channel also satisfies condition (9.91). Hence, using (9.92), we have

$$C_c(\kappa_{d,p}^{\mathrm{era}}) = I_c(\rho_{\mathrm{mix}}, \kappa_{d,p}^{\mathrm{era}}) = H(\kappa_{d,p}^{\mathrm{era}}(\rho_{\mathrm{mix}})) - \min_\rho H(\kappa_{d,p}^{\mathrm{era}}(\rho))$$
$$= -(1-p)\log\frac{1-p}{d} - p\log p - h(p) = (1-p)\log d.$$

Since it is attained by commutative input states, $C_c(\kappa_{d,p}^{\mathrm{era}}) = \tilde{C}_c(\kappa_{d,p}^{\mathrm{era}})$.

Next, we consider the capacity $C_c^e(\kappa_{d,p}^{\mathrm{era}})$. Because

$$(\kappa_{d,p}^{\mathrm{era}})^{\otimes n}(\rho) = \bigoplus_{\{i_1,\ldots,i_k\}\subset\{1,\ldots,n\}} (1-p)^k p^{n-k} \operatorname{Tr}_{i_1,\ldots,i_k} \rho \otimes |u_d\rangle\langle u_d|^{\otimes(n-k)},$$

the minimum entropy $\min_\rho H((\kappa_{d,p}^{\mathrm{era}})^{\otimes n}(\rho))$ is calculated as

$$\min_\rho H((\kappa_{d,p}^{\mathrm{era}})^{\otimes n}(\rho))$$
$$= \min_\rho \sum_{\{i_1,\ldots,i_k\}\subset\{1,\ldots,n\}} (1-p)^k p^{n-k}(-\log(1-p)^k p^{n-k} + H(\operatorname{Tr}_{i_1,\ldots,i_k}\rho))$$
$$= nh(p) + \min_\rho \sum_{\{i_1,\ldots,i_k\}\subset\{1,\ldots,n\}} (1-p)^k p^{n-k} H(\operatorname{Tr}_{i_1,\ldots,i_k}\rho) = nh(p).$$

Hence, from (9.93),

$$C_c^e(\kappa_{d,p}^{\mathrm{era}}) = C_c(\kappa_{d,p}^{\mathrm{era}}) = \tilde{C}_c(\kappa_{d,p}^{\mathrm{era}}) = (1-p)\log d.$$

The entanglement-assisted capacity is calculated as

$$C_{c,e}^e(\kappa_{d,p}^{\mathrm{era}}) = I(\rho_{\mathrm{mix}}, \kappa_{d,p}^{\mathrm{era}})$$
$$= \log d + (1-p)\log\frac{d}{1-p} - p\log p - p\log\frac{d}{p} + (1-p)\log(1-p) \qquad (9.99)$$
$$= 2(1-p)\log d,$$

where we used Exercise 5.13 in (9.99).

From Exercise 5.13, the wiretap channel $(\kappa^{\mathrm{era}}_{d,p}, (\kappa^{\mathrm{era}}_{d,p})^E)$ is a degraded channel. Hence, from (9.96), its quantum capacity is calculated as [45]

$$C_{q,1}(\kappa^{\mathrm{era}}_{d,p}) = I_c(\rho_{\mathrm{mix}}, \kappa^{\mathrm{era}}_{d,p}) = (1-2p)\log d \text{ for } p \le 1/2.$$

9.7.7 Phase-Damping Channel

Any phase-damping channel κ^{PD}_D clearly satisfies

$$C_c(\kappa^{\mathrm{PD}}_D) = \tilde{C}_c(\kappa^{\mathrm{PD}}_D) = C^e_c(\kappa^{\mathrm{PD}}_D) = \log d.$$

Indeed, we can show that the wiretap channel $(\kappa^{\mathrm{PD}}_D, (\kappa^{\mathrm{PD}}_D)^E)$ is a degraded channel as follows. The purification of the output state $\sum_{k,l} d_{k,l}|e_k, e^R_k\rangle\langle e_l, e^R_l|$ of the maximally entangled state $\frac{1}{d}\sum_{k,l}|e_k, e^R_k\rangle\langle e_l, e^R_l|$ is given as

$$\frac{1}{d_A}\sum_{k,k,',l,l'} y_{k,k'}\overline{y_{l,l'}}|e_k, e^R_k, e^E_{k'}\rangle\langle e_l, e^R_l, e^E_{l'}|,$$

where $Y = (y_{k,k'})$ satisfies $Y^*Y = D$. From the condition $X_{k,k} = 1$, the positive semidefinite matrix $\rho^E_k \stackrel{\mathrm{def}}{=} \sum_{k'l'} y_{k,k'}\overline{y_{k,l'}}|e^E_{k'}\rangle\langle e^E_{l'}|$ satisfies the condition of states $\mathrm{Tr}\,\rho^E_k = 1$. Then, by applying (8.30), the channel $(\kappa^{\mathrm{PD}}_D)^E$ to the environment is described as

$$\begin{aligned}
(\kappa^{\mathrm{PD}}_D)^E(\rho) &= \mathrm{Tr}_{R,A}(I_{A,E}\otimes\rho^T)\sum_{k,k,',l,l'} y_{k,k'}\overline{y_{l,l'}}|e_k, e^R_k, e^E_{k'}\rangle\langle e_l, e^R_l, e^E_{l'}| \\
&= \sum_k \rho_{k,k} y_{k,k'}\overline{y_{k,l'}}|e^E_{k'}\rangle\langle e^E_{l'}| = \sum_k \langle e_k|\kappa^{\mathrm{PD}}_D(\rho)|e_k\rangle y_{k,k'}\overline{y_{k,l'}}|e^E_{k'}\rangle\langle e^E_{l'}| \\
&= \sum_k \langle e_k|\kappa^{\mathrm{PD}}_D(\rho)|e_k\rangle \rho^E_k.
\end{aligned}$$

Hence, the wiretap channel $(\kappa^{\mathrm{PD}}_D, (\kappa^{\mathrm{PD}}_D)^E)$ is a degraded channel. Further, the phase-damping channel κ^{PD}_D satisfies the invariance

$$I_c(U_\theta\rho U^*_\theta, \kappa^{\mathrm{PD}}_D) = I_c(\rho, \kappa^{\mathrm{PD}}_D), \quad U_\theta \stackrel{\mathrm{def}}{=} \sum_k e^{i\theta_k}|e_k\rangle\langle e_k|.$$

Hence, using concavity (9.89), we have

$$\begin{aligned}
C_{q,1}(\kappa^{\mathrm{PD}}_D) &= \max_\rho I_c(\rho, \kappa^{\mathrm{PD}}_D) = \max_p I_c(\sum_k p_k|e_k\rangle\langle e_k|, \kappa^{\mathrm{PD}}_D) \\
&= \max_p H(p) - H(\sum_k p_k\rho^E_k) \ge \log d - H(\frac{1}{d}D).
\end{aligned}$$

9.8 Historical Note

Bennett et al. [37] considered the transmission of classical information by using entangled states as input states. After this research, in order to consider the additivity of the classical channel capacity, Nagaoka [305] proposed quantum analogs of the Arimoto–Blahut algorithms [13, 53], and Nagaoka and Osawa [311] numerically analyzed two-tensor product channels in the qubit case with quantum analogs based on this algorithms. In this numerical analysis, all the examined channels κ satisfy $C(\kappa^{\otimes 2}) = 2C(\kappa)$. This numerical analysis strongly suggests Conjecture **HM**. This research was published by Osawa and Nagaoka [325]. Independently, King proved Conditions **HM**, **EM**, and **RM** with κ^1 as a unital channel in the qubit system and κ^2 as an arbitrary channel [253]. Following this result, Fujiwara and Hashizume [133] showed **HM** and **EM** with κ^1 and κ^2 as depolarizing channels. Further, King [254] proved **HM**, **EM**, and **RM** with only κ^1 as a depolarizing channel. Shor [372] also proved **HM** with only κ_1 as an entanglement-breaking channel.

On the other hand, Vidal et al. [403] pointed out that the entanglement of formation is $\log 2$ when the support of the state is contained by the antisymmetric space of $\mathbb{C}^3$. Following this research, Shimono [370] proved **FA** when the supports of ρ_1 and ρ_2 are contained by the antisymmetric space of $\mathbb{C}^3$; Yura [433] proved that $E_f(\rho) = E_c(\rho)$ for this case. Further, using the idea in Vidal et al. [403], Matsumoto et al. [290] introduced the MSW correspondence (9.6) or (9.25). Using this correspondence, they proved **FS** $\Rightarrow$ **HM** and **FS** $\Rightarrow$ **HL**.

Following this result, Shor [374] proved **HL** $\Rightarrow$ **FA** and **HM** $\Rightarrow$ **HL**. Audenaert and Braunstein [17] pointed out the importance of the conjugate function in this problem. Further, Pomeransky [350] proved the equivalence among **FA**, **FC**, and **FS** by employing the idea by Audenaert and Braunstein [17]. Shor also showed **FA** $\Rightarrow$ **FS** independently. He also proved **EM** $\Rightarrow$ **FA** and (**HM** or **FA**) $\Rightarrow$ **EM**. Further, applying this idea, Koashi and Winter [259] obtained relation (8.142). Recently, Matsumoto [287] found short proofs of **EM** $\Rightarrow$ **EL** and **EL** $\Leftrightarrow$ **ML**. In this textbook, based on his idea, we analyze the structure of equivalence among these conditions and derive 14 conditions (Theorem 9.3). Matsumoto [288] also introduced another measure of entanglement and showed that its additivity is equivalent to the additivity of entanglement of formation.

Further, Matsumoto and Yura [291] showed $E_f(\rho) = E_c(\rho)$ for antisymmetric states. Applying the concept of channel states to antisymmetric states, they proved that $C(\kappa^{\otimes n}) = C(\kappa)$ for antisymmetric channels. Indeed, this channel has been proposed by Werner and Holevo [413] as a candidate for a counterexample of Additivity **HM** or **EM** because they showed that it does not satisfy Condition **RM** for sufficiently large s. Vidal et al. implicitly applied the same concept to entanglement-breaking channels and proved **FA** when only ρ_1 satisfies condition (8.117). Following discovery of this equivalence, Datta et al. [90] and Fannes et al. [112] showed **HM** and **EM** when κ_1 and κ_2 are transpose-depolarizing channels. Wolf and Eisert [423], Fukuda [138], and Datta and Ruskai [91] extended the above results to larger classes of channels. Further, Matsumoto [288] obtained further equivalent conditions.

Concerning the channel coding with shared entanglement, Bennett and Wiesner [43] found the effectiveness of shared entanglement. Assuming Theorem 4.1

in the nonorthogonal two-pure-state case,[8] Barenco and Ekert [28] proved the direct part of Theorem 9.4 in the two-dimensional pure-state case. Hausladen et al. [164] independently proved the unitary coding version of Theorem 9.4 in the two-dimensional pure-state case. Bose et al. [56] showed the direct part of Theorem 9.4 in the two-dimensional mixed-state case. Hiroshima [237] showed the unitary coding version of Theorem 9.4 in the general mixed-state case. Bowen [58] independently showed the same fact in the two-dimensional case. Finally, Horodecki et al. [234] and Winter [419] independently proved Theorem 9.4 in the form presented in this book. When the channel has noise, Bennett et al. [41] showed the direct part of Theorem 9.5 in the general case and its converse part in the generalized Pauli case. In this converse part, they introduced the reverse Shannon theorem. Following this result, Bennett et al. [42] and Holevo [220] completed the proof of Theorem 9.5. In this book, we proved this theorem in a way similar to Holevo [220].

Many researchers have treated the capacity of quantum-state transmission via a noisy quantum channel by algebraic methods first [66, 67, 150, 155, 256, 371, 378]. This approach is called quantum error correction. Using these results, Bennett et al. [40] discussed the relation between quantum error correction and entanglement of distillation. Following these studies, Schumacher [361] introduced many information quantities for noisy channels (Sect. 8.2). Barnum et al. [32] showed that a capacity with the error $\varepsilon_2[\Phi^{(n)}]$ is less than $\lim \frac{1}{n} \max_{\rho \in \mathcal{S}(\mathcal{H}_A^{\otimes n})} I_c(\rho, \kappa^{\otimes n})$ if the encoding is restricted to being isometry. Barnum et al. [31] proved the coincidence with two capacities $C_1(\kappa)$ and $C_2(\kappa)$. They also showed that these capacities are less than $\lim \frac{1}{n} \max_{\rho \in \mathcal{S}(\mathcal{H}_A^{\otimes n})} I_c(\rho, \kappa^{\otimes n})$. On the other hand, Lloyd [273] predicted that the bound $I_c(\rho, \kappa)$ could be achieved without a detailed proof, and Shor [373] showed its achievability. Then, the capacity theorem for quantum-state transmission (Theorem 9.10) was obtained. Further, Devetak [95] formulated a capacity theorem for quantum wiretap channels (Theorem 9.8). Applying this discussion, he gave an alternative proof of Theorem 9.10. Here, the bit error of state transmission corresponds to the error of normal receiver in wiretap channels, and the phase error of state transmission corresponds to information obtained by the eavesdropper in a wiretap channel. Indeed, the analysis of information obtained by the eavesdropper is closely related to the channel resolvability. Hence, in this book, we analyze quantum-channel resolvability first. Then we proceed to quantum wiretap channels and quantum-state transmission. Indeed, Devetak [95] also essentially showed the direct part of quantum-channel resolvability (Theorem 9.7) in the tensor product case; however, our proof of it is slightly different from the proof by Devetak. Further, Devetak and Shor [96] studied the asymptotic tradeoff between the transmission rates of transmissions of quantum-state and classical information.

When a channel is degraded, the wiretap capacity can be single-letterized as (9.62). This formula was obtained independently in the original Japanese version of this book and by Devetak and Shor [96]. By applying this relation to quantum capacity, the single-letterized capacity is derived independently in several examples in this English version and Yard [427].

[8] In their paper, it is mentioned that Levitin [265] showed the direct part of Theorem 4.1 in this special case.

10

Source Coding in Quantum Systems

Summary. Recently, compression software used in computers to compress data has become an indispensable computational tool. Why is this compression possible in the first place? Information commonly possesses redundancies. In other words, information possesses some regularity. If one randomly types letters of the alphabet, it is highly unlikely that these will form a meaningful sentence or program. Imagine that we are assigned a task of communicating a sequence of 1000 binary digits via telephone. Assume that the $2n$th and $(2n+1)$th digits of this sequence are the same. Naturally, we would not read out all 1000 digits of the sequence; we would first say that the $2n$th and $(2n+1)$th digits are the same, and then read out the even-numbered (or odd-numbered) digits. We may even check whether there is any further structure in the sequence. In this way, compression software works by changing the input sequence of letters (or numbers) into another sequence of letters that can reproduce the original sequence, thereby reducing the necessary storage. The compression process may therefore be regarded as an encoding. This procedure is called source coding in order to distinguish it from the channel coding examined in Chap. 4.

Applying this idea to the quantum scenario, the presence of any redundant information in a quantum system may be similarly compressed to a smaller quantum memory for storage or communication. However, in contrast to the classical case, we may identify at least two distinct cases.

The task of the first case is saving memory in a quantum computer. This will be relevant when quantum computers are used in practice. In this case, a given quantum state is converted into a state on a system of lower size (dimension). The original state must then be recoverable from the compressed state. Note that the state to be compressed is unknown. The task of the second case is to save the quantum system to be sent for quantum cryptography. In this case, the sender knows what state to send. This provides the encoder with more options for compression. In the decompression stage, there is no difference between the first and second cases, since they are conversions from one quantum system to another. In this chapter, the two types of compression outlined above are discussed in detail.

Table 10.1. Denotations used in Chap. 10

Minimum compression rates

$R_{B,q}(W,p)$	Minimum compression rate in the blind and ensemble setting (10.2)
$R_{V,q}(W,p)$	Minimum compression rate in visble and ensemble setting (10.3)
$R_{P,q}(\rho)$	Minimum compression rate in purification setting (10.13)
$R^{\dagger}_{B,q}(W,p)$	Strong converse compression rate in blind and ensemble setting (10.4)
$R^{\dagger}_{V,q}(W,p)$	Strong converse compression rate in visible and ensemble setting (10.5)
$R^{\dagger}_{P,q}(\rho)$	Strong converse compression rate in purification setting (10.14)
$R_{V,c}(W,p)$	Minimum visible compression rate with classical memory (10.42)
$R_{V,q,r}(W,p)$	Minimum visible compression rate with quantum memory and shared randomness (10.44)
$R_{V,c,r}(W,p)$	Minimum visible compression rate with classical memory and shared randomness (10.45)

Codes

$\mathit{\Psi}$	Blind code
Ψ	Visible code
Ψ_c	Visible code by classical memory
Ψ_r	Visible code with common randomness
$\Psi_{c,r}$	Visible code with common randomness by classical memory

Chanel capacities

$C_{c,r}(W)$	Channel capacity for sending classical information with shared randomness (10.55)
$C^{R}_{c,r}(W)$	Reverse channel capacity for sending classical information with shared randomness (10.54)
$C_{c,e}(W)$	Channel capacity for sending classical information with shared entanglement
$C^{R}_{c,e}(W)$	Reverse channel capacity for sending classical information with shared entanglement
$C^{e}_{c,e}(\kappa)$	Channel capacity for sending classical information with shared entanglement and entangled input
$C^{e,R}_{c,e}(\kappa)$	Reverse channel capacity for sending classical information with shared entanglement and entangled input
$C^{e}_{q,e}(\kappa)$	Channel capacity for sending quantum states with shared entanglement and entangled input
$C^{e,R}_{q,e}(\kappa)$	Reverse channel capacity for sending quantum states with shared entanglement and entangled input

10.1 Four Kinds of Source Coding Schemes in Quantum Systems

As discussed above, source coding can be formulated in two ways. In the encoding process of the first scheme, we perform a state evolution from an original quantum system to a system of lower dimension. In that of the second, the encoder prepares a state in a system of lower dimension depending on the input signal. In the first case, the state is unknown since only the quantum system is given. Hence, the first scheme is called *blind*. In the second case, the state is known, and this scheme is called *visible*. The quality of the compression is evaluated by its compression rate. Of course, a lower dimension of the compressed quantum system produces a better encoding in terms of its compression rate. We may choose the compression rate to be either fixed or dependent on the input state. Coding with a fixed compression rate is called fixed-length coding, while that with a compression rate that depends on the input state is called variable-length coding. Therefore, there exist four schemes for the problem, i.e., fixed-/variable-length and visible/blind coding.

Let us summarize the known results on fixed- and variable-length compression in classical systems. In fixed-length compression, it is not possible to completely recover all input signals. Decoders can erroneously recover some input signals. However, by assuming that the state on the input system follows some probability distribution, a code can be obtained such that the probability of erroneously encoding a state is sufficiently close to zero for a compression rate below a thereshold [160, 366]. This thereshold is called the *minimum admissible rate*. In order to treat this problem precisely, we often assume that input data generate subject to the n-fold independent and identical distribution of a given probability distribution with sufficiently large n.

In variable-length compression, it is always possible to construct a code recovering all input signals perfectly. This is an advantage of variable-length encoding over fixed-length encoding. In this case, since there is no occurrence of errorly decoding, we measure the quality of the variable-length encoding by the coding length. The worst-case scenario in this type of coding occurs when the coding length is greater than the input information. However, if we assume some probability with respect to the input, the average coding length will be shorter than the number of bits in the input. For an independent and identical distribution, it has been shown that the average coding length is equal to its entropy in the optimal case [160, 366].

Let us now turn to quantum systems. As for the classical case, for fixed-length coding, it is possible to construct a coding protocol with an error of sufficiently small size for both visible and blind cases, provided the compression rate is below a certain value [245, 360]. This will be examined in more detail later. In fact such an encoding has already been realized experimentally [292]. For variable-length coding, in many cases there does not exist a code with zero error of a smaller coding length than the size of the input information [258]. However, when we replace the condition "zero error" by "almost zero error," it is possible to construct codes with the admissible compression rate. Therefore, if the information source is a quantum state that is generated by an n-fold independent and identical distribution of a "known" distribution, variable-length encoding does not offer any advantage.

On the other hand, if the probability distribution that generates the quantum state is unknown, the situation is entirely different. In fixed-length coding, since the compression rate is fixed a priori, it is impossible to recover the input state with a small error when the compression rate is less than the minimum admissible rate. In this case, it is preferable to use variable-length encoding wherein the compression rate depends on the input state [93, 277]. However, as a measurement is necessary to determine the compression rate, this will demolish the state after the measurement due to the quantum mechanical nature of the problem. Consider a method in which the encoding method and the compression rate are determined by an approximate estimation of an input state. If the initial state is demolished at a nonnegligible degree, clearly we cannot expect a decoding error that is close to zero. It is therefore necessary to examine the tradeoff between the degree to which the state is demolished due to this measurement and the estimation error of the distribution of the input state, which is required for determining the encoding method. As will be discussed later, both this estimation error and the degree of state demolition can be made to approach zero simultaneously and asymptotically. Therefore, even when the probability distribution of the quantum state is unknown, we can asymptotically construct a variable-length code such that the minimum admissible rate is achieved with a probability close to 1 and the decoding error is almost 0 [184, 185]. Hence, when a given coding protocol is effective for all probability distributions, it is called *universality*; this is an important topic in information theory. Various other types of source compression problems have also been studied [97, 197].

10.2 Quantum Fixed-Length Source Coding

The source of quantum system $\mathcal{H}$ is denoted by

$$W : \mathcal{X} \to \mathcal{S}(\mathcal{H}) \ (x \mapsto W_x) \tag{10.1}$$

(which is the same notation as that in a quantum channel) and a probability distribution p in $\mathcal{X}$. That is, the quantum information source is described by the ensemble $(p_x, W_x)_{x \in \mathcal{X}}$. Let $\mathcal{K}$ be the compressed quantum system. For the blind case, the *encoder* is represented by the TP-CP map τ from $\mathcal{S}(\mathcal{H})$ to $\mathcal{S}(\mathcal{K})$. The *decoder* is represented by a TP-CP map ν from $\mathcal{S}(\mathcal{K})$ to $\mathcal{S}(\mathcal{H})$. The triplet $\mathit{\Psi} \stackrel{\text{def}}{=} (\mathcal{K}, \tau, \nu)$ is then called a blind *code*. In the visible case, the encoder is not as restricted as in the blind case. In this case, the *encoder* is given by a map T from $\mathcal{X}$ to $\mathcal{S}(\mathcal{K})$. Any blind encoder τ can be converted into a visible encoder according to $\tau \circ W$. The triplet $\Psi \stackrel{\text{def}}{=} (\mathcal{K}, T, \nu)$ is then called a visible *code*. That is, the information is stored by a quantum memory. The errors $\varepsilon_p(\mathit{\Psi})$ and $\varepsilon_p(\Psi)$ and *sizes* $|\mathit{\Psi}|$ and $|\Psi|$ of the codes $\mathit{\Psi}$ and Ψ, respectively, are defined as follows:

$$\varepsilon_p(\Psi) \stackrel{\text{def}}{=} \sum_{x \in \mathcal{X}} p_x \left(1 - F^2(W_x, \nu \circ \tau(W_x))\right), \quad |\Psi| \stackrel{\text{def}}{=} \dim \mathcal{K}$$

$$\varepsilon_p(\Psi) \stackrel{\text{def}}{=} \sum_{x \in \mathcal{X}} p_x \left(1 - F^2(W_x, \nu \circ T(x))\right), \quad |\Psi| \stackrel{\text{def}}{=} \dim \mathcal{K}.$$

We used $1 - F^2(\cdot, \cdot)$ in our definition of the decoding error.

Now, let the source be given by the quantum system $\mathcal{H}^{\otimes n}$ and its candidate states be given by $W^{(n)} : \mathcal{X}^{(n)} \to \mathcal{S}(\mathcal{H}^{\otimes n})$ ($x^n = (x_1, \ldots, x_n) \mapsto W^{(n)}_{x^n} \stackrel{\text{def}}{=} W_{x_1} \otimes \cdots \otimes W_{x_n}$). Further, let the probability distribution for these states be given by the nth-order independent and identical distribution of the probability distribution p in $\mathcal{X}$. Denote the blind and visible codes by $\Psi^{(n)}$ and $\Psi^{(n)}$, respectively. Define[1]

$$R_{B,q}(W,p) \stackrel{\text{def}}{=} \inf_{\{\Psi^{(n)}\}} \left\{ \overline{\lim} \frac{1}{n} \log |\Psi^{(n)}| \,\middle|\, \varepsilon_{p^n}(\Psi^{(n)}) \to 0 \right\}, \tag{10.2}$$

$$R_{V,q}(W,p) \stackrel{\text{def}}{=} \inf_{\{\Psi^{(n)}\}} \left\{ \overline{\lim} \frac{1}{n} \log |\Psi^{(n)}| \,\middle|\, \varepsilon_{p^n}(\Psi^{(n)}) \to 0 \right\}, \tag{10.3}$$

$$R^{\dagger}_{B,q}(W,p) \stackrel{\text{def}}{=} \inf_{\{\Psi^{(n)}\}} \left\{ \overline{\lim} \frac{1}{n} \log |\Psi^{(n)}| \,\middle|\, \overline{\lim}\, \varepsilon_{p^n}(\Psi^{(n)}) < 1 \right\}, \tag{10.4}$$

$$R^{\dagger}_{V,q}(W,p) \stackrel{\text{def}}{=} \inf_{\{\Psi^{(n)}\}} \left\{ \overline{\lim} \frac{1}{n} \log |\Psi^{(n)}| \,\middle|\, \overline{\lim}\, \varepsilon_{p^n}(\Psi^{(n)}) < 1 \right\}. \tag{10.5}$$

Since a blind code $\Psi^{(n)}$ can be regarded as a visible code, we have

$$R_{B,q}(W,p) \geq R_{V,q}(W,p), \quad R^{\dagger}_{B,q}(W,p) \geq R^{\dagger}_{V,q}(W,p). \tag{10.6}$$

From the definitions it is also clear that

$$R_{B,q}(W,p) \geq R^{\dagger}_{B,q}(W,p), \quad R_{V,q}(W,p) \geq R^{\dagger}_{V,q}(W,p). \tag{10.7}$$

The following theorem holds with respect to the above.

Theorem 10.1 *If all W_xs are pure states, then the quantities defined above are equal. We have*

$$R_{B,q}(W,p) = R^{\dagger}_{B,q}(W,p) = R_{V,q}(W,p) = R^{\dagger}_{V,q}(W,p) = H(W_p). \tag{10.8}$$

This theorem can be proved by combining the following two lemmas.

Lemma 10.1 (Direct Part) *There exists a sequence of blind codes $\{\Psi^{(n)}\}$ satisfying*

$$\frac{1}{n} \log |\Psi^{(n)}| \leq H(W_p) - \delta \tag{10.9}$$

$$\varepsilon_{p^n}(\Psi^{(n)}) \to 0 \tag{10.10}$$

for arbitrary real number $\delta > 0$.

[1] The subscript q indicates the "quantum" memory.

Lemma 10.2 (Converse Part) *If all W_xs are pure states and the sequence of visible codes $\{\Psi^{(n)}\}$ satisfies*

$$\overline{\lim} \frac{1}{n} \log |\Psi^{(n)}| < H(W_p), \tag{10.11}$$

then

$$\varepsilon_{p^n}(\Psi^{(n)}) \to 1. \tag{10.12}$$

Lemma 10.1 tells us that $R_{B,q}(W,p) \leq H(W_p)$, and Lemma 10.2 tells us that $R^{\dagger}_{V,q}(W,p) \geq H(W_p)$. Using (10.6) and (10.7), we thus obtain (10.8).

Further, we have another fixed-length coding scheme. In this scheme, the state is given as a pure state $|x\rangle\langle x|$ on the composite system $\mathcal{H}_A \otimes \mathcal{H}_R$, and encoder and decoder can treat only the local system $\mathcal{H}_A$. Then, our task is recovering the state $|x\rangle\langle x|$ on the composite system $\mathcal{H}_A \otimes \mathcal{H}_R$. Hence, the code of this scheme is the triplet $\Psi \stackrel{\text{def}}{=} (\mathcal{K}, \tau, \nu)$, which is the same as that of the blind scheme. The error is given as

$$\varepsilon'_{\rho}(\Psi) \stackrel{\text{def}}{=} 1 - \langle x|(\nu \otimes \iota) \circ (\tau \otimes \iota)(|x\rangle\langle x|)|x\rangle = 1 - F_e^2(\rho, \nu \circ \tau),$$

where $\rho = \mathrm{Tr}_R |x\rangle\langle x|$. Recall the definition of the entanglement fidelity (8.18). Hence, the quality depends only on the reduced density ρ. This scheme is called the *purification scheme*, while the former scheme with the visible case and the blind case is called the *ensemble scheme*. Hence, we define the minimum compression rate as

$$R_{P,q}(\rho) \stackrel{\text{def}}{=} \inf_{\{\Psi^{(n)}\}} \left\{ \overline{\lim} \frac{1}{n} \log |\Psi^{(n)}| \,\middle|\, \varepsilon'_{\rho^{\otimes n}}(\Psi^{(n)}) \to 0 \right\}, \tag{10.13}$$

$$R^{\dagger}_{P,q}(\rho) \stackrel{\text{def}}{=} \inf_{\{\Psi^{(n)}\}} \left\{ \overline{\lim} \frac{1}{n} \log |\Psi^{(n)}| \,\middle|\, \overline{\lim}\, \varepsilon'_{\rho^{\otimes n}}(\Psi^{(n)}) < 1 \right\}. \tag{10.14}$$

From inequality (8.22) we have $R_{P,q}(\rho) \leq R_{P,q}(W,p)$ and $R^{\dagger}_{P,q}(\rho) \leq R^{\dagger}_{P,q}(W,p)$ when all W_x states are pure and $\rho = W_p$. The following theorem also holds (Exercise 10.2).

Theorem 10.2

$$R_{P,q}(\rho) = R^{\dagger}_{P,q}(\rho) = H(\rho). \tag{10.15}$$

Exercises

10.1. Show that $\sum_{\boldsymbol{x} \in \mathcal{X}^n} p^n_{\boldsymbol{x}} F(W^{(n)}_{\boldsymbol{x}}, \nu_n(T(\boldsymbol{x}))) \to 1$ is equivalent to $\sum_{\boldsymbol{x} \in \mathcal{X}^n} p^n_{\boldsymbol{x}} F(W^{(n)}_{\boldsymbol{x}}, \nu_n(T(\boldsymbol{x}))) \to 1$.

10.3 Construction of a Quantum Fixed-Length Source Code

Let us construct a blind fixed-length code that attains the minimum compression rate $H(W_p)$ when the quantum state is generated subject to the independent and identical distribution of the probability distribution p. (Since any blind code can be regarded as a visible code, it is sufficient to construct a blind code.) Since formula (8.22) guarantees that

$$\varepsilon_p(\Psi) = \sum_{x \in \mathcal{X}} p(x)(1 - F^2(W_x, \nu \circ \tau(W_x))) \le 1 - F_e^2(W_p, \nu \circ \tau) = \varepsilon'_{W_p}(\Psi),$$

it is sufficient to treat the purification scheme.

Now define $\rho_{\text{mix}}^P \stackrel{\text{def}}{=} \frac{P}{\operatorname{Tr} P}$. Let the encoder $\tau_P : \mathcal{S}(\mathcal{H}) \to \mathcal{S}(\operatorname{Ran} P)$, using the projection P in $\mathcal{H}$, be given by

$$\tau_P(\rho) \stackrel{\text{def}}{=} P\rho P + \operatorname{Tr}[(I - P)\rho]\rho_{\text{mix}}^P. \tag{10.16}$$

Define the encoder ν_P as the natural embedding from $\mathcal{S}(\operatorname{Ran} P)$ to $\mathcal{S}(\mathcal{H})$, where $\operatorname{Ran} A$ is the range of A.

Let x be the purification of ρ. Then,

$$\begin{aligned}
&F_e^2(\rho, \nu_P \circ \tau_P) \\
&= \langle x|(I \otimes P)|x\rangle\langle x|(I \otimes P)|x\rangle \\
&\quad + \langle x|\rho_{\text{mix}}^P \otimes \operatorname{Tr}_{\mathcal{H}} \left[(I \otimes (I - P))|x\rangle\langle x|(I \otimes (I - P))\right]|x\rangle \\
&\le \langle x|(I \otimes P)|x\rangle\langle x|(I \otimes P)|x\rangle = (\operatorname{Tr} P\rho)^2 \\
&= (1 - (1 - \operatorname{Tr} P\rho))^2 \le 1 - 2(1 - \operatorname{Tr} P\rho).
\end{aligned} \tag{10.17}$$

We now define $b(s,R) \stackrel{\text{def}}{=} \frac{R - \psi(s)}{1-s}$, $\psi(s) = \psi(s|\rho)$, $s_0 \stackrel{\text{def}}{=} \operatorname{argmax}_{0 < s < 1} \frac{sR - \psi(s)}{1-s}$ for $R > H(\rho)$ and $0 < s < 1$. We choose P such that

$$P_n \stackrel{\text{def}}{=} \left\{ W_p^{\otimes n} - e^{-nb(s_0,R)} > 0 \right\}. \tag{10.18}$$

Then, from (2.45) and (2.47), the code $\Phi^{(n)} = (\mathcal{K}_n, \tau_{P_n}, \nu_{P_n})$ satisfies

$$\begin{aligned}
\dim \mathcal{K}_n &\stackrel{\text{def}}{=} \operatorname{Ran} P_n = \operatorname{Tr} \left\{ \rho^{\otimes n} - e^{-nb(s,R)} > 0 \right\} \le e^{nR} \\
\varepsilon'_{\rho^{\otimes n}}(\Phi^{(n)}) &\le 2 \operatorname{Tr} \rho^{\otimes n} \left\{ \rho^{\otimes n} - e^{-nb(s_0,R)} \le 0 \right\} \le 2e^{-n\frac{s_0 R - \psi(s_0)}{1-s_0}}.
\end{aligned} \tag{10.19}$$

Thus, (10.9) and (10.10) hold. This proves the existence of a code that attains the compression rate $H(\rho) + \delta$ for arbitrary $\delta > 0$. Note that the code constructed here depends only on the state ρ and the rate R. In order to emphasize this dependence, we denote this encoding and decoding as $\tau_{n,\rho,R}$ and $\nu_{n,\rho,R}$, respectively.

We next show that the code given above still works even when the true density ρ' is slightly different from the predicted density ρ. This property is called *robustness* and is important for practical applications.

Let us consider the case where the true density ρ' is close to the predicted one ρ. Choosing a real number $\alpha > 0$, we have

$$\rho' \leq \rho e^{\alpha}. \tag{10.20}$$

Hence,

$$\operatorname{Tr} \rho'^{\otimes n} \leq e^{n\alpha} \operatorname{Tr} \rho^{\otimes n}.$$

Using the same argument as that in the derivation of (10.19), we obtain

$$\begin{aligned} &\varepsilon'_{\rho'^{\otimes n}}(\Phi^{(n)}) \leq 2 \operatorname{Tr} \rho'^{\otimes n} \{\rho^{\otimes n} - e^{-na} < 0\} \\ \leq &2e^{n\alpha} \operatorname{Tr} \rho^{\otimes n} \{\rho^{\otimes n} - e^{-na} < 0\} \leq 2e^{n(\alpha + \frac{\psi(s_0) - s_0 R}{1-s_0})}. \end{aligned} \tag{10.21}$$

Therefore, if $\alpha < \max_{0 \leq s \leq 1} \frac{sR - \psi(s)}{1-s}$, then $\varepsilon'_{\rho'^{\otimes n}}(\Phi^{(n)}) \to 0$.

Let us now prove the converse part of the theorem, i.e., Lemma 10.2. For a proof of Lemma 10.2, we prepare the following lemma, which is proved in Appendix B.8.

Lemma 10.3 *(Hayashi [175]) Any visible code $\Psi = (\mathcal{K}, T, \nu)$ satisfies*

$$1 - \varepsilon(\Psi) \leq a|\Psi| + \operatorname{Tr} W_p\{W_p - a \geq 0\} \tag{10.22}$$

for $\forall a > 0$.

Proof of Lemma 10.2. The above inequality (10.22) can be regarded as the "dual" inequality of inequality (10.19) given in the proof of the direct part of the theorem. Inequality (10.22) shows that the quality of any code is evaluated by use of $\operatorname{Tr} W_p\{W_p - e^{\lambda} \geq 0\}$. Inequality (10.22) plays the same role as (2.41) in Sect. 2.1.4, and thus any sequence of codes $\{\Psi^{(n)}\}$ satisfies

$$1 - \varepsilon(\Psi^{(n)}) \leq 2e^{n\frac{\psi(s) - sR}{1-s}}. \tag{10.23}$$

Choosing an appropriate $s_0 < 0$, we have $\frac{\psi(s_0) - s_0 R}{1-s_0} < 0$. Therefore, we obtain (10.12), which gives us Lemma 10.2. ∎

In order to construct a code with the compression rate $H(W_p)$, we replace R by $H(W_p) - \frac{1}{n^{1/4}}$ in (10.23). Approximating $\psi(s)$ as $H(W_p)s + \frac{1}{2}\psi''(0)s^2$, we obtain

$$\min_{s<0} \frac{\psi(s) - s(H(W_p) - \frac{1}{n^{1/4}})}{1-s} \cong -\frac{C^2}{\psi''(0)\sqrt{n}}. \tag{10.24}$$

Hence,

$$\varepsilon_{p^n}(\Psi^{(n)}) \le 2e^{-\sqrt{n}\frac{C^2}{\psi''(0)}} \to 0. \tag{10.25}$$

Finally, let us focus on the property of the state on the compressed system when the asymptotic compression rate is $H(W_p)$.

Theorem 10.3 *(Han [162]) When a sequence of codes $\{\Psi_n = (\mathcal{K}_n, T_n, \nu_n)\}$ satisfies*

$$\varepsilon_{p^n}(\Psi_n) \to 0, \quad \frac{1}{n}\log|\Psi_n| \to H(W_p), \tag{10.26}$$

we obtain

$$\frac{1}{n}D(\sum_x p_x^n T_n(x)\|\rho_{\mathrm{mix}}^{\mathcal{K}_n}) \to 0. \tag{10.27}$$

That is, the compressed state is almost completely mixed in the sense of the normalized quantum relative entropy $\frac{1}{n}D(\rho_n\|\sigma_n)$. However, the compressed state is different from the completely mixed state if we focus on the Bures distance, trace norm, or quantum relative entropy. This fact has been shown in the classical case [179].

Proof. From the monotonicity of the transmission information (5.48) we have

$$\begin{aligned}\log|\Phi_n| &\ge H\Big(\sum_x p_x^n T_n(x)\Big) \ge H\Big(\sum_x p_x^n T_n(x)\Big) - \sum_x p_x^n H(T_n(x)) \\ &\ge H\Big(\sum_x p_x^n \nu(T_n(x))\Big) - \sum_x p_x^n H(\nu(T_n(x))).\end{aligned}$$

From condition (10.26) the two conditions in (5.72) yield

$$\lim \frac{1}{n}\left(H\Big(\sum_x p_x^n \nu(T_n(x))\Big) - \sum_x p_x^n H(\nu(T_n(x))\right) = H(W_p).$$

Hence, we obtain

$$\lim \frac{1}{n} H\Big(\sum_x p_x^n T_n(x)\Big) = H(W_p).$$

Since $\mathrm{Tr}\sum_x p_x^n T_n(x)\log\rho_{\mathrm{mix}}^{\mathcal{K}_n} = -\log|\Phi_n|$, relation (10.27) holds. ■

Exercises

10.2. Prove equation (10.15) using the inequality $R^{\dagger}_{P,q}(\rho) \le R^{\dagger}_{P,q}(W,p)$ and (10.17).

10.4 Universal Quantum Fixed-Length Source Codes

The code given in Sect. 10.3 depends on the quantum state W_p. For the classical case, there exists a code that depends on the value $H(p)$ and works when the data are generated subject to the independent and identical information source of p. Such codes are called *universal* codes. To construct such universal codes, we often use the method of types, which is discussed in Sect. 2.5.1 [85]. Similarly, for the quantum case, there exists a code that depends only on the entropy $H(W_p)$ of the average state W_p and works well provided the states are generated according to an independent and identical distribution of p [247]. For this purpose, the projection P_n given by (10.18) should depend only on the compression rate. That is, we have to construct a subspace $\Upsilon_n(R)$ of $\mathcal{H}^{\otimes n}$ depending only on the compression rate R. As the first step, we construct a code depending only on the compression rate and the basis $B = (u_1, \dots, u_d)$ comprising the eigenvectors of W_p. Let us consider a set of types T_n with the probability space $\mathbb{N}_d = \{1, \dots, d\}$. Define the subspace $\Upsilon_n(R, B)$ of the n-fold tensor product space $\mathcal{H}^{\otimes n}$ to be the space spanned by $\cup_{q \in T^n : H(q) \le R} \{u(\boldsymbol{i}_n)\}_{\boldsymbol{i}_n \in T_q^n}$, where $u(\boldsymbol{i}_n) \stackrel{\text{def}}{=} u_{i_1} \otimes \cdots \otimes u_{i_n} \in \mathcal{H}^{\otimes n}$ and $\boldsymbol{i}_n = (i_1, \dots, i_n)$. Let $P_{n,R,B}$ be a projection to $\Upsilon_n(R, B)$. Then, according to the discussion in Sect. 2.5.1,

$$\dim \Upsilon_n(R, B) \le (n+1)^d e^{nR},$$
$$\mathrm{Tr}(I - P_{n,R,B}) W_p^{\otimes n} \le (n+1)^l \exp(-n \inf_{q : H(q) > R} D(q\|r)).$$

Hence, the code $\{(\Upsilon_n(R, B), \nu_n, \tau_{P_{n,R,B}})\}$ almost has the compression rate R. Since $\min_{0 \le s \le 1} \frac{\psi(s) - sR}{1-s} = \min_{q : H(q) \ge R} D(q\|r)$, its entanglement fidelity $F_e(W_p^{\otimes n}, \nu_n \circ \tau_{P_{n,R,B}})$ asymptotically approaches 1. This code is effective when the basis $B = \{u_1, \dots, u_d\}$ is known. However, when the basis B is unknown, it is suitable to use the code constructed from the space $\Upsilon_n(R)$ spanned by $\cup_B \Upsilon_n(R, B)$. The space $\Upsilon_n(R)$ and the projection $P_{n,R}$ to $\Upsilon_n(R)$ satisfy

$$\dim \Upsilon_n(R) \le (n+1)^{d+d^2} e^{nR}, \tag{10.28}$$
$$\mathrm{Tr}(I - P_{n,R}) W_p^{\otimes n} \le (n+1)^d \exp(-n \inf_{q : H(q) > R} D(q\|r)). \tag{10.29}$$

Hence, there exists a code attaining the entropy rate in a manner similar to that for a fixed basis. Since (10.29) follows immediately from $P_{n,R} \ge P_{n,R,B}$, we prove inequality (10.28) as follows. For simplicity, we consider the case of $d = 2$, but this discussion may be easily extended to the general case. First, we fix the basis $B = \{u_1, u_2\}$. Then, an arbitrary basis $B' = \{u_1', u_2'\}$ may be written as $u_1' = a u_1 + b u_2, u_2' = c u_1 + d u_2$ using $d^2 = 4$ complex numbers a, b, c, d. Thus,

$$u_1' \otimes u_2' \otimes \cdots \otimes u_1' = (a u_1 + b u_2) \otimes (c u_1 + d u_2) \otimes \cdots \otimes (a u_1 + b u_2).$$

Choosing an appropriate vector $v_{n_1,n_2,n_3,n_4} \in \mathcal{H}^{\otimes n}$, we have

$$u_1' \otimes u_2' \otimes \cdots \otimes u_1' = \sum_{n_1,n_2,n_3,n_4} a^{n_1} b^{n_2} c^{n_3} d^{n_4} v_{n_1,n_2,n_3,n_4}.$$

The vector v_{n_1,n_2,n_3,n_4} does not depend on a, b, c, and d. Hence, the vector $u_1' \otimes u_2' \otimes \cdots \otimes u_1'$ belongs to the subspace spanned by the vectors v_{n_1,n_2,n_3,n_4} with the condition $n_1 + n_2 + n_3 + n_4 = n$. The dimension of this subspace is at most $(n+1)^{2^2-1} = (n+1)^{d^2-1}$. Since the dimension of the space $\Upsilon_n(R)$ is at most this number multiplied by the dimension of the space $\Upsilon_n(R,B)$ with a fixed basis, we obtain (10.28).

Moreover, combining (10.28) with (2.42) and (2.142), we obtain

$$\begin{aligned}
\operatorname{Tr} P_{n,R} W_p^{\otimes n} &\le 2\exp\left(\min_{s\le 0} \frac{n\psi(s) - s\log\dim \Upsilon_n(R)}{1-s}\right) \\
&\le 2\exp\left(n\frac{\psi(s_0) - s_0 R}{1-s_0} + \frac{-s_0}{1-s_0}(d+d^2)\log(n+1)\right) \\
&= 2(n+1)^{\frac{-s_0}{1-s_0}(d+d^2)} \exp(-n \inf_{q:H(q)\le R} D(q\|r)), \qquad (10.30)
\end{aligned}$$

where $s_0 \stackrel{\text{def}}{=} \operatorname{argmin}_{s\le 0} \frac{\psi(s)-sR}{1-s}$.

10.5 Universal Quantum Variable-Length Source Codes

Let us construct a code that has a sufficiently small error and achieves the entropy rate $H(W_p)$, even though it is unknown, provided the source follows an independent and identical distribution of p. Such codes are called universal quantum variable-length source codes. For these codes, it is essential to determine the compression ratio depending on the input state.[2] If nonorthogonal states are included in the source, then state demolition inevitably occurs during the determination of the compression ratio. Hence, the main problem is to reduce as much state demolition as possible [184, 185].

Let us first construct a measurement that determines the compression ratio using the projection $P_{n,R}$ constructed in the previous section. Consider the projection $E_{n,R} \stackrel{\text{def}}{=} \lim_{\epsilon\to+0}(P_{n,R} - P_{n,R-\epsilon})$. Let $\Omega_n = \{H(p)\}_{p\in T^n}$ be a set of R such that $E_{n,R}$ is nonzero. Then, $\sum_{R\in\Omega_n} E_{n,R} = I$. The probability distribution by the measurement $E_n = \{E_{n,R_i}\}_i$ satisfies

[2] Even though the error of universal quantum variable-length source code converges to 0, its convergence rate is not exponential [184].

$$
\begin{aligned}
&\mathrm{P}^{E_n}_{W_p^{\otimes n}}\{|H(W_p) - R_i| \geq \epsilon\} \\
&\leq 2\max\Bigl\{(n+1)^{d+d^2}\exp(-n \inf_{q:H(q)\leq R-\epsilon} D(q\|r)), \\
&\quad (n+1)^d \exp(-n \inf_{q:H(q)\geq R+\epsilon} D(q\|r))\Bigr\}.
\end{aligned}
$$

We may therefore apply the arguments of Sect. 7.3. Choosing l_n, δ_n so that they satisfy (7.44) and (7.45), and constructing $\boldsymbol{M}^{(n),\delta_n,l_n}$ from E_n, we have

$$
F_e^2(W_p^{\otimes n}, \boldsymbol{\kappa}_{\boldsymbol{M}^{(n),\delta_n,l_n}}) \to 1.
$$

Since the measurement $\boldsymbol{M}^{(n),\delta_n,l_n}$ takes values in $[0, \log d]$ with spacing δ_n, the number of its measurement data is $(\frac{\log d}{\delta_n} + 1)$. Hence, we choose δ_n such that $\frac{1}{n}\log\delta_n \to 0$.

We now construct a universal variable-length code based on this measurement. In the encoding step, we perform a measurement corresponding to the instrument $\boldsymbol{\kappa}_{\boldsymbol{M}^{(n),\delta_n,l_n}}$. When the measurement data are R_i, the resulting state is a state in $\operatorname{Ran} M_i^{(n),\delta_n,l_n}$. The state in the space $\operatorname{Ran} M_i^{(n),\delta_n,l_n}$ is sent with the measurement data R_i. Then, the coding length is $\log\dim\operatorname{Ran} M_i^{(n),\delta_n,l_n} + \log(\frac{\log d}{\delta_n} + 1)$. The compression ratio is this value divided by n. Since the second term converges to zero after the division by n, we only consider the first term. From

$$
\begin{aligned}
\dim M_R^{(n),\delta_n,l_n} &= \sum_{R-\delta_n < R' < R+\delta_n} \operatorname{rank} E_{n,R'} \\
&\leq \dim \Upsilon_n(R+\delta_n) \leq (n+1)^{d+d^2} e^{n(R+\delta_n)}
\end{aligned}
$$

we obtain

$$
\frac{1}{n}\left\{\log\dim\operatorname{Ran} M_i^{(n),\delta_n,l_n} + \log\left(\frac{\log d}{\delta_n} + 1\right)\right\} \leq R.
$$

Therefore, in this protocol, the compression ratio is asymptotically less than the entropy $H(W_p)$ with a probability of approximately 1 [more precisely the compression ratio converges to the entropy $H(W_p)$ in probability]; the error also approaches zero asymptotically.

10.6 Mixed-State Case

So far, we have treated quantum data compression when W_x is pure. That is, in this case, the optimal compression rate in the blind case coincides with that in the visible case. However, when W_x is not pure, these are different. In this problem, one may think that the quantity $H(W_p)$ or $I(p, W)$ is a good candidate for the optimal rate. This intuition is not entirely inaccurate. In the blind scheme, if the ensemble (p_x, W_x) has no trivial redundancy, the optimal compression rate is given as follows [257]:

$$R_{B,q}(W,p) = H(W_p). \tag{10.31}$$

On the other hand, in the visible scheme, the inequality

$$R_{B,q}(W,p) \geq I(p,W) \tag{10.32}$$

holds[Ex. 10.5] [228]. However, it does not give the optimal rate in general:

Theorem 10.4

$$R_{V,q}(W,p) = E_c^{\dashrightarrow}(\tilde{W}_p) = \lim \frac{1}{n} E_p(\tilde{W}_p^{\otimes n}), \quad \tilde{W}_p \stackrel{\text{def}}{=} \sum_x p_x |e_x^A\rangle\langle e_x^A| \otimes W_x. \tag{10.33}$$

In fact, Horodecki [231] focused on the quantity

$$H^{ext}(W,p) \stackrel{\text{def}}{=} \inf_{W_x^{ext}:\text{purification of } W_x} H(\sum_x p_x W_x^{ext})$$

and proved

$$R_{V,q}(W,p) = \lim \frac{H^{ext}(W^{(n)}, p^n)}{n}. \tag{10.34}$$

From the definition of $E_p(\tilde{W}_p)$, we can easily check that $E_p(\tilde{W}_p) \leq H^{ext}(W,p)$. When all W_xs are pure, $R_{V,q}(W,p) = H(W_P)$. This fact fits into (8.141).

The direct part essentially is obtained by the following lemma.

Lemma 10.4 *Let κ be a one-way LOCC operation. There exists a code Ψ such that*

$$\frac{1}{2}\varepsilon_p(\Psi) \leq \varepsilon(\tilde{W}_p, \kappa, L) + \frac{1}{2}\|\tilde{W}_p - \kappa(|\Phi_L\rangle\langle\Phi_L|)\|_1, \tag{10.35}$$

$$|\Psi| = L \cdot \mathrm{CC}(\kappa), \tag{10.36}$$

where $\mathrm{CC}(\kappa)$ is the size of the classical communication of κ.

(Note that any two-way LOCC operation can be simulated by one-way LOCC when the initial state is pure [276].)

Proof of Theorem 10.4. First, we prove the direct part. Using this lemma, we obtain the direct part as follows. Let κ_n be a one-way LOCC operation satisfying

$$\lim F(\tilde{W}_p^{\otimes n}, \kappa_n(|\Phi_{L_n}\rangle\langle\Phi_{L_n}|)) = 1, \quad \frac{\log \mathrm{CC}(\kappa_n)}{n} \to 0,$$

$$\lim \frac{\log L_n}{n} \leq E_c^{\dashrightarrow}(\tilde{W}_p) + \epsilon$$

for any $\epsilon > 0$. Thus, the application of this lemma indicates that there exists a sequence of codes $\{\Psi_n\}$ such that

$$\varepsilon_{p^n}(\Psi_n) \to 0, \ \lim \frac{\log |\Psi_n|}{n} \le E_c^{-\to}(\tilde{W}_p) + \epsilon.$$

Therefore, we obtain

$$R(W,p) \le E_c^{-\to}(\tilde{W}_p).$$

Next, we prove the converse part. For any $\epsilon > 0$, we choose a sequence of codes $\Psi_n = (\mathcal{K}_n, T_n, \nu_n)$ such that

$$R \stackrel{\text{def}}{=} \overline{\lim} \frac{1}{n} \log |\Psi_n| \le R_{V,q}(W,p) + \epsilon, \quad \varepsilon_{p^n}(\Psi_n) \to 0.$$

Since $\tilde{W}_p^{\otimes n} = \sum_{\boldsymbol{x} \in \mathcal{X}^n} p^n_{\boldsymbol{x}} |e^A_{\boldsymbol{x}}\rangle\langle e^A_{\boldsymbol{x}}| \otimes W^{(n)}_{\boldsymbol{x}}$, the state $\tilde{\rho}_n \stackrel{\text{def}}{=} \sum_{\boldsymbol{x} \in \mathcal{X}^n} p^n_{\boldsymbol{x}} |e^A_{\boldsymbol{x}}\rangle\langle e^A_{\boldsymbol{x}}| \otimes T_n(\boldsymbol{x})$ satisfies

$$\begin{aligned} F(\tilde{W}_p^{\otimes n}, \iota_A \otimes \nu_n(\tilde{\rho}_n)) &= \sum_{\boldsymbol{x} \in \mathcal{X}^n} p^n_{\boldsymbol{x}} F(W^{(n)}_{\boldsymbol{x}}, \nu_n \circ T_n(\boldsymbol{x})) \\ &\ge \sum_{\boldsymbol{x} \in \mathcal{X}^n} p^n_{\boldsymbol{x}} F^2(W^{(n)}_{\boldsymbol{x}}, \nu_n \circ T_n(\boldsymbol{x})) \to 1. \end{aligned}$$

From Lemma 8.11 and Condition **E2′** we have

$$\log |\Psi_n| \ge H(\operatorname{Tr}_A \tilde{\rho}_n) \ge E_p(\tilde{\rho}_n) \ge E_p(\iota_A \otimes \nu_n(\tilde{\rho}_n)).$$

Since E_p satisfies Condition **E3** (Exercise 8.50), we have

$$\underline{\lim} \frac{1}{n} \log |\Psi_n| \ge \lim \frac{1}{n} E_p(\tilde{W}_p^{\otimes n}).$$

Hence, using Theorem 8.11, we obtain

$$R_{V,q}(W,p) \ge E_c^{-\to}(\tilde{W}_p) = \lim \frac{1}{n} E_p(\tilde{W}_p^{\otimes n}).$$

∎

Proof of Lemma 10.4.
Construction of code Ψ satisfying (10.35) and (10.36): Assume that operation κ has the form $\kappa = \sum_i \kappa_{A,i} \otimes \kappa_{B,i}$, where $\{\kappa_{A,i}\}_{i=1}^{l_n}$ is an instrument (a TP-CP-map-valued measure) on $\mathcal{H}_A$ and $\kappa_{B,i}$ is a TP-CP map on $\mathcal{H}_B$ for each i. Define the probability q_x

$$\begin{aligned} q_x &\stackrel{\text{def}}{=} \operatorname{Tr} |e^A_x\rangle\langle e^A_x| \otimes I_B \sum_i \kappa_{A,i} \otimes \kappa_{B,i}(|\Phi_L\rangle\langle\Phi_L|) \qquad (10.37) \\ &= \sum_i \operatorname{Tr} \kappa^*_{A,i}(|e^A_x\rangle\langle e^A_x|) \otimes I_B(|\Phi_L\rangle\langle\Phi_L|), \end{aligned}$$

the probability $p_{i,x}$, and the state $\rho_{i,x}$ as

$$p_{i,x} \stackrel{\text{def}}{=} \frac{\operatorname{Tr} \kappa_{A,i}^*(|e_x^A\rangle\langle e_x^A|) \otimes I_B(|\Phi_L\rangle\langle\Phi_L|)}{q_x}$$

$$\rho_{i,x} \stackrel{\text{def}}{=} \frac{\operatorname{Tr}_A \kappa_{A,i}^*(|e_x^A\rangle\langle e_x^A|) \otimes I_B(|\Phi_L\rangle\langle\Phi_L|)}{q_x p_{i,x}}.$$

Now we construct the coding protocol Ψ. When the encoder receives the input signal x, he sends the state $\rho_{i,x}$ with the probability $p_{i,x}$ and sends the classical information i. The decoder performs the TP-CP map $\kappa_{B,i}$ depending on the classical signal i. Then, inequality (10.36) follows from this construction.

Proof of (10.35): First, we have the following inequality:

$$\begin{aligned}
& F^2\Big(\sum_x p_x |e_x^A\rangle\langle e_x^A| \otimes W_x, \sum_i \kappa_{A,i} \otimes \kappa_{B,i}(|\Phi_L\rangle\langle\Phi_L|)\Big) \\
&\le \operatorname{Tr} \sqrt{\sum_x p_x |e_x^A\rangle\langle e_x^A| \otimes W_x} \sqrt{\sum_i \kappa_{A,i} \otimes \kappa_{B,i}(|\Phi_L\rangle\langle\Phi_L|)} \\
&= \operatorname{Tr} \sum_x \sqrt{p_x} |e_x^A\rangle\langle e_x^A| \otimes \sqrt{W_x} \sqrt{\sum_i \kappa_{A,i} \otimes \kappa_{B,i}(|\Phi_L\rangle\langle\Phi_L|)} \\
&= \sum_x \sqrt{p_x} \operatorname{Tr}_B \sqrt{W_x} (\operatorname{Tr}_A |e_x^A\rangle\langle e_x^A| \otimes I_B \sqrt{\sum_i \kappa_{A,i} \otimes \kappa_{B,i}(|\Phi_L\rangle\langle\Phi_L|)}) \\
&\le \sum_x \sqrt{p_x} \operatorname{Tr}_B \sqrt{W_x} \sqrt{(\operatorname{Tr}_A |e_x^A\rangle\langle e_x^A| \otimes I_B \sum_i \kappa_{A,i} \otimes \kappa_{B,i}(|\Phi_L\rangle\langle\Phi_L|))} \\
&= \sum_x \sqrt{p_x q_x} \operatorname{Tr}_B \sqrt{W_x} \sqrt{\sum_i p_{i,x} \kappa_{B,i}(\rho_{i,x})} && (10.38) \\
&= \sum_x p_x \operatorname{Tr}_B \sqrt{W_x} \sqrt{\sum_i p_{i,x} \kappa_{B,i}(\rho_{i,x})} \\
&\quad + \sum_x (\sqrt{p_x q_x} - p_x) \operatorname{Tr}_B \sqrt{W_x} \sqrt{\sum_i p_{i,x} \kappa_{B,i}(\rho_{i,x})}. && (10.39)
\end{aligned}$$

The above relations can be checked as follows: (i) The first inequality follows from a basic inequality $F^2(\rho,\sigma) \le \operatorname{Tr} \sqrt{\rho}\sqrt{\sigma}$. (ii) The second inequality follows from Exercise 1.16 and Condition ② of Theorem A.1 because $\sqrt{t}$ is matrix concave. (iii) Equation (10.38) follows from

$$
\begin{aligned}
& q_x \sum_i p_{i,x} \kappa_{B,i}(\rho_{i,x}) \\
&= \sum_i \kappa_{B,i}(\mathrm{Tr}_A(\kappa^*_{A,i}(|e_x^A\rangle\langle e_x^A|) \otimes I_B)|\Phi_L\rangle\langle\Phi_L|) \\
&= \sum_i \kappa_{B,i}(\mathrm{Tr}_A(|e_x^A\rangle\langle e_x^A| \otimes I_B)(\kappa_{A,i} \otimes \iota_B)(|\Phi_L\rangle\langle\Phi_L|)) \\
&= \mathrm{Tr}_A \, |e_x^A\rangle\langle e_x^A| \otimes I_B \sum_i \kappa_{A,i} \otimes \kappa_{B,i}(|\Phi_L\rangle\langle\Phi_L|).
\end{aligned}
$$

Further, the second term of (10.39) is evaluated by

$$
\begin{aligned}
& \sum_x (\sqrt{p_x q_x} - p_x) \,\mathrm{Tr}_B \sqrt{W_x} \sqrt{\sum_i p_{i,x} \kappa_{B,i}(\rho_{i,x})} \\
&\le \sum_x (\sqrt{p_x q_x} - p_x)_+ = \sum_x \Big(\sqrt{\frac{q_x}{p_x}} - 1\Big)_+ p_x \le \sum_x \Big(\frac{q_x}{p_x} - 1\Big)_+ p_x \\
&= \sum_x (q_x - p_x)_+ = \frac{1}{2}\|q - p\|_1 \le \frac{1}{2}\|\tilde{W}_p - \kappa(|\Phi_L\rangle\langle\Phi_L|)\|_1,
\end{aligned}
\tag{10.40}
$$

where $(t)_+$ is t when t is positive and 0 otherwise. The final inequality follows from the definition of distribution q (10.37).

Concerning the first term of (10.39), the inequality

$$
\begin{aligned}
& \frac{1}{2}(1 - F^2(W_x, \sum_i p_{i,x} \kappa_{B,i}(\rho_{i,x}))) \le 1 - F(W_x, \sum_i p_{i,x} \kappa_{B,i}(\rho_{i,x})) \\
&\le 1 - \mathrm{Tr}_B \sqrt{W_x} \sqrt{\sum_i p_{i,x} \kappa_{B,i}(\rho_{i,x})}
\end{aligned}
\tag{10.41}
$$

holds. Hence, (10.35) follows from (10.39), (10.40), and (10.41). ∎

10.7 Compression by Classical Memory

In the previous section, we treated visible compression by quantum memory. In this section, we consider the compression rate by classical memory. This problem was first discussed by Hayden et al. [197]. In this problem, when the state W_x is to be sent to the decoder, the *encoder* is given by stochastic transition matrix Q with input system $\mathcal{X}$ and output system $\{1, \dots, M\}$. The *decoder* is represented by a c-q channel $\{W'_i\}_{i=1}^M$ with output system $\mathcal{H}$. Hence, our code in this problem is given by triplet $\Psi_c \stackrel{\text{def}}{=} (M, Q, W')$, which can be regarded as a code in the visible scheme. Then, the optimal compression rate is defined as[3]

[3] The subscript c denotes classical memory.

$$R_{V,c}(W,p) \stackrel{\text{def}}{=} \inf_{\{\Psi_c^{(n)}\}} \left\{ \overline{\lim} \frac{1}{n} \log |\Psi_c^{(n)}| \middle| \varepsilon(\Psi_c^{(n)}) \to 0 \right\}. \tag{10.42}$$

Clearly, the inequality

$$R_{V,c}(W,p) \geq R_{V,q}(W,p)$$

holds.

Theorem 10.5

$$R_{V,c}(W,p) = C(\tilde{W}_p) = C_c(\tilde{W}_p). \tag{10.43}$$

Note that the quantities $C(\tilde{W}_p)$ and $C_c(\tilde{W}_p)$ are defined in Sect. 8.9.

Proof. First, for $\epsilon > 0$, we choose local states $\rho_{n,i}^A$ and $\rho_{n,i}^B$ such that

$$\left\| \frac{1}{M_n} \sum_{i=1}^{M_n} \rho_{n,i}^A \otimes \rho_{n,i}^B - \tilde{W}_p^{\otimes n} \right\|_1 \to 0, \quad \overline{\lim} \frac{1}{n} \log M_n \leq C_c(\tilde{W}_p) + \epsilon.$$

Next, we define the distribution q_n on $\mathcal{X}^n$ and two states ρ_n and ρ_n' as

$$q_{n,\boldsymbol{x}} \stackrel{\text{def}}{=} \sum_{i=1}^{M_n} \frac{1}{M_n} \langle e_{\boldsymbol{x}}^A | \rho_{n,i}^A | e_{\boldsymbol{x}}^A \rangle,$$

$$\rho_n \stackrel{\text{def}}{=} \sum_{\boldsymbol{x} \in \mathcal{X}^n} p_{\boldsymbol{x}}^n |e_{\boldsymbol{x}}^A\rangle\langle e_{\boldsymbol{x}}^A| \otimes \sum_{i=1}^{M_n} \frac{1}{M_n} \frac{\langle e_{\boldsymbol{x}}^A | \rho_{n,i}^A | e_{\boldsymbol{x}}^A \rangle}{q_{n,\boldsymbol{x}}} \rho_{n,i}^B,$$

$$\rho_n' \stackrel{\text{def}}{=} \sum_{\boldsymbol{x} \in \mathcal{X}^n} q_{n,\boldsymbol{x}} |e_{\boldsymbol{x}}^A\rangle\langle e_{\boldsymbol{x}}^A| \otimes \sum_{i=1}^{M_n} \frac{1}{M_n} \frac{\langle e_{\boldsymbol{x}}^A | \rho_{n,i}^A | e_{\boldsymbol{x}}^A \rangle}{q_{n,\boldsymbol{x}}} \rho_{n,i}^B.$$

Then, applying the monotonicity of a trace norm to a partial trace and the pinching of PVM $\{|e_{\boldsymbol{x}}^A\rangle\langle e_{\boldsymbol{x}}^A|\}$, we have

$$\|q_n - p^n\|_1 \leq \left\| \rho_n' - \tilde{W}_p^{\otimes n} \right\|_1 \leq \left\| \frac{1}{M_n} \sum_{i=1}^{M_n} \rho_{n,i}^A \otimes \rho_{n,i}^B - \tilde{W}_p^{\otimes n} \right\|_1 \to 0.$$

Hence,

$$\begin{aligned} \left\| \rho_n - \tilde{W}_p^{\otimes n} \right\|_1 &\leq \|\rho_n - \rho_n'\|_1 + \left\| \rho_n' - \tilde{W}_p^{\otimes n} \right\|_1 \\ &= \|q_n - p^n\|_1 + \left\| \rho_n' - \tilde{W}_p^{\otimes n} \right\|_1 \to 0. \end{aligned}$$

That is,

$$\sum_{\boldsymbol{x} \in \mathcal{X}^n} p_{\boldsymbol{x}}^n F\Big(\sum_{i=1}^{M_n} \frac{1}{M_n} \frac{\langle e_{\boldsymbol{x}}^A | \rho_{n,i}^A | e_{\boldsymbol{x}}^A \rangle}{q_{n,\boldsymbol{x}}} \rho_{n,i}^B, W_{\boldsymbol{x}}^{(n)} \Big) = F(\rho_n, \tilde{W}_p^{\otimes n}) \to 1.$$

Now, we define a code $\Psi^{(n)} = (M_n, Q^{(n)}, W'^{(n)})$ as

$$(Q^{(n)})_i^x \stackrel{\text{def}}{=} \frac{1}{M_n} \frac{\langle e_{\boldsymbol{x}}^A | \rho_{n,i}^A | e_{\boldsymbol{x}}^A \rangle}{q_{n,\boldsymbol{x}}}, \quad W'^{(n)}_i \stackrel{\text{def}}{=} \rho_{n,i}^B.$$

Then, its error is calculated by

$$\varepsilon_{p^n}(\Psi^{(n)}) = \sum_{\boldsymbol{x} \in \mathcal{X}^n} p_{\boldsymbol{x}}^n F\Big(\sum_{i=1}^{M_n} \frac{1}{M_n} \frac{\langle e_{\boldsymbol{x}}^A | \rho_{n,i}^A | e_{\boldsymbol{x}}^A \rangle}{q_{n,\boldsymbol{x}}} \rho_{n,i}^B, W_{\boldsymbol{x}}^{(n)} \Big) \to 1.$$

Hence, from Exercise 10.1 we obtain

$$R_{V,c}(W,p) \le C_c(\tilde{W}_p).$$

Now we prove the converse inequality. For any $\epsilon > 0$, we choose a sequence of codes $\Psi_c^{(n)} = (M_n, Q^{(n)}, W'^{(n)})$ such that

$$\overline{\lim} \frac{1}{n} \log |\Psi_c^{(n)}| \le R_{V,c}(W,p) + \epsilon,$$
$$\sum_{\boldsymbol{x} \in \mathcal{X}^n} p_{\boldsymbol{x}}^n F\Big(\sum_{i=1}^{M_n} (Q^{(n)})_i^{\boldsymbol{x}} W'^{(n)}_i, W_{\boldsymbol{x}}^{(n)} \Big) \to 0.$$

Hence, the state $\rho_n'' \stackrel{\text{def}}{=} \sum_{\boldsymbol{x} \in \mathcal{X}^n} |e_{\boldsymbol{x}}^A\rangle\langle e_{\boldsymbol{x}}^A| \otimes \sum_{i=1}^{M_n} (Q^{(n)})_i^{\boldsymbol{x}} W'^{(n)}_i$ satisfies

$$F(\rho_n'', \tilde{W}_p^{\otimes n}) = \sum_{\boldsymbol{x} \in \mathcal{X}^n} p_{\boldsymbol{x}}^n F\Big(\sum_{i=1}^{M_n} (Q^{(n)})_i^{\boldsymbol{x}} W'^{(n)}_i, W_{\boldsymbol{x}}^{(n)} \Big) \to 1.$$

From (8.161) and (8.155),

$$\log M_n \ge C\Big(\sum_{\boldsymbol{x} \in \mathcal{X}^n} |e_{\boldsymbol{x}}^A\rangle\langle e_{\boldsymbol{x}}^A| \otimes \sum_{i=1}^{M_n} (Q^{(n)})_i^{\boldsymbol{x}} |e_i^B\rangle\langle e_i^B| \Big) \ge C(\rho_n'').$$

Therefore, from (8.159),

$$\underline{\lim} \frac{\log M_n}{n} \ge C(\tilde{W}_p),$$

which implies

$$R_{V,c}(W,p) \ge C(\tilde{W}_p).$$

Thus, combining Theorem 8.12, we obtain (10.43). ∎

10.8 Compression by Shared Randomness

Next, we consider the case when the sender and decoder share a common random number a priori. In this problem, the *encoder* is given by a c-q channel T_X with input system $\mathcal{X}$ and output system $\mathcal{K}$, where T_X depends on the common random number X. The *decoder* is represented by a TP-CP map ν_X from $\mathcal{K}$ to $\mathcal{H}$. The TP-CP map T_X also depends on the common random number X. Hence, our code in this problem is given by triplet $\Psi_r \stackrel{\text{def}}{=} (\mathcal{K}, T_X, \nu_X)$. Further, when storage is a classical system, the problem is modified as follows. That is, the *encoder* is given by stochastic transition matrix Q_X with input system $\mathcal{X}$ and output system $\{1, \ldots, M\}$, where Q_X depends on the common random number. The *decoder* is represented by a c-q channel $\{W'_{X,i}\}_{i=1}^M$ with output system $\mathcal{H}$. C-q channel W'_X also depends on the common random number. Hence, our code is given by triplet $\Psi_{c,r} \stackrel{\text{def}}{=} (M, Q_X, W'_X)$.[4] Then, these optimal compression rates are defined as

$$R_{V,q,r}(W,p) \stackrel{\text{def}}{=} \inf_{\{\Psi_r^{(n)}\}} \left\{ \overline{\lim} \frac{1}{n} \log |\Psi_r^{(n)}| \,\middle|\, \varepsilon(\Psi_r^{(n)}) \to 0 \right\}, \tag{10.44}$$

$$R_{V,c,r}(W,p) \stackrel{\text{def}}{=} \inf_{\{\Psi_{c,r}^{(n)}\}} \left\{ \overline{\lim} \frac{1}{n} \log |\Psi_{c,r}^{(n)}| \,\middle|\, \varepsilon(\Psi_{c,r}^{(n)}) \to 0 \right\}. \tag{10.45}$$

Clearly, we have

$$R_{V,c}(W,p) \geq R_{V,c,r}(W,p) \geq R_{V,q,r}(W,p), \quad R_{V,q}(W,p) \geq R_{V,q,r}(W,p). \tag{10.46}$$

Lemma 10.5

$$R_{V,q,r}(W,p) \geq I(p,W) = I_{\tilde{W}_p}(A:B) = C_d^{A\to B}(\tilde{W}_p). \tag{10.47}$$

Proof. Let $\Psi_r^{(n)} \stackrel{\text{def}}{=} (\mathcal{K}_n, T_X^{(n)}, \nu_X^{(n)})$ be a sequence of codes achieving the optimal rate $R_{V,q,r}(W,p)$. From monotonicity (5.30) and joint convexity (5.31) of quantum relative entropy,

$$\begin{aligned}
\log |\Psi_r^{(n)}| &\geq \mathrm{E}_X \sum_{\boldsymbol{x}\in\mathcal{X}^n} p_{\boldsymbol{x}}^n D\Big(T_X^{(n)}(x) \Big\| \sum_{\boldsymbol{x}\in\mathcal{X}^n} p_{\boldsymbol{x}}^n T_X^{(n)}(x)\Big) \\
&\geq \mathrm{E}_X \sum_{\boldsymbol{x}\in\mathcal{X}^n} p_{\boldsymbol{x}}^n D\Big(\nu_X T_X^{(n)}(x) \Big\| \sum_{\boldsymbol{x}\in\mathcal{X}^n} p_{\boldsymbol{x}}^n \nu_X T_X^{(n)}(x)\Big) \\
&\geq \sum_{\boldsymbol{x}\in\mathcal{X}^n} p_{\boldsymbol{x}}^n D\Big(\mathrm{E}_X \nu_X T_X^{(n)}(x) \Big\| \mathrm{E}_X \sum_{\boldsymbol{x}\in\mathcal{X}^n} p_{\boldsymbol{x}}^n \nu_X T_X^{(n)}(x)\Big) \\
&= H\Big(\sum_{\boldsymbol{x}\in\mathcal{X}^n} p_{\boldsymbol{x}}^n \mathrm{E}_X \nu_X T_X^{(n)}(x)\Big) - \sum_{\boldsymbol{x}\in\mathcal{X}^n} p_{\boldsymbol{x}}^n H(\mathrm{E}_X \nu_X T_X^{(n)}(x)) = I_{\rho_n}(A:B),
\end{aligned}$$

[4] The last subscript r denotes shared "randomness."

where $\rho_n \stackrel{\text{def}}{=} \sum_{\boldsymbol{x}\in\mathcal{X}^n} p^n_{\boldsymbol{x}} |e^A_{\boldsymbol{x}}\rangle\langle e^A_{\boldsymbol{x}}| \otimes (\mathrm{E}_X \nu_X T_X^{(n)}(x))$. From the choice of our code we can check that $F(\rho_n, \tilde{W}_p) \to 0$. Hence, using Fannes inequality (Theorem 5.9), we have

$$\log |\Psi_r^{(n)}| \geq I_{\tilde{W}_p}(A:B) = I(p,W).$$

■

In fact, these optimal rates are calculated in the classical case.

Theorem 10.6 *(Bennett et al. [41], Dür et al. [100]) If all W_xs are commutative, we have*

$$R_{V,q,r}(W,p) = R_{V,c,r}(W,p) = I(p,W) = C_d^{A\to B}(\tilde{W}_p). \tag{10.48}$$

Hence, from (10.46) we have

$$R_{V,c}(W,p) \geq R_{V,q}(W,p) \geq R_{V,q,r}(W,p) = R_{V,c,r}(W,p). \tag{10.49}$$

Proof. From Theorem 10.5 it is sufficient to show the inequality $R_{V,c,r}(W,p) \leq I(p,W)$. Now we construct a protocol achieving the rate $R = I(p,W) + \epsilon$ for any $\epsilon > 0$.

First, we take the the expectation of the right-hand side (RHS) of (10.51) in Lemma 10.6 with $M = e^{nR}, C = e^{-nr}$, with a distribution $p^n(\boldsymbol{x})$. Applying (2.121) to $X = \log \frac{p(y|x)}{p(y)}$, we can show that this expectation is less than

$$2(e^{n(\phi(s|R\|S)-R+r)} + e^{-nr}), \quad \text{for } \forall s \leq 0, \tag{10.50}$$

where R and S are as given in (4.31). Note that R is commutative with S. Hence, we obtain (10.48). ■

Lemma 10.6 *For any classical source $\{(P_X(x), P_{Y|X}(y|x))\}$, there exists a code $\Psi_{c,r} \stackrel{\text{def}}{=} (M, Q_X, Q'_X)$ such that*

$$\frac{1}{2}\sum_y \left| \mathrm{E}_X \sum_{i=1}^{M} (Q'_X)^i_y (Q_X)^x_i - p(y|x) \right| \leq \delta + \sum_{y: \frac{1}{M}\frac{P_{Y|X}(y|x)}{P_Y(y)} \geq \delta} P_{Y|X}(y|x) \tag{10.51}$$

for $1 > \forall\delta > 0$, where $P_Y(y) \stackrel{\text{def}}{=} \sum_x P_X(x) P_{Y|X}(y|x)$.

Proof of Lemma 10.6. First, the encoder and the decoder prepare the M i.i.d. common random numbers $Y_1, \ldots, Y_M$ subject to $P_Y(y) \stackrel{\text{def}}{=} \sum_x P_{Y|X}(y|x) P_X(x)$. When the encoder receives the original message x, he sends the signal i obeying the distribution $P(i) \stackrel{\text{def}}{=} \frac{P_{X|Y}(x|Y_i)}{P_{X|Y}(x|Y_1)+\ldots+P_{X|Y}(x|Y_M)}$, where $P_{X|Y}(x|y) \stackrel{\rm def}{=} P_{Y|X}(y|x)\frac{P_X(x)}{P_Y(y)}$. The receiver recovers $y = Y_i$ when he receives i.

In this protocol, when the original message is x, the probability of $i = 1$ and $y = Y_1$ is given as

$$\mathrm{E}_{Y_1,\ldots,Y_M} \frac{P_Y(y)P_{X|Y}(x|y)}{P_{X|Y}(x|y) + P_{X|Y}(x|Y_2) + \ldots + P_{X|Y}(x|Y_M)}.$$

Hence, the recovered signal is equal to y with the probability

$$\mathrm{E}_{Y_1,\ldots,Y_M} \frac{MP_Y(y)P_{X|Y}(x|y)}{P_{X|Y}(x|y) + P_{X|Y}(x|Y_2) + \ldots + P_{X|Y}(x|Y_M)}.$$

Thus, since $\frac{1}{x}$ is concave,

$$\begin{aligned}
&\frac{1}{2}\sum_y \left| P_{Y|X}(y|x) - \mathrm{E}_{Y_1,\ldots,Y_M} \frac{Mp(y)P_{X|Y}(x|y)}{P_{X|Y}(x|y) + \sum_{i=2}^M P_{X|Y}(x|Y_i)} \right| \\
&= \sum_y \left(P_{Y|X}(y|x) - \mathrm{E}_{Y_1,\ldots,Y_M} \frac{MP_Y(y)P_{X|Y}(x|y)}{P_{X|Y}(x|y) + \sum_{i=2}^M P_{X|Y}(x|Y_i)} \right)_+ \\
&\le \sum_y \left(P_{Y|X}(y|x) - \frac{MP_Y(y)P_{X|Y}(x|y)}{P_{X|Y}(x|y) + (M-1)P_X(x)} \right)_+ \\
&= \sum_y P_{Y|X}(y|x) \left(\frac{\frac{1}{M}(\frac{P_{Y|X}(y|x)}{P_Y(y)} - 1)}{1 + \frac{1}{M}(\frac{P_{Y|X}(y|x)}{P_Y(y)} - 1)} \right)_+ && (10.52) \\
&\le \delta + \sum_{x,y: \frac{1}{M}(\frac{P_{Y|X}(y|x)}{P_Y(y)} - 1) \ge \delta} P_X(x)P_{Y|X}(y|x) && (10.53) \\
&\le \delta + \sum_{y: \frac{1}{M}\frac{P_{Y|X}(y|x)}{P_Y(y)} \ge \delta} P_{Y|X}(y|x),
\end{aligned}$$

where (10.52) and (10.53) follow from Exercise 10.4 and Exercise 10.3, respectively. ∎

Exercises

10.3. Prove the inequality $\frac{x}{1+x} \le \min\{x, 1\}$ for any real number $x > -1$.

10.4. Show that

$$P_{Y|X}(y|x) - \frac{MP_Y(y)P_{X|Y}(x|y)}{P_{X|Y}(x|y) + (M-1)P_X(x)} = \frac{\frac{1}{M}(\frac{P_{Y|X}(y|x)}{P_Y(y)} - 1)}{1 + \frac{1}{M}(\frac{P_{Y|X}(y|x)}{P_Y(y)} - 1)}.$$

10.5. Using a discussion similar to the proof of Lemma 10.5, prove (10.32).

10.9 Relation to Channel Capacities

In the above discussion, we consider the average error concerning the prior distribution on the input system. Sometimes it is suitable to treat the worst error concerning the input signal as follows:

$$C_{c,r}^R(W) \stackrel{\text{def}}{=} \inf_{\{\Psi_{c,r}^{(n)}\}} \left\{ \overline{\lim} \frac{1}{n} \log |\Psi_{c,r}^{(n)}| \,\middle|\, \tilde{\varepsilon}(\Psi_{c,r}^{(n)}) \to 0 \right\}, \tag{10.54}$$

$$\tilde{\varepsilon}(\Psi_{c,r}) \stackrel{\text{def}}{=} \max_{x \in \mathcal{X}} 1 - F^2(W_x, E_X (W'_X)_i (Q_X)_i^x).$$

This problem is called reverse Shannon theorem and is closely related to the following modification of channel coding, i.e., channel capacity with shared randomness:

$$C_{c,r}(W) \stackrel{\text{def}}{=} \inf_{\{\Phi_{c,r}^{(n)}\}} \left\{ \overline{\lim} \frac{1}{n} \log |\Phi_{c,r}^{(n)}| \,\middle|\, \tilde{\varepsilon}(\Phi_{c,r}^{(n)}) \to 0 \right\}, \tag{10.55}$$

$$\varepsilon(\Phi_{c,r}) \stackrel{\text{def}}{=} \max_{1 \le i \le M} 1 - E_X \operatorname{Tr}(Y_X)_i W_{\varphi_X(i)},$$

where $\Phi_{c,r}$ is a code with shared randomness, i.e., it is written as the triplet (M, φ_X, Y_X) with shared randomness X. In the above two problems, one is allowed to prepare any amount of shared random numbers. Indeed, the relation between these two problems is similar to that between entanglement distillation and dilution, in which dilution corresponds to $C_{c,r}^R(W)$, distillation corresponds to $C_{c,r}(W)$, and maximally entangled states correspond to noiseless channels. Hence, we can show

$$C_c(W) \le C_{c,r}(W) \le C_{c,r}^R(W).$$

The first inequality above follows from the comparison between the definitions of $C_c(W)$ and $C_{c,r}(W)$. From the discussion in Chap. 4 we have

$$\max_p I(p, W) = C_c(W).$$

Further, the following theorem holds.

Theorem 10.7 *(Bennett et al. [41]) When all W_xs are commutative,*

$$C_c(W) = C_{c,r}(W) = C_{c,r}^R(W) = \max_p I(p, W).$$

It is sufficient to show $C_{c,r}^R(W) \le \max_p I(p, W)$ for this theorem. This inequality follows from the proof of Theorem 10.6 with the distribution $p = \operatorname{argmax}_p I(p, W)$. In this proof, (10.50) is replaced by $2(e^{n(\phi(s|W_x \| W_p) - R + r)} + e^{-nr})$.

Further, we can define $C_{c,e}(W)$ and $C^R_{c,e}(W)$ by replacing the shared randomness by the shared entanglement.[5] Since the shared randomness can be generated from the shared entanglement,

$$C_c(W) \le C_{c,r}(W) \le C_{c,e}(W) \le C^R_{c,e}(W) \le C^R_{c,r}(W).$$

Hence, when all W_xs are commutative,

$$C(W) = C_{c,r}(W) = C_{c,e}(W) = C^R_{c,e}(W) = C^R_{c,r}(W) = \max_p I(p, W).$$

When W is a q-q channel κ, we must simulate entangled inputs. Considering this requirement, we can define the capacity $C^{e,R}_{c,e}(\kappa)$ as the reverse capacity of $C^e_{c,e}(\kappa)$ [41,42]. This capacity can be regarded as the capacity of teleportation through a noisy channel κ. Similarly, we have

$$C^e_{c,e}(\kappa) \le C^{e,R}_{c,e}(\kappa).$$

Recall our treatment of $C^e_{c,e}(\kappa)$ in Sect. 9.3. Originally, the reverse capacities $C^R_{c,e}(W)$ and $C^{e,R}_{c,e}(\kappa)$ were introduced for proving the converse part of $C^e_{c,e}(\kappa)$ by Bennett et al. [41]. They proved the equation $C^e_{c,e}(\kappa^{\mathrm{GP}}_p) = \max_\rho I(\rho, \kappa^{\mathrm{GP}}_p)$ for the generalized Pauli channel κ^{GP}_p by showing the two inequalities

$$\begin{aligned} C^e_{c,e}(\kappa^{\mathrm{GP}}_p) &\ge \max_\rho I(\rho, \kappa^{\mathrm{GP}}_p), \\ C^{e,R}_{c,e}(\kappa^{\mathrm{GP}}_p) &\le \max_\rho I(\rho, \kappa^{\mathrm{GP}}_p) = \log d - H(p), \end{aligned} \tag{10.56}$$

where d is the dimension of the system. They also conjectured [42]

$$C^e_{c,e}(\kappa) = C^{e,R}_{c,e}(\kappa) = \max_\rho I(\rho, \kappa).$$

In addition, when W is a q-c channel, i.e., a POVM, this problem was solved as the compression of POVM by Winter [420] and Massar and Winter [418].

Moreover, replacing the classical noiseless channel by the quantum noiseless channel, we can define the capacities $C^e_{q,e}(\kappa)$, $C^e_{q,r}(\kappa)$, $C^{e,R}_{q,e}(\kappa)$, and $C^{e,R}_{q,r}(\kappa)$. Then, the relations

$$C^e_{q,r}(\kappa) \le C^e_{q,e}(\kappa) \le C^{e,R}_{q,r}(\kappa), \quad C^e_{q,r}(\kappa) \le C^{e,R}_{q,e}(\kappa)$$

hold.

Proof of (10.56). Here, we give a proof for inequality (10.56). Assume that the sender and the receiver share the maximally entangled state $|\Phi_d\rangle\langle\Phi_d|$ on the tensor product $\mathcal{H}_B \otimes \mathcal{H}_C$. When the sender performs the generalized

[5] The last subscript e denotes the shared "entanglement" while the superscript e denotes entangled input.

Bell measurement $\{|u_{i,j}^{A,C}\rangle\langle u_{i,j}^{A,C}|\}_{(i,j)}$ on the composite system between the input system $\mathcal{H}_A$ and the sender's local system $\mathcal{H}_C$, he obtains the data (i,j) subject to the uniform distribution p_{mix,d^2}. In this case, the generalized Pauli channel κ_p^{GP} can be written as

$$\kappa_p^{\mathrm{GP}}(\rho) = \sum_{(i,j)} \sum_{(i',j')} p(i'-i,j'-j)\overline{\mathbf{X}_B^{i'}\mathbf{Z}_B^{j'}} \cdot \mathrm{Tr}_{A,C}(I \otimes |u_{i,j}^{A,C}\rangle\langle u_{i,j}^{A,C}|)(|\Phi_d\rangle\langle\Phi_d| \otimes \rho)\overline{\mathbf{X}_B^{i'}\mathbf{Z}_B^{j'}}^* .$$

Hence, if the classical channel $Q_{(i',j')}^{(i,j)} \stackrel{\mathrm{def}}{=} p(i'-i,j'-j)$ is simulated with the shared randomness, the generalized Pauli channel κ_p^{GP} can be simulated with the shared entanglement. Since $C_{c,r}^R(Q_{(i',j')}^{(i,j)}) = \log d - H(\kappa_p^{\mathrm{GP}})$, we have (10.56). ∎

10.10 Historical Note

First, we briefly treat the pure-state case. The source coding problem in the quantum case was initiated by Schumacher [360]. In this paper, he formulated the blind scheme and derived the direct part and the strong converse part assuming only unitary coding. Jozsa and Schumacher [245] improved this discussion. Barnum et al. [29] introduced the purification scheme and proved the strong converse part without assuming unitary coding. Further, Horodecki [228] introduced the visible scheme as an arbitrary coding scheme and showed the weak converse part. Further, Barnum et al. [30] pointed out that the previous proof by Barnum et al. [29] could be used as the proof of the strong converse part even in the visible scheme. In this book, Lemma 10.3 plays a central role in the proof of the strong converse part. This lemma was proved by Hayashi [175]. Winter [415] also proved the strong converse part using a related lemma. Using this formula, Hayashi [175] derived the optimal rate with an exponential error constraint. When the probability of the information source is unknown, we cannot use the coding protocol based on the prior distribution p. Using the type method, Jozsa et al. [247] constructed a fixed-length universal code achieving the optimal rate. In addition, In the classical case, Han [162] showed that compressed states that achieve the minimum rate are almost uniformly random in the fixed-length scheme. In this book, part of the quantum extension of the above Han's result is proved in as Theorem 10.3.

In the variable-length scheme, the problem is not so easy. In the classical case, we can compress classical data without any loss. However, Koashi and Imoto [258] proved that if all information sources cannot be diagonalized simultaneously, compression without any loss is impossible. Of course, using Schumacher's [360] compression, we can compress quantum information sources with a small error. Further, using Jozsa et al.'s [247] compression, we can compress quantum information sources with a small error based only on the knowledge of the entropy rate $H(W_p)$ if the information is generated by an independent and identical distribution of the distribution p. Hence, universal variable-length compression with a small error is possible if we can estimate the entropy rate $H(W_p)$ with a small state demolition.

For this estimation, the estimation method in Sect. 7.3 can be applied. Using this idea, a variable-length universal compression theorem is constructed in this book. This construction is slightly different from the original construction by Hayashi and Matsumoto [184]. The modified construction by Hayashi and Matsumoto [185] is closer to the construction of this book. Further, Hayashi and Matsumoto [184] showed that the average error of variable-length compression does not approach 0 exponentially in the two-level systemwhen the compression scheme has group covariance and achieves the entropy rate $H(W_p)$. Jozsa and Presnell [248] applied this idea to the Lempel–Ziv method. Bennett et al. [46] considered the complexity of universal variable-length compression.

In the analysis presented in this book, we have only considered probability distributions that satisfy the independent and identical condition for the source. Petz and Mosonyi [346] showed that the optimal compression rate is $\lim \frac{H(W_{p_n})}{n}$ when the information source p_n is stationary. Bjelaković and Szkoła [51] extended this result to the ergodic case. Datta and Suhov [89] treated nonstationary quantum spin systems. Further, Bjelaković et al. [50] extended Bjelaković and Szkoła's result to the quantum lattice system. Nagaoka and Hayashi [310] derived the optimal compression rate without any assumption of the information source based on the quantum information spectrum method. Using Lemma 10.3, they reduced quantum source coding to quantum hypothesis testing. Indeed, it is expected that the above results will be derived based on the asymptotic general formula by Nagaoka and Hayashi [310]. Kaltchenko and Yang [249] showed that this optimal rate can be attained by fixed-length source coding in the ergodic case.

Concerning the mixed-state case, Jozsa's talk [246] was the first study to focus on this problem. Horodecki derived the lower bound $I(p, W)$ (10.32) [228] and derived the optimal rate (10.34) [231] in the visible case. However, our optimal rate (10.33) has a slightly different form. Koashi and Imoto also derived the optimal rate in the blind case (10.31).

When the memory is classical, Bennett and Winter [47] pointed out that the compression problem with commutative mixed states is essentially equivalent to Wyner's [426] problem (Theorem 8.12). Theorem 10.5 can be regarded as its quantum extension. Further, Hayden et al. [197] treated the tradeoff between the sizes of classical memory and quantum memory with the visible scheme in the pure-state case.

Next, let us proceed to compression with shared randomness. Bennett et al. [41] introduced a reverse Shannon theorem (Theorem 10.7) and proved Theorem 10.6 as its corollary. Dür et al. [100] also proved Theorem 10.6 independently. In this book, we prove it via Lemma 10.6. Since this lemma has a general form, it can be extended to a general sequence of channels.

Further, we can consider the tradeoff between the sizes of the classical noiseless channel and the shared randomness as an intermediate problem between $R_{V,c,r}(W, p)$ and $R_{V,c}(W, p)$. Bennett and Winter [47] treated this problem in the commutative case.

In the classical case, Slepian and Wolf [377] considered the compression problem when the information source lies in the composite system and has a correlation. In their problem, the encoder in each system is is divided into two players, who can only perform local operations. However, the decoder is allowed to treat any information processing. Devetak and Winter [97] treated the quantum extension of this problem in the special case with an ensemble scheme. Ahn et al. [7] treated a more general case with the ensemble scheme. Further, Abeyesinghe et al. [1] treated this problem with a purification scheme.

A

Limits and Linear Algebra

A.1 Limits

In this text, frequently we discuss the asymptotic behaviors in several problems when the number n of prepared systems is sufficiently large. In this situation, we often take the limit $n \to \infty$. In this section, we give a brief summary of the fundamental properties of limits. Given a general sequence $\{a_n\}$, the limit $\lim a_n$ does not necessarily exist. For example, a sequence a_n is a counterexample when a_n diverges to $+\infty$ or $-\infty$. In such a case, it is possible to at least denote these limits as $\lim a_n = +\infty$ or $\lim a_n = -\infty$. However, the sequence a_n has no limit as $n \to infinity$, even allowing possibilities such as $+\infty$ or $-\infty$, when a_n is defined to be 0 when n is even and 1 when it is odd. This is caused by its oscillatory behavior. In this case, we can consider the upper limit $\underline{\lim}\, a_n$ and the lower limit $\overline{\lim}\, a_n$, which are given as $\underline{\lim}\, a_n = 0$ and $\overline{\lim}\, a_n = 1$. More precisely, $\underline{\lim}\, a_n$ and $\overline{\lim}\, a_n$ are defined as follows:

$$\underline{\lim}\, a_n \stackrel{\text{def}}{=} \sup\{a | \forall \epsilon > 0, \exists N, \forall n \geq N, a \leq a_n + \epsilon\},$$
$$\overline{\lim}\, a_n \stackrel{\text{def}}{=} \inf\{a | \forall \epsilon > 0, \exists N, \forall n \geq N, a \geq a_n - \epsilon\}.$$

When $\underline{\lim}\, a_n = \overline{\lim}\, a_n$, the limit $\lim a_n$ exists and is equal to $\underline{\lim}\, a_n = \overline{\lim}\, a_n$. The following three lemmas hold concerning limits.

Lemma A.1 *Let sequences* $\{a_n\}_{n=1}^{\infty}$ *and* $\{b_n\}_{n=1}^{\infty}$ *satisfy*

$$a_n + a_m \leq a_{n+m} + b_{n+m}, \quad \sup_n \frac{a_n}{n} < \infty, \quad \lim_n \frac{b_n}{n} = 0.$$

Then, the limit $\lim \frac{a_n}{n}$ *exists and satisfies*

$$\lim \frac{a_n}{n} = \overline{\lim}\, \frac{a_n}{n} = \sup_n \frac{a_n}{n}. \tag{A.1}$$

If $a_n + a_m \geq a_{n+m} - b_{n+m}$ *and* $\inf_n \frac{a_n}{n} > \infty$, $b_n \to 0$, *then similarly* $\lim \frac{a_n}{n} = \underline{\lim}\, \frac{a_n}{n} = \inf_n \frac{a_n}{n}$, *as shown by considering* $-a_n$.

Proof. Fix the integer m. Then, for any integer n, there uniquely exist integers l_n and r_n such that $0 \leq r_n \leq m-1$ and $n = l_n m + r_n$ for each n. Thus, we have

$$\frac{a_n}{n} = \frac{a_{l_n m+r}}{l_n m + r} \geq \frac{a_{l_n m}}{l_n m + r} + \frac{a_r - b_n}{l_n m + r} \geq \frac{l_n a_m}{l_n m + r} + \frac{a_r - b_n - b_{l_n m}}{l_n m + r}.$$

Since $l_n \to \infty$ as $n \to \infty$, taking the limit $n \to \infty$, we have $\underline{\lim} \frac{a_n}{n} \geq \frac{a_m}{m}$ for arbitrary m. Next, taking the limit $m \to \infty$, we have $\underline{\lim} \frac{a_n}{n} \geq \sup_m \frac{a_m}{m} \geq \overline{\lim}_{m\to\infty} \frac{a_m}{m}$. Since $\overline{\lim} \frac{a_n}{n} \geq \underline{\lim} \frac{a_n}{n}$, we obtain (A.1). ■

Lemma A.2 *Let $\{a_n\}$ and $\{b_n\}$ be two sequences of positive real numbers. Then,*

$$\overline{\lim} \frac{1}{n} \log(a_n + b_n) = \max\left\{ \overline{\lim} \frac{1}{n} \log a_n, \overline{\lim} \frac{1}{n} \log b_n \right\}.$$

Proof. Since $(a_n + b_n) \geq a_n, b_n$ and

$$\overline{\lim} \frac{1}{n} \log(a_n + b_n) \geq \overline{\lim} \frac{1}{n} \log a_n, \overline{\lim} \frac{1}{n} \log b_n,$$

we obtain the $\geq$ part of the proof. Since $2 \max\{a_n, b_n\} \geq (a_n + b_n)$, we have

$$\max\left\{ \overline{\lim} \frac{1}{n} \log a_n, \overline{\lim} \frac{1}{n} \log b_n \right\} = \overline{\lim} \frac{1}{n} \log \max\{a_n, b_n\}$$
$$= \overline{\lim} \frac{1}{n} \log 2 \max\{a_n, b_n\} \geq \overline{\lim} \frac{1}{n} \log(a_n + b_n),$$

which gives the reverse inequality. This completes the proof. ■

Lemma A.3 *Let $\{f_n(x)\}$ be a sequence of functions such that $f_n(x) \leq f_n(y)$ if $x \geq y$, and $f_n(x) \to 0$ if $x > 0$. There exists a sequence $\{\epsilon_n\}$ of positive real numbers converging to zero such that $f_n(x) \to 0$.*

Proof. Let N be a positive integer. Choose positive integers $n(N)$ such that $n(N) < n(N+1)$ and $f_n(\frac{1}{N}) \leq \frac{1}{N}$ for $n \geq n(N)$. We also define $\epsilon_n \stackrel{\text{def}}{=} \frac{1}{N}$ for $n(N) \leq n < n(N+1)$. Then, $\epsilon_n \to 0$. If $n \geq n(N)$, then $f_n(\epsilon_n) \leq \frac{1}{N}$. Therefore, $f_n(\epsilon_n) \to 0$. ■

For any two continuous functions f and g on an open subset $X \subset \mathbb{R}^d$, we define

$$[f, g](a) \stackrel{\text{def}}{=} \min_{x \in V} \{f(x) | g(x) \leq a\}. \tag{A.2}$$

Lemma A.4 *When X is closed and bounded, i.e., compact,*

$$[f, g](a) = \lim_{\epsilon \downarrow 0} [f, g](a + \epsilon). \tag{A.3}$$

Proof. From the definition for $\epsilon > 0$, $[f, g](a) \geq [f, g](a + \epsilon)$. Hence, $[f, g](a) \geq \lim_{\epsilon\downarrow 0}[f, g](a + \epsilon)$. From the compactness, for any $\epsilon_1 > 0$ there exists $\epsilon_2 > 0$ such that $\|x - x'\| < \epsilon_2 \Rightarrow |f(x) - f(x')| < \epsilon_1$. Further, from the compactness of X we can choose a small number $\epsilon_3 > 0$ such that $\{x|g(x) \le a + \epsilon_3\} \subset \cup_{x':g(x)\le a} U_{x',\epsilon_2}$. Hence,

$$\min_{x|g(x)\le a+\epsilon_3} f(x) \ge \min_{x\in\cup_{x':g(x)\le a} U_{x',\epsilon_2}} f(x) \ge \min_{x|g(x)\le a} f(x) - \epsilon_1, \tag{A.4}$$

which implies (A.3). ■

A.2 Singular Value Decomposition and Polar Decomposition

Any $d \times d'$ complex-valued matrix X has the form

$$X = U_1 X' U_2^*$$

with isometric matrices U_1 and U_2 and a diagonal matrix X'. This is called a *singular value decomposition* (the matrix U is an *isometric matrix* if U^*U is the identity matrix; U is a *partially isometric matrix* for the partial space $\mathcal{K}$ if it is a projection onto the partial space $\mathcal{K}$). Choosing a $d \times d'$ partially isometric matrix U in the range $\{X^*Xv|v \in \mathbb{C}^d\}$ of X^*X, we have

$$X = U|X|, \quad |X| \stackrel{\text{def}}{=} \sqrt{X^*X}, \tag{A.5}$$

which is called a *polar decomposition.* If X is Hermitian and is diagonalizable according to $X = \sum_i \lambda_i |u_i\rangle\langle u_i|$, then $|X| = \sum_i |\lambda_i||u_i\rangle\langle u_i|$. Since $X^* = |X|U^*$,

$$XX^* = U|X||X|U^* = UX^*XU^*, \quad \sqrt{XX^*} = U\sqrt{X^*X}U^*, \tag{A.6}$$

$$UX^*U = X. \tag{A.7}$$

Therefore,

$$X = \sqrt{XX^*}U. \tag{A.8}$$

If X is a square matrix (i.e., $d = d'$), then U is unitary. If $d \ge d'$, then U can be chosen as an isometric. If $d \le d'$, U can be chosen such that U^* is isometric. We now show that these two decompositions exist.

Since X^*X is Hermitian, we may choose a set of mutually orthogonal vectors $u_1, \dots, u_l$ of norm 1 such that

$$X^*X = \sum_{i=1}^{l} \lambda_i |u_i\rangle\langle u_i|.$$

In the above, we choose $\{\lambda_k\}_{k=1}^l$ such that $\lambda_i \ge \lambda_{i+1} > 0$. Hence, l is not necessarily equal to the dimension of the space because there may exist zero eigenvalues. Defining $v_i \stackrel{\text{def}}{=} \sqrt{\frac{1}{\lambda_i}} X u_i$, we have

$$\langle v_i|v_j\rangle = \sqrt{\frac{1}{\lambda_i}}\sqrt{\frac{1}{\lambda_j}}\langle Xu_i|Xu_j\rangle = \sqrt{\frac{1}{\lambda_i}}\sqrt{\frac{1}{\lambda_j}}\langle u_i|X^*X|u_j\rangle$$
$$= \sqrt{\frac{1}{\lambda_i}}\sqrt{\frac{1}{\lambda_j}}\delta_{i,j}\lambda_j = \delta_{i,j}.$$

Furthermore, from the relation

$$\langle v_i|X|u_j\rangle = \sqrt{\frac{1}{\lambda_i}}\langle Xu_i|X|u_j\rangle = \sqrt{\frac{1}{\lambda_i}}\langle Xu_i|X|u_j\rangle$$
$$= \sqrt{\frac{1}{\lambda_i}}\langle u_i|X^*X|u_j\rangle = \sqrt{\lambda_i}\delta_{i,j}$$

we can show that

$$\sum_i \sqrt{\lambda_i}|v_i\rangle\langle u_i| = \sum_i |v_i\rangle\langle v_i|X\sum_j |u_j\rangle\langle u_j| = X.$$

One may be concerned about the validity of the second equality if X^*X has some eigenvectors u with zero eigenvalue. However, since $\langle u|X^*X|u\rangle = 0$, we have $Xu = 0$. Hence, both sides of the image of vector u coincide with the 0 vector. We define $U_2 \stackrel{\text{def}}{=} (u_i^j)$ and $U_1 \stackrel{\text{def}}{=} (v_i^j)$, which are $l \times d$ and $l \times d'$ isometric matrices, respectively. Let X' be an $l \times l$ diagonal matrix $(\sqrt{\lambda_i}\delta_{i,j})$. This gives us (A.5).

Using the above, we obtain the following lemma.

Lemma A.5 *Let a density matrix ρ be written as*

$$\rho = \sum_{j=1}^{d'} |v_j\rangle\langle v_j|, \tag{A.9}$$

where $\{v_i\}$ is a set of vectors that are not necessarily orthogonal. Let its diagonalization be given by $\rho = \sum_{i=1}^{l} \lambda_i |u_i\rangle\langle u_i|$. Since $\lambda_i > 0$, l is not necessarily equal to the dimension of the space. Then, the vector v_j can be written as $v_j = \sum_{i=1}^{l} w_{j,i}\sqrt{\lambda_i}u_i$ by using an $l \times d'$ isometric matrix $W = (w_{j,i})$ [316].

The set of vectors $\{v_i\}$ satisfying (A.9) is called the *decomposition* of the density matrix ρ.

Proof. Let Y be a $j \times d$ matrix given by (v_i^i). Then,

$$\rho = \sum_{i=1}^{l} \lambda_i |u_i\rangle\langle u_i| = YY^*.$$

Define $w_i \stackrel{\text{def}}{=} \sqrt{\frac{1}{\lambda_i}}Y^*u_i$. Then, $Y^* = \sum_{i=1}^{l} \sqrt{\lambda_i}|w_i\rangle\langle u_i|$. Taking its conjugate, we obtain $Y = \sum_{i=1}^{l} \sqrt{\lambda_i}|u_i\rangle\langle w_i|$. Looking at the jth row, we obtain $|v_j\rangle = \sum_{i=1}^{l}(w_i^j)^*\sqrt{\lambda_i}|u_i\rangle$. Since $\sum_j (w_i^j)^*(w_{i'}^j) = \delta_{i,i'}$, $(w_i^j)^*$ is an isometric matrix. The proof is complete. ∎

Next, we consider the case where X is a real $d \times d$ matrix. Since a real symmetric matrix can be diagonalized by an orthogonal matrix, the unitary matrices U_1 and U_2 may be replaced by orthogonal matrices O_1 and O_2. In fact, we may further restrict the orthogonal matrices to orthogonal matrices with determinant 1 (these are called special orthogonal matrices). However, the following problem occurs. Assume that the determinant of O_i $(i = 1, 2)$ is -1. Then, O_i may be redefined by multiplying it by a diagonal matrix with diagonal elements $-1, 1, \ldots, 1$. The redefined matrix is then a special orthogonal matrix, and $O_1^* X O_2$ is diagonal. Choosing O_1 and O_2 in a suitable way, all the diagonal elements of $O_1^* X O_2$ will be positive if $\det X > 0$. On the other hand, if $\det X < 0$, then it is not possible to make all the diagonal elements of $O_1^* X O_2$ positive for special orthogonal matrices O_1, O_2.

Exercises

A.1. Define $J_{i,j} \stackrel{\mathrm{def}}{=} \langle u_i | u_j \rangle$ for a set of linearly independent vectors $u_1, \ldots, u_k$ in $\mathcal{H}$. Show that

$$\sum_{i,j} (J^{-1})^{j,i} |u_i\rangle\langle u_j| = \sum_{i,j} ((J^{-1})^{j,i})^* |u_i\rangle\langle u_j|.$$

Show that this is a projection to the subspace of $\mathcal{H}$ spanned by $u_1, \ldots, u_k$.

A.2. Using relation (A.5), show that

$$AA^* f(AA^*) = Af(A^*A) = A^*. \tag{A.10}$$

A.3 Norms of Matrices

We often focus on the norm between two matrices as a measure of the difference between them. There are two types of norms, the matrix norm and the trace norm. The matrix norm $\|A\|$ of a matrix A is defined as

$$\|A\| \stackrel{\mathrm{def}}{=} \max_{\|x\|=1} \|Ax\|.$$

Since $\|x\| = \max_{\|y\|=1} |\langle y, x\rangle|$, we have $\|A\| = \max_{\|y\|=\|x\|=1} |\langle y, Ax\rangle|$; therefore, $\|A\| = \|A^*\|$. From the definition we have $\|U_1 A U_2\| = \|A\|$ for unitary matrices U_1 and U_2. Defining

$$w(A) \stackrel{\mathrm{def}}{=} \max_{\|x\|=1} |\langle x, Ax\rangle|, \quad \mathrm{spr}(A) \stackrel{\mathrm{def}}{=} \max\{|\lambda| : \lambda \text{ is the eigenvalue of } A\},$$

we obtain

$$\mathrm{spr}(A) \le w(A) \le \|A\|. \tag{A.11}$$

Assume that A is a Hermitian matrix. Then, it may be diagonalized as $A = \sum_{i=1}^d \lambda_i |u_i\rangle\langle u_i|$. Thus,

$$
\begin{aligned}
|\langle y, Ax\rangle| &= \sum_{i=1}^d |\lambda_i||\langle y|u_i\rangle\langle u_i|x\rangle| \le \max_i |\lambda_i| \sum_{i=1}^d |\langle y|u_i\rangle||\langle u_i|x\rangle| \\
&\le \max_i |\lambda_i| \sqrt{\sum_{i=1}^d |\langle y|u_i\rangle|^2} \sqrt{\sum_{i=1}^d |\langle u_i|x\rangle|^2} = \max_i |\lambda_i| = \operatorname{spr}(A).
\end{aligned}
$$

The above inequality implies the equality sign in (A.11). Since $\|A\|^2 = \max_{\|x\|=1} \langle x|A^*A|x\rangle = \operatorname{spr}(A^*A) = (\operatorname{spr}(\sqrt{A^*A}))^2$, then $\|A\| = \|\sqrt{A^*A}\| = \|A^*\| = \|\sqrt{AA^*}\|$.

On the other hand, the trace norm $\|X\|_1$ of a matrix X is defined as

$$
\|X\|_1 = \max_{U:\text{unitary}} \operatorname{Tr} UX. \tag{A.12}
$$

Choosing a unitary matrix U_X such that $X = U_X|X|$ (i.e., a polar decomposition), we obtain Ex. A.8

$$
\|X\|_1 = \max_{U:\text{unitary}} \operatorname{Tr} UX = \operatorname{Tr} U_X^* X = \operatorname{Tr} |X|. \tag{A.13}
$$

Hence, we also have

$$
\|X^*\|_1 = \max_{U:\text{unitary}} \operatorname{Tr} U^* X^* = \operatorname{Tr} U_X X^* = \operatorname{Tr} |X^*|.
$$

If X is Hermitian, then

$$
\|X\|_1 = \max_{T:-I\le T\le I} \operatorname{Tr} XT = \operatorname{Tr} X(\{X \ge 0\} - \{X < 0\}) = \operatorname{Tr} X(I - 2\{X < 0\}). \tag{A.14}
$$

Exercises

A.3. Show that the trace norm of a Hermitian matrix $\begin{pmatrix} -a & b \\ b^* & a \end{pmatrix}$ is equal to $2\sqrt{|b|^2 + a^2}$.

A.4. Show that

$$
\|X\|_1 \ge \| \operatorname{Tr}_B X\|_1. \tag{A.15}
$$

for a matrix X in $\mathcal{H}_A \otimes \mathcal{H}_B$.

A.5. Let A and B be square matrices of dimension d. Show that the eigenvalues of BA are the same as the eigenvalues of AB including degeneracies if A or B possesses the inverse.

A.6. Show that $\mathrm{spr}(AB) = \mathrm{spr}(BA)$.

A.7. Show that the function $t \mapsto t^{1/2}$ is a matrix monotone function following the steps below.

a Show that $\|A^{1/2}B^{-1/2}\| \le 1$ when the Hermitian matrices B and A satisfy $B \ge A \ge 0$ and B possesses the inverse.
b Show that $1 \le \mathrm{spr}(B^{-1/4}A^{1/2}B^{-1/4})$ under the same conditions as **a**.
c Show that

$$B^{1/2} \ge A^{1/2} \tag{A.16}$$

under the same conditions as **a**.
d Show that (A.16) holds even if B does not possess the inverse.

A.8. Prove (A.13) following the steps below.

a Show that $\max_{v:\|v\|=1}\langle v|\,|X|\,|u_i\rangle = \langle u_i|\,|X|\,|u_i\rangle$ for eigenvectors u_i of $|X|$ of length 1.
b Show that $\max_{U:\mathrm{unitary}}\langle u_i|UX|u_i\rangle = \langle u_i|U_X^* X|u_i\rangle = \langle u_i||X||u_i\rangle$.
c Prove (A.13).

A.9. Show that $\|XY\|_1 \le \|X\|\|Y\|_1$ for two matrices X and Y.

A.10. (*Poincaré inequality*) Let A be a $d \times d$ Hermitian matrix. Let a_i be the eigenvalues of A ordered from largest to smallest. Show that $\min_{x\in\mathcal{K},\|x\|=1} \langle x|A|x\rangle \le a_k$ for any k-dimensional subspace $\mathcal{K}$.

A.11. Show that $\max_{P:\mathrm{rank}\,P=k} \min_{x} \frac{\langle x|PAP|x\rangle}{\langle x|P|x\rangle} = a_k$ under the same conditions as above.

A.12. Let A and B be Hermitian matrices, and let a_i and b_i be their ordered eigenvalues from largest to smallest. Show that $a_i \ge b_i$ if $A \ge B$.

A.4 Convex Functions and Matrix Convex Functions

Linear functions are often used in linear algebra. On the other hand, functions such as x^2 and $\exp(x)$ do not satisfy the linearity property. If we denote such functions by f, then they instead satisfy

$$f(\lambda x_1 + (1-\lambda)x_2) \le \lambda f(x_1) + (1-\lambda)f(x_2), \quad 0 \le \forall\lambda \le 1, \forall x_1, x_2 \in \mathbb{R}.$$

A function is called a *convex function* when it satisfies the above inequality. If $-f$ is a convex function, then f is called a *concave function*. In the above, its domain is restricted to real numbers. However, this restriction is not necessary and may be defined in a more general way. For example, for a vector space, we may define the *convex combination* $\lambda v_1 + (1-\lambda)v_2$ for two vectors v_1 and v_2 with $0 < \lambda < 1$. More generally, a set is called a *convex set* when the convex combination of any two elements is defined. Further, a convex set L is called a *convex cone* if $v \in L$ and $\lambda > 0$ imply $\lambda v \in L$. Therefore, it is possible to define convex and concave functions for functions with a vector space domain and a real number range. Similarly, convex and concave functions may be defined with a convex set domain. Examples of convex sets are the set of probability distributions and the set of density matrices. In particular, an element v of the convex set V is called an *extremal point* if $v_i \in V$ and $v = \lambda v_1 + (1-\lambda)v_2, (0 < \lambda < 1)$ imply $\lambda = 1$ or 0. For example, a pure state is an extremal point in the set of density matrices.

Lemma A.6 *Let f be a convex function on the convex set V. For any element v_0 of V, there exists a linear function g such that*

$$g(v_0) - f(v_0) = \max_{v \in V} g(v) - f(v).$$

When f is differentiable, g coincides with the derivative of f at v_0. Further, for any linear function g and a constant $C_0 \geq 0$, there exists the Lagrange multiplier λ such that

$$\max_{v \in V} f(v) + \lambda g(v) = \max_{v \in V : g(v) \leq C_0} f(v) + \lambda g(v).$$

In this case, λg coincides with the derivative of f at $\operatorname{argmax}_{v \in V : g(v) \leq C_0} f(v)$.

Lemma A.7 *Consider two vector spaces V_1 and V_2 and consider a real-valued function $f(v_1, v_2)$ with the domain $V_1 \times V_2$. If f is convex with respect to v_2 and concave with respect to v_1, then (Chap. VI Prop. 2.3 of [105])*[1]

$$\sup_{v_1 \in S_1} \min_{v_2 \in S_2} f(v_1, v_2) = \min_{v_2 \in S_2} \sup_{v_1 \in S_1} f(v_1, v_2),$$

where S_1 and S_2 are convex subsets of V_1 and V_2.

Next, we focus on the set of probability distributions on $\mathcal{S}(\mathcal{H})$ and denote it by $\mathcal{P}(\mathcal{S}(\mathcal{H}))$. In particular, we consider extremal points of the set $\mathcal{S}(\mathcal{H})$:

$$\mathcal{P}(\rho, \mathcal{S}(\mathcal{H})) \stackrel{\text{def}}{=} \left\{ p \in \mathcal{P}(\mathcal{S}(\mathcal{H})) \middle| \sum_i p_i \rho_i = \rho \right\}.$$

Such extremal points of the above set are characterized as follows.

Lemma A.8 *(Fujiwara and Nagaoka [128]) Let $p \in \mathcal{P}(\rho, \mathcal{S}(\mathcal{H}))$ be an extremal point and $\{\rho_1, \ldots, \rho_k\}$ be the support of p. Then, $\rho_1, \ldots, \rho_k$ are linearly independent. Hence, the number of supports of p is less than* $\dim \mathcal{T}(\mathcal{H}) = (\dim \mathcal{H})^2$.

[1] This relation holds even if V_1 is infinite dimensional, as long as S_2 is a closed and bounded set.

Note that we obtain the same result when we replace $\mathcal{P}(\rho, \mathcal{S}(\mathcal{H}))$ by $\mathcal{P}(\mathcal{S}(\mathcal{H}))$.
Proof. Assume that $\rho_1, \dots, \rho_k$ are linearly dependent. That is, we choose real numbers $\lambda_1, \dots, \lambda_k$ such that $\sum_{i=1}^k \lambda_i \rho_i = 0$ and $\sum_i \lambda_i = 0$. Define two distributions q^+ and q^- with the same support by

$$q_i^{\pm} \stackrel{\text{def}}{=} p_i \pm \epsilon \lambda_i. \tag{A.17}$$

Then, we have $p = \frac{1}{2}q^+ + \frac{1}{2}q^-$ and $q^+ \neq q^-$. It is a contradiction. ∎

Indeed, applying this lemma to ρ_{mix}, we can see that any extremal POVM has at most $(\dim \mathcal{H})^2$ elements. Further, we focus on the cost functions $f_1, \dots, f_l$ on $\mathcal{S}(\mathcal{H})$ and treat the following sets:

$$\mathcal{P}_{=(\leq)c}(\rho, f, \mathcal{S}(\mathcal{H})) \stackrel{\text{def}}{=} \left\{ p \in \mathcal{P}(\rho, \mathcal{S}(\mathcal{H})) \middle| \sum_i p_i f_j(\rho_i) = (\leq) c \forall j = 1, \dots, l \right\}$$

$$\mathcal{P}_{=(\leq)c}(f, \mathcal{S}(\mathcal{H})) \stackrel{\text{def}}{=} \left\{ p \in \mathcal{P}(\mathcal{S}(\mathcal{H})) \middle| \sum_i p_i f_j(\rho_i) = (\leq) c \forall j = 1, \dots, l \right\}.$$

Lemma A.9 *(Fujiwara and Nagaoka [128]) Let p be an extremal point of one of the above sets. Then, the number of supports of p is less than $(l + 1)(\dim \mathcal{H})^2$.*

The concept of "convex function" can be extended to functions of matrices. If a function f with the range $[0, \infty]$ satisfies

$$\lambda f(A) + (1 - \lambda) f(B) \geq f(\lambda A + (1 - \lambda) B),$$

for arbitrary Hermitian matrices A, B with eigenvalues in $[0, \infty]$, it is called a *matrix convex function*. See Sect. 1.5 for the definition of $f(A)$. Also, the function f is called a *matrix concave function* when the function $-f$ is a matrix convex function. The following equivalences are known [49]:

① $f(t)$ is matrix monotone.
② $t/f(t)$ is matrix monotone.
③ $f(t)$ is matrix concave.

Furthermore, it is known that if the function f satisfies one of the above conditions, $1/f(t)$ is matrix convex [49]. Hence, since the functions $t^s, -t^{-s} (s \in [0,1])$, and $\log t$ are matrix monotone, the functions t^s $(s \in [-1, 0] \cup [1, 2])$, $-t^s$ $(s \in [0, 1])$, $-\log t$, and $t \log t$ are matrix convex functions.

Theorem A.1 *The following conditions are equivalent for a function f [49].*

① *$f(t)$ is matrix convex.*
② *When a matrix C satisfies $C^* C = I$, any Hermitian matrix A with eigenvalues in $[0, \infty]$ satisfies $f(C^* A C) \leq C^* f(A) C$.*

③ *When matrices $C_1, \ldots, C_k$ satisfy $\sum_i C_i^* C_i = I$, any Hermitian matrices $A_1, \ldots, A_k$ with eigenvalues in $[0, \infty]$ satisfy $f(\sum_i C_i^* A_i C_i) \leq \sum_i C_i^* f(A_i) C_i$.*

Now, we prove important inequalities.

Proof of (6.11), (5.40), and (5.41). In what follows, we focus on the linear space $\mathcal{M}_A$ of matrices on $\mathcal{H}_A$ and the linear space $\mathcal{M}_A$ of matrices on $\mathcal{H}_A$. For a given density ρ on $\mathcal{H}_A$, we also define the map L_ρ and R_ρ as a linear map on the matrix space $\mathcal{M}_A$:

$$L_\rho(A) \stackrel{\text{def}}{=} \rho A, \quad R_\rho(A) \stackrel{\text{def}}{=} A\rho.$$

The map L_ρ is positive Hermitian under the inner product $\langle Y, X \rangle = \operatorname{Tr} Y^* X$ because

$$\begin{aligned}\langle Y, L_\rho X \rangle_{\rho,r}^{(e)} &= \operatorname{Tr} Y^* \rho \rho X = \operatorname{Tr}(\rho Y)^* \rho X = \langle L_\rho Y, X \rangle_{\rho,r}^{(e)}, \\ \langle X, L_\rho X \rangle &= \operatorname{Tr} X^* \rho^2 X \geq 0.\end{aligned}$$

For another state σ, the map R_σ is positive Hermitian under the inner product $\langle Y, X \rangle_{\rho,r}^{(e)}$. It is checked as follows:

$$\begin{aligned}\langle Y, R_\sigma X \rangle_{\rho,r}^{(e)} &= \operatorname{Tr} Y^* \rho X \sigma = \operatorname{Tr}(Y\sigma)^* \rho X = \langle R_\sigma Y, X \rangle_{\rho,r}^{(e)}, \\ \langle X, R_\sigma X \rangle_{\rho,r}^{(e)} &= \operatorname{Tr} X^* \rho X \sigma \geq 0.\end{aligned}$$

Since L_ρ and R_σ are commutative, $L_\rho^{-1} R_\sigma$ is also positive Hermitian. Now, for any state $\rho^{A,B}$ on $\mathcal{H}_A \otimes \mathcal{H}_B$, and the TP-CP map $\kappa = \operatorname{Tr}_B$, we focus on the map $\kappa_{\rho^{A,B},r}$ from $\mathcal{M}_{A,B}$ to $\mathcal{H}_A$ (this map is defined in Sect. 6.1). The dual map $\kappa^*_{\rho^{A,B},r}$ is written as $\kappa^*_{\rho^{A,B},r}(Y) = Y \otimes I$ because

$$\begin{aligned}&\langle \kappa^*_{\rho^{A,B},r}(Y), X \rangle_{\rho^{A,B},r}^{(e)} = \langle Y, \kappa_{\rho^{A,B},r}(X) \rangle_{\operatorname{Tr}_B \rho^{A,B},r}^{(e)} \\ =& \operatorname{Tr}_A Y^* (\operatorname{Tr}_B \rho^{A,B}) \kappa_{\rho^{A,B},r}(X) = \operatorname{Tr}_A Y^* \operatorname{Tr}_B(\rho^{A,B} X) \\ =& \operatorname{Tr}_{A,B} Y^* \otimes I_B \rho^{A,B} X = \langle Y \otimes I_B, X \rangle_{\rho^{A,B},r}^{(e)}.\end{aligned}$$

Hence, $\kappa_{\rho^{A,B},r} \kappa^*_{\rho^{A,B},r}(Y) = Y$. Now, let f be a matrix convex function. Applying Condition ② in Theorem A.1, we obtain

$$\begin{aligned}&\operatorname{Tr} X^* f(\kappa_{\rho^{A,B},r} L^{-1}_{\rho^{A,B}} R_{\sigma^{A,B}} \kappa^*_{\rho^{A,B},r})(\operatorname{Tr}_B \rho^{A,B}) X \\ =& \langle X, f(\kappa_{\rho^{A,B},r} L^{-1}_{\rho^{A,B}} R_{\sigma^{A,B}} \kappa^*_{\rho^{A,B},r}) X \rangle_{\operatorname{Tr}_B \rho^{A,B},r}^{(e)} \\ \leq& \langle X, \kappa_{\rho^{A,B},r} f(L^{-1}_{\rho^{A,B}} R_{\sigma^{A,B}}) \kappa^*_{\rho^{A,B},r} X \rangle_{\operatorname{Tr}_B \rho^{A,B},r}^{(e)} \\ =& \operatorname{Tr}(X \otimes I)^* f(L^{-1}_{\rho^{A,B}} R_{\sigma^{A,B}}) \rho^{A,B} (X \otimes I).\end{aligned}$$

Thus, substituting $\rho^{A,B}$ $(-x^\lambda)$ into $\sigma^{A,B}$ $(f(x))$, we have

$$
\begin{aligned}
& - \operatorname{Tr}_A X^* (\operatorname{Tr}_B \rho^{A,B})^{1-\lambda} X (\operatorname{Tr}_B \rho^{A,B})^{\lambda} \\
\le & - \operatorname{Tr}_{A,B} (X \otimes I_B)^* (\rho^{A,B})^{1-\lambda} (X \otimes I_B) (\rho^{A,B})^{\lambda}
\end{aligned}
$$

because $f(L^{-1}_{\rho^{A,B}} R_{\sigma^{A,B}})X = (\rho^{A,B})^{-\lambda} X (\sigma^{A,B})^{\lambda}$. Hence, we obtain (6.11). Further, substituting X $(-x^{\lambda})$ into I $(f(x))$, we obtain

$$
- \operatorname{Tr}_A (\operatorname{Tr}_B \rho^{A,B})^{1-\lambda} (\operatorname{Tr}_B \sigma^{A,B})^{\lambda} \le - \operatorname{Tr}_{A,B} (\rho^{A,B})^{1-\lambda} (\sigma^{A,B})^{\lambda}
$$

for $\le \lambda \le 1$, which implies (5.40) in the partial trace case. From the Stinespring representation we obtain (5.40) in the general case. Also, substituting X (x^{λ}) into I $(f(x))$, we obtain

$$
\operatorname{Tr}_A (\operatorname{Tr}_B \rho^{A,B})^{1-\lambda} (\operatorname{Tr}_B \sigma^{A,B})^{\lambda} \le \operatorname{Tr}_{A,B} (\rho^{A,B})^{1-\lambda} (\sigma^{A,B})^{\lambda}, \quad -1 \le \lambda \le 0,
$$

which implies (5.41) in the partial trace case. Therefore, we obtain (5.41) in the general case. ■

Exercises

A.13. Show the concavity of the von Neumann entropy (5.49) using the matrix convexity of $x \log x$.

A.14. Show the joint convexity of (5.31) using the matrix convexity of $-\log x$ and $x \log x$.

A.5 Proof and Construction of Stinespring and Choi–Kraus Representations

In this section, we will prove Theorem 5.1 and construct the Stinespring and Choi–Kraus representations. First, let us consider the following theorem for completely positive maps, without the trace-preserving condition.

Theorem A.2 *Given a linear map κ from the set of Hermitian matrices on the d-dimensional system $\mathcal{H}_A$ to that on the d'-dimensional system $\mathcal{H}_B$, the following conditions are equivalent.*

① *κ is a completely positive map.*
② *κ^* is a completely positive map.*
③ *κ is a $\min\{d,d'\}$-positive map.*
④ *The matrix $K(\kappa)$ on $\mathcal{H}_A \otimes \mathcal{H}_B$ is positive semidefinite.*
⑤ *(Stinespring representation) There exist a Hilbert space $\mathcal{H}_C$ with the same dimension as $\mathcal{H}_B$, a pure state $\rho_0 \in \mathcal{S}(\mathcal{H}_B \otimes \mathcal{H}_C)$, and a matrix W in $\mathcal{H}_A \otimes \mathcal{H}_B \otimes \mathcal{H}_C$ such that $\kappa(X) = \operatorname{Tr}_{A,C} W(X \otimes \rho_0)W^*$.*
⑥ *(Choi–Kraus representation) There exist dd' linear maps $F_1, \ldots, F_{dd'}$ from $\mathcal{H}_A$ to $\mathcal{H}_B$ such that $\kappa(X) = \sum_i F_i X F_i^*$.*

We also define Conditions ⑤′ and ⑥′ by deforming Conditions ⑤ and ⑥ as:

⑤′ *There exist Hilbert spaces $\mathcal{H}_C$ and $\mathcal{H}'_C$, a positive semidefinite state $\rho'_0 \in \mathcal{S}(\mathcal{H}_C)$, and a linear map W from $\mathcal{H}_A\mathcal{H}_C$ to $\mathcal{H}_A\mathcal{H}'_C$ such that $\kappa(X) = \operatorname{Tr}_{C'} W(X \otimes \rho'_0)W^*$.*

⑥′ *There exist linear maps $F_1, \ldots, F_k$ from $\mathcal{H}_A$ to $\mathcal{H}_B$ such that $\kappa(X) = \sum_i F_i X F_i^*$.*

Proof. We now show that ②⇔ ①⇒③⇒④⇒ ⑤⇒⑥⇒⑥′⇒ ① and ⑤⇒⑤′⇒①. Since ①⇒③, ⑤⇒⑤′, and ⑥⇒⑥′ by inspection, we prove the remaining relations.

We first prove ①⇔②. The n-positivity of κ is equivalent to

$$\operatorname{Tr} \kappa \otimes \iota_n(X) \geq 0 \tag{A.18}$$

for any positive semidefinite Hermitian matrix X on $\mathcal{H}_A \otimes \mathbb{C}^n$, which is equivalent to

$$\operatorname{Tr} \kappa \otimes \iota_n(X)Y \geq 0$$

for any positive semidefinite Hermitian matrix X (Y) on $\mathcal{H}_A \otimes \mathbb{C}^n$ ($\mathcal{H}_B \otimes \mathbb{C}^n$). Since $(\kappa \otimes \iota_n)^* = \kappa^* \otimes \iota_n$, we have $\operatorname{Tr} \kappa \otimes \iota_n(X)Y = \operatorname{Tr} X\kappa^* \otimes \iota_n(Y)$. Therefore, the n-positivity of κ is equivalent to the n-positivity of κ^*. Hence, the complete positivity of κ is equivalent to the complete positivity of κ^*.

Next, we derive ③⇒④. We show ④ for $d' \leq d$. Since κ is a d-positive map, $\kappa \otimes \iota_B$ is a positive map (ι_B is the identity map in $\mathcal{T}(\mathcal{H}_B)$). Let X be a positive semidefinite Hermitian matrix on $\mathcal{H}_A \otimes \mathcal{H}_B$. Assume that $X = \sum_{i,k,j,l,} x^{(i,k),(j,l)} |e_i^A \otimes e_k^B\rangle\langle e_j^A \otimes e_l^B|$. Since $(\kappa \otimes \iota_B)(X) \geq 0$, we have

$$\begin{aligned}
0 &\leq \langle I_B|(\kappa \otimes \iota_B)(X)|I_B\rangle \\
&= \sum_{i,j,k,l} x^{(i,k),(j,l)} \langle I_B| \left(\kappa(|e_i^A\rangle\langle e_j^A|) \otimes |e_k^B\rangle\langle e_l^B|\right) |I_B\rangle \\
&= \sum_{i,j,k,l} x^{(i,k),(j,l)} \langle e_k^B|\kappa(|e_i^A\rangle\langle e_j^A|)|e_l^B\rangle \\
&= \sum_{i,j,k,l} x^{(i,k),(j,l)} K(\kappa)^{(j,l),(i,k)} = \operatorname{Tr} XK(\kappa).
\end{aligned} \tag{A.19}$$

Therefore, $K(\kappa) \geq 0$, and we obtain ④. In the above, we denote the vector $\sum_{k=1}^{d'} e_k^B \otimes e_k^B$ in the space $\mathcal{H}_B \otimes \mathcal{H}_B$ by I_B. In the derivation of (A.19), we used the fact that

$$\langle I_B| \left(|e_k^B\rangle\langle e_l^B| \otimes |e_s^B\rangle\langle e_t^B|\right) |I_B\rangle = \langle I_B| \left(|e_k^B \otimes e_s^B\rangle\langle e_l^B \otimes e_t^B|\right) |I_B\rangle = \delta_{k,s}\delta_{l,t}.$$

Using a discussion in the proof of ①⇔②, we can show that Condition ③ implies that κ^* is a d-positive map if $d \leq d'$. From this fact we can derive

Condition ④ for κ^*. This gives us $K(\kappa^*) \geq 0$, which is equivalent to $K(\kappa) \geq 0$. Thus, we obtain ④ for κ.

We now derive ④ ⇒ ⑤. Since $K(\kappa) \geq 0$, $\sqrt{K(\kappa)}$ exists. In what follows, we consider a space $\mathcal{H}_C$ with a basis $e_1^C, \ldots, e_{d'}^C$. Note that the space $\mathcal{H}_C$ is isometric to the space $\mathcal{H}_B$. Defining $U_{C,B} \stackrel{\text{def}}{=} \sum_{k=1}^{d'} e_k^C \otimes e_k^B$, we have

$$\begin{aligned}&\operatorname{Tr} |e_{i'}^A\rangle\langle e_{j'}^A| \otimes (|U_{C,B}\rangle\langle U_{C,B}|)\, |e_j^A \otimes e^C{}_k \otimes e_s^B\rangle\langle e_i^A \otimes e^C{}_l \otimes e_t^B| \\ =&\delta_{j',j}\delta_{i',i}\delta_{l,t}\delta_{k,s},\end{aligned} \tag{A.20}$$

where the order of the tensor product is $\mathcal{H}_A \otimes \mathcal{H}_C \otimes \mathcal{H}_B$. Although $K(\kappa)$ is originally a Hermitian matrix on $\mathcal{H}_A \otimes \mathcal{H}_B$, we often regard it as a Hermitian matrix on $\mathcal{H}_A \otimes \mathcal{H}_C$ because $\mathcal{H}_C$ is isometric to $\mathcal{H}_B$. Using (A.20), we have

$$\begin{aligned}\operatorname{Tr} \kappa(X)Y &= \operatorname{Tr}(X \otimes Y)K(\kappa) = \operatorname{Tr}\left(X \otimes |U_{C,B}\rangle\langle U_{C,B}|\right) K(\kappa) \otimes Y \\ &= \operatorname{Tr}\left(X \otimes |U_{C,B}\rangle\langle U_{C,B}|\right)\left(\sqrt{K(\kappa)} \otimes I_B\right)(I_{A,C} \otimes Y)\left(\sqrt{K(\kappa)} \otimes I_B\right) \\ &= \operatorname{Tr}_B \operatorname{Tr}_{A,C}\left(\left(\sqrt{K(\kappa)} \otimes I_B\right)(X \otimes |U_{C,B}\rangle\langle U_{C,B}|)\left(\sqrt{K(\kappa)} \otimes I_B\right)\right) Y\end{aligned}$$

for $\forall X \in \mathcal{T}(\mathcal{H}_A), \forall Y \in \mathcal{T}(\mathcal{H}_B)$. Therefore, we can show that

$$\kappa(X) = \operatorname{Tr}_{A,C}\left[\left(\sqrt{d'K(\kappa)} \otimes I_B\right)\left(X \otimes \frac{|U_{C,B}\rangle\langle U_{C,B}|}{d'}\right)\left(\sqrt{d'K(\kappa)} \otimes I_B\right)\right]. \tag{A.21}$$

Letting $\rho_0 = \frac{|U_{C,B}\rangle\langle U_{C,B}|}{d'}$ and $W = \sqrt{d'K(\kappa)} \otimes I_B$, we obtain ⑤.

Next, we show that ⑤ ⇒ ⑥. Let ρ_0 be $|x\rangle\langle x|$, P be a projection from $\mathcal{H}_A \otimes \mathcal{H}_B \otimes \mathcal{H}_C$ to $\mathcal{H}_A \otimes |x\rangle$, and $P_{i,k}$ be a projection from $\mathcal{H}_A \otimes \mathcal{H}_B \otimes \mathcal{H}_C$ to $\mathcal{H}_B \otimes |e_i^A \otimes e^C{}_k\rangle$. Using formula (1.26) of the partial trace, we have

$$\begin{aligned}\kappa(X) = \operatorname{Tr}_{A,C} W(X \otimes \rho_0)W^* &= \sum_{i=1}^{d}\sum_{k=1}^{d'} P_{i,k} WPXPW^* P_{i,k} \\ &= \sum_{i=1}^{d}\sum_{k=1}^{d'} (P_{i,k}WP)X(P_{i,k}WP)^*.\end{aligned}$$

We thus obtain ⑥.

Finally, we show ⑤′ ⇒ ①. From Condition ⑤′ any positive semidefinite Hermitian matrix X on $\mathcal{H}_A \otimes \mathbb{C}^n$ satisfies

$$\kappa \otimes \iota_n(X) = \operatorname{Tr}_{\mathcal{H}_A \otimes \mathcal{H}_C}(W \otimes I_n)(X \otimes \rho_0')(W^* \otimes I_n) \geq 0,$$

where I_n is an identity matrix in $\mathbb{C}^n$. From condition (A.18), therefore, κ is an n-positive map for arbitrary n. It follows that κ is a completely positive map from which we obtain ①.

Concerning a proof of ⑥′ ⇒ ①, we have

$$\kappa \otimes \iota_n(X) = \sum_i (F_i \otimes I_n) X (F_i^* \otimes I_n) \geq 0,$$

for a semipositive definite Hermitian matrix X on $\mathcal{H}_A \otimes \mathbb{C}^n$. Thus, we obtain ①. ■

Next, we prove Theorem 5.1. Thanks to Theorem A.2, it is sufficient to show the equivalence of Conditions ① to ⑥′ in Theorem 5.1 when κ is a completely positive map. Indeed, ① ⇒ ③, ⑤ ⇒ ⑤′, and ⑥ ⇒ ⑥′ by inspection. Concerning ⑤′ ⇒ ① and ⑥′ ⇒ ①, it is sufficient to show the trace-preserving property because of Theorem A.2. Therefore, we only show ③ ⇒ ④ ⇒ ⑤ ⇒ ⑥ as follows.

We first show ③ ⇒ ④. From definition (1.23) of the partial trace we obtain

$$\operatorname{Tr}_A \rho = \operatorname{Tr}_B \kappa(\rho) = \operatorname{Tr}_{A,B} (\rho \otimes I_B) K(\kappa) = \operatorname{Tr}_A \rho \left(\operatorname{Tr}_B K(\kappa)\right)$$

for arbitrary $\rho \in \mathcal{S}(\mathcal{H}_A)$. Hence, $\operatorname{Tr}_B K(\kappa) = I_A$, and thus we obtain ④.

Next, we show ④ ⇒ ⑤. Employing the notation used in the proof of ④ ⇒ ⑤ in Theorem A.2, we let P be the projection from $\mathcal{H}_A \otimes \mathcal{H}_B \otimes \mathcal{H}_C$ to $\mathcal{H}_A \otimes |U_{C,B}\rangle$. Since any $\rho \in \mathcal{S}(\mathcal{H}_A)$ satisfies

$$\begin{aligned}\operatorname{Tr} \rho &= \operatorname{Tr}_B \operatorname{Tr}_{A,C} \left(\left(\sqrt{d' K(\kappa)} \otimes I_B\right) P \rho P \left(\sqrt{d' K(\kappa)} \otimes I_B\right)\right) \\ &= \operatorname{Tr}_{A,C,B} \left(\left(\sqrt{d' K(\kappa)} \otimes I_B\right) P \rho P \left(\sqrt{d' K(\kappa)} \otimes I_B\right)\right),\end{aligned}$$

we obtain

$$\operatorname{Tr}_{A,C,B} \left(\left(\sqrt{d' K(\kappa)} \otimes I_B\right) P\right)^* \left(\sqrt{d' K(\kappa)} \otimes I_B\right) P = P.$$

Let $\mathcal{H}_R$ be the range of $\left(\sqrt{d' K(\kappa)} \otimes I_B\right) P$ for $\mathcal{H}_A \otimes |U_{C,B}\rangle$. Then, the dimension of $\mathcal{H}_R$ is equal to that of $\mathcal{H}_A$. $\left(\sqrt{d' K(\kappa)} \otimes I_B\right) P$ can be regarded as a map from $\mathcal{H}_A \otimes |U_{C,B}\rangle$ to $\mathcal{H}_R$.

Let ${\mathcal{H}_R}^\perp$ be the orthogonal complementary space of $\mathcal{H}_R$ in $\mathcal{H}_A \otimes \mathcal{H}_B \otimes \mathcal{H}_C$, and ${\mathcal{H}_A}^\perp$ be the orthogonal complementary space of $\mathcal{H}_A \otimes |U_{C,B}\rangle$. Since the dimension of ${\mathcal{H}_R}^\perp$ is equal to that of ${\mathcal{H}_A}^\perp$, there exists a unitary (i.e., metric-preserving) linear mapping U' from ${\mathcal{H}_R}^\perp$ to ${\mathcal{H}_A}^\perp$. Then, $U_\kappa \stackrel{\text{def}}{=} \left(\sqrt{d' K(\kappa)} \otimes I_B\right) P \oplus U'$ is a unitary linear map from $\mathcal{H}_A \otimes \mathcal{H}_B \otimes \mathcal{H}_C = (\mathcal{H}_A \otimes |U_{C,B}\rangle) \oplus {\mathcal{H}_A}^\perp$ to $\mathcal{H}_A \otimes \mathcal{H}_B \otimes \mathcal{H}_C = \mathcal{H}_R \oplus {\mathcal{H}_R}^\perp$. Therefore, from (A.21) we have $\kappa(\rho) = \operatorname{Tr}_{A,C} U_\kappa \rho \otimes \frac{|U_{C,B}\rangle\langle U_{C,B}|}{d'} U_\kappa$, which gives Condition ⑤.

Next, we show ⑤ ⇒ ⑥ by employing the notation used in the proof of ⑤ ⇒ ⑥ in Theorem A.2. Since

$$\operatorname{Tr}\rho = \operatorname{Tr}\kappa(\rho) = \operatorname{Tr}_B \operatorname{Tr}_{A,C} U_\kappa(\rho\otimes\rho_0)U_\kappa^* = \sum_{i=1}^{d}\sum_{k=1}^{d'} \operatorname{Tr}_B P_{i,k}WP\rho PW^*P_{i,k}$$

$$= \sum_{i=1}^{d}\sum_{k=1}^{d'} \operatorname{Tr}_B (P_{i,k}WP)\rho(P_{i,k}WP)^* = \operatorname{Tr}_A \sum_{i=1}^{d}\sum_{k=1}^{d'} (P_{i,k}WP)^*(P_{i,k}WP)\rho,$$

we obtain $\sum_{i=1}^{d}\sum_{k=1}^{d'}(P_{i,k}WP)^*(P_{i,k}WP) = I_A$. Therefore, we obtain ⑥. Further, from the proof ⑤ ⇒ ⑥, we obtain (5.1).

Finally, we directly construct Stinespring representation ⑤′ from Choi–Kraus representation ⑥′. Define the map W from $\mathcal{H}_A$ to $\mathcal{H}_B \otimes \mathbb{C}^k$ as

$$W(x) \stackrel{\text{def}}{=} \sum_{i=1}^{k} F_i(x) \otimes e_i.$$

Then, W satisfies

$$\operatorname{Tr}_{\mathbb{C}^k} W\rho W^* = \sum_{i=1}^{k} F_i \rho F_i^*.$$

We obtain Condition ⑤′ from ⑥′ in Theorem A.2. In Theorem 5.1, we have to check the unitarity. From the condition $\sum_{i=1}^{k} F_i^* F_i = I$, we obtain $W^*W = I$, i.e., W is an isometry map. Hence, it is possible to deform map W to a unitary map by extending the input space. In this case, the state in the environment $\kappa^E(\rho)$ equals $\operatorname{Tr}_B W\rho W^* = (\operatorname{Tr} F_j^* F_i \rho)_{i,j}$. Thus, we obtain Lemma 5.1.

B

Proofs of Theorems and Lemmas

B.1 Proof of Theorem 3.1

In this proof, we only consider the case in which there exists an element $x \in L$ such that $A(x) = b$.[1] Otherwise, since both sides are equal to $-\infty$, the theorem holds.

When $x \in L$ satisfies $A(x) = b$ and y satisfies $A^T(y) - c \in L^T$, we have $0 \leq \langle A^T(y) - c, x\rangle = \langle y, A(y)\rangle - \langle c, x\rangle = \langle y, b\rangle - \langle c, x\rangle$. Hence, we can check that

$$\max_{x \in V_1}\{\langle c, x\rangle | x \in L, A(x) = b\} \leq \min_{y \in V_2}\{\langle y, b\rangle | A^T(y) - c \in L^T\}. \tag{B.1}$$

Furthermore,

$$\begin{aligned}
&\min_{y \in V_2}\{\langle y, b\rangle | A^T(y) - c \in L^T\} \\
=& \min_{(\mu, y) \in \mathbb{R} \times V_2}\{\mu | \exists y \in V_2, \forall x \in L, \langle y, b\rangle - \langle A^T(y) - c, x\rangle \leq \mu\}.
\end{aligned}$$

This equation can be checked as follows. When $y \in V_2$ satisfies $A^T(y) - c \in L^T$, the real number $\mu = \langle y, b\rangle$ satisfies the condition on the right-hand side (RHS). Hence, we obtain the $\geq$ part. Next, we consider a pair (μ, y) satisfying the condition on the RHS. Then, we can show that $\langle A^T(y) - c, x\rangle$ is greater than zero for all $x \in L$, by reduction to absurdity. Assume that there exists an element $x \in L$ such that $\langle A^T(y) - c, x\rangle$ is negative. By choosing a sufficiently large number $t > 0$, $tx \in L$, but $\langle y, b\rangle - \langle A^T(y) - c, tx\rangle \leq \mu$ does not hold. It is a contradiction. This proves the $\leq$ part.

Let $\eta_0 \stackrel{\text{def}}{=} \max_{x \in V_1}\{\langle c, x\rangle | x \in L, A(x) = b\}$. Then $(\eta_0, 0)$ is a point that lies on the boundary of the convex set $\{(\langle c, x\rangle, A(x) - b)\}_{x \in L} \subset \mathbb{R} \times V_2$. Choosing an appropriate $y_0 \in V_2$ and noting that $(1, -y_0) \in \mathbb{R} \times V_2$, we have

[1] Our proof follows [397].

$$\eta_0 = \eta_0 - \langle y, 0\rangle \geq \langle c, x\rangle - \langle y_0, A(x) - b\rangle, \quad \forall x \in L.$$

From this fact we have

$$\eta_0 \geq \min_{(\mu,y)\in\mathbb{R}\times V_2} \{\mu | \exists y \in V_2, \forall x \in L, \langle y, b\rangle - \langle A^T(y) - c, x\rangle \leq \mu\}.$$

This proves the reverse inequality of (B.1) and completes the proof.

B.2 Proof of Theorem 8.2

We prove this theorem in the following steps: ① ⇒ ② ⇒ ③ ⇒ ①, ② ⇒ ④ ⇒ ①. The proof given here follows from Bhatia [49].

We first show ① ⇒ ② for dimension d by induction. Let $t \stackrel{\text{def}}{=} (y_1 - x_1)/(y_1 - y_2) = (x_2 - y_2)/(y_1 - y_2)$ for $d = 2$. Since $x \preceq y$, we have $0 \leq t \leq 1$. Further, the relation

$$\begin{pmatrix} x_1 \\ x_2 \end{pmatrix} = \begin{pmatrix} 1-t & t \\ t & 1-t \end{pmatrix} \begin{pmatrix} y_1 \\ y_2 \end{pmatrix} \tag{B.2}$$

proves the case for $d = 2$. In the following proof, assuming that the result holds for $d \leq n-1$, we prove the case for $d = n$. Any permutation is expressed by by a product of T transforms. Hence, it is sufficient to show ② when $x_1 \geq x_2 \geq \ldots \geq x_n$ and $y_1 \geq y_2 \geq \ldots \geq y_n$. Since $x \preceq y$, we have $y_n \leq x_1 \leq y_1$. Choosing an appropriate k, we have $y_k \leq x_1 \leq y_{k-1}$. When t satisfies $x_1 = ty_1 + (1-t)y_k$, the relation $0 \leq t \leq 1$ holds. Let T_1 be the T transform among the first and kth elements defined by t. Define

$$x' \stackrel{\text{def}}{=} (x_2, \ldots, x_n)^T, \tag{B.3}$$

$$y' \stackrel{\text{def}}{=} (y_2, \ldots, y_{k-1}, (1-t)y_1 + ty_k, y_{k+1}, \ldots, y_n)^T. \tag{B.4}$$

Then, $T_1 y = (x_1, y')$. Since $x' \preceq y'$ (as shown below), from the assumptions of the induction there exist T transforms $T_f, \ldots, T_2$ such that $T_f \cdots T_2 y' = x'$. Therefore, $T_f \cdots T_2 T_1 y = T_f \cdots T_2(x_1, y') = (x_1, x') = x$, which completes the proof for this part. We now show that $x' \preceq y'$. For an integer m satisfying $2 \leq m \leq k-1$, we have

$$\sum_{j=2}^{m} x_j \leq \sum_{j=2}^{m} y_j. \tag{B.5}$$

If $k \leq m \leq n$, then

$$\begin{aligned}
\sum_{j=2}^{m} {y'}_j &= \left(\sum_{j=2}^{k-1} y_j\right) + (1-t)y_1 + ty_k + \left(\sum_{j=k+1}^{m} y_j\right) \\
&= \left(\sum_{j=1}^{m} y_j\right) - ty_1 + (t-1)y_k = \sum_{j=1}^{m} y_j - x_1 \geq \sum_{j=1}^{m} x_j - x_1 = \sum_{j=2}^{m} x_j,
\end{aligned}$$

which shows that $x' \preceq y'$.

Next, we show ② ⇒ ③. The product of two double stochastic transition matrices A_1 and A_2 is also a double stochastic transition matrix A_1A_2. Since a T transform is a double stochastic transition matrix, we obtain ③.

For proving ③ ⇒ ①, it is sufficient to show that

$$\sum_{t=1}^{k}\sum_{j=1}^{d} x^{i_t,j} a_j \le \sum_{j=1}^{k} a_j^{\downarrow}$$

for an arbitrary integer k and a set of k arbitrary integers $i_1, \dots, i_k$ from 1 to d. This can be shown from the fact that $\sum_{j=1}^{d}\sum_{t=1}^{k} x^{i_t,j} = k$ and $\sum_{t=1}^{k} x^{i_t,j} \le 1$ for each j.

We now show ② ⇒ ④. For simplicity, we consider $d = 2$ and let

$$B = \begin{pmatrix} \frac{(y_1/y_2)^2-(x_2/x_1)(y_1/y_2)}{(y_1/y_2)^2-1} & \frac{(y_1/y_2)(x_1/x_2)-1}{(y_1/y_2)^2-1} \\ \frac{(x_2/x_1)(y_1/y_2)-1}{(y_1/y_2)^2-1} & \frac{(y_1/y_2)^2-(y_1/y_2)(x_1/x_2)}{(y_1/y_2)^2-1} \end{pmatrix}. \tag{B.6}$$

It can be verified that this is a stochastic transition matrix. Since

$$\begin{pmatrix} \frac{(y_1/y_2)^2-(x_2/x_1)(y_1/y_2)}{(y_1/y_2)^2-1} x_1 \\ \frac{(y_1/y_2)(x_1/x_2)-1}{(y_1/y_2)^2-1} x_2 \end{pmatrix} = \frac{(y_1/y_2)x_1 - x_2}{(y_1/y_2)^2 - 1} \begin{pmatrix} (y_1/y_2) \\ 1 \end{pmatrix}$$

$$\begin{pmatrix} \frac{(x_2/x_1)(y_1/y_2)-1}{(y_1/y_2)^2-1} x_1 \\ \frac{(y_1/y_2)^2-(y_1/y_2)(x_1/x_2)}{(y_1/y_2)^2-1} x_2 \end{pmatrix} = \frac{(y_1/y_2)x_2 - x_1}{(y_1/y_2)^2 - 1} \begin{pmatrix} 1 \\ (y_1/y_2) \end{pmatrix},$$

we observe that $B^1 \circ x \approx B^2 \circ x \approx y$.

Let T_0 be a T transform defined with respect to t between kth and lth elements ($k < l$), and define B^1 and B^2 as

$$\begin{pmatrix} b^{1,k} & b^{1,l} \\ b^{2,k} & b^{2,l} \end{pmatrix} = \begin{pmatrix} \frac{(y_k/y_l)^2-(x_l/x_k)(y_k/y_l)}{(y_k/y_l)^2-1} & \frac{(y_k/y_l)(x_k/x_l)-1}{(y_k/y_l)^2-1} \\ \frac{(x_l/x_k)(y_k/y_l)-1}{(y_k/y_l)^2-1} & \frac{(y_k/y_l)^2-(y_k/y_l)(x_k/x_l)}{(y_k/y_l)^2-1} \end{pmatrix}$$

$$b^{1,i} = \frac{(y_k/y_l)x_k - x_l}{(y_k/y_l) - 1}, \quad b^{2,i} = \frac{(y_k/y_l)x_l - x_k}{(y_k/y_l) - 1} \quad \text{if } i \neq k, l.$$

Then, $B^1 \circ x \approx B^2 \circ x \approx y$, if $x = T_0 y$.

Further, if two stochastic transition matrices B, C satisfy $y \approx (B^j)^* \circ x$ and $z \approx (C^i)^* \circ y$ for arbitrary integers i and j, then there exists an appropriate substitution $s(j)$ such that

$$y \propto s(j)((B^j)^* \circ x),$$

where we identify the permutation $s(j)$ and the matrix that represents it. Since $s(j)^* = (s(j))^{-1}$,

$$\begin{aligned} z &\approx (C^i)^* \circ y = s(j)\left(((s(j))^{-1}(C^i)^*) \circ (s(j))^{-1} y\right) \\ &\propto s(j)\left(((s(j))^{-1}(C^i)^*) \circ (B^j)^* \circ x\right) = s(j)\left((C^i s(j))^* \circ (B^j)^* \circ x\right) \\ &= s(j)\left((C^i s(j))^* \circ (B^j)^* \circ x\right) \approx (C^i s(j))^* \circ (B^j)^* \circ x. \end{aligned}$$

Therefore,

$$\sum_{i,j}(C^i s(j))^* \circ (B^j)^* = \sum_j (\sum_i C^i s(j))^* \circ (B^j)^*$$
$$= \sum_j (e^* s(j))^* \circ (B^j)^* = \sum_j e \circ (B^j)^* = \sum_j (B^j)^* = e.$$

When we define the matrix D by $(D^{i,j})^* \stackrel{\text{def}}{=} (C^i s(j))^* \circ (B^j)^*$ (note that the pair i, j refers to one column), this matrix is a stochastic transition matrix and satisfies

$$(D^{i,j})^* \circ x = z. \tag{B.7}$$

Using this and the previous facts, we obtain ② ⇒ ④.

Finally, we show ④ ⇒ ①. It is sufficient to show the existence of a d-dimensional vector $c = (c_i)$ with positive real elements such that

$$c_i \le 1, \quad \sum_{i=1}^d c_i = k, \quad \sum_{j=1}^k y_{i_j} \ge \sum_{j=1}^k x_j^\downarrow \tag{B.8}$$

for arbitrary k. For this purpose, we choose k different integers $i_1, \dots, i_k$ such that

$$\sum_{j=1}^k x_j^\downarrow = \sum_{j=1}^k x_{i_j}.$$

For each j, we choose the permutation $s(j)$ and the positive real number d_j such that $(B^j)^* \circ x = d_j s(j) y$. Note that $\sum_{j=1}^d d_j = 1$. Since

$$\sum_{j=1}^d b^{j,i} x_i = x_i,$$

we have

$$\sum_{j=1}^k x_{i_j} = \sum_{t=1}^d \sum_{j=1}^k b^{t,i_j} x_{i_j} = \sum_{t=1}^d \sum_{j=1}^k d_t (s(t)y)_{i_j} = \sum_{t=1}^d \sum_{j=1}^k \sum_{l=1}^d d_t s(t)_{i_j,l} y_l.$$

Since

$$\sum_{j=1}^k s(t)_{i_j,l} \le 1, \quad \sum_{l=1}^d \sum_{j=1}^k s(t)_{i_j,l} = k,$$

we obtain

$$\sum_t \sum_{j=1}^k d_t s(t)_{i_j,l} \le 1, \quad \sum_{l=1}^d \sum_t \sum_{j=1}^k s(t)_{i_j,l} = k$$

where we used $\sum_t d_t = 1$. This shows the existence of a vector $c = (c_i)$ satisfying (B.8).

B.3 Proof of Theorem 8.3

Let ρ be a separable state on $\mathcal{H}_A \otimes \mathcal{H}_B$. We can choose an appropriate set of vectors $\{u_i\}_i$ in $\mathcal{H}_A$ and $\{v_i\}_i$ in $\mathcal{H}_B$ such that $\rho = \sum_i |u_i \otimes v_i\rangle\langle u_i \otimes v_i| = \sum_j \lambda_j |e_j\rangle\langle e_j|$, where the RHS is the diagonalized form of ρ. From Lemma A.5 we can take an isometric matrix $W = (w_{i,j})$ such that $u_i \otimes v_i = \sum_j w_{i,j}\sqrt{\lambda_j} e_j$. Since $W^* W = I$, we have

$$\sum_i w_{i,j}^* u_i \otimes v_i = \sqrt{\lambda_j} e_j. \tag{B.9}$$

Similarly, we diagonalize $\mathrm{Tr}_B\, \rho$ such that $\mathrm{Tr}_B\, \rho = \sum_k \lambda'_k |f_k\rangle\langle f_k|$. Then, we can take an isometric matrix $W' = (w'_{i,k})$ such that $u_i = \sum_k w'_{i,k}\sqrt{\lambda'_k} f_k$.

Substituting this into (B.9), we obtain

$$\sqrt{\lambda_j} e_j = \sum_i \sum_k w'_{i,k} w_{i,j}^* \sqrt{\lambda'_k} f_k \otimes v_i.$$

Taking the norm on both sides, we have

$$\lambda_j = \sum_k D_{j,k} \lambda'_k, \quad D_{j,k} \stackrel{\text{def}}{=} \left(\sum_{i,i'} w'_{i,k} w_{i,j}^* (w'_{i',k})^* w_{i',j} \langle v_{i'} | v_i \rangle \right).$$

If we can show that $D_{j,k}$ is a double stochastic transition matrix, Condition ③ in Theorem 8.2 implies $(\lambda'_k) \preceq (\lambda_j)$. Since

$$\left(\sum_{i,i'} w'_{i,k} w_{i,j}^* (w'_{i',k})^* w_{i',j} \langle v_{i'} | v_i \rangle \right) = \left\langle \sum_{i'} w'_{i',k} w_{i',j}^* v_{i'} \middle| \sum_i w'_{i,k} w_{i,j}^* v_i \right\rangle \geq 0$$

and $W'^* W' = I, W^* W = I$, we obtain

$$\sum_k \left(\sum_{i,i'} w'_{i,k} w_{i,j}^* (w'_{i',k})^* w_{i',j} \langle v_{i'} | v_i \rangle \right) = \sum_{i,i'} \delta_{i,i'} w_{i,j}^* w_{i',j} \langle v_{i'} | v_i \rangle$$
$$= \sum_i w_{i,j}^* w_{i,j} = 1.$$

We may similarly show that $\sum_j D_{j,k} = 1$. Hence, $D_{j,k}$ is a double stochastic transition matrix.

B.4 Proof of Theorem 8.8 for Mixed States

We show the $\leq$ part of (8.87) for a general state ρ. Let $\{E_{A,i} \otimes E_{B,i}\}_i$ be the Choi–Kraus representation of an S-TP-CP map κ. Then,

$$\kappa(|\Phi_L\rangle\langle\Phi_L|) = \sum_i (E_{A,i} \otimes E_{B,i}) |\Phi_L\rangle\langle\Phi_L| (E_{A,i} \otimes E_{B,i})^* .$$

Now, choose y_i such that

$$\begin{aligned} p'_i &\stackrel{\mathrm{def}}{=} \mathrm{Tr}\, (E_{A,i} \otimes E_{B,i}) |\Phi_L\rangle\langle\Phi_L| (E_{A,i} \otimes E_{B,i})^* \\ p'_i|y_i\rangle\langle y_i| &= (E_{A,i} \otimes E_{B,i}) |\Phi_L\rangle\langle\Phi_L| (E_{A,i} \otimes E_{B,i})^* . \end{aligned}$$

From Corollary 8.1 there exists a probabilistic decomposition $\{(p_i, x_i)\}$ of ρ such that

$$F(\kappa(|\Phi_L\rangle\langle\Phi_L|), \rho) = \sum_i \sqrt{p_i p'_i} |\langle x_i|y_i\rangle|.$$

Since the Schmidt rank of y_i is at most L,

$$|\langle x_i|y_i\rangle| \le \sqrt{P(x_i, L)}. \tag{B.10}$$

From the Schwarz inequality,

$$\begin{aligned} F(\kappa(|\Phi_L\rangle\langle\Phi_L|), \rho) &\le \sum_i \sqrt{p_i p'_i} \sqrt{P(x_i, L)} \\ &\le \sqrt{\sum_i p'_i} \sqrt{\sum_i p_i P(x_i, L)} = \sqrt{\sum_i p_i P(x_i, L)}. \end{aligned} \tag{B.11}$$

Thus, we obtain the $\le$ part of (8.87).

Conversely, if there exists a vector y_i with a Schmidt rank of at most L that satisfies the equality in (B.10) and

$$p'_i = \frac{p_i P(x_i, L)}{\sum_j p_j P(x_j, L)},$$

the equality in (B.11) holds. Therefore, according to Theorem 8.4, there exists a one-way LOCC satisfying the RHS of (8.85).

B.5 Proof of Theorem 8.9 for Mixed States

B.5.1 Proof of Direct Part

The second equality in (8.94) holds according to (8.84) and Lemma A.1. We therefore show the $\le$ part of the first equality. Let us first show that

$$\min_{(p_i, x_i)} \left\{ \sum_i p_i (1 - P(x_i, [e^{nR}])) \middle| \sum_i p_i |x_i\rangle\langle x_i| = \rho^{\otimes n} \right\}$$

converges to zero exponentially for $R > E_f(\rho)$. The convergence of this expression to zero is equivalent to that of the value inside the $\sqrt{\ }$ on the RHS of (8.87) to one. Hence, we consider the latter quantity, i.e., the value inside the $\sqrt{\ }$. Choose a decomposition $\{(p_i, x_i)\}$ such that $R > \sum_i p_i E(|x_i\rangle\langle x_i|)$. Let $\rho_i \stackrel{\text{def}}{=} \mathrm{Tr}_B |x_i\rangle\langle x_i|$. From (8.88),

$$\sum_i p_i P(x_i, [e^R]) \le \sum_i p_i e^{\frac{\log \mathrm{Tr}\, \rho_i^{1-s} - sR}{1-s}}.$$

In particular, since $\frac{\log \mathrm{Tr}(\rho_i \otimes \rho_j)^{1-s} - 2sR}{1-s} = \frac{\log \mathrm{Tr}\, \rho_i^{1-s} - sR}{1-s} + \frac{\log \mathrm{Tr}\, \rho_j^{1-s} - sR}{1-s}$, we obtain

$$\begin{aligned}\sum_{i^n} p_{i^n}^n P(x_{i^n}^n, [e^{nR}]) &\le \sum_{i^n} p_{i^n}^n e^{\frac{\log \mathrm{Tr}(\rho_{i^n}^n)^{1-s} - snR}{1-s}} \\ &= \left(\sum_i p_i e^{\frac{\log \mathrm{Tr}\, \rho_i^{1-s} - sR}{1-s}}\right)^n, \end{aligned} \tag{B.12}$$

where we define $x_{i^n}^n \stackrel{\text{def}}{=} x_{i_1} \otimes \cdots \otimes x_{i_n}$, $\rho_{i^n}^n \stackrel{\text{def}}{=} \rho_{i_1} \otimes \cdots \otimes \rho_{i_n}$ with respect to $i^n \stackrel{\text{def}}{=} (i_1, \ldots, i_n)$, and p^n is the independent and identical distribution of p. Further, we obtain

$$\begin{aligned}\lim_{s \to 0} \frac{1}{s} \log\left(\sum_i p_i e^{\frac{\log \mathrm{Tr}\, \rho_i^{1-s} - sR}{1-s}}\right) &= \frac{d}{ds} \log\left(\sum_i p_i e^{\frac{\log \mathrm{Tr}\, \rho_i^{1-s} - sR}{1-s}}\right)\Bigg|_{s=0} \\ &= \sum_i p_i (H(\rho_i) - R) < 0. \end{aligned}$$

Note that the inside of the logarithmic on the left-hand side (LHS) of the above equation is equal to 1 when $s = 0$. Taking an appropriate $1 > s_0 > 0$, we have $\log\left(\sum_i p_i e^{\frac{\log \mathrm{Tr}\, \rho_i^{1-s_0} - s_0 R}{1-s_0}}\right) < 0$. Thus, the RHS of (B.12) exponentially converges to zero. Therefore, we obtain $E_c^{\rightarrow}(\rho) \le E_f(\rho)$. Similarly, $E_c^{\rightarrow}(\rho^{\otimes k}) \le E_f(\rho^{\otimes k})$.

Next, we choose a sequence $\{m_n\}$ such that $(m_n - 1)k \le n \le m_n k$ with respect to n. Denote the partial trace of $(\mathcal{H}_A \otimes \mathcal{H}_B)^{\otimes m_n k - n}$ by C_n. Then,

$$F(\rho^{\otimes m_n k}, \kappa_{m_n}(|\Phi_{L_{m_n}}\rangle\langle\Phi_{L_{m_n}}|)) \ge F(\rho^{\otimes n}, C_n \circ \kappa_{m_n}(|\Phi_{L_{m_n}}\rangle\langle\Phi_{L_{m_n}}|)) \tag{B.13}$$

for κ_m, L_m. Therefore, if the LHS of (B.13) converges to zero, then the RHS also converges to zero. Since

$$\overline{\lim} \frac{1}{n} \log L_{m_n} = \frac{1}{k} \overline{\lim}_{m \to \infty} \frac{1}{m} \log L_m,$$

C_n is a local quantum operation, and $E_c^{\rightarrow}(\rho^{\otimes k}) \le E_f(\rho^{\otimes k})$, and we have

$$E_c^{\to}(\rho) \leq \frac{E_c^{\to}(\rho^{\otimes k})}{k} \leq \frac{E_f(\rho^{\otimes k})}{k}.$$

Considering $\inf_k$, we obtain the $\leq$ part of (8.94).

B.5.2 Proof of Converse Part

Let us first consider the following lemma as a preparation.

Lemma B.1 *Let p be a probability distribution $p = \{p_i\}_{i=1}^d$. Then,*

$$\sum_{i=L+1}^{d} p_i^{\downarrow} \geq \frac{H(p) - \log L - \log 2}{\log(d-L) - \log L}. \tag{B.14}$$

Proof. By defining the double stochastic transition matrix $A = (a_{i,j})$

$$a_{i,j} \stackrel{\text{def}}{=} \begin{cases} \frac{1}{L} & \text{if } i,j \leq L \\ \frac{1}{d-L} & \text{if } i,j > L \\ 0 & \text{otherwise,} \end{cases}$$

the image Ap satisfies $(Ap)_i = \begin{cases} \frac{P(p,L)}{L} & \text{if } i \leq L \\ \frac{1-P(p,L)}{d-L} & \text{if } i > L \end{cases}$. From Condition ③ in Theorem 8.2 we have $Ap \preceq p$. Therefore,

$$H(p) \geq H(Ap) = -P(p,L)\log\frac{P(p,L)}{L} - (1 - P(p,L))\log\frac{1-P(p,L)}{d-L}.$$

Since the binary entropy $h(x)$ is less than $\log 2$, we have

$$\begin{aligned} &P^c(p,L)(\log(d-L) - \log L) + \log 2 \\ \geq &P^c(p,L)(\log(d-L) - \log L) + h(P(p,L)) \geq H(p) - \log L. \end{aligned}$$

We thus obtain (B.14). ∎

We now show the $\geq$ part of Theorem 8.9 by using Lemma B.1. Consider the sequence of S-TP-CP maps $\{\kappa_n\}$ and the sequence of maximally entangled states $\{|\Phi_{L_n}\rangle\}$ satisfying

$$F(\kappa_n(|\Phi_{L_n}\rangle\langle\Phi_{L_n}|), \rho^{\otimes n}) \to 1. \tag{B.15}$$

Combining (8.86) and (8.87) in Theorem 8.8 and Lemma B.1, we have

$$\begin{aligned} &1 - F^2(\kappa_n(|\Phi_{L_n}\rangle\langle\Phi_{L_n}\||), \rho^{\otimes n}) \\ &\geq \min_{(p_i,x_i)} \left\{ \sum_i p_i P^c(x_i, L_n) \middle| \sum_i p_i |x_i\rangle\langle x_i| = \rho^{\otimes n} \right\} \\ &\geq \min_{(p_i,x_i)} \left\{ \sum_i p_i \frac{E(|x_i\rangle\langle x_i|) - \log L_n - \log 2}{\log(d^n - L_n) - \log L_n} \middle| \sum_i p_i |x_i\rangle\langle x_i| = \rho^{\otimes n} \right\} \\ &= \frac{E_f(\rho^{\otimes n}) - \log L_n - \log 2}{\log(d^n - L_n) - \log L_n} = \frac{\frac{E_f(\rho^{\otimes n})}{n} - \frac{\log L_n}{n} - \frac{\log 2}{n}}{\frac{\log(d^n - L_n)}{n} - \frac{\log L_n}{n}}. \end{aligned}$$

Using (B.15) and Lemma A.1, we obtain

$$0 = \lim \left(\frac{\log(d^n - L_n)}{n} - \frac{\log L_n}{n} \right) \left(1 - \left(F(\kappa_n(|\Phi_{L_n}\rangle\langle\Phi_{L_n}|), \rho^{\otimes n})\right)^2\right)$$
$$\geq \overline{\lim} \left(\frac{E_f(\rho^{\otimes n})}{n} - \frac{\log L_n}{n} - \frac{\log 2}{n} \right) = \lim \frac{E_f(\rho^{\otimes n})}{n} - \underline{\lim} \frac{\log L_n}{n}.$$

Thus, we obtain

$$\lim \frac{E_f(\rho^{\otimes n})}{n} \leq \underline{\lim} \frac{\log L_n}{n} \leq \overline{\lim} \frac{\log L_n}{n} \leq E_c(\rho),$$

which completes the proof of Theorem 8.9.

B.6 Proof of Theorem 9.3

First we prove Lemma 9.1.

Proof of Lemma 9.1.
S ⇒ A: From (9.26), **S** implies (9.30), *i.e.*, **A**.
S ⇒ L: From (9.27),

$$\begin{aligned}
&\min_{\rho^{1,2}:\operatorname{Tr}\rho^{1,2}(X^1+X^2)\leq K} f^{1,2}(\rho^{1,2}) \\
&\geq \min_{\rho^{1,2}:\operatorname{Tr}\rho^{1,2}(X^1+X^2)\leq K} f^1(\rho^1) + f^2(\rho^2) \\
&= \min_{\rho^1,\rho^2:\operatorname{Tr}\rho^1 X^1+\operatorname{Tr}\rho^2 X^2\leq K} f^1(\rho^1) + f^2(\rho^2) \\
&= \min_{0\leq\lambda\leq 1} \min_{\rho^1:\operatorname{Tr}\rho^1 X^1\leq\lambda K} f^1(\rho^1) + \min_{\rho^2:\operatorname{Tr}\rho^2 X^2\leq(1-\lambda)K} f^2(\rho^2).
\end{aligned}$$

On the other hand, since $f^1(\rho^1) + f^2(\rho^2) \geq f^{1,2}(\rho^1 \otimes \rho^2)$, we have

$$\begin{aligned}
&\min_{\rho^1,\rho^2:\operatorname{Tr}\rho^1 X^1+\operatorname{Tr}\rho^2 X^2\leq K} f^1(\rho^1) + f^2(\rho^2) \\
&\geq \min_{\rho^1,\rho^2:\operatorname{Tr}\rho^1 X^1+\operatorname{Tr}\rho^2 X^2\leq K} f^{1,2}(\rho^1 \otimes \rho^2) \\
&\geq \min_{\rho^{1,2}:\operatorname{Tr}\rho^{1,2}(X^1+X^2)\leq K} f^{1,2}(\rho^{1,2}).
\end{aligned}$$

Hence, we obtain (9.29).
L ⇒ C: Choose $\rho_0^{1,2}$ such that $\operatorname{Tr}\rho_0^{1,2}(X^1 + X^2) - f^{1,2}(\rho_0^{1,2}) = \max_{\rho^{1,2}} \operatorname{Tr}\rho^{1,2}(X^1+X^2) - f^{1,2}(\rho^{1,2})$. Then, the real number $K \stackrel{\text{def}}{=} \operatorname{Tr}\rho_0^{1,2}(X^1+X^2)$ satisfies

$$\begin{aligned}
&\max_{\rho^{1,2}} \operatorname{Tr} \rho^{1,2}(X^1+X^2) - f^{1,2}(\rho^{1,2}) \\
&= \max_{\rho^{1,2}:\operatorname{Tr}\rho^{1,2}(X^1+X^2)\ge K} \operatorname{Tr} \rho^{1,2}(X^1+X^2) - f^{1,2}(\rho^{1,2}) \\
&= K + \max_{\rho^{1,2}:\operatorname{Tr}\rho^{1,2}(X^1+X^2)\ge K} -f^{1,2}(\rho^{1,2}) \\
&= K + \max_{\rho^1,\rho^2:\operatorname{Tr}\rho^1 X^1+\operatorname{Tr}\rho^2X^2\ge K} -f^1(\rho^1) - f^2(\rho^2) \\
&= \max_{\rho^1,\rho^2:\operatorname{Tr}\rho^1 X^1+\operatorname{Tr}\rho^2X^2\ge K} \operatorname{Tr}\rho^1 X^1 - f^1(\rho^1) + \operatorname{Tr}\rho^2 X^2 - f^2(\rho^2) \\
&\le \max_{\rho^1,\rho^2} \operatorname{Tr}\rho^1 X^1 - f^1(\rho^1) + \operatorname{Tr}\rho^2 X^2 - f^2(\rho^2).
\end{aligned}$$

Conversely, from (9.26),

$$\begin{aligned}
&\max_{\rho^{1,2}} \operatorname{Tr} \rho^{1,2}(X^1+X^2) - f^{1,2}(\rho^{1,2}) \\
&\ge \max_{\rho^1,\rho^2} \operatorname{Tr} \rho^{1,2}(X^1+X^2) - f^{1,2}(\rho^1\otimes\rho^2) \\
&\ge \max_{\rho^1,\rho^2} \operatorname{Tr}\rho^1 X^1 - f^1(\rho^1) + \operatorname{Tr}\rho^2 X^2 - f^2(\rho^2).
\end{aligned}$$

Hence, we obtain (9.28).
C ⇒ S: For any $\rho_0^{1,2}$, from Lemma A.6, we choose Hermitian matrices X^1 and X^2 such that $\operatorname{Tr}\rho_0^i X^i - f^i(\rho_0^i) = \max_{\rho^i}\operatorname{Tr}\rho^i X^i - f^i(\rho^i)$. Hence,

$$\begin{aligned}
&\sum_{i=1}^2 \operatorname{Tr}\rho_0^i X^i - f^i(\rho_0^i) = \sum_{i=1}^2 \max_{\rho^i}\operatorname{Tr}\rho^i X^i - f^i(\rho^i) \\
&= \max_{\rho^{1,2}} \operatorname{Tr} \rho^{1,2}(X^1+X^2) - f^{1,2}(\rho^{1,2}) \ge \operatorname{Tr}\rho_0^{1,2}(X^1+X^2) - f^{1,2}(\rho_0^{1,2}).
\end{aligned}$$

Since $\operatorname{Tr}\rho_0^{1,2}(X^1+X^2) = \operatorname{Tr}\rho_0^1 X^1 + \operatorname{Tr}\rho_0^2 X^2$, we have (9.27). ∎

Proof of HM ⇒ HC. First, we assume that there exists a channel $\kappa_{X,p}$ for any channel κ, any positive semi-definite Hermitian matrix X on the input system $\mathcal{H}_A$, and any probability p, such that

$$C_c(\kappa_{X,p}) = \max_{\rho}(1-p)(\chi_\kappa(\rho)) + p\operatorname{Tr} H\rho, \tag{B.16}$$

and

$$\begin{aligned}
&C_c(\kappa^1_{X^1,p}\otimes\kappa^2_{X^2,p}) \\
&= \max_{\rho}\Big((1-p)^2\chi_{\kappa^1\otimes\kappa^2}(\rho) + (1-p)p(\chi_{\kappa^1}(\rho^1) + \operatorname{Tr} X^2\rho^2) \\
&\quad + (1-p)p(\chi_{\kappa^2}(\rho^2) + \operatorname{Tr} X^1\rho^1) + p^2(\operatorname{Tr} X^1\rho^1 + \operatorname{Tr} X^2\rho^2)\Big).
\end{aligned} \tag{B.17}$$

The channel $\kappa_{X,p}$ is called *Shor Extension* [374] of κ. Apply Condition **HM** to the channel $\kappa^1_{\frac{1}{p}X^1,p}\otimes\kappa^2_{\frac{1}{p}X^2,p}$, then we have

$$\max_{\rho^{1,2}}\Big((1-p)^2\chi_{\kappa^1\otimes\kappa^2}(\rho^{1,2})+(1-p)p(\chi_{\kappa^1}(\rho^1)+\operatorname{Tr}\frac{1}{p}X^2\rho^2)$$
$$+(1-p)p(\chi_{\kappa^2}(\rho^2)+\operatorname{Tr}\frac{1}{p}X^1\rho^1)+p^2(\operatorname{Tr}\frac{1}{p}X^1\rho^1+\operatorname{Tr}\frac{1}{p}X^2\rho^2)\Big)$$
$$\le\max_{\rho^1}(1-p)(\chi_{\kappa^1}(\rho^1)+\operatorname{Tr}\frac{1}{p}X^1\rho^1)+\max_{\rho^2}(1-p)(\chi_{\kappa^2}(\rho^2)+\operatorname{Tr}\frac{1}{p}X^2\rho^2).$$

Taking the limit $p\to 0$, we obtain

$$\max_{\rho^{1,2}}\chi_{\kappa^1\otimes\kappa^2}(\rho^{1,2})+\operatorname{Tr}(X^1+X^2)\rho^{1,2}$$
$$\le\max_{\rho^1}(\chi_{\kappa^1}(\rho^1)+\operatorname{Tr}X^1\rho^1)+\max_{\rho^2}(\chi_{\kappa^2}(\rho^2)+\operatorname{Tr}X^2\rho^2),$$

which implies Condition **HC**.

Next, we define the channel $\kappa_{X,p}$ with the input system $\mathcal{H}_A\otimes\mathbb{C}^k$, where $k\ge\|X\|$, and check (B.17). First, we generate one-bit random number X with probability $P_0=1-p$ and $P_1=p$. When $X=0$, the output state is $\kappa(\operatorname{Tr}_{\mathbb{C}^k}\rho)$ for the input state ρ. Otherwise, we perform the measurement of the spectral decomposition of X, and send the receiver the state $\tilde{\kappa}^y(\operatorname{Tr}_A\rho)$ depending on its measured data y, which is eigenvalue of X. Here, we defined the channel $\tilde{\kappa}^y$ and the stochastic transition matrix Q_l^j such that

$$\tilde{\kappa}^y(\sigma)=\sum_l\sum_j Q_l^j|u_l\rangle\langle u_l|\langle u_j|\sigma|u_j\rangle,\quad y=C_c(Q)=I(p_{\mathrm{mix}},Q).$$

In this case, we assume that the receiver received the information X and y. Then, the relation (B.16) holds. From these discussions, we can check the equation (B.17). ■

Proof of EM ⇒ EC. We define the channel $\tilde{\kappa}_{H,p}$ with the input system $\mathcal{H}_A$ as follows First, we generate one-bit random number X with probabilities $P_0=1-p$ and $P_1=p$. When $X=0$, the output state is $\kappa(\operatorname{Tr}_{\mathbb{C}^k}\rho)$ for the input state ρ. When $X=1$, we perform the measurement of the spectral decomposition of H, and obtain the eigenvalue y of H. Then, the output state is ρ_y, where ρ_y satisfies $H(\rho_y)=y$. In this case, the receiver is assumed to receive the information X and y. Then, the output entropy of the channel $\tilde{\kappa}_{H,p}$ can be calculated as

$$H(\tilde{\kappa}_{X,p}(\rho))=(1-p)H(\kappa(\rho))+p\operatorname{Tr}X\rho+h(p)-pH(\mathrm{P}_\rho^{E_X}).$$

Further,

$$H(\tilde{\kappa}^1_{X^1,p}\otimes\tilde{\kappa}^2_{X^2,p}(\rho))$$
$$=(1-p)^2(H(\kappa^1\otimes\kappa^2(\rho)))+p(1-p)(\operatorname{Tr}X^1\rho^1+H(\kappa^2(\rho^2)))$$
$$+p(1-p)(\operatorname{Tr}X^2\rho^2+H(\kappa^1(\rho^1)))+p^2(\operatorname{Tr}X^1\rho^1+\operatorname{Tr}X^2\rho^2)+2h(p)$$
$$-pH(\mathrm{P}_{\rho^1}^{E_{X^1}})-pH(\mathrm{P}_{\rho^2}^{E_{X^2}}).$$

Condition **EM** implies

$$\min_{\rho^{1,2}} H_{\tilde{\kappa}^1_{\frac{1}{p}X^1,p} \otimes \tilde{\kappa}^2_{\frac{1}{p}X^2,p}}(\rho^{1,2}) = \min_{\rho^1} H_{\tilde{\kappa}^1_{\frac{1}{p}X^1,p}}(\rho^1) + \min_{\rho^2} H_{\tilde{\kappa}^2_{\frac{1}{p}X^2,p}}(\rho^2).$$

Since $H(\mathrm{P}_{\rho^2}^{E_{X^2}}) \le \log d_A d_B$, taking the limit $p \to 0$, we have

$$\begin{aligned} H(\tilde{\kappa}_{\frac{1}{p}X,p}, \rho) &\to H_\kappa(\rho) + \operatorname{Tr} X\rho \\ H_{\tilde{\kappa}^1_{\frac{\lambda}{p}X^1,p} \otimes \tilde{\kappa}^2_{\frac{1}{p}X^2,p}}(\rho) &\to H_{\kappa^1 \otimes \kappa^2}(\rho) + (\operatorname{Tr} X^1\rho^1 + \operatorname{Tr} X^2\rho^2). \end{aligned}$$

Since the set of density matrices is compact, we obtain

$$\begin{aligned} &\min_{\rho^{1,2}} H_{\kappa^1\otimes\kappa^2}(\rho^{1,2}) + \operatorname{Tr}(X^1 + X^2)\rho^{1,2} \\ &\quad = \min_{\rho^1,\rho^2} (H_{\kappa^1}(\rho^1) + H_{\kappa^2}(\rho^2) + \operatorname{Tr} X^1\rho^1 + \operatorname{Tr} X^2\rho^2, \end{aligned}$$

which implies **EC**. ∎

Proof of FA ⇒ FS. See Pomeransky [350].

B.7 Proof of Lemma 9.4

In this section, we prove Lemma 9.4 from Lemma B.2 given below. For this proof, we use arguments similar to that in the proof of Theorem 9.7.

Following these arguments, we can choose ML elements $\{x_{m,l}\}_{(m,l)\in\{1,\dots,M\}\times\{1,\dots,L\}}$ of $\mathcal{X}$ and a POVM $\{Y_{m,l}\}$ (the elements $x_{m,l}$ are not necessarily different from each other) such that

$$\frac{8}{ML}\sum_{m,l} \epsilon_{B,m,l} + \frac{1}{M}\sum_{m=1}^{M} \epsilon_{E,m} \le (\text{RHS of (9.83)})$$

$$\epsilon_{B,m,l} \overset{\text{def}}{=} 1 - \operatorname{Tr} W_{B,x_{m,j}} Y_{m,l}, \quad \epsilon_{E,m} \overset{\text{def}}{=} \left\| \frac{1}{L}\sum_{l=1}^{L} W_{E,x_{m,l}} - \sum_x p_x W_{E,x} \right\|. \tag{B.18}$$

In what follows, we prove Lemma 9.4 using the above argument and Lemma B.2. The chosen element $x_{m,l}$ may in general contain some degeneracies. However, to apply Lemma B.2, no degeneracies should be present. In order to resolve this problem, we define new ML elements $x'_{m,l}$ satisfying the following conditions: (1) the elements $x'_{m,l}$ have no degeneracies. (2) $\{x_{m,l}\} \subset \{x'_{m,l}\}$. (3) For any element $x_{m,l}$, there exits an index (m_1,l_1) such that $x_{m_1,l_1} = x'_{m_1,l_1}$ among indices (m_1,l_1) satisfying $x_{m_1,l_1} = x_{m,l}$. The POVM $\{Y_{m,l}\}$ is also rewritten such that

$$Y'_{m,l} \stackrel{\text{def}}{=} \begin{cases} \sum_{x_{m',l'}=x_{m,l}} Y_{m',l'} & x_{m,l} = x'_{m,l} \\ 0 & x_{m,l} \neq x'_{m,l}. \end{cases}$$

Denoting the error probability by $\epsilon'_{B,m,l}$ in a manner similar to (B.18), we have

$$\sum_{m,l} \epsilon_{B,m,l} = \sum_{m,l} \epsilon'_{B,m,l}.$$

The number of elements in $\{x_{m,l} \neq x'_{m,l}\}$ is less than $\sum_{m,l} \epsilon'_{B,m,l}$. Therefore,

$$\frac{1}{M}\sum_{m=1}^{M} \left\| \frac{1}{L}\sum_{l=1}^{L} W^E_{x_{m,l}} - \frac{1}{L}\sum_{l=1}^{L} W^E_{x'_{m,l}} \right\| \leq \frac{2}{ML}\sum_{m,l} \epsilon_{B,m,l}.$$

Define $\epsilon'_{E,m}$ with respect to the pair $\{x'_{m,l}\}$ in a manner similar to (B.18). Then, we have

$$\begin{aligned} \frac{6}{ML}\sum_{m,l} \epsilon'_{B,m,l} + \frac{1}{M}\sum_{m=1}^{M} \epsilon'_{E,m} &\leq \frac{8}{ML}\sum_{m,l} \epsilon_{B,m,l} + \frac{1}{M}\sum_{m=1}^{M} \epsilon_{E,m} \\ &\leq (\text{RHS of (9.83)}). \end{aligned}$$

Hence, Lemma 9.4 can be obtained by applying Lemma B.2 to $u_{x'_{m,l}}$.

Lemma B.2 *Let $\{u^A_{m,l}\}_{(m,l)\in\{1,\ldots,M\}\times\{1,\ldots,L\}}$ be ML mutually orthogonal normalized vectors in $\mathcal{H}_A$, $U^{B,E}_A$ be an isometric map from $\mathcal{H}_A$ to $\mathcal{H}_B \otimes \mathcal{H}_E$, $Y = \{Y_{m,l}\}_{(m,l)\in\{1,\ldots,M\}\times\{1,\ldots,L\}}$ be a POVM in $\mathcal{H}_B$, and κ be a TP-CP map from $\mathcal{H}_A$ to $\mathcal{H}_B$ defined according to $U^{B,E}_A$. We define*

$$\epsilon^B_{m,l} \stackrel{\text{def}}{=} 1 - \operatorname{Tr} U^{B,E}_A |u^A_{m,l}\rangle\langle u^A_{m,l}| (U^{B,E}_A)^* (Y_{m,l} \otimes I_E)$$

$$\epsilon^E_m \stackrel{\text{def}}{=} \left\| \left(\frac{1}{L}\sum_{l=1}^{L} W^E_{m,l} \right) - W^E \right\|_1, \quad W^E_{m,l} \stackrel{\text{def}}{=} \operatorname{Tr}_B U^{B,E}_A |u^A_{m,l}\rangle\langle u^A_{m,l}| (U^{B,E}_A)^*$$

with respect to a state W^E in $\mathcal{H}_E$.

Now let $\mathcal{H}_C \stackrel{\text{def}}{=} \langle u^C_1, \ldots, u^C_M \rangle$ and $\mathcal{H}_D \stackrel{\text{def}}{=} \langle u^D_1, \ldots, u^D_M \rangle$ be spaces with a dimension of M. When we choose an encoding τ_α from $\mathcal{H}_C$ to $\mathcal{H}_A$ and a decoding ν_α from $\mathcal{H}_B$ to $\mathcal{H}_D$, depending on the random variable $\alpha = (\alpha_1, \ldots, \alpha_M)$ by following the method below, we have

$$1 - \mathrm{E}_\alpha F_e(\rho_{\mathrm{mix},C}, \nu_\alpha \circ \kappa \circ \tau_\alpha) \leq \frac{6}{ML}\sum_{m,l} \epsilon^B_{m,l} + \frac{1}{M}\sum_{M} \epsilon^E_m, \tag{B.19}$$

where α_m $(m = 1, \ldots, M)$ are independent random variables subject to the uniform distribution on the integers $0, \ldots, L-1$, and $\mathcal{H}_C$ and $\mathcal{H}_D$ are identified due to the natural correspondence $u^C_m \mapsto u^D_m$.

The encoder τ_α *and the decoder* ν_α *are constructed as follows. Define the isometric map* $U^A_{C,\alpha}$ *from* $\mathcal{H}_C$ *to* $\mathcal{H}_A$ *by*

$$U^A_{C,\alpha} u^C_m \stackrel{\text{def}}{=} \left(u^A_{m,\alpha} \stackrel{\text{def}}{=} \sum_{l=1}^{L} e^{(2l\alpha_m \pi i)/L} u^A_{m,l} \right).$$

We also define τ_α *according to* $\tau_\alpha(\rho) \stackrel{\text{def}}{=} U^A_{C,\alpha}\rho(U^A_{C,\alpha})^*$*. Next, we choose the space* $\mathcal{H}_{B'}$*, containing* $\mathcal{H}_B$*, to be sufficiently large such that the purification* $\hat{u}$ *of* W^E *can be taken as an element of* $\mathcal{H}_E \otimes \mathcal{H}_{B'}$*. We also choose the isometric map* $U^{B,D}_B$ *from* $\mathcal{H}_B$ *to* $\mathcal{H}_B \otimes \mathcal{H}_D$*, the unitary matrix* $U^{B'}_{m,\alpha}$ *in* $\mathcal{H}_{B'}$*, and the unitary matrix* $U^{B',D}_\alpha$ *in* $\mathcal{H}_{B'} \otimes \mathcal{H}_D$ *such that*

$$\operatorname{Tr} \rho Y_m = \langle u^D_m | U^{B,D}_B \rho (U^{B,D}_B)^* | u^D_m \rangle, \quad Y_m \stackrel{\text{def}}{=} \sum_{l=1}^{L} Y_{m,l}, \tag{B.20}$$

$$U^{B'}_{m,\alpha} \stackrel{\text{def}}{=} \operatorname*{argmax}_U |\langle u^D_m \otimes (U^* \otimes I_E)\hat{u} | U^{B,D}_B U^{B,E}_A u^A_{m,\alpha} \rangle|,$$

$$U^{B',D}_\alpha \stackrel{\text{def}}{=} \sum_{m=1}^{M} |u^D_m\rangle\langle u^D_m| \otimes U^{B'}_{m,\alpha}.$$

The first condition (B.20) is the condition that $U^{B,D}_B$ *gives a Naĭmark extension of* $\{Y_m\}_{m=1}^M$*. We choose* $U^{B'}_{m,\alpha}$ *such that* $\langle u^D_m \otimes ((U^{B'}_{m,\alpha})^* \otimes I_E)\hat{u} | U^{B,D}_B U^{B,E}_A u^A_{m,\alpha}\rangle$ *is a nonnegative real number. Let* ν_α *be given by*

$$\nu_\alpha(\rho) \stackrel{\text{def}}{=} \operatorname{Tr}_{B'} U^{B',D}_\alpha U^{B,D}_B \rho (U^{B,D}_B)^* (U^{B',D}_\alpha)^*. \tag{B.21}$$

This discussion gives the construction of the encoder τ_α *and the decoder* ν_α*.*

Proof. Let $\mathcal{H}_R \stackrel{\text{def}}{=} \langle u^R_1, \dots, u^R_M \rangle$ be the environment of $\mathcal{H}_C$. Then,

$$\begin{aligned}
&\mathrm{E}_\alpha F_e(\rho_{\mathrm{mix},C}, \nu_\alpha \circ \kappa \circ \tau_\alpha) \\
&= \mathrm{E}_\alpha F\left(\frac{1}{\sqrt{M}} \sum_{m=1}^{M} u^R_m \otimes u^C_m, \nu_\alpha \circ \kappa \circ \tau_\alpha \left(\frac{1}{\sqrt{M}} \sum_{m=1}^{M} u^R_m \otimes u^C_m \right)\right) \\
&\le F\left(\left(\frac{1}{\sqrt{M}} \sum_{m=1}^{M} u^R_m \otimes u^C_m \right) \otimes \hat{u}, \frac{1}{\sqrt{M}} \sum_{m=1}^{M} u^R_m \otimes U^{B',D}_\alpha U^{B,D}_B U^{B,E}_A u^A_{m,\alpha} \right) \\
&= \left| \left\langle \frac{1}{\sqrt{M}} \sum_{m'=1}^{M} u^R_{m'} \otimes u^C_{m'} \otimes \hat{u} \right| \frac{1}{\sqrt{M}} \sum_{m=1}^{M} u^R_m \otimes U^{B',D}_\alpha U^{B,D}_B U^{B,E}_A u^A_{m,\alpha} \right\rangle \right| \\
&= \left| \frac{1}{M} \sum_{m=1}^{M} \left\langle u^C_m \otimes \hat{u} \left| U^{B',D}_\alpha U^{B,D}_B U^{B,E}_A u^A_{m,\alpha} \right\rangle \right. \right|
\end{aligned}$$

$$= \frac{1}{M} \sum_{m=1}^{M} \left\langle u_m^C \otimes \hat{u} \left| U_\alpha^{B',D} U_B^{B,D} U_A^{B,E} u_{m,\alpha}^A \right.\right\rangle$$
$$= \frac{1}{M} \sum_{m=1}^{M} F(U_B^{B,D} U_A^{B,E} u_{m,\alpha}^A, u_m^C \otimes (U_{m,\alpha}^{B'})^* \hat{u}). \tag{B.22}$$

Next, we evaluate $1 - F(U_B^{B,D} U_A^{B,E} u_{m,\alpha}^A, u_m^C \otimes (U_{m,\alpha}^{B'})^* \hat{u})$. Note that the relation $b^2(\rho, \sigma) = 1 - F(\rho, \sigma)$ will be used frequently in what follows. For this evaluation, we choose a sufficiently large space $\mathcal{H}_{B''}$ containing $\mathcal{H}_{B'}$ and a PVM $E^m = \{E_l^m\}_{l=1}^L$ on $\mathcal{H}_{B''}$ such that

$$\begin{aligned}
&\operatorname{Tr} \rho Y_{m,l} (= \operatorname{Tr} \sqrt{Y_m} \rho \sqrt{Y_m} (\sqrt{Y_m})^{-1} \rho Y_{m,l} (\sqrt{Y_m})^{-1}) \\
&\quad = \operatorname{Tr}(|u_m^D\rangle\langle u_m^D| \otimes E_l^m) U_B^{B,D} \rho (U_B^{B,D})^* \\
&\quad\quad \left(= \operatorname{Tr}(I_D \otimes E_l^m)(|u_m^D\rangle\langle u_m^D| \otimes I_B) U_B^{B,D} \rho (U_B^{B,D})^* (|u_m^D\rangle\langle u_m^D| \otimes I_B) \right).
\end{aligned}$$

Define a normalized vector $u_{m,l}^{B'',E}$ in $\mathcal{H}_{B'',E}$ by

$$u_m^D \otimes u_{m,l}^{B'',E} \stackrel{\text{def}}{=} \frac{1}{\sqrt{1 - \epsilon_{m,l}^E}} (|u_m^D\rangle\langle u_m^D| \otimes E_l^m) U_B^{B,D} U_A^{B,E} u_{m,l}^A$$

and a unitary $U_{m,\alpha}^{B''}$ by

$$U_{m,\alpha}^{B''} \stackrel{\text{def}}{=} \operatorname*{argmax}_U F\left(\frac{1}{\sqrt{L}} \sum_{l=1}^{L} e^{(2l\alpha_m \pi i)/L} u_m^D \otimes u_{m,l}^{B'',E}, u_m^C \otimes (U)^* \hat{u} \right).$$

Then,

$$\begin{aligned}
&b^2(U_B^{B,D} U_A^{B,E} u_{m,\alpha}^A, u_m^C \otimes (U_{m,\alpha}^{B'})^* \hat{u}) \\
&\quad \le b^2(U_B^{B,D} U_A^{B,E} u_{m,\alpha}^A, u_m^C \otimes (U_{m,\alpha}^{B''})^* \hat{u}) \\
&\quad \le 2b^2 \left(\frac{1}{\sqrt{L}} \sum_{l=1}^{L} e^{(2l\alpha_m \pi i)/L} u_m^D \otimes u_{m,l}^{B'',E}, u_m^C \otimes (U_{m,\alpha}^{B''})^* \hat{u} \right) \\
&\quad\quad + 2b^2 \left(U_B^{B,D} U_A^{B,E} u_{m,\alpha}^A, \frac{1}{\sqrt{L}} \sum_{l=1}^{L} e^{(2l\alpha_m \pi i)/L} u_m^D \otimes u_{m,l}^{B'',E} \right).
\end{aligned} \tag{B.23}$$

Now we evaluate the first term in (B.23). Since

$$
\begin{aligned}
&\mathrm{Tr}_{B''}\left|\frac{1}{\sqrt{L}}\sum_{l=1}^{L} e^{(2l\alpha_m\pi i)/L} u_{m,l}^{B'',E}\right\rangle\left\langle\frac{1}{\sqrt{L}}\sum_{l'=1}^{L} e^{(2l'\alpha_m\pi i)/L} u_{m,l'}^{B'',E}\right| \\
&= \mathrm{Tr}_{B''}\sum_{l''=1}^{L}(E_{l''}^{m}\otimes I_E) \\
&\quad \left(\frac{1}{L}\sum_{l=1}^{L}\sum_{l'=1}^{L} e^{(2(l-l')\alpha_m\pi i)/L}\left|u_{m,l}^{B'',E}a\right\rangle\left\langle u_{m,l'}^{B'',E}\right|\right)(E_{l''}^{m}\otimes I_E) \\
&= \mathrm{Tr}_{B''}\sum_{l=1}^{L}\frac{1}{L}\left|u_{m,l}^{B'',E}\right\rangle\left\langle u_{m,l}^{B'',E}\right|,
\end{aligned}
$$

from the definition of $U_{m,\alpha}^{B''}$, we obtain

$$
\begin{aligned}
&b^2\left(\frac{1}{\sqrt{L}}\sum_{l=1}^{L} e^{(2l\alpha_m\pi i)/L} u_m^D\otimes u_{m,l}^{B'',E}, u_m^C\otimes (U_{m,\alpha}^{B''})^*\hat{u}\right) \\
&= b^2\left(\frac{1}{\sqrt{L}}\sum_{l=1}^{L} e^{(2l\alpha_m\pi i)/L} u_{m,l}^{B'',E}, (U_{m,\alpha}^{B''})^*\hat{u}\right) \\
&= b^2\left(\mathrm{Tr}_{B''}\sum_{l=1}^{L}\frac{1}{L}\left|u_{m,l}^{B'',E}\right\rangle\left\langle u_{m,l}^{B'',E}\right|, W_E\right) \\
&\le 2b^2\Big(\mathrm{Tr}_{B''}\sum_{l=1}^{L}\frac{1}{L}\left|u_{m,l}^{B'',E}\right\rangle\left\langle u_{m,l}^{B'',E}\right|, \\
&\qquad \sum_{l=1}^{L}\frac{1}{L}\mathrm{Tr}_{B,D}\left|U_B^{B,D}U_A^{B,E}u_{m,l}^{A}\right\rangle\left\langle U_B^{B,D}U_A^{B,E}u_{m,l}^{A}\right|\Big) \\
&\quad + 2b^2\left(\sum_{l=1}^{L}\frac{1}{L}W_{m,l}^{E}, W_E\right),
\end{aligned}
\tag{B.24}
$$

where we use Lemma 8.2 and the fact that $\mathrm{Tr}_{B,D}\left|U_B^{B,D}U_A^{B,E}u_{m,l}^{A}\right\rangle$ $\left\langle U_B^{B,D}U_A^{B,E}u_{m,l}^{A}\right| = W_{m,l}^{E}$. For the second term in (B.24), inequality (5.42) yields

$$
b^2\left(\sum_{l=1}^{L}\frac{1}{L}W_{m,l}^{E}, W_E\right) \le \frac{1}{2}\left\|\sum_{l=1}^{L}\frac{1}{L}W_{m,l}^{E} - W_E\right\|_1 \le \frac{1}{2}\epsilon_m^E. \tag{B.25}
$$

The first term in (B.24) is evaluated as

$$
\begin{aligned}
& b^2\Big(\mathrm{Tr}_{B''} \sum_{l=1}^{L} \frac{1}{L} \left|u_{m,l}^{B'',E}\right\rangle \left\langle u_{m,l}^{B'',E}\right|, \\
& \qquad \sum_{l=1}^{L} \frac{1}{L} \mathrm{Tr}_{B,D} \left|U_B^{B,D} U_A^{B,E} u_{m,l}^A\right\rangle \left\langle U_B^{B,D} U_A^{B,E} u_{m,l}^A\right|\Big) \\
& \le \sum_{l=1}^{L} \frac{1}{L} b^2\Big(\mathrm{Tr}_{B''} \left|u_{m,l}^{B'',E}\right\rangle \left\langle u_{m,l}^{B'',E}\right|, \\
& \qquad \mathrm{Tr}_{B'',D} \left|U_B^{B,D} U_A^{B,E} u_{m,l}^A\right\rangle \left\langle U_B^{B,D} U_A^{B,E} u_{m,l}^A\right|\Big) \\
& \le \sum_{l=1}^{L} \frac{1}{L} b^2 \left(u_m^D \otimes u_{m,l}^{B'',E}, U_B^{B,D} U_A^{B,E} u_{m,l}^A\right) \\
& = \sum_{l=1}^{L} \frac{1}{L} \left(1 - \sqrt{1 - \epsilon_{m,l}^B}\right) \le \sum_{l=1}^{L} \frac{1}{L} \epsilon_{m,l}^B,
\end{aligned} \tag{B.26}
$$

where we use the fact that $\mathrm{Tr}_{B''} \left|u_{m,l}^{B'',E}\right\rangle \left\langle u_{m,l}^{B'',E}\right| = \mathrm{Tr}_{B'',D} \left|u_m^D \otimes u_{m,l}^{B'',E}\right\rangle \left\langle u_m^D \otimes u_{m,l}^{B'',E}\right|$. The second term of (B.23) can be evaluated as

$$
\begin{aligned}
& \mathrm{E}_\alpha F\left(U_B^{B,D} U_A^{B,E} u_{m,\alpha}^A, \sum_{l=1}^{L} \frac{e^{(2l\alpha_m \pi i)/L}}{\sqrt{L}} u_m^D \otimes u_{m,l}^{B'',E}\right) \\
& = \mathrm{E}_\alpha \left\langle \sum_{l=1}^{L} \frac{e^{(2l\alpha_m \pi i)/L}}{\sqrt{L}} U_B^{B,D} U_A^{B,E} u_{m,l}^A, \sum_{l=1}^{L} \frac{e^{(2l\alpha_m \pi i)/L}}{\sqrt{L}} u_m^D \otimes u_{m,l}^{B'',E}\right\rangle \\
& = \frac{1}{L} \sum_{l=1}^{L} \sum_{l'=1}^{L} (\mathrm{E}_\alpha e^{(2(l-l')\alpha_m \pi i)/L}) \left\langle U_B^{B,D} U_A^{B,E} u_{m,l'}^A, u_m^D \otimes u_{m,l}^{B'',E}\right\rangle \\
& = \frac{1}{L} \sum_{l=1}^{L} \langle U_B^{B,D} U_A^{B,E} u_{m,l}^A, u_m^D \otimes u_{m,l}^{B'',E}\rangle \\
& = \frac{1}{L} \sum_{l=1}^{L} \left\langle U_B^{B,D} U_A^{B,E} u_{m,l}^A, \frac{1}{\sqrt{1-\epsilon_{m,l}^E}} (|u_m^D\rangle\langle u_m^D| \otimes E_l^m) U_B^{B,D} U_A^{B,E} u_{m,\alpha}^A\right\rangle \\
& = \frac{1}{L} \sum_{l=1}^{L} \sqrt{1-\epsilon_{m,l}^E} \ge \frac{1}{L} \sum_{l=1}^{L} (1-\epsilon_{m,l}^E).
\end{aligned} \tag{B.27}
$$

Combining (B.23)–(B.27) with (B.22), we obtain (B.19). ■

B.8 Proof of Lemma 10.3

We first prove the following lemma.

Lemma B.3 *A visible encoder may be represented by a map from $\mathcal{X}$ to $\mathcal{S}(\mathcal{K})$. Consider the convex combination of codes T and T':*

$$(\lambda T + (1-\lambda)T')(x) \stackrel{\text{def}}{=} \lambda T(x) + (1-\lambda)T'(x), \quad 0 < \forall\lambda < 1.$$

Then, the set of visible encoders is a convex set, and the set of extremal points (see Sect. A.4 for the definition of an extremal point) is equal to

$$\{T\,|T(x) \text{ is a pure state } \forall x \in \mathcal{X}\}. \tag{B.28}$$

Proof. When $T(x)$ is a pure state for every input x, T is therefore an extremal point because it is impossible to represent the encoder T as a convex combination of other encoders. Hence, to complete the proof, it is sufficient to show that an arbitrary visible encoder $T(x) = \sum_{j_x} s_{j_x}|\phi_{j_x}\rangle\langle\phi_{j_x}|$ can be represented as a convex combination of encoders satisfying the condition in (B.28). Define a visible encoder $T(j_1, j_2, \cdots, j_n)$ by

$$T(j_1, j_2, \cdots, j_n|i) = |\phi_{j_x}\rangle\langle\phi_{j_x}|.$$

Then, this encoder belongs to the set (B.28). Since $T = \sum_{j_1,j_2,\cdots,j_n} s_{j_1}s_{j_2}\cdots s_{j_n}$ $T(j_1, j_2, \cdots, j_n)$, the proof is completed. ∎

We also require the following lemma for the proof of Lemma 10.3. This lemma is equivalent to Theorem 8.3, which was shown from the viewpoint of entanglement in Sect. 8.4.

Lemma B.4 *Let $\rho \in \mathcal{S}(\mathcal{H}_A \otimes \mathcal{H}_B)$ be separable. Then,*

$$\begin{aligned} &\max\{\operatorname{Tr} P\rho_A | P : \text{Projection in } \mathcal{H}_A \text{ with rank } k\} \\ &\quad \geq \max\{\operatorname{Tr} P\rho | P : \text{Projection in } \mathcal{H}_A \otimes \mathcal{H}_B \text{ with rank } k\} \end{aligned}$$

holds for any integer k.

Proof of Lemma 10.3. According to Lemma B.3, it is sufficient to show (10.22) for a visible encoder T in (B.28). From Condition ⑥ in Theorem 5.1, there exist a space $\mathcal{H}'$ with the same dimension of $\mathcal{H}$, a pure state ρ_0 in $\mathcal{H}'\otimes\mathcal{H}$, and a unitary matrix U in $\mathcal{K}\otimes\mathcal{H}'\otimes\mathcal{H}$ such that $\nu(\rho) = \operatorname{Tr}_{\mathcal{K},\mathcal{H}'} U(\rho\otimes\rho_0)U^*$, and the state

$$\rho_x \stackrel{\text{def}}{=} \frac{(W_x \otimes I)\,U\,(T(x)\otimes\rho_0)\,U^*\,(W_x\otimes I)}{\operatorname{Tr} U\,(T(x)\otimes\rho_0)\,U^*\,(W_x\otimes I)} \in \mathcal{S}(\mathcal{K}\otimes\mathcal{H}\otimes\mathcal{H}')$$

is a pure state. Since $UT(x)\otimes\rho_0 U^*$ is a pure state and $(W_x\otimes I)$ is a projection, we have

$$\operatorname{Tr} \nu(T(x))W_x = \operatorname{Tr} UT(x) \otimes \rho_0 U^* (W_x \otimes I) = \operatorname{Tr} U(T(x) \otimes \rho_0) U^* \rho_x. \qquad \text{(B.29)}$$

Since $\operatorname{Tr}_{\mathcal{K},\mathcal{H}'} \rho_x = W_x$, we may write $\rho_x = W_x \otimes \sigma_x$ by choosing an appropriate pure state $\sigma_x \in \mathcal{S}(\mathcal{K} \otimes \mathcal{H}')$. Hence, the state $\rho_p \stackrel{\text{def}}{=} \sum_{i \in \mathcal{X}} p(x) \rho_x = \sum_{x \in \mathcal{X}} p(x) W_x \otimes \sigma_x$ is separable and satisfies $W_p = \operatorname{Tr}_{\mathcal{H},\mathcal{K}'} \rho_p$. Since $I_{\mathcal{K}} \geq T(x)$, we have $U(I_{\mathcal{K}} \otimes \rho_0) U^* \geq U(T(x) \otimes \rho_0) U^*$. Thus, from (B.29) we have

$$\begin{aligned} \sum_{x \in \mathcal{X}} p(x) \operatorname{Tr} \nu(T(x)) W_x &= \sum_{x \in \mathcal{X}} p(x) \operatorname{Tr}_{\mathcal{H}} \operatorname{Tr}_{\mathcal{K} \otimes \mathcal{H}'} U(T(x) \otimes \rho_0) U^* \rho_x \\ &\leq \sum_{x \in \mathcal{X}} p(x) \operatorname{Tr} U(I_{\mathcal{K}} \otimes \rho_0) U^* \rho_x = \operatorname{Tr} U(I_{\mathcal{K}} \otimes \rho_0) U^* \rho_p. \end{aligned} \qquad \text{(B.30)}$$

According to $I \geq U(I_{\mathcal{K}} \otimes \rho_0) U^* \geq 0$ and $\operatorname{Tr} U(I_{\mathcal{K}} \otimes \rho_0) U^* = \operatorname{Tr} I_{\mathcal{K}} = \dim \mathcal{K}$, we obtain

$$\begin{aligned} \operatorname{Tr} U(I_{\mathcal{K}} \otimes \rho_0) U^* \rho_p &\leq \max \left\{ \operatorname{Tr} P \rho_p \,\middle|\, \begin{array}{l} P : \text{Projection in } \mathcal{K} \otimes \mathcal{H} \otimes \mathcal{H}', \\ \operatorname{rank} P = \dim \mathcal{K} \end{array} \right\} \\ &\leq \max\{\operatorname{Tr} PW_p | P : \text{Projection in } \mathcal{H}, \operatorname{rank} P = \dim \mathcal{K}\}. \end{aligned} \qquad \text{(B.31)}$$

(B.31) may be obtained from Lemma B.4 and the separability of ρ_p. The projection P on $\mathcal{H}$ satisfies

$$\operatorname{Tr}(W_p - a)P \leq \operatorname{Tr}(W_p - a)\{W_p - a \geq 0\}.$$

If the rank of P is $\dim \mathcal{K}$ (i.e., if $\operatorname{Tr} P = \dim \mathcal{K}$), then

$$\operatorname{Tr} W_p P \leq a \dim \mathcal{K} + \operatorname{Tr} W_p \{W_p - a \geq 0\}. \qquad \text{(B.32)}$$

From (B.30)–(B.32),

$$\begin{aligned} 1 - \varepsilon(\Psi) &= \sum_{x \in \mathcal{X}} p(x) \operatorname{Tr} \nu(T(x)) W_x \\ &\leq \max\{\operatorname{Tr} PW_p | P : \text{ Projection in } \mathcal{H}, \ \operatorname{rank} P = \dim \mathcal{K}\} \\ &\leq a \dim \mathcal{K} + \operatorname{Tr} W_p \{W_p - a \geq 0\}. \end{aligned}$$

We therefore obtain (10.22). ∎

C

Hints and Brief Solutions to Exercises

Exercise 1.1. Use the fact that the discriminant of $\langle u + rcv | u + rcv \rangle$ concerning r is negative.

Exercise 1.3. Note that $X^T = \sum_{i,j} \overline{x^{i,j}} |u_i\rangle\langle u_j|$. Show that $\overline{\langle x|Xx\rangle} = \langle \overline{x}|X^T\overline{x}\rangle \geq 0$, where $x = \sum_i x^i u_i$ and $\overline{x} = \sum_i \overline{x^i} u_i$.

Exercise 1.4. Consider the derivative of the product of the matrix elements.

Exercise 1.13. **d**: Take the partial trace on A or B.

Exercise 1.19. A matrix cannot be semipositive definite if its determinant is negative.

Exercise 1.23. Show $\mathrm{Tr}\{Y \geq 0\}X\{Y \geq 0\} \geq \mathrm{Tr}\{Y \geq 0\}Y\{Y \geq 0\}$ and $\mathrm{Tr}\{Y < 0\}X\{Y < 0\} \geq -\mathrm{Tr}\{Y < 0\}Y\{Y < 0\}$.

Exercise 2.2. Consider the case $\mathrm{P}_Y(1) = \lambda, \mathrm{P}_Y(0) = 1 - \lambda, \mathrm{P}_{X|Y=1} = p, \mathrm{P}_{X|Y=0} = p'$.

Exercise 2.6. Apply a stochastic transition matrix of rank 1 to Theorem 2.1.

Exercise 2.8. Use the fact that $\sum_j \sum_i Q_j^i |p_i - q_i| \geq \sum_j |\sum_i Q_j^i (p_i - q_i)|$.

Exercise 2.9. Consider the $x \geq y$ and $x < y$ cases separately.

Exercise 2.10. **a**: Use $|p_i - q_i| = |\sqrt{p_i} - \sqrt{q_i}||\sqrt{p_i} + \sqrt{q_i}|$. **b**: Use $p_i + q_i \geq 2\sqrt{p_i}\sqrt{q_i}$.

Exercise 2.11. Assume that the datum i generates with the probability distribution p_i. Apply Jensen's inequality to the random variable $\sqrt{q_i/p_i}$ and the convex function $-\log x$.

Exercise 2.13. Check that $\phi'(s|p\|q) = \dfrac{\sum_i p_i^{1-s} q_i^s (\log q_i - \log p_i)}{\sum_i p_i^{1-s} q_i^s}$.

Exercise 2.14. Check that $\phi''(s|p\|q) =$

$$\frac{(\sum_i p_i^{1-s} q_i^s)(\sum_i p_i^{1-s} q_i^s (\log q_i - \log p_i)^2) - (\sum_i p_i^{1-s} q_i^s (\log q_i - \log p_i))^2}{(\sum_i p_i^{1-s} q_i^s)^2}.$$

Next, use Schwarz's inequality between two vectors 1 and $(-\log p_i + \log q_i)$.

Exercise 2.18. Consider the spectral decomposition $\boldsymbol{M}$ of X and apply Jensen's inequality to $\mathrm{P}_\rho^{\boldsymbol{M}}$.

Exercise 2.21. Note that $2(x^2+y^2) \geq (x+y)^2$.

Exercise 2.22. **b**: Take the average by integrating between $[0, 2\pi]$ for each θ_i. Note that $\langle u|v\rangle$ is continuous for each θ_i.

Exercise 2.24. **d**: Use Schwarz's inequality with respect to the inner product $\operatorname{Tr} XY^*$ with two vectors $\rho^{(1-s)/2}(\log\sigma - \log\rho)\sigma^{s/2}$ and $\rho^{(1-s)/2}\sigma^{s/2}$.

Exercise 2.25. **a**: Use the Schwarz inequality. **c**: Choose U such that $|\rho^{1/2}\sigma^{1/2}| = U\rho^{1/2}\sigma^{1/2}$. Note that $|\operatorname{Tr} U\rho^{1/2}M_i\sigma^{1/2}| \geq \operatorname{Tr} U\rho^{1/2}M_i\sigma^{1/2}$.

Exercise 2.26. Note that the spectral decomposition $\sum_i \lambda M_i$ of $\rho^{1/2}U^*\sigma^{-1/2}$ satisfies $M_i^{1/2}\sigma^{1/2} = \lambda_i M_i^{1/2}\rho^{1/2}U^*$.

Exercise 2.27. **b**: See the hint for Exercise 2.26.

Exercise 2.31. Use the approximation
$\sqrt{p_{\theta+\epsilon}(\omega)} \cong \sqrt{p_\theta(\omega)}\sqrt{1 + l_\theta(\omega)\epsilon + \frac{1}{2}\frac{d^2 p_\theta(\omega)}{d\theta^2}\epsilon^2}$.

Exercise 2.34. **c**: Use $\frac{d\theta}{d\eta} = J_\eta$. **g**: Use the uniqueness of the solution of the differential equation.

Exercise 2.35. $D(q_{\eta_j = X_j(\omega)}\|p_{\theta'}) - D(q_{\eta_j = X_j(\omega)}\|p_\theta)$

$$
\begin{aligned}
&= \textstyle\sum_{\omega'} q_{\eta_j=X_j(\omega)}(\omega')\left((\log q_{\eta_j=X_j(\omega)}(\omega') - \log p_{\theta'}(\omega'))\right.\\
&\quad \left. - (\log q_{\eta_j=X_j(\omega)}(\omega') - \log p_\theta(\omega'))\right)\\
&= \textstyle\sum_{\omega'} q_{\eta_j=X_j(\omega)}(\omega')(\log p_\theta(\omega') - \log p_{\theta'}(\omega'))\\
&= \textstyle\sum_{\omega'} q_{\eta_j=X_j(\omega)}(\omega')\left(\sum_i(\theta^i - \theta'^i)X_i(\omega') + \mu(\theta') - \mu(\theta')\right)\\
&= \left(\textstyle\sum_i(\theta^i - \theta'^i)X_i(\omega) + \mu(\theta') - \mu(\theta')\right) = \log p_\theta(\omega) - \log p_{\theta'}(\omega).
\end{aligned}
$$

Exercise 2.36. Show that $\frac{p_\theta(\omega)}{d\theta} = 0$ if and only if $\eta(\theta) = X(\omega)$.

Exercise 2.37. Combine (2.9) and (2.77).

Exercise 2.38. The case of $n \geq m$ can be obtained from $n, n-1, \ldots, m+1 \geq m$. The $n < m$ case may be obtained from $\frac{1}{m}, \frac{1}{m-1}, \ldots, \frac{1}{n+1} \leq \frac{1}{n}$.

Exercise 2.41. **o**: Replace $H(p)$ and $\psi(s)$ with $-D(q\|p')$ and $\phi(s|p\|p')$, respectively.

Exercise 3.1. Note that $\operatorname{Tr}|X|$ is equal to the sum of the absolute values of the eigenvalues of X.

Exercise 3.3. $\|\rho_{\mathrm{mix}} - \rho\|_1 = 2\operatorname{Tr}(\rho_{\mathrm{mix}} - \rho)\{\rho_{\mathrm{mix}} - \rho \geq 0\}$. Hence, $2 - \|\rho_{\mathrm{mix}} - \rho\|_1 = 2 - 2\operatorname{Tr}(\rho_{\mathrm{mix}} - \rho)\{\rho_{\mathrm{mix}} - \rho \geq 0\} \leq 2 - 2\operatorname{Tr}\rho_{\mathrm{mix}}\{\rho_{\mathrm{mix}} - \rho \geq 0\} = 2\operatorname{Tr}\rho_{\mathrm{mix}}\{\rho_{\mathrm{mix}} - \rho < 0\} \leq 2\operatorname{Tr}\rho_{\mathrm{mix}}\{0 < \rho\} = 2\frac{\operatorname{rank}\rho}{d}$.

Exercise 3.4. $\sum_{i=1}^k \frac{1}{k}\operatorname{Tr}\rho_i M_i \leq \sum_{i=1}^k \frac{1}{k}\operatorname{Tr} M_i = \frac{1}{k}\operatorname{Tr}\sum_{i=1}^k M_i = \frac{1}{k}\operatorname{Tr} I = \frac{d}{k}$. Further, $\sum_{i=1}^k \frac{1}{k}\operatorname{Tr}\rho_i M_i \leq \sum_{i=1}^k \frac{1}{k}\|M_i\|_1\|\rho_i\| = \frac{1}{k}\max_{i'}\|\rho_{i'}\|\sum_{i=1}^k \|M_i\|_1 = \frac{1}{k}\max_{i'}\|\rho_{i'}\|\sum_{i=1}^k \operatorname{Tr} M_i = \frac{d}{k}\max_i\|\rho_i\|$.

Exercise 3.5. Use Cramér's theorem with $X = -\log \frac{p(\omega)}{\bar{p}(\omega)}$, $\theta = s$, $x = 0$, and show that $\lim \frac{-1}{n} \log p^n \left\{ \frac{-1}{n} \log \left(\frac{p^n(x^n)}{\bar{p}^n(x^n)} \right) \geq 0 \right\} = \max_{s \geq 0} -\phi(s)$ and $\lim \frac{-1}{n} \log \bar{p}^n \left\{ \frac{-1}{n} \log \left(\frac{p^n(x^n)}{\bar{p}^n(x^n)} \right) \leq R \right\} = \max_{s \leq 1} -\phi(s)$.

Exercise 3.9. Use the convexity of $\phi(s)$ and the symmetry $\phi(s) = \phi(1-s)$.

Exercise 3.11. **a**: For the derivations of (3.20) and (3.21), substitute $\theta = s$ and $X = -\log \left(\frac{p(x)}{\bar{p}(x)} \right)$ in (2.117) and (2.119), respectively. For the derivation of (3.22), substitute $\theta = s - 1$ $X = -\log \left(\frac{p(x)}{\bar{p}(x)} \right)$ in (2.119). **e**: Lemma 3.1 guarantees that the optimal test is always a likelihood test. When $R \leq \phi'(s_r)$, (3.22) $\geq r$, and when $R = \phi'(s_r)$, (3.20) $= \sup_{0 \leq s \leq 1} \frac{-sr - \phi(s|p\|\bar{p})}{1-s}$.

Exercise 3.12.

$$
\begin{aligned}
& D\left(\mathrm{P}^{\boldsymbol{M}^n}_{\rho^{\otimes n}} \middle\| \mathrm{P}^{\boldsymbol{M}^n}_{\sigma^{\otimes n}}\right) \\
&= \sum_{\omega_n} \left(\prod_{k=1}^n \operatorname{Tr} M^n_{k,\omega_n} \rho \right) \log \left(\prod_{k=1}^n \operatorname{Tr} M^n_{k,\omega_n} \rho \right) - \log \left(\prod_{k=1}^n \operatorname{Tr} M^n_{k,\omega_n} \sigma \right) \\
&= \sum_{\omega_n} \left(\prod_{k=1}^n \operatorname{Tr} M^n_{k,\omega_n} \rho \right) \sum_{k=1}^n \left(\log \operatorname{Tr} M^n_{k,\omega_n} \rho - \log \operatorname{Tr} M^n_{k,\omega_n} \sigma \right) \\
&= \sum_{\omega_n} \sum_{k=1}^n \operatorname{Tr} a_{k,\omega_n} M^n_{k,\omega_n} \rho \left(\log \operatorname{Tr} a_{k,\omega_n} M^n_{k,\omega_n} \rho - \log \operatorname{Tr} a_{k,\omega_n} M^n_{k,\omega_n} \sigma \right) \\
&= \sum_{k=1}^n \sum_{\omega_n} \operatorname{Tr} a_{k,\omega_n} M^n_{k,\omega_n} \rho \left(\log \operatorname{Tr} a_{k,\omega_n} M^n_{k,\omega_n} \rho - \log \operatorname{Tr} a_{k,\omega_n} M^n_{k,\omega_n} \sigma \right) \\
&= \sum_{k=1}^n D\left(\mathrm{P}^{\boldsymbol{M}^{n,k}}_{\rho} \middle\| \mathrm{P}^{\boldsymbol{M}^{n,k}}_{\sigma}\right).
\end{aligned}
$$

Exercise 3.13. **a**: Use the fact that $\{\sigma^{\otimes n} \leq e^{-na}\} \sigma^{\otimes n} \{\sigma^{\otimes n} \leq e^{-na}\} \leq \{\sigma^{\otimes n} \leq e^{-na}\} (\sigma^{\otimes n})^s e^{-n(1-s)a} \{\sigma^{\otimes n} \leq e^{-na}\}$ for (3.36). Apply Lemma 3.6 for (3.35). **b**: Use the fact that $\frac{-sr - \phi(s)}{1-s} = -(\phi(s) + sa)$ if $r = -(\phi(s) - (1-s)a)$. **d**: Show that $(\kappa_{\sigma^{\otimes n}}(\rho^{\otimes n}))^{-s} \leq (n+1)^d (\rho^{\otimes n})^{-s}$. **e**: See the hint for **b**.

Exercise 3.15. Lemma 3.5 guarantees that $\lim \operatorname{Tr} \tau^{\otimes n}(I - T_n) = 1$ because $D(\tau\|\sigma) \leq \underline{\lim} \frac{-1}{n} \log \operatorname{Tr} \sigma^{\otimes n}(I - T_n)$. Hence, applying Lemma 3.5 again, we have $\underline{\lim} \frac{-1}{n} \log \operatorname{Tr} \rho^{\otimes n} T_n \leq D(\tau\|\rho)$.

Exercise 3.16. **a**: Join equation (3.17) and inequality (3.40). **c**: Consider the case $r = r_s$ in **a**.

Exercise 3.17. **a**: Use (3.17) with $\kappa_{\sigma^{\otimes n}}$ and $\sigma^{\otimes n}$. **b**: Show that the function $s \mapsto \frac{-sr - \phi(s|p\|\bar{p})}{1-s}$ is monotone increasing in $(-\infty, s_r)$ and is monotone decreasing in (s_r, ∞). **c**: Use (3.43).

Exercise 4.1. Since $H(p|u\rangle\langle u|+(1-p)|v\rangle\langle v|) = H((1-p)|u\rangle\langle u|+p|v\rangle\langle v|)$, the concavity implies $H(1/2|u\rangle\langle u| + 1/2|v\rangle\langle v|) \geq H((1-p)|u\rangle\langle u| + p|v\rangle\langle v|)$. Show that the larger eigenvalue of $1/2|u\rangle\langle u| + 1/2|v\rangle\langle v|$ is $\frac{1+|\langle v|u\rangle|}{2}$.

Exercise 4.2.

$$
\begin{aligned}
& I(p_A, W^A) + I(p_B, W^B) - I(p, W^A \otimes W^B) \\
&= \sum_{x_A, x_B} p_A(x_A) p_B(x_B) D(W^A_{x_A} \otimes W^B_{x_B} \| W^A_{p_A} \otimes W^B_{p_B}) \\
&\quad - \sum_{x_A, x_B} p(x_A, x_B) D(W^A_{x_A} \otimes W^B_{x_B} \| W^A_{p_A} \otimes W^B_{p_B}) \\
&\quad + \sum_{x_A, x_B} p(x_A, x_B) D(W^A_{x_A} \otimes W^B_{x_B} \| W^A_{p_A} \otimes W^B_{p_B}) \\
&\quad - \sum_{x_A, x_B} p(x_A, x_B) D(W^A_{x_A} \otimes W^B_{x_B} \| \left(W^A \otimes W^B\right)_p) \\
&= \sum_{x_A, x_B} (p_A(x_A) p_B(x_B) - p(x_A, x_B)) \left(D(W^A_{x_A} \| W^A_{p_A}) + D(W^B_{x_B} \| W^B_{p_B})\right) \\
&\quad + \sum_{x_A, x_B} p(x_A, x_B) \Big(- \operatorname{Tr} \left(W^A_{x_A} \otimes W^B_{x_B}\right) \log \left(W^A_{p_A} \otimes W^B_{p_B}\right) \\
&\quad + \operatorname{Tr} \left(W^A_{x_A} \otimes W^B_{x_B}\right) \log \left(W^A \otimes W^B\right)_p \Big) \\
&= D\left(W^A_{p_A} \otimes W^B_{p_B} \middle\| \left(W^A \otimes W^B\right)_p\right) \geq 0.
\end{aligned}
$$

Exercise 4.3. The $\leq$ part in (4.19) follows from $I(\boldsymbol{M}, p, W) \leq I(p, W)$. Use the Fano inequality noting the definition of $C(W)$ for the proof of the $\geq$ part.

Exercise 4.4. Use the method of Lagrange multipliers.

Exercise 4.5. From $(A - cB)^*(A - cB) \geq 0$ we have $A^*B + B^*A \leq c^{-1} A^*A + c B^*B$.

Exercise 4.6. Consider the case of $c = \sqrt{\beta/(\alpha + \beta)}$.

Exercise 4.7. Apply Lemma 3.6 to the first term on the RHS of (4.35).

Exercise 4.9. Let $\nu_i = W_{\varphi(i)}$ in (4.39).

Exercise 4.10. Apply lemma 3.1 to the RHS of (4.35).

Exercise 4.11. **a**: Define $A = -nR$. **b**: Define $T_n = \{\kappa_{S^{\otimes n}}(R^{\otimes n}) - N_n S^{\otimes n} \geq 0\}$ in (4.30) and use the arguments in **c,d** of Exercise 3.13.

Exercise 4.12. Order the N_n signals from smallest to largest, and note that the error probability of the first $N_n/2$ signals is less than twice the average error probability.

Exercise 4.13. First, show that

$$\mathrm{P}_X\Big\{\varepsilon[\Phi_X] > M \sum_{x\in\mathcal{X}} p(x)\big(2\,\mathrm{Tr}\, W_x^i\,\{W_x^i - 2NW_p^i \le 0\} + 4N\,\mathrm{Tr}\, W_p^i\,\{W_x^i - 2NW_p^i > 0\}\big)\Big\} < \frac{1}{M}.$$

Use this inequality.

Exercise 4.15. **a**: Apply the Markov inequality to the uniform distribution on the message set $\{1,\dots,N_n\}$.

Exercise 5.1. Let $\mathcal{H}_D$ be $\mathcal{H}_C \otimes \mathcal{H}_B$, and consider the unitary matrix corresponding to the replacement $W : u\otimes v \mapsto v\otimes u$ in $\mathcal{H}_A\otimes\mathcal{H}_B$, and define $V \stackrel{\mathrm{def}}{=} (W\otimes I_C)U$.

Exercise 5.3. Use Exercise 5.2.

Exercise 5.5. Consider the Hilbert space $\mathcal{H}_B$ produced by $|\omega\rangle$, and apply Condition ⑤ of Theorem 5.1 to the entanglement-breaking channel $W_\omega = |\omega\rangle\langle\omega|$. Finally, consider the measurement $\{|\omega\rangle\langle\omega|\otimes I_C\}$.

Exercise 5.17. When we choose the coordinate $u_1 = u_1^A\otimes u_1^B, u_2 = u_1^A\otimes u_2^B, u_3 = u_2^A\otimes u_1^B, u_4 = u_2^A\otimes u_2^B$, we have $\mathrm{Inv}_\lambda\otimes\iota_{\mathbb{C}^2}(|\Phi_2\rangle\langle\Phi_2|) = \begin{pmatrix} 1-\lambda & 0 & 0 & 1-2\lambda \\ 0 & \lambda & 0 & 0 \\ 0 & 0 & \lambda & 0 \\ 1-2\lambda & 0 & 0 & 1-\lambda \end{pmatrix}$. This matrix is positive if and only if $(1-\lambda)^2-(1-2\lambda)^2\ge 0$, i.e., $\frac{2}{3}\ge\lambda\ge 0$.

Exercise 5.19. See the proof of $d_1(\rho,\sigma)\ge b^2(\rho,\sigma)$.

Exercise 5.20. Let $\boldsymbol{M}$ be a POVM that satisfies the equality in (2.60). Applying (2.18) to $\mathrm{P}_\rho^{\boldsymbol{M}}$ and $\mathrm{P}_\sigma^{\boldsymbol{M}}$, we obtain $d_1(\mathrm{P}_\rho^{\boldsymbol{M}},\mathrm{P}_\sigma^{\boldsymbol{M}})\ge b^2(\mathrm{P}_\rho^{\boldsymbol{M}},\mathrm{P}_\sigma^{\boldsymbol{M}})$. Finally, adding (2.61), we obtain $d_1(\rho,\sigma)\ge b^2(\rho,\sigma)$.

Exercise 5.23. **c**: Apply the information-processing inequality of the quantum relative entropy to the two-valued measurement $\{P, I-P\}$.

Exercise 5.24. For any POVM $\boldsymbol{M}=\{M_i\}$ on $\mathcal{H}^{\otimes n}$, $\{(\kappa^{\otimes n})^*(M_i)\}$ is also a POVM. Using this fact, show that $B^*(r|\rho\|\sigma)\le B^*(r|\kappa(\rho)\|\kappa(\sigma))$. Next, choose r such that $B^*(r|\kappa(\rho)\|\kappa(\sigma)) = \frac{-sr-\bar\phi(s|\kappa(\rho)\|\kappa(\sigma))}{1-s}$ for any $s\le 0$.

Exercise 5.27. Let ρ_{mix} be a completely mixed state in $\mathcal{H}_B$. Consider the relative entropy $D(\rho_{A,B,C}\|\rho_{\mathrm{mix}}\otimes\rho_{A,C})$ and the partial trace of $\mathcal{H}_C$.

Exercise 5.30. Consider the state $\begin{pmatrix} p_1\rho_1 & & 0 \\ & \ddots & \\ 0 & & p_k\rho_k \end{pmatrix}$ in $\mathcal{H}_A\otimes\mathcal{H}_B\otimes\mathcal{H}_C$.

Exercise 5.31. Let the purification of $\rho_{A,B}$ be $\rho_{A,B,C}$ for a reference state $\mathcal{H}_R$. From the subadditivity (5.58),

$$H(\rho_{A,B}) - H(\rho_A) + H(\rho_B) = H(\rho_C) - H(\rho_{B,C}) + H(\rho_B) \geq 0.$$

Exercise 5.29. a: Use (5.52).

Exercise 5.35. a: Differentiate both sides with respect to ϵ.

Exercise 5.36. a: Use **b** of Exercise 5.35. **b**: Note that $|\eta(a_1) - \eta(b_1)| \leq \eta(1/e) = 1/e$. **c**: Use $\frac{1}{2}\log 2 = 0.34658 < 1/e = 0.36792$.

Exercise 5.37. Note that $I(p, W) = \sum_x p(x)(H(W_p) - H(W_x))$ and η_0 is concave.

Exercise 5.43. Show this inequality when κ_A and κ_B are partial traces using (5.75). Next, show it in the general case.

Exercise 5.44. Use (8.47) and Lemma 3.7.

Exercise 6.4. Assume Condition ①. Since **d** of Exercise 6.1 implies $[\rho, A] = 0$, we have $A = \rho X' = X'\rho$ for a suitable X'. We can choose $A = \rho \circ X$ for a Hermitian matrix X that commutes with each M_i. Since $\rho \circ (X - X') = 0$, $\rho(X - X') = (X - X')\rho = 0$. Therefore, $A = \rho X = X\rho$.

Exercise 6.5. Note that there exists a unitary matrix U_m such that $\|X\rho Y\|_1 = \operatorname{Tr} X\rho Y U_m$.

Exercise 6.7. Letting $\sigma = \operatorname{Tr}_{\mathcal{H}'} |y\rangle\langle y|$, we have $(\|X \otimes I_{\mathcal{H}'}\|^{(e)}_{|y\rangle\langle y|,b})^2 = |\operatorname{Tr} X\sigma|^2 = |\langle I, X\rangle^{(e)}_{\sigma,b}|^2 \leq (\|X\|^{(e)}_{\sigma,b})^2$. The equality holds only when $X = I$.

Exercise 6.10. Apply Theorem 6.2 to the entanglement-breaking channel $\rho \mapsto \sum_i (\operatorname{Tr} M_i \rho)|u_i\rangle\langle u_i|$ with the CONS $\{u_i\}$.

Exercise 6.12. a: Combine **a** and **d** of Exercise 6.2. **b**: Combine Exercise 6.11 and **a**. **c**: Use the fact that $\kappa_{\boldsymbol{M},\rho_\theta,s}$ is a projection, as shown in Exercise 6.2.

Exercise 6.13. **b**: Use $\exp(X(\theta)) = \sum_{n=0}^{\infty} \frac{X(\theta)^n}{n!}$ and **a**.

Exercise 6.14. First, show that $\frac{d\rho_\theta^{\otimes n}}{d\theta} = \sqrt{n}\left(\frac{d\rho_\theta}{d\theta}\right)^{(n)}$. Use this and (6.9).

Exercise 6.16. Use the fact that $\boldsymbol{J}_{\theta,s;i,j} = \operatorname{Tr} L^*_{\theta,i,s} \frac{\partial \rho_\theta}{\partial \theta^j}$.

Exercise 6.18. Use $L_{\theta,b} = \frac{d \log \rho_\theta}{d\theta}$.

Exercise 6.19. Compare the e and m representations of the derivative of the quantum state family $\{\rho_\theta = e^{-i\theta Y}\rho e^{i\theta Y}\}$ in Exercise 6.18.

Exercise 6.20. **b**: Use (6.14) and (5.23).

Exercise 6.21. Consider the TP-CP map $\lambda\rho_\theta^1 \oplus (1-\lambda)\rho_\theta^2 \to \lambda\rho_\theta^1 + (1-\lambda)\rho_\theta^2$.

Exercise 6.23. First, for a given SLD geodesic $\Pi^\theta_{L,s}\sigma$, choose a unitary matrix U_1 such that ULU^* is equal to the constant times of S_1. Then, the SLD geodesic $U\Pi^\theta_{L,s}\sigma U^*$ has the form given in Exercise 6.21. Next, choose another unitary matrix U_2 such that

$$U_2 S_1 U_2^* = S_1, \; U_2(x_2 S_2 + x_3 S_3) U_2^* = \sqrt{x_2^2 + x_3^2} S_3. \tag{C.1}$$

Exercise 6.24. **b**: $\int_0^1 \operatorname{Tr}(\sigma - \rho)^2 (\rho + t(\sigma - \rho))^{-1} dt$
$= \int_0^1 \operatorname{Tr} \rho (\sqrt{\rho}^{-1} \sigma \sqrt{\rho}^{-1} - I)(I + t(\sqrt{\rho}^{-1}\sigma\sqrt{\rho}^{-1} - I))^{-1} (\sqrt{\rho}^{-1}\sigma\sqrt{\rho}^{-1} - I) dt$
$= \operatorname{Tr} \rho \left((\sqrt{\rho}^{-1}\sigma\sqrt{\rho}^{-1} - I) - \log(I + (\sqrt{\rho}^{-1}\sigma\sqrt{\rho}^{-1} - I)) \right)$
$= - \operatorname{Tr} \rho \log(\sqrt{\rho}^{-1}\sigma\sqrt{\rho}^{-1}) = \operatorname{Tr} \rho \log(\sqrt{\rho}\sigma^{-1}\sqrt{\rho})$.

Exercise 6.26. **a**: Use the partial integration formula twice. **c**: Use (6.22). **d**: Use **a**,**b**,**c** and the fact that $\frac{d^2 \rho_\theta}{d\theta^2} = 0$.

Exercise 6.27. Show the equivalence of ② and ③ based on Exercise 6.4.

Exercise 6.28. Show that $\lim \frac{1}{n} D_x^{(m)}((\rho_1 \otimes \rho_2)^{\otimes n} \| (\sigma_1 \otimes \sigma_2)^{\otimes n}) = \lim \frac{1}{n} D_x^{(m)}(\rho_1^{\otimes n} \| \sigma_1^{\otimes n}) + \lim \frac{1}{n} D_x^{(m)}(\rho_2^{\otimes n} \| \sigma_2^{\otimes n})$. Use (6.50) and (5.36).

Exercise 6.30. Consider a state family where $A = \frac{d\rho_\theta}{d\theta}$ and $\rho_{\theta_0} = \rho$, and let κ be given by a POVM M. From property (6.17) and $\|\rho^{-1}\|_{\rho,x}^{(m)} = 1$, we have $J_{\theta_0}^M = \|\kappa(A)\|_{\kappa(\rho),x}^{(m)} \le \|A\|_{\rho,x}^{(m)}$. Then, use Exercise 6.29.

Exercise 6.31. See the hint for Exercise 2.34.

Exercise 6.32. **a**: Let K be the difference K between $O(\boldsymbol{M}, \hat{\theta}) - \theta$ and $\frac{1}{J_{\theta,s}} L_{\theta,s}$. Then, $K\rho + \rho K = 0$ when Condition ① holds. **b**: Use Condition (6.68).

Exercise 6.35. **b**: Define $B_0 \stackrel{\text{def}}{=} \left\{ \hat{\theta} \le \theta \right\}$, $B_i \stackrel{\text{def}}{=} \left\{ \theta + \frac{\delta(i-1)}{m} \le \hat{\theta} \le \theta + \frac{\delta i}{m} \right\}$, $(i = 1, \ldots, m)$, and $B_{m+1} \stackrel{\text{def}}{=} \left\{ \theta + \delta \le \hat{\theta} \right\}$, and consider a POVM $M_i \stackrel{\text{def}}{=} MB_i$ comprising $m+1$ measurements.

Exercise 6.38. **a**: Show that $\frac{\langle M(\omega), M(\omega) \rangle_{\rho_\theta,s}^{(e)}}{\langle M(\omega), I \rangle_{\theta,s}^{(e)}} = \operatorname{Tr} M(\omega)$. **c**: Use Exercise A.1. **d**: Use Exercise 2.37.

Exercise 6.39. If $(\boldsymbol{M}, \hat{\theta})$ is a locally unbiased estimator, then show that $\boldsymbol{V}_\theta(\boldsymbol{M}, \hat{\theta}) \ge (\boldsymbol{J}_\theta^{\boldsymbol{M}})^{-1}$. Then, show that for each POVM $\boldsymbol{M}$ there exists a function $\hat{\theta}$ such that $(\boldsymbol{M}, \hat{\theta})$ is a locally unbiased estimator and $\boldsymbol{V}_\theta(\boldsymbol{M}, \hat{\theta}) = (\boldsymbol{J}_\theta^{\boldsymbol{M}})^{-1}$.

Exercise 6.40. Use the method of Lagrange multipliers.

Exercise 6.41. Show that $\left(\|L(u)\|_{\rho_\theta,s}^{(e)} \right)^2 = \langle u | \boldsymbol{J}_{\theta,s} | u \rangle$. Then, show that $\langle x | \boldsymbol{J}_\theta^{\boldsymbol{M}^u} | x \rangle = \langle L(x) | \kappa_{\boldsymbol{M},\rho_\theta,s} | L(x) \rangle_{\rho_\theta,s}^{(e)} \ge \langle L(x) | \frac{|L(u)\rangle_{\rho_\theta,s}^{(e)} \langle L(u)|}{\langle u | \boldsymbol{J}_{\theta,s} | u \rangle} | L(x) \rangle_{\rho_\theta,s}^{(e)}$ for $x \in \mathbb{R}^d$.

Exercise 6.44. Let $u_1, \ldots, u_d$ be the eigenvectors of $\boldsymbol{J}_{\theta,s}$, and let p_i be the eigenvalues of $\frac{1}{\operatorname{tr} \boldsymbol{J}_{\theta,s}^{-\frac{1}{2}}} \boldsymbol{J}_{\theta,s}^{-\frac{1}{2}}$. Then, the RHS of (6.94) is equal to $\frac{1}{\operatorname{tr} \boldsymbol{J}_{\theta,s}^{-\frac{1}{2}}} \boldsymbol{J}_{\theta,s}^{\frac{1}{2}}$.

Exercise 6.45. Choose the new coordinate such that the SLD Fisher information matrix is $\sqrt{G}^{-1} \boldsymbol{J}_{\theta,s} \sqrt{G}^{-1}$.

Exercise 6.52. c: See **b** of Exercise 6.50. **e**: Consider the estimator $(\boldsymbol{M}, \hat{\theta})$ in the extended system given in **b** for the set of vectors (y^i). Let P be the projection to the original space. Consider the POVM $\{PM_kP\}$ for $\boldsymbol{M} = \{M_k\}$. **g**: Note that the set of vectors (u^i) satisfying $\langle u^i|x_j\rangle = \delta^i_j$ is restricted to $u^i = ((\mathbf{Re}\, J)^{-1})^{i,j} x_j$.

Exercise 7.1. Consider the unitary matrix $(\sum_\omega |u_\omega\rangle\langle u_\omega| \otimes U_{\kappa'_\omega})(I_{B,C} \otimes U)$ and $\rho_1 \otimes \rho_0$.

Exercise 7.2. Let $\mathcal{H}_D$ be $\mathcal{H}_C \otimes \mathcal{H}_B$, consider the unitary matrix corresponding to the replacement $W : u \otimes v \mapsto v \otimes u$ in $\mathcal{H}_A \otimes \mathcal{H}_B$, and define $V \stackrel{\text{def}}{=} (W \otimes I_C)U$.

Exercise 7.3. Equation (5.4) yields that $\operatorname{Tr} \rho M'_{\omega,\omega} = \operatorname{Tr} \rho \kappa^*_\omega(M_{\omega'}) = \operatorname{Tr} \kappa^*_\omega(\rho) M_{\omega'}$.

Exercise 7.5. **b**: Let $\sum_i p_i \rho_i^A \otimes \rho_i^B = (\kappa \otimes \iota_{A'})(|x_M\rangle\langle x_M|)$. Then, $K(\kappa) = d \sum_i p_i \rho_i^A \otimes \rho_i^B$ from (5.3). From the definition of $K(\kappa)$, we have $\operatorname{Tr} \kappa(\rho)\sigma = \operatorname{Tr} K(\kappa) \rho \otimes \sigma = d \sum_i p_i (\operatorname{Tr} \rho \rho_i^A)(\operatorname{Tr} \sigma \rho_i^B)$. Thus, we obtain $\kappa(\rho) = d \sum_i p_i (\operatorname{Tr} \rho \rho_i^A) \rho_i^B$.

Exercise 7.7. Expand the RHS of (7.17) and rearrange.

Exercise 7.8. Similarly, expand the RHS of (7.17) and rearrange.

Exercise 7.13. Use the fact that $\Delta_1(O(\boldsymbol{M}) - X, \rho) \le \Delta_3(\boldsymbol{M}, X, \rho)$, problem 7.11, and $\sqrt{x^2 + y^2} \le x + y$ for $x, y \ge 0$.

Exercise 7.14. Choose a 2×2 orthogonal matrix $(a_{i,j})$ such that the two matrices $\tilde{X} \stackrel{\text{def}}{=} a_{1,1}X + a_{1,2}Y$ and $\tilde{Y} \stackrel{\text{def}}{=} a_{2,1}X + a_{2,2}Y$ satisfy $\operatorname{Cov}_\rho(\tilde{X}, \tilde{Y}) = 0$. Show (7.36) for $\tilde{X}, \tilde{Y}$. Finally, use the fact that both sides of (7.36) are invariant under the orthogonal matrix transformation $(X, Y) \mapsto (\tilde{X}, \tilde{Y})$.

Exercise 7.15. **g**: Note that both sides of (7.38) become their $\det(b_{i,j})$ times value when we perform the transformation $(\boldsymbol{x}, \boldsymbol{y}) \mapsto (\tilde{\boldsymbol{x}}, \tilde{\boldsymbol{y}})$.

Exercise 7.16. Show that $\Delta_3(\boldsymbol{M}_{X,Y,\rho}, O^1(\boldsymbol{M}_{X,Y,\rho}), \rho) = \frac{1-p}{p} \Delta_1(X, \rho)$, $\Delta_3(\boldsymbol{M}_{X,Y,\rho}, O^2(\boldsymbol{M}_{X,Y,\rho}), \rho) = \frac{1-q}{q} \Delta_1(Y, \rho)$, and note that $\frac{1-p}{p}\frac{1-q}{q} = 1$.

Exercise 7.19. Use Exercise 6.21.

Exercise 7.20. Note that the case of $p = 1$ has been shown in the main body. Use Exercise 7.19.

Exercise 8.2. Note that d_1 is monotone concerning the partial trace.

Exercise 8.3. Let u and v be purifications of ρ and σ such that $F(\rho, \sigma) = F(|u\rangle\langle u|, |v\rangle\langle v|) = \operatorname{Tr} \sqrt{|u\rangle\langle u|}\sqrt{|v\rangle\langle v|} = |\langle u|v\rangle|^2 = F^2(|u\rangle\langle u|, |v\rangle\langle v|)$. Using the monotonicity of $\phi(1/2, \rho, \sigma)$, we have $F^2(\rho, \sigma) \le \operatorname{Tr} \sqrt{\rho}\sqrt{\sigma}$.

In addition, $F^2(\rho, \sigma) = \operatorname{Tr} |\sqrt{\rho}\sqrt{\sigma}| \ge \operatorname{Tr} \sqrt{\rho}\sqrt{\sigma}$.

Exercise 8.4. $F_e^2(\rho, \kappa) = \sum_i \langle x|E_i \otimes I|x\rangle\langle x|E_i \otimes I|x\rangle$.

Exercise 8.5. **a**: Consider the singular value decomposition of the matrix $\{\operatorname{Tr} E_i A_j \rho\}_{i,j}$, and retake the Choi–Kraus representation. Use Exercise 5.2. **b**: Applying **a** and defining $p_i \stackrel{\text{def}}{=} \operatorname{Tr} A_i \rho A_i$, $A'_i \stackrel{\text{def}}{=} A_i / \sqrt{p_i}$, we have $F_e^2(\rho, \kappa \circ$

$\kappa') = \sum_i p_i |\operatorname{Tr} E_i A_i' \rho|^2$. **c**: Note that $EA = (U|E|^{1/2})(|E|^{1/2}A)$, and use the Schwarz inequality for the inner product $\operatorname{Tr} X^* Y \rho$. **d**: Note that $F_e^2(\rho, \kappa \circ \kappa_U) \geq |\operatorname{Tr} E_1 U^* \rho|^2$.

Exercise 8.6. Show that $|\operatorname{Tr}(A_1 + A_2 i)\rho|^2 = (\operatorname{Tr} A_1 \rho)^2 + (\operatorname{Tr} A_2 \rho)^2$, where A_1 and A_2 are Hermitian matrices.

Exercise 8.12. Use ② or (8.19).

Exercise 8.14. Denote the input and output system of κ by $\mathcal{H}_A$ and $\mathcal{H}_B$, respectively. Let $|x\rangle\langle x|$ be a purification of ρ_{mix}. Choose the probalistic decomposition as $\kappa \otimes \iota_R(|x\rangle\langle x|) = \sum_i p_i |y_i\rangle\langle y_i|$. Then, the Schmidt rank of $|y_i\rangle$ is less than d'. That is, the rank of $\operatorname{Tr}_B |y_i\rangle\langle y_i|$ is less than d' Therefore, $\operatorname{Tr}_R \sqrt{\operatorname{Tr}_A |y_i\rangle\langle y_i|} \leq \sqrt{d'}$. Thus,

$$\begin{aligned}
&\langle x|(\kappa' \circ \kappa) \otimes \iota_R(|x\rangle\langle x|)|x\rangle = \langle x|\sum_i p_i (\kappa' \otimes \iota_R)(|y_i\rangle\langle y_i|)|x\rangle \\
\leq &\sum_i p_i F^2(\operatorname{Tr}_A(\kappa' \otimes \iota_R)(|y_i\rangle\langle y_i|), \operatorname{Tr}_A |x\rangle\langle x|) \\
= &\sum_i p_i F^2(\operatorname{Tr}_A |y_i\rangle\langle y_i|, \rho_{\mathrm{mix},R}) = \sum_i p_i (\operatorname{Tr}_R |\sqrt{\operatorname{Tr}_A |y_i\rangle\langle y_i|}\sqrt{\rho_{\mathrm{mix},R}}|)^2 \\
= &\sum_i p_i \frac{(\operatorname{Tr}_R \sqrt{\operatorname{Tr}_A |y_i\rangle\langle y_i|})^2}{d} \leq \frac{d'}{d}.
\end{aligned}$$

Exercise 8.15. Since the final state on $\mathcal{H}_R \otimes \mathcal{H}_E \otimes \mathcal{H}_B$ is a pure state, $H(\rho)$ is equal to the entropy of the final state on the reference system $\mathcal{H}_R$. $H(\kappa(\rho))$ is equal to the entropy of the final state on $\mathcal{H}_R \otimes \mathcal{H}_E$. $H_e(\rho, \kappa)$ is therefore equal to the entropy of the final state on the environment $\mathcal{H}_E$.

Exercise 8.16. Note the second inequality in (5.42) and the monotonicity of the trace norm concerning the partial trace on the reference system.

Exercise 8.17. Note that the entropy of $U(\rho \otimes |u\rangle\langle u|)U$ is equal to the entropy of ρ.

Exercise 8.20. Consider the Stinespring represntation of κ', and consider the partial trace with respect to the environment of κ' after the state evolution.

Exercise 8.22. Since x' is a pure state on $\mathcal{H}_{A'} \otimes \mathcal{H}_{E'} \otimes \mathcal{H}_R$, $H_{x'}(A'R) = H_{x'}(E'), H_{x'}(R) = H_{x'}(A'E')$.

Exercise 8.25. **b**: Use (5.52) for the last inequality in (8.54).

Exercise 8.28. Use the concavity of the entropy and (5.52) for the pinching of a PVM that commutes with $|u\rangle\langle u|$.

Exercise 8.29. Consider the unitary matrix

$$\begin{pmatrix} S_0 & 0 & 0 & 0 \\ 0 & S_1 & 0 & 0 \\ 0 & 0 & S_2 & 0 \\ 0 & 0 & 0 & S_3 \end{pmatrix} \begin{pmatrix} \sqrt{p_0} I & * & * & * \\ \sqrt{p_1} I & * & * & * \\ \sqrt{p_2} I & * & * & * \\ \sqrt{p_3} I & * & * & * \end{pmatrix}$$

as a Stinespring representation in $\mathbb{C}^2 \otimes \mathbb{C}^4$, where the elements $*$ of the second matrix are choosen appropriately to preserve unitarity.

Exercise 8.30. Note that this map is a double stochastic matrix.

Exercise 8.32. This follows immediately from Exercise 8.31.

Exercise 8.33. Use Theorem 2.3.

Exercise 8.35. Use inequality (2.140). Note that $\frac{-1}{n} \log L(\kappa_n) = R$ in (8.78).

Exercise 8.36. Put $1 - t = s$ and $\sigma = \sum_i \lambda_i |u_i^A \otimes u_i^B\rangle\langle u_i^A \otimes u_i^B|$ in the proof of Theorem 8.7.

Exercise 8.38. First, consider the case of a pure state in $\{v \otimes u - u \otimes v | u, v \in \mathbb{C}^3\}$.

Exercise 8.39. Show the pure-state case using Theorem 8.4. Next, extend this fact to the mixed-state case.

Exercise 8.40. Show that the RHS of (8.85) in Theorem 8.8 approaches 0 exponentially when $R > E(\rho)$, $L = [e^{nR}]$.

Exercise 8.41. Let $\rho = \sum_i p_i |x_i\rangle\langle x_i|$, and consider a separable state σ_i, where $E_f(|x_i\rangle\langle x_i|) = D(|x_i\rangle\langle x_i| \| \sigma_i)$. Use the joint convexity of the relative entropy.

Exercise 8.42. **a:** First, show that $\| |x_n\rangle\langle x_n| - |y_n\rangle\langle y_n| \|_1 \to 0$ when $\|\rho_n - \sigma_n\|_1 \to 0$. Next, give a probabilistic decomposition using a POVM $\boldsymbol{M}^n = \{M_i^n\}$ on the reference system $\mathcal{H}_R$ (Lemma 8.3). Let $\boldsymbol{M}$ be the POVM giving the decomposition minimizing the average entropy of σ_n. Then, the average entropy $\sum_x p_x^i H(\mathrm{Tr}_B \rho_x^i)$ on $\mathcal{H}_A$ is equal to $H_{\hat{\kappa}_{\boldsymbol{M}^n} \otimes \iota_A(\mathrm{Tr}_B |x_n\rangle\langle x_n|)}(A|R)$. From monotonicity, show that $\|\hat{\kappa}_{\boldsymbol{M}^n} \otimes \iota_A(\mathrm{Tr}_B |x_n\rangle\langle x_n|) - \hat{\kappa}_{\boldsymbol{M}^n} \otimes \iota_A(\mathrm{Tr}_B |y_n\rangle\langle y_n|)\|_1 \le \|\rho_n - \sigma_n\|_1$. Finally, use (5.71).

Exercise 8.43. **a:** First, show that $\| |x_n\rangle\langle x_n| - |y_n\rangle\langle y_n| \|_1 \to 0$ when $\|\rho_n - \sigma_n\|_1 \to 0$. Next, choose a system $\mathcal{H}_E$ as a subsystem of an extended space of the reference system $\mathcal{H}_R$. Note that extenstions of ρ_n and σ_n can be given as the reduced densities on $\mathcal{H}_A \otimes \mathcal{H}_B \otimes \mathcal{H}_E$. Now, we choose the subsystem $\mathcal{H}_E$ giving $E_{sq}(\sigma_n)$. Finally, use (5.73).

Exercise 8.46.

$$
\begin{aligned}
I(AA' : BB'|E) &= I(A : BB'|E) + I(A : BB'|EA) \\
&= I(A : B|E) + I(A : B'|BE) + I(A : B'|EA) + I(A : B|EAB') \\
&\ge I(A : B|E) + I(A : B'|BE).
\end{aligned}
$$

Exercise 8.52. First, show that $I^{A \to B}(|u\rangle\langle u|) \le H(\mathrm{Tr}_A |u\rangle\langle u|)$. Next, show its equality.

Exercise 8.57. **a**: Use Theorem 5.8. **b**: $I_{(\kappa_{\boldsymbol{M}} \otimes \iota_{AB})(\rho^{ABE})}(AB : E) \le I_{\rho^{ABE}}(AB : E)$.

Exercise 9.3. Refer to the proof of Theorem 4.2, and use Fano's inequality for the converse part of the theorem.

Exercise 9.4. Using the monotonicity of the quantum relative entropy and (5.58), we obtain

$$
\begin{aligned}
I(X:Y) \le & \frac{1}{N_n}\sum_{i=1}^{N_n} D\Big(\Big(\varphi_e^{(n)}(i)\otimes \iota_B^{\otimes n}\Big)(\rho_{A,B}^{\otimes n})\Big\| \frac{1}{N_n}\sum_{i=1}^{N_n}\Big(\varphi_e^{(n)}(i)\otimes \iota_B^{\otimes n}\Big)(\rho_{A,B}^{\otimes n})\Big) \\
= & H\Big(\frac{1}{N_n}\sum_{i=1}^{N_n}\Big(\varphi_e^{(n)}(i)\otimes \iota_B^{\otimes n}\Big)(\rho_{A,B}^{\otimes n})\Big) - \frac{1}{N_n}\sum_{i=1}^{N_n} H\Big(\Big(\varphi_e^{(n)}(i)\otimes \iota_B^{\otimes n}\Big)(\rho_{A,B}^{\otimes n})\Big) \\
\le & H\Big(\operatorname{Tr}_{A'_n}\frac{1}{N_n}\sum_{i=1}^{N_n}\Big(\varphi_e^{(n)}(i)\otimes \iota_B^{\otimes n}\Big)(\rho_{A,B}^{\otimes n})\Big) \\
& + H\Big(\operatorname{Tr}_{B}\frac{1}{N_n}\sum_{i=1}^{N_n}\Big(\varphi_e^{(n)}(i)\otimes \iota_B^{\otimes n}\Big)(\rho_{A,B}^{\otimes n})\Big) - \min_{\kappa} H((\kappa\otimes \iota_B^{\otimes n})(\rho_{A,B}^{\otimes n})).
\end{aligned}
$$

Using

$$
\operatorname{Tr}_{A'_n}\Big(\varphi_e^{(n)}(i)\otimes \iota_B^{\otimes n}\Big)(\rho_{A,B}^{\otimes n}) = \operatorname{Tr}_A \rho_{A,B}^{\otimes n}
$$

$$
H\Big(\operatorname{Tr}_{B}\frac{1}{N_n}\sum_{i=1}^{N_n}\Big(\varphi_e^{(n)}(i)\otimes \iota_B^{\otimes n}\Big)(\rho_{A,B}^{\otimes n})\Big) \le \log\dim \mathcal{H}_{A'_n},
$$

we obtain

$$
\begin{aligned}
I(X:Y) \le & H(\operatorname{Tr}_A \rho_{A,B}^{\otimes n}) + \log\dim\mathcal{H}_{A'_n} - \min_{\kappa} H((\kappa\otimes \iota_B^{\otimes n})(\rho_{A,B}^{\otimes n})) \\
= & nH(\operatorname{Tr}_A \rho_{A,B}) + \log\dim\mathcal{H}_{A'_n} - \min_{\kappa} H((\kappa\otimes \iota_B^{\otimes n})(\rho_{A,B}^{\otimes n})).
\end{aligned}
$$

Exercise 9.5. The LHS of (9.42) can be rewritten as

$$
\begin{aligned}
& \sum_j p_j D((\kappa\circ\varphi_e(j)\otimes \iota_R)(\rho_{A',R})\| \sum_j p_j(\kappa\circ\varphi_e(j)\otimes\iota_R)(\rho_{A',R})) \\
& \le \sum_j p_j D((\kappa\circ\varphi_e(j)\otimes \iota_R)(\rho_{A',R})\|(\sum_j p_j(\kappa\circ\varphi_e(j))\rho_{A'})\otimes\rho_R) \\
& = H(\sum_j p_j(\kappa\circ\varphi_e(j))\rho_{A'}) + H(\rho_R) - \sum_j p_j H((\kappa\circ\varphi_e(j)\otimes\iota_R)\rho_{A',R}) \\
& = H(\kappa(\sum_j p_j\varphi_e(j)(\rho_{A'})) + \sum_j p_j \tilde{I}_c(\rho_{A'},\kappa\circ\varphi_e(j)) \\
& \le H(\kappa(\sum_j p_j\varphi_e(j)(\rho_{A'}))) + \tilde{I}_c(\sum_j p_j\varphi_e(j)(\rho_{A'}),\kappa) \\
& = I(\sum_j p_j\varphi_e(j)(\rho_{A'}),\kappa),
\end{aligned}
$$

from (4.6) and (8.50). Since $\rho_{A',R}$ is a pure state, we may write $H(\rho_{A'}) = H(\rho_R)$.

Exercise 9.8. First, show that

$$J(\rho,\sigma,\kappa) \le J\Big(\int_\theta U_\theta \rho U_\theta^* d\theta, \sigma, \kappa\Big) = J(P_1\rho P_1 + P_2\rho P_2, \sigma, \kappa),$$

where $U_\theta = P_1 + e^{i\theta} P_2$.

Next, show that

$$J(\lambda\rho_1 \oplus (1-\lambda)\rho_2, \sigma, \kappa) = \lambda J(\rho_1,\sigma,\kappa) + (1-\lambda) J(\rho_2,\sigma,\kappa).$$

Finaly, using (9.40), we obtain

$$\begin{aligned}
\max_\rho I(\rho,\kappa) &= \max_\rho \min_\sigma J(\rho,\sigma,\kappa) = \min_\sigma \max_\rho J(\rho,\sigma,\kappa) \\
&= \min_\sigma \max_{\lambda,\rho_1,\rho_2} J(\lambda\rho_1 \oplus (1-\lambda)\rho_2, \sigma, \kappa) \\
&= \min_\sigma \max_{\lambda,\rho_1,\rho_2} \lambda J(\rho_1,\sigma,\kappa) + (1-\lambda) J(\rho_2,\sigma,\kappa) \\
&\le \max_{\lambda,\rho_1,\rho_2} \lambda J(\rho_1, \kappa_1(\rho_{\max,1}), \kappa) + (1-\lambda) J(\rho_2, \kappa_1(\rho_{\max,1}), \kappa) \\
&= \max_\lambda \lambda C_{c,e}^e(\kappa_1) + (1-\lambda) C_{c,e}^e(\kappa_2).
\end{aligned}$$

Exercise 9.9. Applying (2.63) to the two-valued POVM $\{\{\kappa_{W_p}(W_x) - CW_p \ge 0\}, \{\kappa_{W_p}(W_x) - CW_p < 0\}\}$, we obtain

$$\begin{aligned}
&\left(\operatorname{Tr} W_x\{\kappa_{W_p}(W_x) - CW_p \ge 0\}\right)^{1-s} \left(\operatorname{Tr} W_p\{\kappa_{W_p}(W_x) - CW_p \ge 0\}\right)^s \\
&\quad + \left(\operatorname{Tr} W_x\{\kappa_{W_p}(W_x) - CW_p < 0\}\right)^{1-s} \left(\operatorname{Tr} W_p\{\kappa_{W_p}(W_x) - CW_p < 0\}\right)^s \\
&\le \operatorname{Tr} W_x^{1-s} W_p^s
\end{aligned}$$

for $0 \ge s$. Since $\operatorname{Tr} W_x\{\kappa_{W_p}(W_x) - CW_p \ge 0\} = \operatorname{Tr} \kappa_{W_p}(W_x)\{\kappa_{W_p}(W_x) - CW_p \ge 0\} \ge C \operatorname{Tr} W_p\{\kappa_{W_p}(W_x) - CW_p \ge 0\}$, we have

$$\begin{aligned}
&\operatorname{Tr} W_x\{\kappa_{W_p}(W_x) - CW_p \ge 0\} \\
&\le C^s \left(\operatorname{Tr} W_x\{\kappa_{W_p}(W_x) - CW_p \ge 0\}\right)^{1-s} \left(\operatorname{Tr} W_p\{\kappa_{W_p}(W_x) - CW_p \ge 0\}\right)^s,
\end{aligned}$$

and thus we obtain (9.53).

Exercise 9.10. Solve $\frac{-\phi(s)+s(R-r)}{2} = \frac{r}{2}$ with respect to r.

Exercise 9.11. In this case, we can replace (9.56) by $\|P_x W_x\|_1 \le \operatorname{Tr} W_x P_x$.

Exercise 9.13. Use the fact that

$$\begin{aligned}
&I(p, W_B) + I(p', W_B') - I(q, W_B \otimes W_B') \\
&= D\left(\sum_{x,x'} q(x,x') W_{B,x} \otimes W'_{B,x'} \middle\| \left(\sum_x p(x) W_{B,x}\right) \otimes \left(\sum_{x'} p'(x') W'_{B,x'}\right)\right).
\end{aligned}$$

Exercise 9.17. First, note that

$$\begin{aligned}\varepsilon_{E,a}[\Phi] &= \sum_i \sum_{j \neq i} \frac{1}{M(M-1)} d_1((W^E Q)_i, (W^E Q)_j) \\ &\geq \frac{1}{M} \sum_i d_1((W^E Q)_i, \frac{1}{M} \sum_{j=1} (W^E Q)_j).\end{aligned}$$

From Fannes' inequality (5.64) and the concavity and monotonicity (Exercise 5.35) of η_0, we have

$$\begin{aligned}I_E(\Phi) &= \frac{1}{M} \sum_{i=1}^{M} H(\frac{1}{M} \sum_{j=1} (W^E Q)_j)) - H(W^E Q)_i) \\ \leq & \frac{1}{M} \sum_{i=1}^{M} |H(\frac{1}{M} \sum_{j=1} (W^E Q)_j)) - H(W^E Q)_i)| \\ \leq & \frac{1}{M} \sum_{i=1}^{M} d_1 \left((W^E Q)_i, \frac{1}{M} \sum_{j=1} (W^E Q)_j \right) \log d \\ & + \eta_0 \left(d_1 \left((W^E Q)_i, \frac{1}{M} \sum_{j=1} (W^E Q)_j \right) \right) \\ \leq & \varepsilon_{E,a}[\Phi] \log d + \eta_0(\varepsilon_{E,a}[\Phi]).\end{aligned}$$

Exercise 9.18. First, show that there exists a code such that $\sum_i \frac{1}{M} d_1((W^E Q)_i, \frac{1}{M} \sum_{j=1} (W^E Q)_j)$ converges to 0. Next, choose the smallest $M/2$ values $i_1, \ldots, i_{M/2}$ concerning $d_1((W^E Q)_i, \frac{1}{M} \sum_{j=1} (W^E Q)_j)$. Finally, show this exercise based on the fact that

$$\sup_i \sup_j d_1((W^E Q)_i, (W^E Q)_j) \leq 2 \sup_i d_1((W^E Q)_i, \frac{1}{M} \sum_{j=1} (W^E Q)_j).$$

Exercise 9.20. Consider, for example, $a = 1/2$.

Exercise A.4. Compare the maximizations in definition (A.13) of $\|X\|_1$ and $\| \operatorname{Tr}_B X\|_1$.

Exercise A.6. If A possesses the inverse, use Exercise A.5. If A does not possess the inverse, choose an invertible matrix approximating A.

Exercise A.7. **a**: Use $B^{-1/2} A B^{-1/2} = (A^{1/2} B^{-1/2})^* (A^{1/2} B^{-1/2})$ and (1.29). **d**: Consider $B + \epsilon I$ and take the limit $\epsilon \to 0$.

Exercise A.10. Consider the $(d-k+1)$-dimensional subspace $\mathcal{K}'$ spanned by the eigenvectors according to the eigenvalues $a_k, \ldots, a_d$. Evaluate $\langle x|A|x\rangle$ when $\|x\| = 1$ and $x \in \mathcal{K} \cap \mathcal{K}'$.

Exercise A.11. Apply Exercise A.10.

Exercise A.12. Apply Exercise A.11.

Postface to Japanese Version

My research on quantum information theory started in October of 1994, when I was a first year master's student. At that time, although Shor's paper on factorization had already been published, I was still unaware of his work. Nor was the field of quantum information theory very well known at that time. The following is a brief summary of how I began working in the field of quantum information theory. Although this is merely a personal account of my experiences, it might be of some interest to those thinking of starting postgraduate studies and pursuing a career in research.

I began my university studies at Kyoto University by studying both mathematics and physics, thanks to their policy that allows students to complete their graduation without electing their major. Although I was mainly interested in physics, I decided to study both physics and mathematics because I was not entirely comfortable with the common thinking manner in physics; I was more naturally inclined towards a mathematical way of thought. As a result, during my undergraduate years although I had a reasonable understanding of mathematics, I could not understand physics sufficiently. Particularly, I could not catch the essence of statistical mechanics, in which "physics thinking" appears most prominently, more seriously. In my fourth year of my undergraduate degree, I realized that based on my understanding of physics at the time, I would probably not pass the entrance exams for a postgraduate course in physics. Therefore, I decided to apply for a postgraduate course in mathematics (which I just barely passed). In particular, while I elected the early universe in the cosmology group to my main research in the undergraduate course, its outcome was rather hopeless due to my poor knowledge of statistical mechanics. In fact, when I told a professor of physics that I would work as a casual teacher of cram school for high school physics in the next year, he said to me "I would never let you teach physics to anyone." Indeed, my physics knowledge was lacking in such a extent at that time. This was a particularly depressing event for me. Although I was still able to graduate, it had hardly felt like a time for celebration.

The following April, I began my postgraduate studies in twistor theory [411][1] under Professor Ueno, who is a professor in mathematics in Kyoto University. I chose this topic because it is related to relativity theory, which I was interested in at that time. However, as is the case with many topics in mathematical physics, although its basic is rooted in physics, it was essentially mathematical. I also found it very difficult to understand the physics behind this mathematical concepts. Ultimately, I realized that it did not suit my interests, and therefore I searched for another topic. Although I was capable of thinking in a mathematical way, I was not interested in mathematics itself. Therefore it was impossible for me to concentrate the research only of pure mathematics. Meanwhile, by teaching high school physics as a casual teacher in cram school during my postgraduate years, I felt that I could truly understand physics for the first time. Since I usually concerned difficult mathematical structure in physics, I realized for the first time that it is important to understand physics based on fundamental concepts.

When I searched for my new research topic, I met Dr. Akio Fujiwara, who came to Osaka University as an assistant professor at that time. He advised me to study Holevo's textbook [216], and I decided that I would start research in quantum information theory. Until this point, I had mainly studied abstract mathematics with little physical connection. I was particularly impressed by the way Holevo's textbook expressed the fundamental concepts of quantum mechanics without high levels of abstraction. although Holevo's textbook is not a particularly easy book to read from the current viewpoint, it was not so difficult for me to read because I had read more difficult books on mathematics.

In retrospect, it might be fortunate that I did not proceed to a postgraduate course in physics because physics community had an implicit strong stress to never try the measurement problem in quantum mechanics due to its philosophical aspect in Japan at that time. Therefore, while I appeared to take a rather indirect path during my years for my undergraduate and master courses, my career may have been the most direct path.

However, I faced a problem at starting my research. Since I had only studied physics and mathematics until this point, I was completely ignorant of subjects in information science such as mathematical statistics. In particular, despite having the opportunity to take these subjects, I had not studied these at all. During my undergraduate years, I regarded statistics to be a rather lightweight subject, as compared with physics, which examines the true nature of reality. I considered statistics to be only a convenient subject not an essential subject. This perception has been changed on reading Holevo's text. The reason is that it is impossible to quantitatively evaluate the information obtained by an observer without a statistical viewpoint because the measurement data is inherently probabilistic under the mathematical formulation of quantum mechanics. Ultimately, I was forced to study subjects such as math-

[1] Professor Richard Jozsa also studied twistor theory in his graduate course.

ematical statistics and information theory, which should be studied during the undergraduate course. In the end, the research for my master's thesis has been completed with a rather insufficient knowledge of mathematical statistics.

Further, as another problem, I lacked researchers to discuss my research, nearby, due to my choose of this research area. Hence, I had to arrange the chance to discuss with researchers located long distances away. Moreover, since I was also financially quite unstable during the first half of my doctor course, I had kept my research time only between casual teaching work at high school and cram school at that time. In particular, in the first six months of my doctoral course, my research progress was rather slow due to little opportunity for discussion about my research interest. Henceforth, the Quantum Computation Society in Kansai opened in November 1996, and it gave me a chance to talk about topics closely related to my interest. Hence, I could continue my research. During this period, I also had many helpful discussions via telephone with Keiji Matsumoto, who was a research associate at University of Tokyo at that time. Thus, I could learn statistics, and I am deeply indebted to him. I am also grateful to Professor Kenji Ueno, who accepted me as a graduate student until my employment at RIKEN.

In retrospect, in less than 10 years, the situation around quantum information theory has changed completely in Japan. The following are my thoughts and opinions on the future of quantum information theory.

Recently, sophisticated quantum operations have become a reality, and some quantum protocols have been realized. I believe that it is necessary to propose protocols that are relatively easy to implement. This is important not only to motivate further research, but also to have some feedback for the foundations of physics. In particular, I believe that the techniques developed in information theory via quantum information theory will be useful to the foundations of physics.

Thanks to the efforts of many researchers, the field of quantum information theory has become a quite well-known field. However, I feel that many universities in Japan have a trouble to internalize quantum information theory in the current organization of disciplines. Of course, scientific study should have no boundaries in themselves. Hence, It is my presumption that we can construct a more constructive research and educational environment through the treatment for fields such as quantum information theory, which transcend the current framework of disciplines.

My hope is that this book will make those unsatisfied with existing fields to be interested in quantum information theory and inspire them to become active researchers in this or any of its associated fields.

References

1. A. Abeyesinghe, I. Devetak, P. Hayden, A. Winter, “Quantum coherent Slepian-Wolf coding,” in preparation.
2. A. Acín, E. Jané, G. Vidal, “Optimal estimation of quantum dynamics,” *Phys. Rev. A*, **64**, 050302(R) (2001); quant-ph/0012015 (2000).
3. C. Adami, N. J. Cerf, “On the von Neumann capacity of noisy quantum channels,” *Phys. Rev. A*, **56**, 3470 (1997); quant-ph/9610005 (1996).
4. D. Aharonov, A. Kitaev, N. Nisan, “Quantum Circuits with Mixed States,” *Proceedings of the 30th Annual ACM Symposium on Theory of Computation (STOC)*, 20–30 (1997); quant-ph/9806029 (1998).
5. R. Ahlswede, G. Dueck, “Identification via channels,” *IEEE Trans. Inf. Theory*, **35**, 15–29 (1989).
6. R. Ahlswede, A. Winter, “ Strong converse for identification via quantum channels,” *IEEE Trans. Inf. Theory*, **48**, 569–579 (2002); quant-ph/0012127 (2000).
7. C. Ahn, A. Doherty, P. Hayden, A. Winter, “On the distributed compression of quantum information,” quant-ph/0403042 (2004).
8. R. Alicki, M. Fannes, “Continuity of quantum mutual information,” quant-ph/0312081 (2003).
9. R. Alicki, M. Fannes, “Note on multiple additivity of minimal Renyi entropy output of the Werner–Holevo channels,” quant-ph/0407033 (2004).
10. S. Amari, *Differential-Geometrical Methods in Statistics*, Lecture Notes in Statistics, Vol. 28 (Springer, Berlin Heidelberg New York, 1985).
11. S. Amari, H. Nagaoka, *Methods of Information Geometry*, (AMS & Oxford University Press, Oxford, 2000).
12. H. Araki, E. Lieb, “Entropy inequalities,” *Comm. Math. Phys.*, **18**, 160–170 (1970).
13. S. Arimoto, “An algorithm for computing the capacity of arbitrary discrete memoryless channels,” *IEEE Trans. Inf. Theory*, **18**, 14–20 (1972).
14. E. Arthurs, M. S. Goodman, “Quantum correlations: a generalized Heisenberg uncertainty relation,” *Phys. Rev. Lett.*, **60**, 2447–2449 (1988).
15. E. Arthurs, J. L. Kelly, Jr., “On the simultaneous measurement of a pair of conjugate observables,” *Bell Syst. Tech.*, **44**, 725–729 (1965).

16. L. M. Artiles, R. D. Gill, M. I. Guţă, "An invitation to quantum tomography (II)," *J. R. Stat. Soc. Ser. B*, (Stat. Methodol.), **67**, 109–134 (2005); math.ST/0405595 (2004).
17. K. M. R. Audenaert, S. L. Braunstein, "On strong superadditivity of the entanglement of formation," *Comm. Math. Phys.*, **246**(3), 443–452 (2004); quant-ph/0303045 (2003).
18. K. Audenaert, J. Eisert, E. Jané, M. B. Plenio, S. Virmani, B. De Moor, "The asymptotic relative entropy of entanglement," *Phys. Rev. Lett.*, **87**, 217902 (2001); quant-ph/0103096 (2001).
19. K. Audenaert, M. B. Plenio, J. Eisert, "Entanglement cost under positive-partial-transpose-preserving operations," *Phys. Rev. Lett.*, **90**, 027901 (2003); quant-ph/0207146 (2002).
20. E. Bagan, M. A. Ballester, R. D. Gill, A. Monras, R. Muñoz-Tapia, O. Romero-Isart, "Parameter estimation of qubit mixed states," *Proc. ERATO Conference on Quantum Information Science (EQIS) 2005*, 35–36 (2005).
21. E. Bagan, M. Baig, R. Muñoz-Tapia, "Entanglement-assisted alignment of reference frames using a dense covariant coding," *Phys. Rev. A*, **69**, 050303(R) (2004); quant-ph/0303019 (2003).
22. E. Bagan, M. Baig, R. Muñoz-Tapia, "Quantum reverse-engineering and reference-frame alignment without nonlocal correlations," *Phys. Rev. A*, **70**, 030301(R) (2004); quant-ph/0405082 (2004).
23. R. R. Bahadur, "On the asymptotic efficiency of tests and estimates," *Sankhyā*, **22**, 229 (1960).
24. R. R. Bahadur, *Ann. Math. Stat.*, **38**, 303 (1967).
25. R. R. Bahadur, *Some limit theorems in statistics,* Regional Conference Series in Applied Mathematics, no. 4, SIAM, Philadelphia (1971).
26. M. A. Ballester, "Estimation of unitary quantum operations," *Phys. Rev. A*, **69**, 022303 (2004); quant-ph/0305104 (2003).
27. M. Ban, K. Kurokawa, R. Momose, O. Hirota, "Optimum measurements for discrimination among symmetric quantum states and parameter estimation," *Int. J. Theor. Phys.*, **36**, 1269 (1997).
28. A. Barenco, A. K. Ekert, "Dense coding based on quantum entanglement," *J. Mod. Opt.*, **42**, 1253 (1995).
29. H. Barnum, C. A. Fuchs, R. Jozsa, B. Schumacher, "A general fidelity limit for quantum channels," *Phys. Rev. A*, **54**, 4707–4711 (1996); quant-ph/9603014 (1996).
30. H. Barnum, C. M. Caves, C. A. Fuchs, R. Jozsa, B. Schumacher, "On quantum coding for ensembles of mixed states," *J. Phys. A Math. Gen.*, **34**, 6767–6785 (2001); quant-ph/0008024.
31. H. Barnum, E. Knill, M. A. Nielsen, "On quantum fidelities and channel capacities," *IEEE Trans. Inf. Theory*, **46**, 1317–1329 (2000); quant-ph/9809010 (1998).
32. H. Barnum, M. A. Nielsen, B. Schumacher, "Information transmission through a noisy quantum channel," *Phys. Rev. A*, **57**, 4153–4175 (1997); quant-ph/9702049 (1997).
33. H. Bechmann-Pasquinucci, N. Gisin, "Incoherent and coherent eavesdropping in the six-state protocol of quantum cryptography," *Phys. Rev. A*, **59**, 4238–4248 (1999).

34. V. P. Belavkin, "Generalized uncertainty relations and efficient measurements in quantum systems," *Teor. Mat. Fiz.*, **26**, 3, 316–329 (1976); quant-ph/0412030 (2004).
35. C. H. Bennett, G. Brassard, "Quantum cryptography: public key distribution and coin tossing," *Proc. IEEE International Conference on Computers, Systems and Signal Processing*, (Bangalore, India), pp. 175–179 (1984).
36. C. H. Bennett, G. Brassard, C. Crepeau, R. Jozsa, A. Peres, W. K. Wootters, "Teleporting an unknown quantum state via dual classical and Einstein-Podolsky-Rosen channels," *Phys. Rev. Lett.*, **70**, 1895 (1993).
37. C. H. Bennett, C. A. Fuchs, J. A. Smolin, "Entanglement-enhanced classical communication on a noisy quantum channel," in *Quantum Communication, Computing, and Measurement*, O. Hirota, A. S. Holevo, C. M. Caves (eds.), pp. 79–88 (Plenum, New York, 1997).
38. C. H. Bennett, H. J. Bernstein, S. Popescu, B. Schumacher, "Concentrating partial entanglement by local operations," *Phys. Rev. A*, **53**, 2046 (1996); quant-ph/9511030 (1995).
39. C. H. Bennett, "Quantum cryptography using any two nonorthogonal states," *Phys. Rev. Lett.*, **68**, 3121–3124 (1992).
40. C. H. Bennett, D. P. DiVincenzo, J. A. Smolin, W. K. Wootters, "Mixed state entanglement and quantum error correction," *Phys. Rev. A*, **54**, 3824–3851 (1996); quant-ph/9604024 (1996).
41. C. H. Bennett, P. W. Shor, J. A. Smolin, A. V. Thapliyal, "Entanglement-assisted classical capacity of noisy quantum channels," *Phys. Rev. Lett.*, **83**, 3081 (1999); quant-ph/9904023 (1999).
42. C. H. Bennett, P. W. Shor, J. A. Smolin, A. V. Thapliyal, "Entanglement-assisted capacity of a quantum channel and the reverse Shannon theorem," *IEEE Trans. Inf. Theory*, **48**, (10), 2637–2655 (2002); quant-ph/0106052.
43. C. H. Bennett and S. J. Wiesner, "Communication via one- and two-particle operators on Einstein-Podolsky-Rosen states," *Phys. Rev. Lett.*, **69**, 2881 (1992).
44. C. H. Bennett, I. Devetak, P. W. Shor, J. A. Smolin, "Inequalities and separations among assisted capacities of quantum channels," quant-ph/0406086 (2004).
45. C. H. Bennett, D. P. DiVincenzo, J. A. Smolin, "Capacities of quantum erasure channels," quant-ph/9701015 (1997).
46. C. H. Bennett, A. W. Harrow, S. Lloyd, "Universal quantum data compression via gentle tomography," quant-ph/0403078 (2000).
47. C. H. Bennett, A. Winter, Private Communication (2005).
48. A. Ben-Tal, A. Nemirovski, *Lectures on Modern Convex Optimization*, (SIAM/MPS, Philadelphia, 2001).
49. R. Bhatia, *Matrix analysis.* (Springer, Berlin Heidelberg New York, 1997).
50. I. Bjelaković, T. Kruger, R. Siegmund-Schultze, A. Szkoła, "The Shannon-McMillan theorem for ergodic quantum lattice systems," Invent. Math. **155**, 203–222 (2004); math.DS/0207121 (2002).
51. I. Bjelaković, A. Szkoła, "The data compression theorem for ergodic quantum information sources," *Quant. Inf. Process.*, **4**, 49–63 (2005); quant-ph/0301043 (2003).
52. D. Blackwell, L. Breiman, A.J. Thomasian, "The capacity of a class of channels," *Ann. Math. Stat.*, **30**, 1229–1241 (1959).

53. R. Blahut, "Computation of channel capacity and rate-distortion functions," *IEEE Trans. Inf. Theory*, **18**, 460–473 (1972).
54. G. Blakely, "Safeguarding cryptographic keys," *Proc. AFIPS*, **48**, 313 (1979).
55. N. A. Bogomolov, "Minimax measurements in a general statistical decision theory," *Teor. Veroyatnost. Primenen.*, **26**, 798–807 (1981); (English translation: *Theory Probab. Appl.*, **26**, 4, 787–795 (1981))
56. S. Bose, M. B. Plenio, B. Vedral, "Mixed state dense coding and its relation to entanglement measures," *J. Mod. Opt.*, **47**, 291, (2000); quant-ph/9810025 (1998).
57. D. Bouwmeester, J.-W. Pan, K. Mattle, M. Eibl, H. Weinfurter, A. Zeilinger, "Experimental quantum teleportation," *Nature*, **390**, 575–579 (1997).
58. G. Bowen, "Classical information capacity of superdense coding," *Phys. Rev. A*, **63**, 022302 (2001); quant-ph/0101117 (2001).
59. D. C. Brody, L. P. Hughston, *R. Soc. Lond. Proc. A*, **454**, 2445–2475 (1998).
60. D. Bruß, "Optimal eavesdropping in quantum cryptography with six states," *Phys. Rev. Lett.*, 81, 3018–3021 (1998).
61. D. Bruß, A. Ekert, C. Machiavello, "Optimal universal quantum cloning and state estimation," *Phys. Rev. Lett.*, **81**, 2598–2601, (1998). (also appeared as Chap. 24 of *Asymptotic Theory of Quantum Statistical Inference,* M. Hayashi eds.)
62. J. A. Bucklew, *Large Deviation Techniques in Decision, Simulation, and Estimation,* (Wiley, New York, 1990).
63. M. V. Burnashev, A. S. Holevo, "On reliability function of quantum communication channel," *Prob. Inf. Trans.*, **34**, 97–107 (1998); quant-ph/9703013 (1997).
64. P. Busch, M. Grabowski, P. J. Lahti, *Operational quantum physics*, Lecture Notes in Physics, vol. 31, (Springer, Berlin Heidelberg New York, 1997).
65. V. Bužek, R. Derka, and S. Massar, "Optimal quantum clocks," *Phys. Rev. Lett.*, **82**, 2207 (1999); quant-ph/9808042 (1998).
66. A. R. Calderbank, E. M. Rains, P. W. Shor, N. J. A. Sloane, "Quantum error correction and orthogonal geometry," *Phys. Rev. Lett.*, **78**, 405 (1996).
67. A.R. Calderbank, P.W. Shor: "Good quantum error-correcting codes exist," *Phys. Rev. A*, **54**, 1098 (1996); quant-ph/9512032 (1995).
68. N. J. Cerf, C. Adami: "Negative entropy and information in quantum mechanics," *Phys. Rev. Lett.*, **79**, 5194 (1997); quant-ph/9512022.
69. N. J. Cerf, C. Adami, R. M. Gingrich, "Reduction criterion for separability," *Phys. Rev. A*, **60**, 898 (1999).
70. A. Chefles, "Condition for unambiguous state discrimination using local operations and classical communication," *Phys. Rev. A*, **69**, 050307(R) (2004).
71. Y.-X. Chen, D. Yang, "Distillable entanglement of multiple copies of Bell states," *Phys. Rev. A*, **66**, 014303 (2002).
72. H. Chernoff, "A measure of asymptotic efficiency for tests of a hypothesis based on the sum of observations," *Ann. Math. Stat.*, **23**, 493–507 (1952).
73. G. Chiribella, G. M. D'Ariano, P. Perinotti, M. F. Sacchi, "Efficient use of quantum resources for the transmission of a reference frame," *Phys. Rev. Lett.*, **93**, 180503 (2004); quant-ph/0405095 (2004).
74. G. Chiribella, G. M. D'Ariano, P. Perinotti, M. F. Sacchi, Covariant quantum measurements which maximize the likelihood *Phys. Rev. A*, **70**, 061205 (2004); quant-ph/0403083 (2004).

75. G. Chiribella, G. M. D'Ariano, M. F. Sacchi, "Optimal estimation of group transformations using entanglement," *Phys. Rev. A*, **72**, 042338 (2005); quant-ph/0506267 (2005).
76. G. Chiribella, G. M. D'Ariano, P. Perinotti, M. F. Sacchi, "Maximum likelihood estimation for a group of physical transformations," quant-ph/0507007 (2005).
77. M.-D. Choi, "Completely positive linear maps on complex matrices," *Lin. Alg. Appl.*, **10**, 285–290 (1975).
78. M. Christandl, A. Winter, "Squashed entanglement"–an additive entanglement measure," *J. Math. Phys.*, **45**, 829–840 (2004); quant-ph/0308088 (2003).
79. M. Christandl, A. Winter, "Uncertainty, monogamy, and locking of quantum correlations," *IEEE Trans Inf Theory*, **51**, 3159–3165 (2005); quant-ph/0501090 (2005).
80. J. I. Cirac, W. Dür, B. Kraus, M. Lewenstein, "Entangling operations and their implementation using a small amount of entanglement," *Phys. Rev. Lett.*, **86**, 544 (2001); quant-ph/0007057 (2000).
81. R. Cleve, D. Gottesman, H.-K. Lo, "How to share a quantum secret," *Phys. Rev. Lett.*, **82**, 648 (1999); quant-ph/9901025 (1999).
82. T. Cover, J. Thomas, *Elements of Information Theory*, (Wiley, New York, 1991).
83. H. Cramér, "Sur un nouveaux theoorème-limite de la théorie des probabilités," in *Actualités Scientifiques et Industrielles*, no. 736 in *Colloque consacré à la thèorie des probabilités*, 5–23, Hermann, Paris, 1938.
84. I. Csiszár, "Information type measures of difference of probability distribution and indirect observations," *Studia Scient. Math. Hungar.*, **2**, 299–318 (1967).
85. I. Csiszár, J. Körner, *Information Theory: Coding Theorems for Discrete Memoryless Systems*, (Academic, 1981).
86. G. M. D'Ariano, "Homodyning as universal detection," in *Quantum Communication, Computing, and Measurement*, O. Hirota, A. S. Holevo, C. A. Caves (eds.), Plenum, New York, pp. 253–264 (1997); quant-ph/9701011 (1997). (also appeared as Chap. 33 of *Asymptotic Theory of Quantum Statistical Inference,* M. Hayashi eds.)
87. G. M. D'Ariano, P. L. Presti, "Imprinting Complete Information about a Quantum Channel on its Output State," *Phys. Rev. Lett.*, **91**, 047902 (2003).
88. G. M. D'Ariano, P. Lo Presti, M. F. Sacchi, "Bell measurements and observables," *Phys. Lett. A*, **272**, 32 (2000).
89. N. Datta, Y. Suhov, "Data compression limit for an information source of interacting qubits," *Quant. Inf. Process.*, **1**, (4), 257–281 (2002); quant-ph/0207069 (2002).
90. N. Datta, A. S. Holevo, Y. Suhov, "Additivity for transpose depolarizing channels," quant-ph/0412034 (2004).
91. N. Datta, M. B. Ruskai, "Maximal output purity and capacity for asymmetric unital qudit channels," quant-ph/0505048 (2005).
92. E. B. Davies, J. T. Lewis, "An operational approach to quantum probability," *Comm. Math. Phys.*, **17**, 239 (1970).
93. L. D. Davisson, "Comments on "Sequence time coding for data compression," *Proc. IEEE*, **54**, 2010 (1966).
94. A. Dembo, O. Zeitouni, *Large Deviation Techniques and Applications*, (Springer, Berlin Heidelberg New York, 1997).

95. I. Devetak, "The private classical capacity and quantum capacity of a quantum channel," *IEEE Trans. Inf. Theory*, **51**, 44–55 (2005); quant-ph/0304127 (2003).
96. I. Devetak, P. W. Shor, "The capacity of a quantum channel for simultaneous transmission of classical and quantum information," quant-ph/0311131 (2003).
97. I. Devetak, A. Winter, "Classical data compression with quantum side information," *Phys. Rev. A*, **68**, 042301 (2003); quant-ph/0209029 (2002).
98. I. Devetak, A. Winter, "Distillation of secret key and entanglement from quantum states," *Proc. R. Soc. Lond. A*, **461**, 207–235 (2005); quant-ph/0306078 (2003).
99. D. DiVincenzo, P. Shor, J. Smolin, "Quantum channel capacities of very noisy channels," *Phys. Rev. A*, **57**, 830–839 (1998); quant-ph/9706061 (1997).
100. W. Dür, G. Vidal, J. I. Cirac, "Visible compression of commuting mixed state," *Phys. Rev. A*, **64**, 022308 (2001); quant-ph/0101111 (2001).
101. W. Dür, J. I. Cirac, M. Lewenstein, D. Bruß, "Distillability and partial transposition in bipartite systems," *Phys. Rev. A*, **61**, 062313 (2000).
102. M. J. Donald, M. Horodecki, "Continuity of relative entropy of entanglement," *Phys. Lett. A*, **264**, 257–260 (1999).
103. M. Donald, M. Horodecki, and O. Rudolph, "The uniqueness theorem for entanglement measures," *J. Math. Phys.*, **43**, 4252–4272 (2002).
104. A. Einstein, R. Podolsky, N. Rosen: "Can quantum-mechanical descriptions of physical reality be considered complete?," *Phys. Rev.*, **47**, 777–780 (1935).
105. I. Ekeland, R. Téman, *Convex Analysis and Variational Problems*, (North-Holland, Amsterdam, 1976); (SIAM, Philadelphia, 1999).
106. R. Ellis, "Large deviations for a general class of random vectors," *Ann. Probab.*, **12**, 1, 1–12 (1984). *Entropy, Large Deviations and Statistical Mechanics*, (Springer, Berlin Heidelberg New York, 1985).
107. H. Fan, "Distinguishability and indistinguishability by local operations and classical communication," *Phys. Rev. Lett.*, **92**, 177905 (2004).
108. H. Fan, "A note on quantum entropy inequalities and channel capacities," *J. Phys. A Math. Gen.*, **36**, 12081–12088 (2003).
109. H. Fan, K. Matsumoto, M. Wadati, "Quantum cloning machines of a d-level system," *Phys. Rev. A*, **64**, 064301 (2001); quant-ph/0103053 (2001).
110. H. Fan, K. Matsumoto, X. Wang, M. Wadati, "Quantum cloning machines for equatorial qubits," *Phys. Rev. A*, **65**, 012304 (2002); quant-ph/0101101 (2001).
111. M. Fannes, "A continuity property of the entropy density for spin lattice systems," *Comm. Math. Phys.*, **31**, 291–294 (1973).
112. M. Fannes, B. Haegeman, M. Mosonyi, D. Vanpeteghem, "Additivity of minimal entropy output for a class of covariant channels," quant-ph/0410195 (2004).
113. R. M. Fano, *Transmission of Information: A Statistical Theory of Communication*, (Wiley, New York, 1961).
114. A. Feinstein: "A new basic theorem of information theory," *IRE Trans. PGIT*, **4**, 2–22 (1954).
115. D. G. Fischer, H. Mack, M. A. Cirone, M. Freyberger, "Enhanced estimation of a noisy quantum channel using entanglement," *Phys. Rev. A*, **64**, 022309 (2001); quant-ph/0103160 (2001).

116. G. D. Forney, Jr., S. M. Thesis, MIT 1963 (unpublished).
117. J. C. Fu, "On a theorem of Bahadur on the rate of convergence of point estimators," *Ann. Stat.*, **1**, 745 (1973).
118. C. A. Fuchs, "Distinguishability and accessible information in quantum theory," quant-ph/9601020 (1996).
119. A. Fujiwara, *Statistical Estimation Theory for Quantum States*, master's thesis, Department of Mathematical Engineering and Information Physics, Graduate School of Engineering, University of Tokyo, Japan (1993) (in Japanese).
120. A. Fujiwara, *A Geometrical Study in Quantum Information Systems*, Ph.D. thesis, Department of Mathematical Engineering and Information Physics, Graduate School of Engineering, University of Tokyo, Japan (1995).
121. A. Fujiwara, private communication to H. Nagaoka (1996).
122. A. Fujiwara, "Geometry of quantum information systems," in *Geometry in Present Day Science*, O. E. Barndorff-Nielsen, E. B. V. Jensen (eds.), (World Scientific, Singapore, 1998), pp. 35–48.
123. A. Fujiwara, "Quantum birthday problems: geometrical aspects of quantum randomcoding," *IEEE Trans. Inf. Theory*, **47**, 2644–2649 (2001).
124. A. Fujiwara, "Quantum channel identification problem," *Phys. Rev. A*, **63**, 042304 (2001).
125. A. Fujiwara, "Estimation of SU(2) operation and dense coding: an information geometric approach," *Phys. Rev. A*, **65**, 012316 (2002).
126. A. Fujiwara, H. Nagaoka, "Quantum Fisher metric and estimation for pure state models," *Phys. Lett.*, **201A**, 119–124 (1995).
127. A. Fujiwara, H. Nagaoka, "Coherency in view of quantum estimation theory," in *Quantum Coherence and Decoherence*, K. Fujikawa, Y. A. Ono (eds.), (Elsevier, Amsterdam, 1996), pp. 303–306.
128. A. Fujiwara, H. Nagaoka, "Operational capacity and pseudoclassicality of a quantum channel," *IEEE Trans. Inf. Theory*, **44**, 1071–1086 (1998).
129. A. Fujiwara, H. Nagaoka, "An estimation theoretical characterization of coherent states," *J. Math. Phys.*, **40**, 4227–4239 (1999).
130. A. Fujiwara, P. Algoet, "One-to-one parametrization of quantum channels," *Phys. Rev. A*, **59**, 3290–3294 (1999).
131. A. Fujiwara, "Quantum channel identification problem," *Phys. Rev. A*, **63**, 042304 (2001).
132. A. Fujiwara, "Estimation of SU(2) operation and dense coding: an information geometric approach," *Phys. Rev. A*, **65**, 012316 (2002).
133. A. Fujiwara, T. Hashizume, "Additivity of the capacity of depolarizing channels," *Phys. Lett A*, **299**, 469–475 (2002).
134. A. Fujiwara and H. Imai, "Quantum parameter estimation of a generalized Pauli channel," *J. Phys. A Math. Gen.*, **36**, 8093–8103 (2003).
135. A. Fujiwara, "Estimation of a generalized amplitude-damping channel," *Phys. Rev. A*, **70**, 012317 (2004).
136. A. Fujiwara, "Mathematics of quantum channels," *Suurikagaku*, 474, 28–35 (2002) (in Japanese).
137. M. Fujiwara, M. Takeoka, J. Mizuno, M. Sasaki, "Exceeding classical capacity limit in quantum optical channel," *Phys. Rev. Lett.*, **90**, 167906 (2003); quant-ph/0304037 (2003).
138. M. Fukuda, "Extending additivity from symmetric to asymmetric channels," *J. Phys. A Math. Gen.*, **38**, L753–L758 (2005); quant-ph/0505022 (2005).

139. A. Furusawa, J. L. Sørensen, S. L. Braunstein, C. A. Fuchs, H. J. Kimble, E. J. Polzik, "Unconditional quantum teleportation," *Science*, **282**, 706 (1998).
140. R. G. Gallager, *Information Theory and Reliable Communication*, (Wiley, New York, 1968).
141. J. Gärtner, "On large deviations from the invariant measure," *Theory Probabil. Appl.*, **22**, 24–39 (1977).
142. R. Gill, S. Massar, "State estimation for large ensembles," *Phys. Rev. A*, **61**, 042312 (2000); quant-ph/9902063 (1999).
143. R. D. Gill, "Asymptotic information bounds in quantum statistics," math.ST/0512443 (2005).
144. R. D. Gill, B. Y. Levit, "Applications of the van Tree inequality: a Bayesian Cramér-Rao bound" *Bernoulli*, **1**, 59–79 (1995).
145. N. Giri, W. von Waldenfels, "An algebraic version of the central limit theorem," *Z. Wahrscheinlichkeitstheorie Verw. Gebiete*, **42**, 129–134 (1978).
146. N. Gisin, contribution to the Torino Workshop, 1997.
147. S. Ghosh, G. Kar, A. Roy, D. Sarkar, "Distinguishability of maximally entangled states," *Phys. Rev. A*, **70**, 022304 (2004).
148. J. P. Gordon, *Proc. IRE*, **50**, 1898–1908 (1962).
149. J. P. Gordon, "Noise at optical frequencies; information theory," in *Quantum Electronics and Coherent Light, Proc. Int. School Phys. "Enrico Fermi," Course XXXI*, P. A. Miles (ed.). (Academic, New York, 1964), pp. 156–181.
150. D. Gottesman, "Class of quantum error-correcting codes saturating the quantum Hamming bound," *Phys. Rev. A*, **54**, 1862 (1996).
151. D. Gottesman, "On the theory of quantum secret sharing," *Phys. Rev. A*, **61**, 042311 (2000); quant-ph/9910067 (1999).
152. J. Gruska, H. Imai, "Power, puzzles and properties of entanglement," in *achines, Computations, and Universality, Proc. 3rd International Conference, MCU 2001*, Lecture Notes in Computer Science, vol. 2055, (Springer, Berlin Heidelberg New York, 2001), pp. 25–68.
153. J. Gruska, H. Imai, K. Matsumoto, "Power of quantum entanglement," in *Foundations of Information Technology in the Era of Networking and Mobile Computing, IFIP 17th World Computer Congress - TC1 Stream / 2nd IFIP International Conference on Theoretical Computer Science*, vol. 223 of IFIP Conference Proceedings, pp. 3–22, 2001.
154. M. Hamada, "Lower bounds on the quantum capacity and highest error exponent of general memoryless channels," *IEEE Trans. Inf. Theory*, **48**, 2547–2557 (2002); quant-ph/0112103 (2002).
155. M. Hamada, "Notes on the fidelity of symplectic quantum error-correcting codes," *Int. J. Quant. Inf.*, **1**, 443–463 (2003).
156. T. S. Han, "Hypothesis testing with the general source," *IEEE Trans. Inf. Theory*, **46**, 2415–2427 (2000).
157. T. S. Han, "The reliability functions of the general source with fixed-length coding," *IEEE Trans. Inf. Theory*, **46**, 2117–2132 (2000).
158. T. S. Han: *Information-Spectrum Methods in Information Theory*, (Springer, Berlin Heidelberg New York, 2002) (originally appeared in Japanese in 1998).
159. T. S. Han, K. Kobayashi, "The strong converse theorem for hypothesis testing," *IEEE Trans. Inf. Theory*, **35**, 178–180 (1989).
160. T. S. Han, K. Kobayashi, *Mathematics of Information and Encoding*, (American Mathematical Society, 2002) (originally appeared in Japanese in 1999).

161. T. S. Han, S. Verdú, "Approximation theory of output statistics," *IEEE Trans. Inf. Theory,* **39**, 752–772 (1993).
162. T. S. Han, "Folklore in source coding: information-spectrum approach," *IEEE Trans. Inf. Theory,* **51**, (2), 747–753 (2005).
163. A. Harrow, H. K. Lo, "A tight lower bound on the classical communication cost of entanglement dilution," *IEEE Trans. Inf. Theory,* **50**, 319–327 (2004); quant-ph/0204096 (2002).
164. P. Hausladen, R. Jozsa, B. Schumacher, M. Westmoreland, W. Wooters, "Classical information capacity of a quantum channel," *Phys. Rev. A*, **54**, 1869–1876 (1996).
165. M. Hayashi, *Minimization of deviation under quantum local unbiased measurements,* master's thesis, Department of Mathematics, Graduate School of Science, Kyoto University, Japan (1996).
166. M. Hayashi, "A linear programming approach to attainable cramer-rao type bound and randomness conditions," Kyoto-Math 97–08; quant-ph/9704044 (1997).
167. M. Hayashi, "A linear programming approach to attainable Cramer–Rao type bound," in *Quantum Communication, Computing, and Measurement*, O. Hirota, A. S. Holevo, C. M. Caves (eds.), (Plenum, New York, 1997), pp. 99–108. (Also appeared as Chap. 12 of *Asymptotic Theory of Quantum Statistical Inference,* M. Hayashi eds.)
168. M. Hayashi, "Asymptotic estimation theory for a finite dimensional pure state model," *J. Phys. A Math. Gen.*, **31**, 4633–4655 (1998); quant-ph/9704041 (1997). (Also appeared as Chap. 23 of *Asymptotic Theory of Quantum Statistical Inference*, M. Hayashi eds.).
169. M. Hayashi, "Asymptotics of quantum relative entropy from a representation theoretical viewpoint," *J. Phys. A Math. Gen.*, **34**, 3413–3419 (2001); quant-ph/9704040 (1997).
170. M. Hayashi, "Asymptotic quantum estimation theory for the thermal states family," in *Quantum Communication, Computing, and Measurement 2*, P. Kumar, G. M. D'ariano, O. Hirota (eds.), (Plenum, New York, 2000) pp. 99–104; quant-ph/9809002 (1998). (Also appeared as Chap. 14 of *Asymptotic Theory of Quantum Statistical Inference,* M. Hayashi eds.).
171. M. Hayashi, "Asymptotic large deviation evaluation in quantum estimation theory," *Proc. Symposium on Statistical Inference and Its Information-Theoretical Aspect*, pp. 53–82 (1998) (in Japanese).
172. M. Hayashi, "Simultaneous measurements of non-commuting physical quantities," *RIMS koukyuroku Kyoto Univiversity*, No. 1099, 96–118 (1999) (in Japanese).
173. M. Hayashi, "Quantum hypothesis testing for the general quantum hypotheses," *Proc. 24th Symposium on Information Theory and Its Applications (SITA)*, pp. 591–594 (2001).
174. M. Hayashi, "Optimal sequence of POVMs in the sense of Stein's lemma in quantum hypothesis," *J. Phys. A Math. Gen.*, **35**, 10759–10773 (2002); quant-ph/0107004 (2001).
175. M. Hayashi, "Exponents of quantum fixed-length pure state source coding," *Phys. Rev. A*, **66**, 032321 (2002); quant-ph/0202002 (2002).
176. M. Hayashi, "General formulas for fixed-length quantum entanglement concentration," appear in *IEEE Trans. Infor. Theory*; quant-ph/0206187 (2002).

177. M. Hayashi, "Two quantum analogues of Fisher information from a large deviation viewpoint of quantum estimation," *J. Phys. A Math. Gen.*, **35**, 7689–7727 (2002); quant-ph/0202003 (2002). (Also appeared as Chap. 28 of *Asymptotic Theory of Quantum Statistical Inference,* M. Hayashi eds.).
178. M. Hayashi, "Quantum estimation and quantum central limit theorem," *Sugaku,* **55**, 4, 368–391 (2003) (in Japanese).
179. M. Hayashi, "Second order asymptotics in fixed-length source coding and intrinsic randomness," cs.IT/0503089 (2005).
180. M. Hayashi, "Characterization of several kinds of quantum analogues of relative entropy," quant-ph/0510181 (2005); *Proc. 9th Quantum Information Technology Symposium (QIT13),* Touhoku University, Sendai, Miyagi, Japan, 24–25 November 2005, pp. 137–142.
181. M. Hayashi, M. Koashi, K. Matsumoto, F. Morikoshi A. Winter, "Error exponents for entangle concentration," *J. Phys. A Math. Gen.*, **36**, 527–553 (2003); quant-ph/0206097 (2002).
182. M. Hayashi, K. Matsumoto, "Statistical model with measurement degree of freedom and quantum physics," *RIMS koukyuroku Kyoto Univiversity,* **1055**, 96–110 (1998) (in Japanese). (Also appeared as Chap. 13 of *Asymptotic Theory of Quantum Statistical Inference,* M. Hayashi eds.)
183. M. Hayashi, K. Matsumoto, "Variable length universal entanglement concentration by local operations and its application to teleportation and dense coding," quant-ph/0109028 (2001); K. Matsumoto, M. Hayashi, "Universal entanglement concentration," quant-ph/0509140 (2005).
184. M. Hayashi, K. Matsumoto, "Quantum universal variable-length source coding," *Phys. Rev. A*, **66**, 022311 (2002); quant-ph/0202001 (2002).
185. M. Hayashi, K. Matsumoto, " Simple construction of quantum universal variable-length source coding," *Quant. Inf. Comput.*, **2**, Special Issue, 519–529 (2002); quant-ph/0209124, (2002).
186. M. Hayashi, H. Nagaoka: "General formulas for capacity of classical-quantum channels," *IEEE Trans. Inf. Theory,* **49**, 1753–1768 (2003); quant-ph/0206186 (2002).
187. M. Hayashi, F. Sakaguchi, "Subnormal operators regarded as generalized observables and compound-system-type normal extension related to su(1,1)," *J. Phys. A Math. Gen.*, **33**, 7793–7820 (2000); quant-ph/0003079 (2000).
188. M. Hayashi, K. Matsumoto, "Asymptotic performance of optimal state estimation in quantum two level system," quant-ph/0411073 (2004).
189. M. Hayashi, "Parallel Treatment of Estimation of SU(2) and Phase Estimation," appear in *Phys. Lett. A*. quant-ph/0407053 (2004).
190. M. Hayashi, "Estimation of SU(2) action by using entanglement," *Proc. 9th Quantum Information Technology Symposium (QIT9),* NTT Basic Research Laboratories, Atsugi, Kangawa, Japan, 11–12 December 2003, pp. 9–13 (in Japanese); *ibid, Proc. 7th International Conference on Quantum Communication, Measurement and Computing,* Glasgow, UK, 25–29 July 2004 (AIP), pp. 269–272.
191. M. Hayashi, D. Markham, M. Murao, M. Owari, S.Virmani, "Bounds on Multipartite Entangled Orthogonal State Discrimination Using Local Operations and Classical Communication," *Phys. Rev. Lett.*, **96**, 040501 (2006).
192. M. Hayashi, "On the second order asymptotics for pure states family," *IEICE Trans.*, **E88-JA**, 903–916 (2005) (in Japanese).

193. M. Hayashi, H. Imai, K. Matsumoto, M. B. Ruskai, T. Shimono, "Qubit channels which require four inputs to achieve capacity: implications for additivity conjectures," *Quant. Inf. Comput.*, **5**, 13–31 (2005); quant-ph/040317.
194. M. Hayashi (eds.), *Asymptotic Theory of Quantum Statistical Inference: Selected Papers,* (World Scientific, Singapore, 2005).
195. M. Hayashi, "Hypothesis Testing of maximally entangled state" in preparation.
196. P. M. Hayden, M. Horodecki, B. M. Terhal, "The asymptotic entanglement cost of preparing a quantum state," *J. Phys. A Math. Gen.*, **34**, 6891–6898 (2001); quant-ph/0008134 (2000).
197. P. Hayden, R. Jozsa, A. Winter, "Trading quantum for classical resources in quantum data compression," *J. Math. Phys.*, **43**, 4404–4444 (2002); quant-ph/0204038 (2002).
198. P. Hayden, A. Winter, "On the communication cost of entanglement transformations," *Phys. Rev. A*, **67**, 012326 (2003); quant-ph/0204092 (2002).
199. P. Hayden, R. Jozsa, D. Petz, A. Winter, "Structure of states which satisfy strong subadditivity of quantum entropy with equality," *Comm. Math. Phys.*, **246**, 359–374 (2004).
200. W. Heisenberg, "Über den anschaulichen Inhalt der quantentheoretischen Kinematik und Mechanik," *Z. Phys.*, **43**, 172–198 (1927).
201. C. W. Helstrom, "Detection theory and quantum mechanics," *Inf. Contr.*, **10**, 254–291 (1967).
202. C. W. Helstrom, "Minimum mean-square error estimation in quantum statistics," *Phys. Lett.*, **25A**, 101–102 (1976).
203. C. W. Helstrom, *Quantum Detection and Estimation Theory,* (Academic, New York, 1976).
204. C. W. Helstrom, *Int. J. Theor. Phys.*, **11**, 357 (1974).
205. L. Henderson, V. Vedral, "Classical, quantum and total correlations," *J. Phys. A Math. Gen.*, **34**, 6899 (2001); quant-ph/0105028 (2001).
206. F. Hiai, D. Petz, "The proper formula for relative entropy and its asymptotics in quantum probability," *Comm. Math. Phys.*, **143**, 99–114 (1991).
207. F. Hiai, D. Petz, "The golden-thompson trace inequality is complemented," *Lin. Alg. Appl.*, **181**, 153–185 (1993).
208. T. Hiroshima, "Majorization criterion for distillability of a bipartite quantum state," *Phys. Rev. Lett.*, **91**, 057902 (2003); quant-ph/0303057 (2003).
209. T. Hiroshima, M. Hayashi, "Finding a maximally correlated state-Simultaneous Schmidt decomposition of bipartite pure states," *Phys. Rev. A*, **70**, 030302(R) (2004).
210. W. Hoeffding, "Asymptotically optimal test for multinomial distributions," *Ann. Math. Stat.*, **36**, 369–400 (1965).
211. A. S. Holevo, "An analog of the theory of statistical decisions in noncommutative theory of probability," *Trudy Moskov. Mat. Obšč.*, **26**, 133–149 (1972) (in Russian). (English translation: *Trans. Moscow Math. Soc.*, **26**, 133–149 (1972)).
212. A. S. Holevo, "Bounds for the quantity of information transmitted by a quantum communication channel," *Problemy Peredachi Informatsii*, **9**, 3–11 (1973) (in Russian). (English translation: *Probl. Inf. Transm.*, **9**, 177–183 (1975)).
213. A. S. Holevo, "Some statistical problems for quantum Gaussian states," *IEEE Trans. Inf. Theory*, **21**, 533–543 (1975).

214. A. S. Holevo, "On the capacity of quantum communication channel," *Problemly Peredachi Informatsii*, **15**, 4, 3–11 (1979) (in Russian). (English translation: *Probl. Inf. Transm.*, **15**, 247–253 (1979).)
215. A. S. Holevo, "Covariant measurements and uncertainty relations," *Rep. Math. Phys.*, **16**, 385–400 (1979).
216. A. S. Holevo, *Probabilistic and Statistical Aspects of Quantum Theory*, (North-Holland, Amsterdam, 1982); originally published in Russian (1980).
217. A. S. Holevo, "The capacity of the quantum channel with general signal states," *IEEE Trans. Inf. Theory*, **44**, 269 (1998); quant-ph/9611023 (1996).
218. A. S. Holevo, "On quantum communication channels with constrained inputs," quant-ph/9705054 (1997).
219. A. S. Holevo, "Coding theorems for quantum channels," *Tamagawa University Res. Rev.*, 4, (1998); quant-ph/9809023 (1998).
220. A. S. Holevo: "On entanglement-assisted classical capacity," *J. Math. Phys.*, **43**, 4326–4333 (2002); quant-ph/0106075 (2001).
221. A. S. Holevo, *Statistical structure of quantum theory*, Lecture Notes in Physics, vol. 67, (Springer, Berlin Heidelberg New York, 2001).
222. A. S. Holevo, M. E. Shirokov, "On Shor's channel extension and constrained channels," *Commun. Math. Phys.*, **249**, 417–430, (2004); quant-ph/0306196 (2003).
223. A. S. Holevo, "Covariant Measurements and Optimality," Chap. IV of *Probabilistic and Statistical Aspects of Quantum Theory*, (North-Holland, Amsterdam, 1982). Originally published in Russian (1980): "Bounds for generalized uncertainty of the shift parameter," Lecture Notes in Mathematics, **1021**, 243–251 (1983).
224. R. Horodecki, P. Horodecki, "Quantum redundancies and local realism," *Phys. Lett.*, **A 194**, 147–152 (1994).
225. M. Horodecki, P. Horodecki, R. Horodecki, "Separability of mixed states: necessary and sufficient conditions," *Phys. Lett.*, **A 223**, 1–8 (1996).
226. M. Horodecki, P. Horodecki, R. Horodecki, "Inseparable two-spin 1/2 density matrices can be distilled to a singlet form," *Phys. Rev. Lett.*, **78**, 574 (1997).
227. M. Horodecki, P. Horodecki, R. Horodecki, "Mixed-state entanglement and distillation: is there a "bound" entanglement in nature?" *Phys. Rev. Lett.*, **80**, 5239 (1998); quant-ph/9801069 (1998).
228. M. Horodecki, "Limits for compression of quantum information carried by ensembles of mixed states," *Phys. Rev. A*, **57**, 3364–3369 (1998); quant-ph/9712035 (1997).
229. M. Horodecki, P. Horodecki, "Reduction criterion of separability and limits for a class of distillation protocols," *Phys. Rev. A*, **59**, 4206 (1999); quant-ph/9708015 (1997).
230. M. Horodecki, P. Horodecki, R. Horodecki, "General teleportation channel, singlet fraction and quasi-distillation," *Phys. Rev. A*, **60**, 1888 (1999); quant-ph/9807091 (1998).
231. M. Horodecki, "Optimal compression for mixed signal states," *Phys. Rev. A*, **61**, 052309 (2000); quant-ph/990508 v3 (2000).
232. M. Horodecki, P. Horodecki, R. Horodecki, "Unified approach to quantum capacities: towards quantum noisy coding theorem," *Phys. Rev. Lett.*, **85**, 433–436 (2000).

233. M. Horodecki, P. Horodecki, R. Horodecki, "Mixed-state entanglement and quantum communication," in *Quantum Information: An Introduction to Basic Theoretical Concepts and Experiments*, (Springer Tracts in Modern Physics, 173), G. Alber, T. Beth, M. Horodecki, P. Horodecki, R. Horodecki, M. Rotteler, H. Weinfurter, R. Werner, A. Zeilinger (eds.), (Springer, Berlin Heidelberg New York, 2001).
234. M. Horodecki, P. Horodecki, R. Horodecki, D. W. Leung, B. M. Terhal, "Classical capacity of a noiseless quantum channel assisted by noisy entanglement," *Quant. Inf. Comput.*, **1**, 70–78 (2001); quant-ph/0106080 (2001).
235. M. Horodecki, P. Shor, M. B. Ruskai, "Entanglement breaking channels," *Rev. Math. Phys.*, **15**, 1–13 (2003); quant-ph/0302031 (2003).
236. P. Horodecki, "Separability criterion and inseparable mixed states with positive partial transposition," *Phys. Lett.*, **A232**, 333 (1997); quant-ph/9703004 (1997).
237. T. Hiroshima, "Optimal dense coding with mixed state entanglement," *J. Phys. A Math. Gen.*, **34**, 6907–6912 (2001); quant-ph/0009048 (2000).
238. H. Imai, M. Hachimori, M. Hamada, H. Kobayashi, K. Matsumoto, "Optimization in quantum computation and information," *Proc. 2nd Japanese-Hungarian Symposium on Discrete Mathematics and Its Applications*, Budapest, Hungary (2001).
239. S. Ishikawa, "Uncertainty relations in simultaneous measurements for arbitrary observables," *Rep. Math. Phys.*, **29**, 257–273 (1991).
240. S. Ishizaka, T. Hiroshima, "Maximally entangled mixed states under nonlocal unitary operations in two qubits," *Phys. Rev. A*, **62**, 022310 (2000).
241. S. Ishizaka, "Binegativity and geometry of entangled states in two states," *Phys. Rev. A*, **69**, 020301(R) (2004); quant-ph/0308056 (2003).
242. S. Ishizaka, "Bound entanglement provides convertibility of pure entangled states," *Phys. Rev. Lett.*, **93**, 190501 (2004); quant-ph/0403016 (2004).
243. A. Jamiołkowski, "Linear transformations which preserve trace and positive semidefiniteness of operators," *Rep. Math. Phys.*, **3**, 275–278 (1972).
244. R. Jozsa, "Fidelity for mixed quantum states," *J. Mod. Opt.*, **41(12)**, 2315–2323 (1994).
245. R. Jozsa, B. Schumacher: "A new proof of the quantum noiseless coding theorem," *J. Mod. Opt.*, **41(12)**, 2343–2349 (1994).
246. R. Jozsa, "Quantum noiseless coding of mixed states," Talk given at *3rd Santa Fe Workshop on Complexity, Entropy, and the Physics of Information*, May 1994.
247. R. Jozsa, M. Horodecki, P. Horodecki, R. Horodecki, "Universal quantum information compression," *Phys. Rev. Lett.*, **81**, 1714 (1998); quant-ph/9805017 (1998).
248. R. Jozsa, S. Presnell, "Universal quantum information compression and degrees of prior knowledge," quant-ph/0210196 (2002).
249. A. Kaltchenko, E.-H. Yang, "Universal compression of ergodic quantum sources," *Quant. Inf. Comput.*, **3**, 359–375 (2003); quant-ph/0302174 (2003).
250. A. Kent, N. Linden, S. Massar, "Optimal entanglement enhancement for mixed states," *Phys. Rev. Lett.*, **83**, 2656 (1999).
251. M. Keyl, R. F. Werner, "Optimal cloning of pure states, judging single clones," *J. Math. Phys.*, **40**, 3283–3299 (1999); quant-ph/9807010 (1998).
252. M. Keyl, R. F. Werner, "Estimating the spectrum of a density operator," *Phys. Rev. A*, **64**, 052311 (2001); quant-ph/0102027 (2001).

253. C. King, "Additivity for a class of unital qubit channels," *J. Math. Phys.*, **43**, 4641–4653 (2002); quant-ph/0103156 (2001).
254. C. King, "The capacity of the quantum depolarizing channel," *IEEE Trans. Infor. Theory*, **49**, 221–229, (2003); quant-ph/0204172 (2002).
255. E. Klarreich, "Quantum cryptography: Can you keep a secret?" *Nature*, **418**, 270–272 (2002).
256. E. Knill, R. Laflamme, "Theory of quantum error-correcting codes," *Phys. Rev. A*, **55**, 900 (1997).
257. M. Koashi, N. Imoto, "Compressibility of mixed-state signals," *Phys. Rev. Lett.*, **87**, 017902 (2001); quant-ph/0103128, (2001).
258. M. Koashi, N. Imoto, "Quantum information is incompressible without errors," *Phys. Rev. Lett.*, **89**, 097904 (2002); quant-ph/0203045, (2002).
259. M. Koashi, A. Winter, "Monogamy of quantum entanglement and other correlations," *Phys. Rev. A*, **69**, 022309 (2004).
260. H. Kosaka, A. Tomita, Y. Nambu, N. Kimura, K. Nakamura, "Single-photon interference experiment over 100 km for quantum cryptography system using a balanced gated-mode photon detector," *Electron. Lett.*, **39**, 16, 1199–1201 (2003).
261. K. Kraus, *States, Effects, and Operations*, Lecture Notes in Physics, vol. 190, (Springer, Berlin Heidelberg New York, 1983).
262. P. G. Kwiat, A. G. White, et al., *Phys. Rev.* A **60**, R773 (1999), *Phys. Rev. Lett.*, **83**, 3103 (1999), *Sience*, **290**, 498 (2000), *Nature*, **409**, 1014 (2001), quant-ph/0108088 (2001).
263. E. L. Lehman, G. Casella, *Theory of Point Estimation*, (Springer, Berlin Heidelberg New York, 1998).
264. L. B. Levitin, "On quantum measure of information," in *Proc. 4th All-Union Conference on Information Transmission and Coding Theory*, pp. 111–115 (Tashkent, 1969) (in Russian). English translation: *Information, Complexity and Control in Quantum Physics*, A. Blaquiere, S. Diner, G. Lochak (eds.), (Springer, Berlin Heidelberg New York, 1987), pp. 15–47.
265. L. B. Levitin, *Information, Complexity and Control in Quantum Physics*, A. Blaquière, S. Diner, G. Lochak (eds.), (Springer, Vienna), pp. 15–47.
266. E. Lieb, *Bull. Am. Math. Soc.*, **81**, 1–13 (1975).
267. E. Lieb, "Convex trace functions and the Wigner-Yanase-Dyson conjecture," *Adv. Math.*, **11**, 267–288 (1973).
268. E. Lieb, M. B. Ruskai, "A fundamental property of quantum mechanical entropy," *Phys. Rev. Lett.*, **30**, 434–436 (1973).
269. E. Lieb, M. B. Ruskai, "Proof of the strong subadditivity of quantum mechanical entropy," *J. Math. Phys.*, **14**, 1938–1941 (1973).
270. G. Lindblad, "Completely positive maps and entropy inequalities," *Comm. Math. Phys.*, **40**, 147–151 (1975).
271. G. Lindblad, "Expectations and entropy inequalities for finite quantum systems," *Comm. Math. Phys.*, **39**, 111–119 (1974).
272. N. Linden, S. Massar, S. Popescu, "Purifying noisy entanglement requires collective measurements," *Phys. Rev. Lett.*, **81**, 3279 (1998).
273. S. Lloyd, "The capacity of the noisy quantum channel," *Phys. Rev. A*, **56**, 1613 (1997); quant-ph/9604015 (1996).
274. H.-K. Lo, "Proof of unconditional security of six-state quantum key distribution scheme," *Quant. Inf. Comput.*, **1**, 81–94 (2001).

275. H.-K. Lo, S. Popescu, "Classical communication cost of entanglement manipulation: is entanglement an interconvertible resource?" *Phys. Rev. Lett.*, **83**, 1459 (1999).
276. H.-K. Lo, S. Popescu, "Concentrating entanglement by local actions: Beyond mean values," *Phys. Rev. A*, **63**, 022301 (2001); quant-ph/9707038 (1997).
277. T. J. Lynch, "Sequence time coding for data compression," *Proc. IEEE*, **54**, 1490–1491 (1966).
278. S. Massar, S. Popescu, "Optimal extraction of information from finite quantum ensembles," *Phys. Rev. Lett.*, **74**, 1259 (1995).
279. F. De Martini, A. Mazzei, M. Ricci, and G. M. D'Ariano, "Pauli Tomography: complete characterization of a single qubit device," *Fortschr. Phys.*, **51**, 342 (2003); quant-ph/0207143 (2002).
280. L. Masanes, "All entangled states are useful for information processing," quant-ph/0508071.
281. K. Matsumoto, *Geometry of a Quantum State*, master's thesis, Department of Mathematical Engineering and Information Physics, Graduate School of Engineering, University of Tokyo, Japan (1995) (in Japanese).
282. K. Matsumoto: "A new approach to the Cramér–Rao type bound of the pure state model," *J. Phys. A Math. Gen.*, **35**, 3111–3123 (2002); quant-ph/9704044 (1997).
283. K. Matsumoto, "Uhlmann's parallelism in quantum estimation theory," quant-ph/9711027 (1997).
284. K. Matsumoto, *A Geometrical Approach to Quantum Estimation Theory*, Ph.D. thesis, Graduate School of Mathematical Sciences, University of Tokyo (1997).
285. K. Matsumoto, "The asymptotic efficiency of the consistent estimator, Berry-Uhlmann' curvature and quantum information geometry," in *Quantum Communication, Computing, and Measurement 2*, P. Kumar, G. M. D'ariano, O. Hirota (eds.), pp. 105–110 (Plenum, New York, 2000).
286. K. Matsumoto, Seminar notes (1999).
287. K. Matsumoto, private communication (2005).
288. K. Matsumoto, "Yet another additivity conjecture," *Physics Letters A*, **350**, 179–181, (2006); quant-ph/0506052 (2005).
289. K. Matsumoto, "Reverse estimation theory, complementarity between RLD and SLD, and monotone distances," quant-ph/0511170 (2005); *Proc. 9th Quantum Information Technology Symposium (QIT13)*, Touhoku University, Sendai, Miyagi, Japan, 24–25 November 2005, pp. 81–86.
290. K. Matsumoto, T. Shimono, A. Winter, "Remarks on additivity of the Holevo channel capacity and of the entanglement of formation," *Comm. Math. Phys.*, **246(3)**, 427–442 (2004); quant-ph/0206148 (2002).
291. K. Matsumoto, F. Yura, "Entanglement cost of antisymmetric states and additivity of capacity of some quantum channel," *J. Phys. A: Math. Gen.*, **37** L167-L171, (2004); quant-ph/0306009 (2003).
292. Y. Mitsumori, J. A. Vaccaro, S. M. Barnett, E. Andersson, A. Hasegawa, M. Takeoka, M. Sasaki, "Experimental demonstration of quantum source coding," *Phys. Rev. Lett.*, **91**, 217902 (2003); quant-ph/0304036 (2003).
293. A. Miyake, "Classification of multipartite entangled states by multidimensional determinants," *Phys. Rev. A*, **67**, 012108 (2003); quant-ph/0206111 (2002).

294. F. Morikoshi, "Recovery of entanglement lost in entanglement manipulation," *Phys. Rev. Lett.*, **84**, 3189 (2000); quant-ph/9911019 (1999).
295. F. Morikoshi, M. Koashi, "Deterministic entanglement concentration," *Phys. Rev. A*, **64**, 022316 (2001); quant-ph/0107120 (2001).
296. M. A. Morozowa, N. N. Chentsov, *Itogi Nauki i Tekhniki*, **36**, 289–304 (1990) (in Russian).
297. M. Murao, D. Jonathan, M. B. Plenio, V. Vedral, "Quantum telecloning and multiparticle entanglement," *Phys. Rev. A*, **59**, 156–161 (1999); quant-ph/9806082 (1998).
298. H. Nagaoka, "On Fisher information of quantum statistical models," in *Proc. 10th Symposium on Information Theory and Its Applications (SITA)*, Enoshima, Kanagawa, Japan, 19–21 November 1987, pp. 241–246. (Originally in Japanese; also appeared as Chap. 9 of *Asymptotic Theory of Quantum Statistical Inference,* M. Hayashi eds.).
299. H. Nagaoka, "An asymptotically efficient estimator for a one-dimensional parametric model of quantum statistical operators," *Proc. 1988 IEEE International Symposium on Information Theory*, 198 (1988).
300. H. Nagaoka, "On the parameter estimation problem for quantum statistical models," *Proc. 12th Symposium on Information Theory and Its Applications (SITA)*, (1989), pp. 577–582. (Also appeared as Chap. 10 of *Asymptotic Theory of Quantum Statistical Inference*, M. Hayashi eds.).
301. H. Nagaoka, "A generalization of the simultaneous diagonalization of Hermitian matrices and its relation to quantum estimation theory," *Trans. Jpn. Soc. Ind. Appl. Math.*, **1**, 43–56 (1991) (Originally in Japanese; also appeared as Chap. 11 of *Asymptotic Theory of Quantum Statistical Inference*, M. Hayashi eds.).
302. H. Nagaoka: Private communication to A. Fujiwara (1991).
303. H. Nagaoka, "On the relation between Kullback divergence and Fisher information—from classical systems to quantum systems," *Proc. Joint Mini-Workshop for Data Compression Theory and Fundamental Open Problems in Information Theory*, (1992), pp. 63–72. (Originally in Japanese; also appeared as Chap. 27 of *Asymptotic Theory of Quantum Statistical Inference*, M. Hayashi eds.).
304. H. Nagaoka, "Differential geometrical aspects of quantum state estimation and relative entropy," in *Quantum Communications and Measurement*, V. P. Belavkin, O. Hirota, R. L. Hudson (eds.), (Plenum, New York, 1995), pp. 449–452.
305. H. Nagaoka, "Algorithms of Arimoto-Blahut type for computing quantum channel capacity," *Proc. 1998 IEEE International Symposium on Information Theory*, 354 (1998).
306. H. Nagaoka, "Information spectrum theory in quantum hypothesis testing," *Proc. 22th Symposium on Information Theory and Its Applications (SITA)*, (1999), pp. 245–247 (in Japanese).
307. H. Nagaoka, "Strong converse theorems in quantum information theory," *Proc. ERATO Conference on Quantum Information Science (EQIS) 2001*, 33 (2001). (also appeared as Chap. 3 of *Asymptotic Theory of Quantum Statistical Inference*, M. Hayashi eds.).
308. H. Nagaoka, "Limit theorems in quantum information theory," *Suurikagaku*, 456, 47–55 (2001) (in Japanese).

309. H. Nagaoka, "The world of quantum information geometry," *IEICE Trans.*, **J88-A**, 8, 874–885 (2005) (in Japanese).
310. H. Nagaoka, M. Hayashi, "An information-spectrum approach to classical and quantum hypothesis testing," quant-ph/0206185 (2002).
311. H. Nagaoka, S. Osawa, "Theoretical basis and applications of the quantum Arimoto-Blahut algorithms," *Proc. 2nd Quantum Information Technology Symposium (QIT2)*, (1999), pp. 107–112.
312. H. Nagaoka, "A new approach to Cramér–Rao bound for quantum state estimation," *IEICE Tech. Rep.*, **IT 89-42**, 228, 9–14 (1989).
313. M. A. Nielsen, "Conditions for a class of entanglement transformations," *Phys. Rev. Lett.*, **83**, 436 (1999); quant-ph/9811053 (1998).
314. M. A. Nielsen, "Continuity bounds for entanglement," *Phys. Rev. A*, **61**, 064301 (2000).
315. M. A. Naĭmark, *Comptes Rendus (Doklady) de l'Acadenie des Sience de l'URSS*, **41**, 9, 359, (1943).
316. M. A. Nielsen, I. L. Chuang, *Quantum Computation and Quantum Information*, (Cambridge University Press, Cambridge, 2000).
317. M. A. Nielsen, J. Kempe, "Separable states are more disordered globally than locally," *Phys. Rev. Lett.*, **86**, 5184–5187 (2001); quant-ph/0011117 (2000).
318. T. Ogawa, *A study on the asymptotic property of the hypothesis testing and the channel coding in quantum mechanical systems*, Ph.D. thesis, Graduate School of Information System, University of Electro-Communication (2000).
319. T. Ogawa, H. Nagaoka, "Strong converse to the quantum channel coding theorem," *IEEE Trans. Inf. Theory*, **45**, 2486–2489 (1999); quant-ph/9808063 (1998).
320. T. Ogawa, H. Nagaoka, "Strong converse and Stein's lemma in quantum hypothesis testing," *IEEE Trans. Inf. Theory*, **46**, 2428–2433 (2000); quant-ph/9906090 (1999).
321. T. Ogawa, H. Nagaoka, "A new proof of the channel coding theorem via hypothesis testing in quantum information theory," *Proc. 2002 IEEE International Symposium on Information Theory*, 73 (2002); quant-ph/0208139 (2002).
322. T. Ogawa, M. Hayashi, "On error exponents in quantum hypothesis testing," *IEEE Trans. Inf. Theory*, **50**, 1368–1372 (2004); quant-ph/0206151 (2002).
323. M. Ohya, D. Petz, *Quantum Entropy and Its Use,* (Springer, Berlin Heidelberg New York, 1993).
324. M. Ohya, D. Petz, N. Watanabe, "On capacities of quantum channels," *Prob. Math. Stat.*, **17**, 179–196 (1997).
325. S. Osawa, H. Nagaoka, "Numerical experiments on the capacity of quantum channel with entangled input states," *IEICE Trans.*, **E84-A**, 2583–2590 (2001); quant-ph/0007115 (2000).
326. M. Owari, M. Hayashi, "Local copying and local discrimination as a study for non-locality of a set," quant-ph/0509062.
327. M. Ozawa, "Quantum measuring processes of continuous observables," *J. Math. Phys.*, **25**, 79 (1984).
328. M. Ozawa, "Quantum limits of measurements and uncertainty principle, in *Quantum Aspects of Optical Communications*, Lecture Notes in Physics, vol. 378, C. Bendjaballah, O. Hirota, S. Reynaud (eds.), (Springer, Berlin Heidelberg New York, 1991), pp. 3–17.

329. M. Ozawa, "Quantum state reduction and the quantum Bayes principle," in *Quantum Communication, Computing, and Measurement*, O. Hirota, A. S. Holevo, C. M. Caves (eds.), (Plenum, New York, 1997), pp. 233–241.
330. M. Ozawa, "An operational approach to quantum state reduction," *Ann. Phys.*, **259**, 121–137 (1997); quant-ph/9706027 (1997).
331. M. Ozawa, "Quantum state reduction: an operational approach," *Fortschr. Phys.*, **46**, 615–625 (1998); quant-ph/9711006 (1997).
332. M. Ozawa, "Operational characterization of simultaneous measurements in quantum mechanics," *Phys. Lett. A*, **275**, 5–11 (2000); quant-ph/9802039 (1998).
333. M. Ozawa, "Measurements of nondegenerate discrete observables," *Phys. Rev. A*, **63**, 062101 (2000); quant-ph/0003033 (2000).
334. M. Ozawa, "Operations, disturbance, and simultaneous measurability," *Phys. Rev. A*, **62**, 032109 (2001); quant-ph/0005054 (2000).
335. M. Ozawa, "Universally valid reformulation of the Heisenberg uncertainty principle on noise and disturbance in measurement," *Phys. Rev. A*, **67**, 042105 (2003); quant-ph/0207121 (2002).
336. M. Ozawa, "Physical content of Heisenberg's uncertainty relation: limitation and reformulation," *Phys. Lett. A*, **318**, 21–29 (2003); quant-ph/0210044 (2002).
337. M. Ozawa, "Uncertainty principle for quantum instruments and computing," *Int. J. Quant. Inf.*, **1**, 569–588 (2003); quant-ph/0310071 (2003).
338. M. Ozawa, "Uncertainty relations for noise and disturbance in generalized quantum measurements," *Ann. Phys.*, **311**, 350–416 (2004); quant-ph/0307057 (2003).
339. M. Ozawa, "On the noncommutative theory of statistical decisions," *Research Reports on Information Sciences*, **A-74** (1980).
340. M. Owari, K. Matsumoto, M. Murao, "Entanglement convertibility for infinite dimensional pure bipartite states," *Phys. Rev. A*, **70**, 050301 (2004); quant-ph/0406141; "Existence of incomparable pure bipartite states in infinite dimensional systems," quant-ph/0312091 (2003).
341. J.-W. Pan, S. Gasparoni, M. Aspelmeyer, T. Jennewein, A. Zeilinger, "Experimental realization of freely propagating teleported qubits," *Nature*, **421**, 721–725 (2003).
342. J.-W. Pan, S. Gasparoni, R. Ursin, G. Weihs, A. Zeilinger, "Experimental entanglement purification of arbitrary unknown states," *Nature*, **423**, 417–422 (2003).
343. D. Petz, D. Petz, "Quasi-entropies for finite quantum systems," *Rep. Math. Phys.*, **23**, 57–65 (1986).
344. D. Petz, *An Invitation to the Algebra of Canonical Commutation Relations*, Leuven Notes in Mathematical and Theoretical Physics, vol. 2 (1990).
345. D. Petz, "Monotone metrics on matrix spaces," *Lin. Alg. Appl.*, **224**, 81–96 (1996).
346. D. Petz, M. Mosonyi: "Stationary quantum source coding," *J. Math. Phys.*, **42**, 48574864 (2001); quant-ph/9912103 (1999).
347. D. Petz, C. Sudár, "Extending the Fisher metric to density matrices," in *Geometry in Present Day Science*, O. E. Barndorff-Nielsen, E. B. V. Jensen (eds.), 21 (World Scientific, Singapore, 1998).
348. D. Petz, G. Toth, *Lett. Math. Phys.*, **27**, 205 (1993).

349. D. Petz, "Monotonicity of quantum relative entropy revisited," *Rev. Math. Phys.*, **15**, 29–91 (2003).
350. A. A. Pomeransky, "Strong superadditivity of the entanglement of formation follows from its additivity," *Phys. Rev. A*, **68**, 032317 (2003); quant-ph/0305056 (2003).
351. H. P. Robertson, "The uncertainty principle," *Phys. Rev.*, **34**, 163 (1929).
352. E. M. Rains, "Bound on distillable entanglement," *Phys. Rev. A*, **60**, 179–184 (1999).
353. E. M. Rains: "A semidefinite program for distillable entanglement," *IEEE Trans. Inf. Theory*, **47**, 2921–2933 (2001); quant-ph/0008047 (2000).
354. M. B. Ruskai, "Beyond strong subadditivity? improved bounds on the contraction of generalized relative entropy," *Rev. Math. Phys.*, **6**, 1147–1161 (1994).
355. M. B. Ruskai, S. Szarek, E. Werner, "An analysis of completely-positive trace-preserving maps on 2×2 matrices," *Lin. Alg. Appl.*, **347**, 159–187 (2002); quant-ph/0101003 (2001).
356. M. B. Ruskai, "Qubit entanglement breaking channels," *Rev. Math. Phys.*, **15**, 643–662 (2003); quant-ph/0302032 (2003).
357. J. J. Sakurai, *Modern Quantum Mechanics*, (Addison-Wesley, Reading, MA, 1985).
358. I. N. Sanov, "On the probability of large deviations of random variables," *Mat. Sbornik*, **42**, 11–44 (1957) (in Russian). English translation: *Selected Translat. Math. Stat.*, **1**, 213–244 (1961).
359. M. Sasaki, M. Ban, S. M. Barnett, "Optimal parameter estimation of a depolarizing channel," *Phys. Rev. A*, **66**, 022308 (2002); quant-ph/0203113 (2002).
360. B. Schumacher, "Quantum coding," *Phys. Rev. A*, **51**, 2738–2747 (1995).
361. B. Schumacher, "Sending quantum entanglement through noisy channels," *Phys. Rev. A*, **54**, 2614–2628 (1996); quant-ph/9604023 (1996).
362. B. Schumacher, M. A. Nielsen, "Quantum data processing and error correction," *Phys. Rev. A*, **54**, 2629 (1996); quant-ph/9604022 (1996).
363. B. Schumacher, M. D. Westmoreland, "Sending classical information via noisy quantum channels," *Phys. Rev. A*, **56**, 131 (1997).
364. B. Schumacher, M. D. Westmoreland, "Optimal signal ensembles," *Phys. Rev. A*, **63**, 022308 (2001).
365. A. Shamir, "How to share a secret," *Commun. ACM*, **22**, 612 (1979).
366. C. E. Shannon, "A mathematical theory of communication," *Bell Syst. Tech. J.*, **27**, 623–656 (1948).
367. C. E. Shannon, "Certain results in coding theory for noisy channels," *Inf. Control*, **1**, 6–25 (1957).
368. C. Gobby, Z. L. Yuan, A. J. Shields, "Quantum key distribution over 122 km of standard telecom fiber," *Appl. Phys. Lett.*, **84**, 3762–3764 (2004); quant-ph/0412171 (2004).
369. T. Shimono, H. Fan, *Proc. ERATO Conference on Quantum Information Science (EQIS) 2003*, 119–120 (2003).
370. T. Shimono, "Additivity of entanglement of formation of two three-level-antisymmetric states," *Int. J. Quant. Inf.*, **1**, 259–268 (2003); quant-ph/0301011 (2003).
371. P. W. Shor, "Scheme for reducing decoherence in quantum computer memory," *Phys. Rev. A*, **52**, 2493 (1995).

372. P. W. Shor, "Additivity of the classical capacity of entanglement-breaking quantum channels," *J. Math. Phys.*, **43**, 4334–4340 (2002); quant-ph/0201149 (2002).
373. P. W. Shor, "The quantum channel capacity and coherent information," *Lecture Notes, MSRI Workshop on Quantum Computation* (2002). Avaliable at http://www.msri.org/publications/ln/msri/2002/quantumcrypto/shor/1/
374. P. W. Shor, "Equivalence of additivity questions in quantum information theory," *Comm. Math. Phys.*, **246**, 3, 453–473 (2004); quant-ph/0305035 (2003).
375. P. W. Shor, "Capacities of quantum channels and how to find them," *Math. Programm.*, **97**, 311–335 (2003); quant-ph/0304102 (2003).
376. P. W. Shor, J. Preskill, "Simple proof of security of the BB84 quantum key distribution protocol," *Phys. Rev. Lett.*, **85**, 441–444 (2000); quant-ph/0003004 (2000).
377. D. Slepian, J. K. Wolf, "Noiseless coding of correlated information sources," *IEEE Trans. Inf. Theory*, **19**, 471 (1973).
378. A. M. Steane, "Multiple particle interference and quantum error correction," *Proc. R. Soc. Lond. A*, **452**, 2551 (1996); quant-ph/9601029 (1996).
379. W. F. Stinespring, "Positive functions on C-algebras," *Proc. Am. Math. Soc.*, **6**, 211 (1955).
380. R. L. Stratonovich, *Izvest. VUZ Radiofiz.*, **8**, 116–141 (1965).
381. R. L. Stratonovich, "The transmission rate for certain quantum communication channels," *Problemy Peredachi Informatsii*, **2**, 45–57 (1966). (in Russian). English translation: *Probl. Inf. Transm.*, **2**, 35–44 (1966.)
382. R. L. Stratonovich, A. G. Vantsjan, *Probl. Control Inf. Theory*, **7**, 161–174 (1978).
383. D. Stucki, N. Gisin, O. Guinnard, G. Ribordy, H. Zbinden, "Quantum key distribution over 67 km with a plug & play system," *New J. Phys.*, **4**, 41.1–41.8 (2002).
384. F. Tanaka, *Investigation on Fisher metric on the classical statistical models and the quantum ones*, master's thesis, Department of Mathematical Informatics, Graduate School of Information Science and Technology, University of Tokyo, Japan (2004) (in Japanese).
385. B. M. Terhal, P. Horodecki, "A Schmidt number for density matrices," *Phys. Rev. A*, **61**, 040301(R)(2000); quant-ph/9911114 (1999).
386. B. M. Terhal, K. G. H. Vollbrecht, "Entanglement of formation for isotropic states," *Phys. Rev. Lett.*, **85**, 2625.
387. B. M. Terhal, M. Horodecki, D. W. Leung, D. P. DiVincenzo, "The entanglement of purification," *J. Math. Phys.*, **43**, 4286 (2002).
388. Y. Tsuda, K. Matsumoto, "Quantum estimation for non-differentiable models," *J. Phys. A Math. Gen.*, **38**, 7, 1593–1613 (2005); quant-ph/0207150 (2002).
389. Y. Tsuda, K. Matsumoto, M. Hayashi, "Hypothesis testing for a maximally entangled state," quant-ph/0504203 (2005).
390. Y. Tsuda, B.S. Shi, A. Tomita, M. Hayashi, K. Matsumoto, Y.K. Jiang, "Hypothesis testing for an entangled state produced by spontaneous parametric down conversion," *ERATO conference on Quantum Information Science 2005 (EQIS 05)*, JST, Tokyo, pp. 57–58 (2005).
391. Y. Tsuda, "Estimation of polynomial of complex amplitude of quantum Gaussian states" *Proc. Annual Meeting of the Japan Statistical Society (2005)* (in Japanese).

392. A. Uhlmann, “The ‘transition probability’ in the state space of *-algebra,” *Rep. Math. Phys.*, **9**, 273–279 (1976).
393. A. Uhlmann, “Relative entropy and the Wigner-Yanase-Dyson-Lieb concavity in an interpolation theory,” *Comm. Math. Phys.*, **54**, 21–32 (1977).
394. A. Uhlmann, “Density operators as an arena for differential geometry,” *Rep. Math. Phys.*, **33**, 253–263 (1993).
395. K. Usami, Y. Nambu, Y. Tsuda, K. Matsumoto, K. Nakamura, “Accuracy of quantum-state estimation utilizing Akaike’s information criterion,” *Phys. Rev. A*, **68**, 022314 (2003); quant-ph/0306083 (2003).
396. A. W. van der Vaart, *Asymptotic Statistics*, (Cambridge University Press, Cambridge, 1998).
397. R. M. Van Slyke, R. J. -B. Wets, “A duality theory for abstract mathematical programs with applications to optimal control theory,” *J. Math. Anal. Appl.*, **22**, 679–706 (1968).
398. H. L. van Trees, *Detection, Estimation and Modulation Theory, Part 1*, (Wiley, New York, 1968).
399. V. Vedral, M. B. Plenio, “Entanglement Measures and Purification Procedures,” *Phys. Rev. A*, **57**, 822 (1998); quant-ph/9707035 (1997).
400. S. Verdú, T. S. Han, “A general formula for channel capacity,” *IEEE Trans. Inf. Theory*, **40**, 1147–1157 1994.
401. F. Verstraete, J. Dehaene, B. DeMorr, “Local filtering operations on two qubits,” *Phys. Rev. A*, **64**, 010101(R) (2001).
402. G. Vidal, “Entanglement of pure states for a single copy,” *Phys. Rev. Lett.*, **83**, 1046–1049 (1999); quant-ph/9902033 (1999).
403. G. Vidal, W. Dür, J. I. Cirac, “Entanglement cost of antisymmetric states,” quant-ph/0112131v1 (2001).
404. G. Vidal, D. Jonathan, M. A. Nielsen, “Approximate transformations and robust manipulation of bipartite pure state entanglement,” *Phys. Rev. A*, **62**, 012304 (2000); quant-ph/9910099 (1999).
405. S. Virmani, M. Sacchi, M.B. Plenio, D. Markham, “Optimal local discrimination of two multipartite pure states,” *Phys. Lett. A*, **288**, 62 (2001).
406. S. Virmani, M. B. Plenio, “Construction of extremal local positive-operator-valued measures under symmetry,” *Phys. Rev. A*, **67**, 062308 (2003).
407. K. G. H. Vollbrecht, R. F. Werner, “Entanglement measures under symmetry,” quant-ph/0010095 (2000).
408. J. von Neumann, *Mathematical Foundations of Quantum Mechanics*, (Princeton University Press, Princeton, NJ, 1955). (Originally appeared in German in 1932).
409. J. Walgate, A. J. Short, L. Hardy, V. Vedral, “Local distinguishability of multipartite orthogonal quantum states,” *Phys. Rev. Lett.*, **85**, 4972 (2000).
410. X. Wang, H. Fan, “Non-post-selection entanglement concentration by ordinary linear optical devices,” *Phys. Rev. A*, **68**, 060302(R) (2003); quant-ph/0302105 (2003).
411. R. S. Ward, R. O. Wells, Jr., *Twistor Geometry and Field Theory*, (Cambridge University Press, Cambridge, 1991).
412. R. F. Werner, “Optimal cloning of pure states,” *Phys. Rev. A*, **58**, 1827 (1998).
413. R.F. Werner and A.S. Holevo, “Counterexample to an additivity conjecture for output purity of quantum channels,” *J. Math. Phys.*, **43**, 4353 (2002); quant-ph/0203003 (2002).

414. H. Weyl, *The Classical Groups, Their Invariants and Representations*, (Princeton University Press, Princeton, NJ, 1939).
415. A. Winter, "Schumacher's Quantum Coding Revisited," Preprint 99–034, Sonder forschungsbereich 343 "Diskrete Strukturen in der Mathematik," Universität Bielefeld (1999).
416. A. Winter, "Coding theorem and strong converse for quantum channels," *IEEE Trans. Inf. Theory*, **45**, 2481–2485 (1999).
417. A. Winter, *Coding Theorems of Quantum Information Theory*, Ph.D. dissertation, Universität Bielefeld (2000); quant-ph/9907077 (1999).
418. A. Winter, S. Massar, "Compression of quantum measurement operations," *Phys. Rev. A*, **64**, 012311 (2001); quant-ph/0012128.
419. A. Winter, "Scalable programmable quantum gates and a new aspect of the additivity problem for the classical capacity of quantum channels," *J. Math. Phys.*, **43**, 4341–4352 (2002); quant-ph/0108066 (2001).
420. A. Winter, "Extrinsic" and "intrinsic" data in quantum measurements: asymptotic convex decomposition of positive operator valued measures," *Comm. Math. Phys.*, **244**, (1) 157–185 (2004); quant-ph/0109050.
421. A. Winter, "Secret, public and quantum correlation cost of triples of random variables," *Proc. 2005 IEEE International Symposium on Information Theory*, 2270 (2005).
422. A. Winter, private communication (2005).
423. M. M. Wolf, J. Eisert, "Classical information capacity of a class of quantum channels," *New J. Phys.*, **7**, 93 (2005).
424. W. K. Wootters, "Entanglement of Formation of an Arbitrary State of Two Qubits," *Phys. Rev. Lett.*, **80**, 2245 (1998).
425. A. D. Wyner, "The wire-tap channel," *Bell. Syst. Tech. J.*, **54**, 1355–1387 (1975).
426. A. D. Wyner, "The common information of two dependent random variables," *IEEE Trans. Inf. Theory*, **21**, 163–179 (1975).
427. J. Yard, in preparation.
428. T. Yamamoto, M. Koashi, Ş. Özdemir, N. Imoto, "Experimental extraction of an entangled photon pair from two identically decohered pairs," *Nature*, **421**, 343–346 (2003).
429. T. Y. Young, "Asymptotically efficient approaches to quantum-mechanical parameter estimation," *Inf. Sci.*, **9**, 25–42 (1975).
430. D. Yang, M. Horodecki, R. Horodecki, B. Synak-Radtke: "Irreversibility for all bound entangled state," *Phys. Rev. Lett.*, **95**, 190501 (2005); quant-ph/0506138 (2005).
431. H. P. Yuen, R. S. Kennedy, M. Lax, "Optimum testing of multiple hypotheses in quantum detection theory," *IEEE Trans. Inf. Theory*, 125–134 (1975).
432. H. P. Yuen, M. Lax, "Multiple-parameter quantum estimation and measurement of non-selfadjoint observables," *IEEE Trans. Inf. Theory*, **19**, 740 (1973).
433. F. Yura, "Entanglement cost of three-level antisymmetric states," *J. Phys. A Math. Gen.*, **36**, L237–L242 (2003); quant-ph/0302163 (2003).

Index